Body and Cosmos

Sir Henry Wellcome Asian Series

VOLUME 20

The titles published in this series are listed at *brill.com/was*

Body and Cosmos

Studies in Early Indian Medical and Astral Sciences in Honor of Kenneth G. Zysk

Edited by

Toke Lindegaard Knudsen
Jacob Schmidt-Madsen
Sara Speyer

BRILL

LEIDEN | BOSTON

Cover illustration: Palmistry chart on dispersed leaf from unknown manuscript. Kenneth G. Zysk Indological Manuscript Collection, University of Copenhagen. © Jacob Schmidt-Madsen.

Library of Congress Cataloging-in-Publication Data

Names: Knudsen, Toke Lindegaard, editor. | Schmidt-Madsen, Jacob, editor. | Speyer, Sara, 1968- editor. | Zysk, Kenneth G., honouree.
Title: Body and cosmos : studies in early Indian medical and astral sciences in honor of Kenneth G. Zysk / edited by Toke Lindegaard Knudsen, Jacob Schmidt-Madsen, Sara Speyer.
Other titles: Sir Henry Wellcome Asian series (Brill Academic Publishers) ; v. 20.
Description: Leiden ; Boston : Brill, [2020] | Series: Sir Henry Wellcome Asian series, 1570-1484 ; volume 20 | Includes bibliographical references and index.
Identifiers: LCCN 2020031603 | ISBN 9789004436695 (hardback) | ISBN 9789004438224 (ebook)
Subjects: LCSH: Medicine, Ayurvedic. | Medicine–India–History. | Astronomy–India–History. | Astronomy, Ancient.
Classification: LCC R605 .B648 2020 | DDC 610.954–dc23
LC record available at https://lccn.loc.gov/2020031603

Typeface for the Latin, Greek, and Cyrillic scripts: "Brill". See and download: brill.com/brill-typeface.

ISSN 1570-1484
ISBN 978-90-04-43669-5 (hardback)
ISBN 978-90-04-43822-4 (e-book)

Contents

PART 4
Interactions across Culture and Time

Preface

It is our great pleasure and privilege to present this volume to the scholarly community on the occasion of the 70th birthday of Kenneth G. Zysk, Professor Emeritus of Indology at the University of Copenhagen in Denmark. Ken, as we are wont to call him, has had a long and esteemed career spanning five decades and several continents. It began under the auspices of Gerald Larson at the University of California, Santa Barbara, continued under A.L. Basham at the Australian National University, Canberra, and found its way via various positions in Europe and the US to the University of Copenhagen, from which Ken retired as Professor and Head of Indology on 1 March, 2020.

Ken's work has always been driven by a great love for India and a great interest in its early scientific traditions. He started out in the field of medicine (Āyurveda), and gradually expanded into the fields of sexology (Kāmaśāstra) and the astral sciences (Jyotiḥśāstra). His focus is not restricted to the sciences themselves, but also takes into account the contexts in which they appear, and the intercultural transmissions which helped support and develop them. In later years he has been particularly interested in exploring possible relations between ancient Mesopotamian, Greek, and Indian traditions of physiognomy and omenology.

His willingness to engage with ever-new fields of study has never compromised his strict adherence to the use of primary sources. As Dominik Wujastyk remarks in his contribution to the present volume, Ken has always been fond of philology, and anyone who has ever worked with him will know this to be true. Over the years, he has built up a sizable collection of Indic manuscripts, which he has now generously donated to the Section of Indology at the University of Copenhagen.[1] The collection consists of around 3,000 manuscripts dating from the 16th to the 19th centuries and covering a wide array of subjects and genres. Steps have been taken to begin cataloguing and digitizing the manuscripts in an effort to make them available to the scholarly community at large.

Another and equally important aspect of his work has been his frequent trips to India and his openness toward cultural and intellectual traditions outside his own. He has been able to form lasting personal and professional bonds with numerous Indian scholars, and, as lovingly told by Kamal Sheel in the present volume, has come to be adopted into the family of the late Indian historian

1 The cover illustration for the present volume was taken from this collection, now officially named the Kenneth G. Zysk Indological Manuscript Collection.

A.K. Narain (1925–2013). Ken has never forgotten his position as an outsider in relation to the region and the people whose intellectual traditions he has tirelessly devoted himself and his work to. His approach has always been one of respect and humility, and, as he himself once said, he has never gone to India without learning something new that he could never have learned from afar.

We have tried to capture all of the above, and a lot more besides, in grouping together the present collection of articles celebrating Ken's life and work under the title *Body and Cosmos*. The body has always stood at the center of Ken's work, but its homology with the cosmos, as laid out, for example, in the early medical treatise *Carakasaṃhitā*,[2] has never been forgotten. This is clearly borne out by two of Ken's best-known publications issued nearly a quarter century apart. The first, *Asceticism and Healing in Ancient India* (New York, 1991), deals with the body as the locus of health and disease, while the second, *The Indian System of Human Marks* (Leiden, 2015), deals with the body as the locus of inscription and prediction.

The volume begins with an introductory section with contributions from friends and colleagues about Ken's qualities as a person and as a scholar. It continues with three sections pertaining to the main focus areas of Ken's work: history of medicine, astral sciences, and interactions across culture and time. Each section contains a number of articles written by Ken's peers, all of whom are among the finest and foremost scholars within their respective fields. The volume also contains a (hopefully) comprehensive list of Ken's publications to date and an index of authors, titles, and keywords referenced throughout.

We, as editors, would like to say a heartfelt thank you to all of the contributors who made this volume possible. The final editorial work was done during the height of the worldwide corona crisis in the spring of 2020, and yet you managed to keep up communications and deliver your articles in due time. For that we remain truly and deeply grateful.

Finally, as friends, students, and colleagues of Ken during his time at the University of Copenhagen, each of us would like to write a few words about our own experiences at the feet of our gracious and capable mentor.

Toke Lindegaard Knudsen: I first met Ken more than 20 years ago when I was a master's student in Mathematics. Ever since our initial meeting, Ken has been a

2 "This human being is coincident with the world: whatever kinds of entity are to be found in the world those same are to be found in the human being and whatever are to be found in the human being are to be found in the world." *Carakasaṃhitā* 5.3 as translated in Dominik Wujastyk, "Interpreting the Image of the Human Body in Premodern India," *International Journal of Hindu Studies* 13, no. 2 (2009): 189–228.

mentor to me, always nurturing my interests and encouraging me to be ambitious and take risks to pursue them. His help has been invaluable to me throughout my career. Ken was the one to introduce me to Indic manuscripts. It was Ken who helped me build the connections to do a PhD, later serving on my PhD committee. He was my supervisor for a Marie Skłodowska-Curie fellowship at the University of Copenhagen. Ken's energy and passion for Indology has always, and is still, propelling me forward. For that I will always be thankful.

Jacob Schmidt-Madsen: Ken was there for me when I first learned the Sanskrit alphabet, and he was still there for me when I defended my PhD dissertation many years later. That about says it all. His positive attitude and encouraging remarks have always carried me forward, and his support has never wavered, not even when I embarked on dubious adventures, such as traveling in the footsteps of the cloud from Kālidāsa's *Meghadūta* or choosing to specialize in the history of Indian board games. If there is one thing that Ken has taught me above all else, it is that there really is a place in this world for those who persevere, no matter how obscure their pursuits.

Sara Speyer: Ken has time for any student who shows genuine interest and is willing to do the work, regardless of who they are. Studying Indology at a relatively high age, with a very alternative background, turned out to be completely unproblematic for me with Ken as teacher and supervisor. His extensive knowledge and open mind make him a brilliant mentor, always pointing in the right direction, but still leaving room for one's own process of discovery.

Figures and Tables

Figures

Tables

PART 1

Introduction

Professor Kenneth G. Zysk. Copenhagen, 2020

CHAPTER 1

Kenneth G. Zysk and the University of Copenhagen

Erik Reenberg Sand

Most Indologists will be aware of Professor Kenneth Gregory Zysk and his excellent contributions to Indology and Indian studies. This is not the place to evaluate his contributions to Indological research. Suffice it to say, the bulk of his scholarly work lies within Indian scientific literature, starting with a focus on ancient Indian medicine (Āyurveda), followed by Kāmaśāstra, and ending with his recent and large study of Indian human marks in 2016. As an offshoot of this latest work, Ken has lately taken an interest in the similarities and possible relations between Indian, Greek, and Mesopotamian sciences of prognostication, especially those based on physiognomy. If one tries to give an overall characterization of Ken's research, one could say that it is everywhere characterized as being original, fundamental, and accomplished with great skill and expertise. It would be wrong to say that he is a great friend of theories; on the contrary, his work is the result of hard and patient work with the original sources, often in the form of unedited works in manuscript form, and most often in hitherto understudied areas.

1 Career in Brief

Ken started his academic career as a premedical student in Oregon and continued as a student of American National Studies at California State University, Fullerton. Here, he started taking an interest in Sanskrit studies, which again led him to study Asian studies with Professor Gerald Larson at the University of California, Santa Barbara. He finished his MA in 1976.

From 1977 to 1981 he studied for his PhD on Vedic medicine at the Australian National University, Canberra, under the guidance of Professor A.L. Basham, whose posthumous *Classical Hinduism* he later edited. After returning from Australia in 1981, Ken had a long career of mostly temporary positions and scholarships, within both Asian and Religious Studies, mainly in the United States but also in Canada and France, until, finally, in 1997, he ended up being appointed as Associate Professor of Indology in the Asian Department of the University of Copenhagen.

 | DOI:10.1163/9789004438224_002

2 Ken's Time in Copenhagen

Before Ken acceded the position as head of Indology, the milieu had, for almost two decades, been influenced by internal disagreements among the staff and not much student activity. For a couple of years, there had been no full-time professor of Indology employed in the department. The reannouncement of a position in Indology was probably seen as the last chance for the discipline to survive, and considering this, it was good to get 'new blood' from the outside; and new blood we certainly got. I was in the History of Religions Department and remember the astonishment of having a colleague whose expertise was Indian natural science, thinking, "Is this not a humanistic faculty?" Since the early days of Viggo Fausbøll, Søren Sørensen, Dines Andersen, and later Poul Tuxen, Indology in Copenhagen had been well known for its studies in Buddhism, Indian mythology, and Indian religion and philosophy, so what were we going to do with a colleague studying Indian medicine? Later, after I got affiliated with Indology and came to know Ken better, I became wiser and got to know about the breadth and scope of his Indological knowledge.

If my own preconceived notions—and perhaps those of others as well—were among Ken's challenges, there also were other, more important challenges, too. One was the language barrier. Although most of the Danish students and colleagues have quite good English, the position as Associate Professor at a Danish university still requires you to master the language within a few years, and to such an extent that you can teach a class in Danish. In Ken's case, this did not prove to be a problem. Thanks to his linguistic skills, he quickly picked up sufficient Danish to make it the language of his courses, although, considering the fact that dictionaries and teaching materials are mostly in English, the students have often preferred this language.

Another challenge must have been the mentality of his Danish colleagues in the department. Contrary to how we like to think of ourselves, we are not very open toward foreigners who settle here, and we rarely socialize with them privately. Luckily for Ken, there were other foreign scholars working in the department who were in the same position, and with whom it was easier for him to socialize. Then, there were the difficulties of getting into and understanding the Danish educational system and the rules governing it. Furthermore, just before his arrival, the curriculum of Indology had been changed in order to fit changes made to the rules of the other Asian disciplines. As a result of this attempt to make the Asian disciplines more uniform in structure, Indology was supposed to have Hindi as a compulsory element. As Ken was the only full-time employee in the discipline, it was decided instead to make Hindi, along with Pāli, elective elements in the BA program, thus giving it a more classical focus.

Shortly after Ken's arrival, the faculty managed to get an extra propaedeutic year for the discipline, giving the students an extra year for their studies, and Ken was now able to limit the first year of Indology to introducing students to Sanskrit script, grammar, and syntax based on an elementary introduction. On top of this, there was a compulsory general introduction to Indology, followed by two parallel tracks of elective language and content study elements. Later, when the Asian studies disciplines were more fully integrated, the Indology program was simplified, and Pāli and Hindi were completely dropped.

3 Ken as a Teacher

It goes without saying that as the only full-time employee in Indology, Ken has broad experience, teaching Indology on all academic levels, as well as teaching both Sanskrit and content courses. He is a most competent teacher and also well-liked and respected by his students. However, after I started teaching Indological courses in 2012, he was happy to leave the more elementary content courses to me. This allowed him to concentrate his efforts on teaching the advanced language courses and more specialized courses, which required a more robust knowledge of Sanskrit. An exception to this was when—in order to generate more interest in the discipline—he started offering Open University courses on yoga, and later Āyurveda, to a broader general public. This was an excellent opportunity for Danish yogis, yoginis, and others to become acquainted with his expert knowledge, and for the discipline to recruit new students.

In addition to this, Ken has always been very helpful in supervising the students, going to great lengths to assist them with their papers. The same can, of course, also be said of his work with his PhD students.

4 Ken as a Colleague

At the University of Copenhagen, Indology is part of the rather complex Department of Cross-Cultural and Regional Studies, which is a merger of the former Carsten Niebuhr Department, Department of Asian Studies, Department of History of Religions, and Department of Eskimology. In addition to the disciplines of religious studies, minority studies and cross-cultural studies, this department deals with a wide range of regions, from Japan to Mesoamerica.

In such an environment, it is important to be open to cooperation and cross-disciplinary approaches, especially when you represent a very small discipline like Indology, and Ken has certainly been able to navigate this ocean, always willing to cooperate with other disciplines in the department in connection with teaching and possible common study programs. Perhaps, this is a skill he developed early on when he was teaching courses in Religious Studies departments.

The author of this introductory piece is himself a good example of Ken's open-mindedness. Since I showed him a manuscript of a critical edition of a Sanskrit Māhātmya text several years ago, he has been supportive and shown confidence in my work; something that finally resulted in my formal affiliation with Indology. He let me take over the introductory courses in Indology and later also let me teach the courses on Epic and Purāṇic Sanskrit. He has been a wonderful, trusting, and faithful colleague. Whatever decisions had to be made, he always included me and took note of my opinion.

Another example of Ken's openness is the enthusiasm and energy with which he lobbied for developing teaching programs in Hindi and Modern Indian Studies. This would later turn into a separate BA studies program in the Department of Regional and Cross-Cultural Studies.

5 Networking Activities

Right from the start, during the days of his PhD studies, Ken has been actively involved in networking activities. Thus, already in 1977, he was the one to come up with the idea for the International Association for the Study of Traditional Asian Medicine, which was founded in 1979 and of which he is still an honorary member. He has also, in various capacities, been involved in editorial work. Here one may mention the *Critical Pāli Dictionary*, the *Indian Journal of Buddhist Studies*, the *Indian Journal of the History of Science*, and the Indian Medical Tradition series published by Motilal Banarsidass. Ken's international orientation has also resulted in several guest appointments during his tenure at the University of Copenhagen. Thus, in May 2005 he was Directeur d'Études Associé at École Pratique des Hautes Études at Sorbonne, France; from January to March 2007 he was Senior Fellow at the French Institute in Pondicherry, India; from April to August 2007 he was Visiting Senior Fellow at Université de Lausanne, Switzerland; from February to April 2014 he was Visiting Professor at the University of Pune and Deccan College Post Graduate Research Institute, Pune, India; and from January to April 2017 and January to March 2018 he was Khyentse Professor at the Department of Pali, Savitribai Phule Pune University, Pune, India. Similarly, Ken is cooperating with several international research partners. Here one may mention the research project on early Sanskrit medical literature housed at the University of Vienna, the editing and translation of Jajjaṭa's *Nirantarapadavyākhyā*, a commentary on the *Carakasaṃhitā* together with Professor Tsutoma Yamashita, Kyoto Gakuen University, Japan, and his recent establishment of a research group, together with Marko Geslani, University of South Carolina; Bill Mak, Kyoto University; and Michio Yano, Kyoto Sangyo University, which aims to prepare a critical edition of the *Gargasaṃhitā*.

In later years, Ken has also spent his energy and resources trying to build up networks and cooperation with foreign institutions, especially the Deccan College Post Graduate and Research Institute and the Department of Pali and Buddhist Studies, both located in Pune, India.

His greatest effort, however, is the recent establishment of The Centre for the Study of Indian Science (CSIS), which focuses on the three traditional branches of Jyotiṣa (*saṃhitā*, *gaṇita*, and *horā*) along with classical Āyurveda. In addition to Ken, the center currently includes a Marie Skłodowska-Curie fellow and a PhD student, but the idea is to also invite foreign guest lecturers and arrange seminars. The center is collaborating with the History of Astronomical and Mathematical Sciences in India (HAMSI) work group at the University of Canterbury, Christchurch, New Zealand.

This initiative is taking place as the future of Indology in Copenhagen is coming under serious threat after years of financial cuts to the Danish educational system. Like most other small Asian disciplines at the University of Copenhagen, Indology has been threatened with closure for decades. Since the middle of the 1980s, it has, however, like other small disciplines, been supported by a special grant from the Danish Government. But, within the last 6 to 7 years, the government has generally tried to speed up the students' studies, threatening to punish the universities if students fall behind in their scheduled study plans. Along with a demand from the Ministry of Education that the universities offer a minimum of 12 hours of weekly teaching to all BA students, this has turned out to be a severe threat to several of the small Asian disciplines at the Humanistic Faculty in Copenhagen.

In 2016 the dean of the Humanistic Faculty decided to stop the intake of new students in several small disciplines, including Southeast Asian studies, Thai, Tibetology, and Indology. In addition to this decision, the Associate Professors of Thai and Tibetology were dismissed. Ken was the only tenured professor in these disciplines who survived, and in 2017, he even got a well-earned promotion to Full Professor after an evaluation by a committee of international Indologists.

Sanskrit has been taught at the University of Copenhagen since 1841, but at present it is only being offered via an Open University program available to the general public, and the special government funding for Indology has gone away. Only time will tell whether this is going to change and what the future of Indology in Copenhagen is going to look like. In case Indology, like Southeast Asian studies, Thai, and Tibetology, is finally closed down, one thing is certain: Ken has done everything humanly possible to keep it from happening.

6 Postscript

After this introduction was written at the end of 2018, Ken retired at the beginning of 2020 and is now Professor Emeritus. In the autumn of 2019, a 3-year postdoc and a new Marie Skłodowska-Curie fellow were hired, thus strengthening the CSIS. Ken has generously donated his large collection of about 3,000 Indic, mainly Sanskrit, manuscripts to the Department of Cross-Cultural and Regional Studies, and an Indological research center has been built around it. It is hoped that funds will be found to catalogue and digitize this collection so that it may be made available to the scholarly community. Hopefully, these initiatives will also secure Indology's strong presence at the University of Copenhagen in the future.

CHAPTER 2

Kenneth G. Zysk as the Narains Remember Him

Kamal Sheel

It has been our pleasure to know and interact with Professor Kenneth G. Zysk (Ken) since his graduate school days at the University of Wisconsin-Madison in 1976. With his prior background in Asian studies at the University of California, he came to the University of Wisconsin-Madison, one of the leading centers of South Asian studies in the United States. He chose to focus on ancient Indian history and culture with my father, Professor A.K. Narain (AK), as his guide and mentor. I was also a graduate student there at the time in the Department of History. Over the semester, Ken impressed all of us with his intelligence, honesty, and integrity. His familial behavior soon made him an extended part of our family. I guess it was Indian food cooked by my mother, Usha Narain, that cemented the bond between Zysk and the Narains. Food in a then alien land was the great cultural assimilator and disseminator. AK's students and friends often visited our home, had food, bridged the hierarchical relationship between teacher and student, and became close to each other, discussing the problems and intricacies of South Asian studies, and thus they gained the extra knowledge necessary to confidently pursue higher studies in their chosen areas. This was my first introduction to Ken.

After completing his master's at the University of Wisconsin-Madison, Ken decided to go to Australia to work with Professor A.L. Basham, the iconic professor who had been responsible for training a long list of first-rate South Asianists since the 1950s. While AK was among the first batch of his students, Ken belonged to one of the last ones. His work on ancient Asian / Indian medicine that was completed there is still one of the pioneering studies in the field. It was his academic courage that propelled him to take up a field of research that was still in its infancy and that some even doubted had connections with the regular study of history and culture. The research for the work made him gain a deeper knowledge of Sanskrit and other ancient Indian languages. The work is now part of all good and standard books on ancient Indian history and culture.

Completing his PhD at the Australian National University, Ken initially settled in New York, teaching at New York University. During this time, AK often visited New York and even stayed there for a semester for his research work. Ken and AK once again began to frequently meet and exchange ideas. Due to Ken's strong background in ancient Indian texts, AK respected his academic opinions

 | DOI:10.1163/9789004438224_003

and judgments. Our family ties were thus further strengthened. This was continued later by my younger sister and her husband who had by then settled in New York. They too formed a bond with Ken.

In 1997, when Ken was offered senior fellowships by the Fulbright Program and the AIIS (American Institute of Indian Studies), he decided to come to Varanasi/Kashi, the most ancient living city in India and the abode of traditional schools of knowledge. AK, who had by then taken early retirement from the University of Wisconsin-Madison and settled in Varanasi, invited him to stay at our big house there. Ken and his lovely wife Susana lived with us for several months. I had joined the Banaras Hindu University as an Associate Professor in Chinese, and my wife and our two school-age children were also living there. We remember tall Ken in his colorful, local cotton kurta and pajamas in the hot and humid summertime diligently going from one library to the other on a tall bicycle in search of his desired texts. His dedication to the work was amazing. We used to often meet in the afternoon at the university's computer center, which was then the only resource for us to send e-mails. He received his offer to join Copenhagen University when he was in Varanasi. Having been to Copenhagen recently, I too, to some extent, contributed to his decision to join the faculty there. Later, he even negotiated a Memorandum of Understanding (MoU) between Copenhagen and Banaras universities to further "Indological research" and a closer connection between the two places. This, however, could not fully come to fruition due to reasons beyond his control. His vision was global.

At home in Varanasi, to my wife Ranjana, he was an adopted brother. He fulfilled the role like a true Indian brother, getting the colorful *rakhi* tied on his hand and enthusiastically participating in the *rakhi* ceremony—which is part of an Indian festival highlighting the bond between brothers and sisters. He was a great uncle to my son, who was then applying to universities in the United States for a bachelor's degree in engineering, and with whom he often engaged in long discussions about the merits and demerits of one or another place. It was really a memorable time for all of us. After that, whenever he visited India, and Pune in particular, as a visiting professor, he tried to come to Varanasi to meet and stay with us. Our home was always open to him as it would be for any family member.

After the death of AK, when his friends and students decided to produce a commemoration volume in his honor, Ken's name as one of the three editors was unanimously approved. As an editor, he worked hard to invite articles from scholars from all around the world and spent a great deal of time thoroughly editing each and every article. The volume appeared in 2017 and was released at the International Association of Buddhist Studies Conference held in Toronto, Canada.

Besides being a serious, conscientious scholar, Ken is a jolly person, who easily mixes with both young and old from all walks of life. In January 2018, he and his loving and caring wife Lykke honored us by participating in our daughter's wedding in Varanasi. Both of them were in a most exhilarating mood and greatly enjoyed each and every ritual and added color to the festivities.

Ken, I am very happy to contribute to this Festschrift, which you richly deserve. We wish you a long and productive life with more opportunities to interact as you superannuate.

PART 2

History of Medicine

CHAPTER 3

Agni and *Soma* Revisited: A Primordial Āyurvedic Concept?

Vitus Angermeier

1 Introduction*

The key concept which binds together the body of Āyurvedic medicine is the doctrine of the three *doṣa*s or morbific factors: bile, wind and phlegm. They fulfill necessary functions in the healthy body and, should they become imbalanced, give rise to ailments. Beyond the *doṣa*s, there are further important sets of constituent elements which describe the composition and the functions of the human body, such as the system of the seven bodily constituents (*dhātu*s)[1] or the widely-known concept of the five elements (*mahābhūta*s)—earth, fire, water, wind and space/ether. Additionally, the compendiums of ancient Indian medicine[2] contain notions of less developed concepts, ideas borrowed from other disciplines or from common knowledge, and transitional constructions that build bridges linking together the aforementioned sets.

* Part of the research for this study was undertaken during my participation in the FWF project P23330-G15, financed by the Austrian Science Fund (FWF). Furthermore, I must thank Karin Preisendanz and Dominik Haas for valuable hints as well as Sarah Kiehne for improving my English.

1 These are nutritious fluid (*rasa*), blood (*rakta*), flesh (*māṃsa*), fat (*medas*), bones (*asthi*), marrow (*majjā*), and semen (*śukra*).

2 This study considers the two oldest foundational medical compendiums, the *Carakasaṃhitā* (CS) and the *Suśrutasaṃhitā* (SS), both roughly dating to the period between 250 BC and 150 AD, as well as the *Bhelasaṃhitā* (BhS), which originated around the same time. All three were revised and partially enhanced in the centuries that followed. The BhS is in the worst state, only transmitted in a single manuscript. Further sources are the younger compendiums, ascribed to Vāgbhaṭa, the *Aṣṭāṅgasaṃgraha* (AS) and the *Aṣṭāṅgahṛdayasaṃhitā* (AHS), which came into existence around 700 AD and are heavily dependent on the two older foundational works. Where advisable, the commentaries on these texts were also consulted. For further information on these works, their contents, authors, and dates cf. Gerrit Jan Meulenbeld, *A History of Indian Medical Literature*, vol. 1A (Groningen: Forsten, 1999) concerning CS, SS, AS and AHS as well as Gerrit Jan Meulenbeld, *A History of Indian Medical Literature*, vol. 2A (Groningen: Forsten, 2000), 13–24 concerning the BhS. For a synopsis of the section names of the examined works cf. table 3.4 below.

DOI:10.1163/9789004438224_004

TABLE 3.1 Correlation of elemental doctrines

Morbific factors	*Agni/Soma*	Five elements
phlegm (*śleṣman, kapha*)	*soma*	water
bile (*pitta*)	*agni*	fire
wind (*vāta, vāyu, …*)		wind
		earth
		space/ether

One of the more peculiar concepts is that of *agni* and *soma*. These two essential, idolized elements from the Vedic ritual make several dubious appearances in the compendiums of early Āyurvedic medicine. They occur most visibly in compounds like *agnīṣomīya*, *agnīṣoma* and *agnīṣomātmaka*, which, in most cases, serve to present the world according to a dualistic principle, dividing it into fiery and watery parts.

In his 2004 study, *Agni and Soma: A Universal Classification*, Dominik Wujastyk collected and analyzed evidence on these compounds.[3] He explains their relations to the Vedic ritual, where Agni and Soma are both main constituents of the sacrifice and essential divinities invoked during the ritual. In addition, he refers to Frauwallner's depiction of the water, breath, and fire doctrines as expressed in the Upaniṣads[4] and sees in them a potential connection between an older dualistic world view and the doctrine of the three morbific factors. In order to substantiate this theory, he examines occurrences of the aforementioned compounds in the *Suśrutasaṃhitā* and in both medical compendiums ascribed to Vāgbhaṭa, the *Aṣṭāṅgasaṃgraha* and the *Aṣṭāṅgahṛdayasaṃhitā*. The following topics are addressed in these cases and discussed by Wujastyk:

- According to some medical authorities, the six flavors are partly watery (sweet, bitter, and astringent), and partly fiery (pungent, sour, and salty)[5]
- According to some medical authorities, potency (*vīrya*) is either watery or fiery[6]

3 Dominik Wujastyk, "Agni and Soma: A Universal Classification," *Studia Asiatica* 4–5 (2004): 347–369.

4 Cf. Erich Frauwallner, *Geschichte Der Indischen Philosophie*, vol. 1 (Salzburg: Otto Müller, 1953), 49–80.

5 See SS 1.42.7 and Wujastyk, "Agni and Soma," 251 f.

6 See SS 1.40.5 and Wujastyk, "Agni and Soma," 352 f.

- Relating to their harvest, drugs, as well as their terrain and season, are defined as either watery or fiery[7]
- There is discussion as to whether the nutritious fluid or chyle (*rasa*) is watery or fiery; due to its other properties, it is defined as watery[8]
- Lethal points (*marmans*) can be watery or fiery[9]
- The whole year is subdivided into watery and fiery halves[10]

I will come back to these cases in the ensuing sections of this study.

In the following, Wujastyk addresses information from younger sources, ethnographic evidence, comparable ancient European sources and preceding research such as Kirfel's 1951 study on fire and water.[11] His conclusion is bipartite:

> The certain conclusion concerns the fundamental importance of the *agni/soma* polarity as a conscious organising principle for understanding the relationships of many categories of the world. This polarity is a clear theme in the āyurvedic literature and beyond. It can be traced backward to the very earliest layer of Indian thought in the Vedas, and forward to the present in the widespread contemporary custom of classifying foods and medicines as expressing heating or cooling powers.[12]

Furthermore, he suggests two conjectures:

> The first conjecture is that the Agni/Soma polarity expressed itself as a two-humor fire-water medical theory that is older than the classical three-humor doctrine in Āyurveda. I have also referred very briefly to the fact that in earliest Greek medical tradition a similar doctrine of hot/cold polarity was, for a period at least, proposed as the principle explanation for disease. The second conjecture is that the Indian and Greek two-

7 See SS 1.36.5 and Wujastyk, "Agni and Soma," 353 f.

8 See SS 1.14.3 and Wujastyk, "Agni and Soma," 354–356.

9 See SS 3.6.16 and Wujastyk, "Agni and Soma," 356 f.

10 See AHS 1.3.2–8 and Wujastyk, "Agni and Soma," 357–360.

11 Willibald Kirfel, *Die fünf Elemente, insbesondere Wasser und Feuer: Ihre Bedeutung für den Ursprung altindischer und altmediterraner Heilkunde* (Walldorf: Vorndran, 1951). Kirfel already suggested that there existed an old fire/water duality, which provided the basis for the medical three-humor doctrine. In order to validate his points, he draws on evidence from ancient Mediterranean sources and from the Balinese medical tradition, which in his view preserved very ancient Indian ideas. However, he does not address the topic of *agni* and *soma* directly.

12 Wujastyk, "Agni and Soma," 366.

> humor theories may have a common origin, either Indo-European or possibly Indo-Iranian.[13]

While Wujastyk's essay has a wider scope, discussing the concept's development throughout South-Asian history and including thoughts on ethnographic evidence and parallels from ancient Greece, in the domain of Āyurveda it concentrates on the above-mentioned compound words which contain both terms and occur in three of the most important compendiums. In this study, I want to take a closer look at Wujastyk's first conjecture and further investigate how this basically Vedic concept emerged in the early Āyurvedic compendiums. To achieve this, I included two further sources, the *Carakasaṃhitā* and the *Bhelasaṃhitā*, and collected and analyzed references of both terms occurring in these five works in all possible forms.[14] Furthermore, I took into account how the relevant passages are embedded in the surrounding context and endeavored to understand the implications of their interaction with connected topics. This enables us to understand how deeply this concept was integrated in these works and how closely it is linked to other medical concepts. This should shed some light on the question of whether the *agni/soma* polarity is a primordial concept of Indian medicine, predating the doctrine of the three morbific factors, or whether it was included at a later date. Through these means, we can obtain a clearer picture of the development of the *agni/soma* concept within early Āyurvedic tradition. As a first step, however, a brief discussion of the meanings of the words *agni* and *soma* is necessary.

2 Word Meaning

The word *agni* quite unambiguously denotes fire (as a natural phenomenon, as an element, and as a deity) and—especially in Āyurvedic contexts—figuratively represents the digestive fire. The meanings of *soma* are much more diverse. Basically, the word refers to a plant and a deity, well known from the Vedic ritual and extensively discussed in secondary literature.[15] At the time the

13 Wujastyk, "Agni and Soma," 366.

14 See fn. 2.

15 Concerning *soma* as a deity, sacrificial offering, and plant, cf. Thomas Oberlies, *Die Religion des Ṛgveda*, 2 vols., Publications of the De Nobili Research Library 27 (Wien: Gerold, 1998–1999), Heinrich Lüders, *Varuṇa*, ed. Ludwig Alsdorf, vol. 1 (Göttingen: Vandenhoeck & Ruprecht, 1951), 202–275, and H. Lommel, "König Soma," *Numen* 2, no. 3 (1955): 196–205. The most favored identifications for Vedic *soma* are a species of Ephedra and the fly

oldest surviving Āyurveda works were created, this plant was no longer available. Thus, the name became associated with several plants which today we cannot identify.[16] In addition, *soma* is already used in the *Ṛgveda* as a synonym for water as well as for other nourishing liquids[17] and—from the Brāhmaṇas onward—for the moon.[18] Because of this wide range of meanings for *soma*, the adjective *saumya* cannot be translated uniformly in the examined works. It includes meanings such as "watery," "cool," "cooling," and "referring to the moon." Furthermore, the plant background should not be forgotten. This, combined with the water and moon identifications, points toward an interconnection with topics like growth and nourishment. Finally, the adjective is broadly used to describe lovely, often moist and cool, objects or scenes. And in the CS, *saumya* is a popular salutatory address for a pupil.

In the following, I translate *saumya* as watery, because in the cited contexts the main aspect is its opposition to *āgneya*. However, the reader should keep in mind that the other above-mentioned meanings of *saumya* are important for a thorough understanding of its meaning and are relevant in specific cases.

---

agaric (*Amanita muscaria* (L.) Lam.). For arguments in favor of Ephedra cf. Harry Falk, "Soma I and II," *Bulletin of the School of Oriental and African Studies, University of London* 52 (1989): 77–90, Harry Falk, "Decent Drugs for Decent People: Further Thoughts on the Nature of Soma," *Orientalia Suecana* 51–52 (2003): 141–155, and Jan E.M. Houben, "The Soma–Haoma Problem: Introductory Overview and Observations on the Discussion," *Electronic Journal of Vedic Studies* 9, no. 1a (2003). For arguments in favor of the fly agaric, cf. Rainer Stuhrmann, "Ṛgvedische Lichtaufnahmen: Soma botanisch, pharmakologisch, in den Augen der Kavis," *Electronic Journal of Vedic Studies* 13, no. 1 (2006): 1–93. For a different approach suggesting that the *soma* drink was composed of a variety of plants, cf. Matthew Clark, *The Tawny One: Soma, Haoma and Ayahuasca* (London: Muswell Hill Press, 2017).

16 See especially SS 4.29.3–9. There, *soma* gets introduced as an exceptional immortality nectar (*amṛta*), made by Brahma. It is described as differentiated into 24 species "according to place, name, form, potency and distinctiveness." The names of these species are *aṃśumān, muñjavān, candramā, rajataprabha, dūrvāsoma, kanīyān, śvetākṣa, kanakaprabha, pratānavān, tālavṛnta, karavīra, aṃśavān, svayamprabha, mahāsoma, garuḍāhṛta, gāyatra, traiṣṭubha, pāṅkta, jāgata, śākvara, agniṣṭoma, raivata, gāyatryā tripadā yukta*, and *uḍupati*. All of these plants are recommended as substitutes for *soma* and described in SS 4.29.20–26. The subsequent passage 27–31 informs about regions where these plants grow and 32 reveals that all of these plants are only visible for certain people.

17 Cf. Oberlies, *Die Religion des Ṛgveda*, vol. 2, 31–50.

18 Regarding the identification of *soma* with the moon, which is not established firmly in the *Ṛgveda*, cf. Eva Wilden, *Der Kreislauf der Opfergaben im Veda*, Alt- und Neu-Indische Studien 51 (Stuttgart: Steiner, 2000) and Heinrich Lüders, *Varuṇa*, ed. Ludwig Alsdorf, vol. 2 (Göttingen: Vandenhoeck & Ruprecht, 1959), 698–701.

3 Cosmological Contexts

Already in the first lesson of the SS, the author postulates that the world is dual in nature: fiery and watery—"according to their own relative dominance" (*tadbhūyastvāt*).[19] Here, supposedly "their own" (*tad-*) aims toward the gross elements of fire and water, or even toward the deities Agni and Soma. As Wujastyk mentions, the polarity of *agni* and *soma* is visible in the CS, but less well-established than in the SS and the works of Vāgbhaṭa. However, in order to examine the development of the *agni/soma* concept within the Āyurvedic corpus, it is essential to also consult references from this work. For example, in the CS lesson on the seasons, the year is segmented into watery and fiery halves. We find the same notion in the corresponding lesson of the considerably younger AS and rudimentarily in the AHS. This segmentation comes about via the dominance of the moon in the watery phase beginning with the rainy season and ending with winter, while the sun prevails in the fiery phase, starting with a second cold season named *śiśira* and coming to an end with summer.[20] The terms *āgneya* and *saumya* are not mentioned in the corresponding lessons of the SS, although the division into two phases, governed by the sun and the moon, is discussed here.[21] The mention in the CS and subsequent omission from the AHS suggest that the *agni/soma* concept was introduced in the earlier work but later omitted for whatever reasons.

In another context—concerning the collection of drugs—the model of a watery and a fiery phase is preferred to a more seasonal concept in the SS. Drugs

19 See SS 1.1.22. This passage deals with the definition of the human being (*puruṣa*) and its relationship with its environment in the context of medicine. The human being is described as consisting of five gross elements and a mental constituent (*śarīrin*). It is the only target of medical treatment. Moreover, it is the seat or basis of medical treatment to the extent that the world is divided into stationary and mobile entities. Furthermore, the world is by nature fiery and watery. She, the world, can also be defined as consisting of five elements. The beings that populate her are distinguishable according to their way of giving birth. The human being is defined as the most important among these beings. (*asmin śāstre pañcamahābhūtaśarīrisamavāyaḥ puruṣa ity ucyate. tasmin kriyā so 'dhiṣṭhānaṃ kasmāt lokasya dvaividhyāt loko hi dvividhaḥ sthāvaro jaṅgamaś ca dvividhātmaka evāgneyaḥ saumyaś ca tadbhūyastvāt pañcātmako vā tatra caturvidho bhūtagrāmaḥ saṃsvedajarāyujāṇḍajodbhijjasaṃjñaḥ tatra puruṣaḥ pradhānaṃ tasyopakaraṇam anyat tasmāt puruṣo 'dhiṣṭhānam.*)

20 See CS 1.5.6, AS 1.4.5–6 and AHS 1.3.2–6. The Āyurvedic year knows six seasons: the rainy season (*varṣā*), autumn (*śarad*), winter (*hemanta*), frost (*śiśira*), spring (*vasanta*), and summer (*grīṣma*). Regarding the halves of the year, cf. Vitus Angermeier, "Regenzeiten, Feuchtgebiete, Körpersäfte: Das Wasser in der klassischen Indischen Medizin" (Dissertation, Universität Wien, 2016), 24–29.

21 See SS 1.6.7–8.

with watery and fiery qualities should each be collected during the corresponding phase. The competing idea of collecting specific substances in specific seasons is discarded. Here once more, the world is described as consisting of water and fire.[22] This time, the concept is mentioned in the SS and AS, but, again, not in the AHS.[23]

The BhS also refers to this elementary dualism of the world in a discussion of the digestive fire, including a rather cryptic passage on extended non-eating:

> *rasaṃ ca śoṇitaṃ caiva medo māṃsam athāpi ca* // 18
> *mahatyanaśane nṝṇāṃ sarvāṇy etāni khādati* /
> *agnīṣomātmakaṃ sarvaṃ jagat sthāvarajaṅgamam* // 19
> *agnīṣomātmakāḥ sarve dehinas tu caturvidhāḥ* /
> *sūryātmakāni sarvāṇi*[24] *tathā somātmakāni ca* // 20
> *mahaty anaśane nṝṇāṃ tenāntrāṇi sa*[25] *khādati* /
> BhS 4.4.18cd–21ab

> The nutritious fluid as well as blood, fat, and then also flesh, all of these are consumed [by the digestive fire] during extended non-eating of the people. The whole world, the stationary and the mobile, is of the nature of fire and water. Of the nature of fire and water, again, are all the creatures, which are of four kinds;[26] all of them have the nature of the sun as well as that of the moon. During extended non-eating of the people, it (i.e. the digestive fire, which of course is related to the sun) consumes the intestines.

Apparently, the fact that creatures have a fiery and a watery nature is seen as the reason for disease through imbalance. When the nourishing principle—the fluid or food corresponding to the moon—is missing, the opposite consuming

---

22 See SS 1.36.5. Cf. AS 5.8.4–5 where both models—harvesting according to seasons and according to watery and fiery phases—are presented as different positions without one having preference over the other.

23 The relation of AS and AHS is still under discussion. By now we cannot be sure which of them came first and if they are really by the same author. Cf. Angermeier, "Regenzeiten, Feuchtgebiete, Körpersäfte," 9–11. Both of the above-mentioned cases suggest that the AS follows the older compendiums more literally while the AHS deviates more strongly from its predecessors.

24 Unemended reading of the manuscript: *cāndrāṇi*.

25 Unemended reading of the manuscript: *na*.

26 This refers to the four ways of giving birth as described, for example, in SS 1.1.22 (see fn. 19).

principle—the digestive fire corresponding to the sun—prevails and, in want of fuel, commences consumption of bodily constituents. This passage has no parallels in the other examined works.

The idea of a bipolar world governed by fire and water is also visible in a discussion in the SS on the taste (*rasa*), the potency (*vīrya*), and the post-digestive taste (*vipāka*) of substances. There, some authorities claim that potency is the most important among these three properties. Others, again, argue that this potency has two qualities, hot (*uṣṇa*) and cold (*śīta*), because the whole world is fiery and watery. Their opponents, however, are convinced that there are eight qualities of potency: hot and cold, unctuous (*snigdha*) and dry (*rūkṣa*), clear (*viśada*) and mucilaginous (*picchila*), as well as soft (*mṛdu*) and sharp/intensive (*tīkṣṇa*). Here, the author does not show preference for either of the two theories,[27] though in general potency is seen as eightfold in accordance with the second theory. The AS also mentions the disagreement between the two theories. In that work, however, the world as a whole is not seen as fiery and watery. Instead, the theories refer to qualities such as light and heavy, etc., and to the two divisions of the year, all of which are seen either as fiery or watery. As a result, it is argued, potency must also be twofold, i.e. hot and cold. Substances with hot potency induce a burning sensation, digestion, sweating, and dissolution, as well as pacification of the morbific factors wind and phlegm. Cold potency, on the other hand, brings about content, stability, and stimulation, as well as the pacification of blood and bile. Once again, the author is undecided as to which of the suggested theories should be regarded as valid.[28] It seems that, in the case of potency, different theories were accepted and applied in parallel. As for the *agni/soma* concept, the development from the SS to the AS suggests that the general idea of a dualistic world, introduced in the SS, was adapted by the author of the AS to better mesh with the pre-established medical theories.

27 See SS 1.40.5. For further information on *rasa*, *vīrya*, and *vipāka*, cf. Gerrit Jan Meulenbeld, "Reflections on the Basic Concepts of Indian Pharmacology," in *Studies on Indian Medical History: Papers Presented at the International Workshop on the Study of Indian Medicine Held at the Wellcome Institute for the History of Medicine, 2–4 September 1985*, eds. Gerrit Jan Meulenbeld and Dominik Wujastyk (Groningen: Forsten, 1987), 1–17.

28 See AS 1.17.20–23. The same topic is discussed more briefly in AHS 1.9.17–19.

4 Physiology and Etiology

It is not only the substances of the outside world that are sometimes understood as governed by fire and water; bodily constituents are also discussed under this aspect. As already mentioned by Wujastyk, the SS contemplates the question of whether the nutritious fluid (*rasa*), produced from the ingested food, has a fiery (*taijasa*) or a watery (*saumya*) nature. Ultimately, the fluid is defined as watery because it is known to be oily, stimulating, nourishing, maintaining, etc.[29] The nature of blood, which originates from the nutritious fluid, however, remains somehow uncertain in the subsequent passage.[30] Menstrual blood, as opposed to semen, is definitely seen to be fiery, thus securing that the embryo is provided with both essential elements.[31] How this fieriness comes about is not explained. The polarity of menstrual blood and semen, however is clearly phrased in the Śārīrasthāna of the SS.[32]

In the lesson on the treatment of fever in the CS, the disease is categorized according to several criteria. For instance, it is called twofold insofar as it can be watery or fiery. The AS and AHS break down this clear hot/cold polarity in their respective lessons by talking of mild (*saumya*) and intensive (*tīkṣṇa*) fever.[33] Again, the concept is present in its plain form in the older work and was dissolved and adapted in the AS and AHS. Some lessons later, however, when discussing excessive alcohol consumption, both mention fiery and watery fevers as symptoms of this disease.[34] The CS lesson on fever provides further relevant information. In the following, the morbific factors are used to explain the physical processes. A few lines later, however, the importance of wind is relativized and fire/heat (here: *tejas*) and water come once again to the fore:

29 See SS 1.14.3; cf. Wujastyk, "Agni and Soma," 354–356.

30 The description of blood as nutritious fluid dyed red in SS 1.14.4–5 suggests that it is seen as watery as well. Also the *tu* ("however") in 1.14.7, which introduces the discussion of menstrual blood (seen as fiery), speaks in favor of a watery blood (cf. the subsequent footnote). AS 1.36.5, however, describes blood as fiery and watery at the same time.

31 See SS 1.14.7: Menstrual blood, however, is fiery because the embryo has the nature of *agni* and *soma* (*ārtavaṃ śoṇitaṃ tv āgneyam agnīṣomīyatvād garbhasya*). So, the quality of menstrual blood must follow from the firm assumption that semen is watery. This view is explicitly expressed in AS 2.1.4.

32 Cf. SS 3.3.3–4.

33 See CS 6.3.32 and AS 3.2.46 as well as AHS 3.2.46–47ab.

34 See AS 3.6.15 and AHS 3.6.15.

yogavāhaḥ paraṃ vāyuḥ saṃyogād ubhayārthakṛt // 38
dāhakṛt tejasā yuktaḥ śītakṛt somasaṃśrayāt /
CS 6.3.38cd–39ab

The wind is a "co-carrier"[35] to the highest degree, insofar as it achieves the goals of both [other morbific factors] through connection [to them]. It causes a burning sensation if connected with heat, it causes a cold sensation if it is combined with *soma*.

Here again we clearly have a strong water/fire polarity, while the wind plays a separate role.[36] The stanza appears to have the task of moderating between the *agni/soma* concept and the doctrine of the morbific factors, explaining how a dualistic and a tripartite concept can be reconciled. A passage in the SS, in a lesson on the tastes, aims in the same direction. After an explanation of the influence of the six tastes on the morbific factors bile, wind, and phlegm, the nature of these three factors is characterized in a nutshell:

tatra vāyor ātmaivātmā, pittam āgneyaṃ, śleṣmā saumya iti /
SS 1.42.5

In this case the nature of the wind is exactly its own nature, bile is fiery, phlegm is watery.

Immediately afterward, the author suggests that some authorities distinguish the tastes into two groups. Sweet, bitter and astringent tastes are seen as watery while pungent, sour and salty taste would be fiery.[37] It is noteworthy that nei-

35 The commentator Cakrapāṇidatta explains: *yogavāha* means that, because of the connection, it carries the property of the connected (*yogād yogino guṇaṃ vahatīti yogavāhaḥ*). Similarly, in CS 1.27.249 and SS 1.45.142, honey is described as a strong *yogavāhin*, because it is made of diverse substances and transports their properties. Also, long pepper (*pippalī, Piper longum* L.) is a *yogavāhin* according to CS 3.1.16 and should not be used excessively. So *yogavāha/yogavāhin* seems to be an undesired phenomenon because these substances are complicated in medical application and can even be injurious to health if combined with some unhealthy substance bearing and boosting the influence of its properties.

36 For further evidence on the disparity between wind on one side, and bile and phlegm on the other, cf. Wujastyk, "Agni and Soma," 363–365.

37 See SS 1.42.7: *kecid āhur agnīṣomīyatvāj jagato rasā dvividhāḥ saumyā āgneyāś ca. tatra madhuratiktakaṣāyāḥ saumyāḥ kaṭvamlalavaṇā āgneyāḥ. tatra madhurāmlalavaṇāḥ snigdhā guravaś ca kaṭutiktakaṣāyā rūkṣā laghavaś ca saumyāḥ śītā āgneyāś coṣṇāḥ*. Here, some further properties are attributed to the different tastes. Thus, sweet, sour, and salty

ther the AS nor the AHS, in their respective lessons on taste, mentions *agni* and *soma* or hints about them, even though the lessons correspond closely to the cognate lesson in the SS, which was obviously used as a template. Instead, they rely completely on the doctrine of the five elements, intentionally omitting the *agni/soma* concept.[38]

The BhS also uses the words *agni* and *soma* in the discussion on the disease of the organ *grahaṇī*.[39] There, the excitation of the digestive fire (*agni*), leading to a disease called *atyagni* (too intense fire), is traced back to a diminishment of *soma*.[40] Though *soma* here could be a synonym of *rasa* (nutritious fluid), or could be understood as simply meaning bodily water, the direct juxtaposition with *agni* brings to mind Kirfel's presumption that there is an underlying, archaic conception in Indian medicine, dividing the human body into two parts, governed by fire and water respectively.[41]

---

are described as oily and heavy, while pungent, bitter, and astringent are described as dry and light; furthermore, *saumya* is defined as cold and *āgneya* as hot. However, the application of these properties appear to be inconsistent: Though oily and heavy are definitely properties of the element of water, here sour and salty, while tagged as oily and heavy, are described as fiery; dry and light, on the other hand, are definitely connected to fire, but here bitter and astringent tastes, tagged as dry and light, are defined as watery. For a better visualization of the incongruencies, cf. table 2 below. The description of the different tastes that follows after the above explain the inconsistency to some extent. For example, the tastes of sour and salty are seen as digestive and thereby connected to the digestive fire; bitter and astringent, on the other hand, have purifying and healing properties, which relate them to water (cf. SS 1.42.10).

TABLE 3.2 Tastes and their properties in SS 1.42.7

<table>
<tr><td>sweet (madhura)</td><td>sour (amla)</td><td>salty (lavaṇa)</td><td>pungent (kaṭu)</td><td>bitter (tikta)</td><td>astringent (kaṣāya)</td></tr>
<tr><td>watery (saumya)</td><td colspan="3">fiery (āgneya)</td><td colspan="2">watery (saumya)</td></tr>
<tr><td colspan="3">oily (snigdha) and heavy (guru)</td><td colspan="3">dry (rūkṣa) and light (laghu)</td></tr>
</table>

38 Cf. AS 1.18 and AHS 1.10. In each case, the interrelations between the six tastes and the five elements are discussed at the beginning of the lesson, i.e. AS 1.18.3 and AHS 1.10.1.

39 The *grahaṇī* is the bodily receptacle in which the digestive fire "cooks" the ingested food. If the food or the fire is somehow impaired, the *grahaṇī* itself can become damaged. Such ills are summarized as *grahaṇīdoṣa* or simply *grahaṇī*. Cf. CS 6.15 and AHS 2.3.49–53.

40 Cf. BhS 6.11.15.

41 Cf. Kirfel, *Die fünf Elemente*, 20–22.

5 Substances

The SS describes caustics (*kṣāra*) as both fiery and watery at the same time because they convey properties linked to both elements. This is explained by the fact that caustics are made from many different plants.[42] Such a claim is made neither in the AS nor in the AHS.

Those two works, however, do classify the medicinal plants as fiery and watery in one case at the end of their Kalpasthānas, where a strange remark on geography is also found. While the works examined here usually demonstrate a good awareness of the entire South-Asian subcontinent, this passage describes the world as being mainly covered by the Himālaya and Vindhya mountains. Plants from the first mountain range are referred to as watery; plants from the second as fiery.[43] Since the stanza is poorly connected to the previous passage and concludes the final lesson of the Sthāna, it may be an interpolation.

6 Fire, Water, Wind

In addition to the previously mentioned references, one can find several passages that include more than two dualistic elements without directly hinting at the classic system of five elements. In these cases, it is always wind which is added as the third element.

According to the CS, you should not defecate or urinate in view of wind, fire, or water; moon or sun; or Brahmins or teachers.[44] This clearly indicates a certain respect for these three elements. On the other hand, if you were not allowed to defecate or urinate in the view of the remaining two elements, earth and space/ether, life would become rather complicated.

Also in the CS, at the beginning of the *Nidānasthāna*, the bodily diseases are classified as watery, fiery, and windy.[45] Here we can assume that *saumya* and *āgneya* are in fact synonymous with *ślaiṣmika* (caused by phlegm) and *paittika* (caused by bile). This exact correlation, however, might hint at a historical change from a bipolar water/fire system to a tripolar system of morbific factors; or, perhaps rather, at reconciliation between the two systems. As previ-

42 See SS 1.11.5. Regarding the caustics in the SS, cf. Prafulla Chandra Ray, *A History of Hindu Chemistry from the Earliest Times to the Middle of the Sixteenth Century, A.D.* (London: Williams and Norgate, 1902), 17–22.

43 See AS 5.8.33 and AHS 5.6.29.

44 See CS 1.8.21.

45 See CS 2.1.4.

ously mentioned on p. 23, wind, fire, and water (*soma*) play a certain role in the explanations of fever in the CS. The respective lesson states that fever, which is localized in semen, sets free the semen, destroys the vital breath (*prāṇa*), and, together with wind, fire, and water (again *soma*), spreads everywhere in the body.[46] Here, the elements are likely poetic synonyms for the morbific factors, again reconciling the two systems.

As mentioned by Wujastyk, the SS and AS both classify the lethal points (*marmans*) as fiery and watery.[47] The relevant passages, however, also mention wind-related (*vāyava*) lethal points. Therefore, the classification should not be understood as dualistic, but rather as referring to three different elements, and thus again as pointing toward the doctrine of the three morbific factors.[48]

Finally, small animals such as insects and worms (*kīṭa*)[49] are likewise classified in the SS and AS. The stings and bites of wind-related creatures excite the bodily wind and cause diseases related to it. The stings and bites of fiery and watery creatures affect the respective morbific factors in the same way.[50] Furthermore, the AS analyzes spiders in the same manner in a separate lesson.[51]

46 See CS 6.3.82.

47 Cf. Wujastyk, "Agni and Soma," 356 f.

48 Cf. SS 3.6.16 and 35cd–36ab, as well as AS 2.7.15–19. Injuries to fiery lethal points lead to a quick death, while injuries to wind-related points are fatal only if they set free the bodily wind. Injuries to watery points pose no direct mortal danger. Besides these lethal points, there are others with mixed characteristics. Injuries to watery-fiery points lead to death after awhile, and injuries to fiery-windy points cause severe pain. It is conspicuous that, despite there being five types of lethal points, instead of all five classic elements, only three are applied—namely those directly connected to the morbific factors. Here, it is also important to note that the doctrine of the five elements is apparently not a primordial feature of Āyurvedic medicine. Cf. Angermeier, "Regenzeiten, Feuchtgebiete, Körpersäfte," 172–180.

49 *kīṭa* refers to worms, insects, and other similarly small creatures. Cf. Gerrit Jan Meulenbeld, *The Mādhavanidāna and Its Chief Commentary: Chapters 1–10* (Leiden: Brill, 1974), 455 and Albrecht Wezler, "Epische Streifzüge / Roaming Through the Indian Epics I. Plaintive Cry of a *Kīṭa* Endangered by Traffic: How Are We to React to It?" in *Expanding and Merging Horizons: Contributions to South Asian and Cross-Cultural Studies in Commemoration of Wilhelm Halbfass*, ed. Karin Preisendanz (Wien: Verlag der Österreichischen Akademie der Wissenschaften, 2007), 641–652.

50 See SS 5.8.12cd–15ab and AS 6.43.3–11.

51 Cf. AS 6.44.6–10.

7 Conclusions

In summary, it is noticeable that the division of the world into fiery and watery parts, as well as the application of the *agni/soma* dualism, are frequently presented as alternative opinions or doctrines. In other cases, the concerned topics are not quite central to Āyurveda, or we find alternative doctrines on the same topics elsewhere in the examined works (cf. table 3.3). This suggests that we might be dealing with a concept from another scientific or religious tradition. Or, even more likely, this dualistic conception is rooted in a less scientific, folk medicinal background. Probably it found acceptance in the Āyurvedic corpus by means of its popularity (or by pressure), even though it was in competition with the already established assumptions.

Furthermore, the suggestion—implicitly already made by Kirfel[52] and more obviously by Wujastyk—that this (definitely old) model precedes the doctrine of the morbific factors in an organic way, has been proven more unlikely by the present study. Though many of the passages which mention fire, water/*soma*, and wind as a group of three and, in some cases, connect them to the morbific factors, show an effort to harmonize the various concepts, this is no argument for an organically grown doctrine, but rather represents an attempt at bringing together different conceptions retrospectively. This is also achieved by poetic equations between the dualistic *agni/soma* doctrine, the doctrine of the five classic elements, and the doctrine of the three morbific factors.

As mentioned in the beginning, Wujastyk stated in the conclusion of his study that "[t]his polarity is a clear theme in the āyurvedic literature and beyond. It can be traced backward to the very earliest layer of Indian thought in the Vedas, […]"[53] However, he does not address the question of the migration of the concept from Veda to Āyurveda, nor that of its development within the history of the Āyurvedic discipline. Today we know—not least thanks to the research of Kenneth Zysk—that Āyurveda originated in the surroundings of the Śramaṇa movement, quite separate from the Vedic tradition.[54] This fact, coupled with the circumstance that the *agni/soma* concept is strongly rooted in the Vedic tradition, clearly suggests that the incorporation of this concept was part of the Brāhmaṇization of this scientific tradition. What we observe in the texts today might well be the remains of a struggle between Brāhmaṇic

52 Cf. Kirfel, *Die fünf Elemente*, 42–44.

53 Wujastyk, "Agni and Soma," 366.

54 Cf. Kenneth G. Zysk, *Asceticism and Healing in Ancient India: Medicine in the Buddhist Monastery* (New York: Oxford University Press, 1991) and Johannes Bronkhorst, *Greater Magadha: Studies in the Culture of Early India* (Leiden: Brill, 2007), 56–60.

TABLE 3.3 *Agni* and *soma* in context. The different quotation styles indicate whether *agni/soma* is mentioned as A) a doctrine among two or more equal alternatives; A+) a doctrine preferred above other alternatives; W) a doctrine without any alternative; or as W~) a doctrine without any alternative in the indicated passage, but with an alternative elsewhere

Occurrence and context	Quotation style
SS 1.1.22 (cosmological)	A
CS 1.6.5, AS 1.4.5–6, AHS 1.3.2–6 (year)	W
SS 1.36.5 (seasons for collection of plants)	A+
AS 5.8.4–5 (seasons for collection of plants)	A
BhS 4.4.18cd–21ab (cosmological/dietetic)	W
SS 1.40.5, AS 1,17,20–23, AHS 1.9.17–19 (cosmological/tastes)	A
SS 1.14.3–7, SS 3.3.32 (physiological/nutritious fluid and blood)	W
BhS 4.5.1 (physiological/metabolism)	W~
CS 6.3.32 and 38–39ab, AS 3.6.15 (physiological/fever)	W~
SS 1.42.7 (physiological/tastes)	A
BhS 6.11.15 (physiological/*grahaṇī*)	W~
SS 1.11.5 (substances/caustics)	W
AS 5.8.33, AHS 5.6.29 (substances/medicinal plants)	W~

and original Āyurvedic convictions; one trying to impregnate these works with their ideology, the other struggling to keep their own laboriously compiled theories intact. These diverging modifications could also trace back to single individuals, protecting their knowledge sources against Brāhmaṇic assaults by the sporadic inclusion of Brāhmaṇic ideas.[55] The originally Vedic concept was added to the earlier compendiums (the CS, SS, and BhS) essentially in its pure form. However, when the Brāhmaṇic pressure subsided, authors of later works (the AS and AHS), as well as redactors, dared to integrate it more freely or to dissolve already integrated occurrences of the concept and harmonize it with other medical theories. The BhS, which was less popular and consequentially less extensively revised, still retains the most singular insertions of the concept, without any parallels in the other compendiums.

55 About this struggle, cf. Debiprasad Chattopadhyaya. *Science and Society in Ancient India*, Philosophical Currents 22 (Amsterdam: B.R. Grüner, 1977). In the last chapter (pp. 363–426), he examines this process using the CS as an example; however, the *agni/soma* concept is not addressed there.

There clearly exists a much older and much more deeply rooted concept of hot/cold in these Āyurvedic works than the *agni/soma* concept. As Kirfel has already suggested, this conception could well have been the basis for the doctrine of the morbific factors.[56] However, the *agni/soma* concept is not directly related to it.

TABLE 3.4 Arrangement of the sections (*sthānas*) in the examined works

	CS, BhS	SS	AS, AHS
1	Sūtra	Sūtra	Sūtra
2	Nidāna	Nidāna	Śārīra
3	Vimāna	Śārīra	Nidāna
4	Śārīra	Cikitsā	Cikitsā
5	Indriya	Kalpa	Kalpa
6	Cikitsā	Uttara	Uttara
7	Kalpa		
8	Siddhi		

Primary Sources

AHS Kuṇṭe, Aṇṇā Moreśwar, and Kṛṣṇa Rāmchandra Śāstrī Navre, eds. *The Aṣṭāṅgahṛdayasaṃhitā of Vāgbhaṭa, with the Commentaries Sarvāṅgasundarā of Aruṇadatta & Āyurvedarasāyana of Hemādri*, edited by Sadāśiva Śāstrī Parāḍakara. Sixth edition. Bombay: Nirṇaya Sāgar Press, 1935.

AS Āṭhavale, Anaṃt Dāmodar, ed. *Śrīmad vṛddhavāgbhaṭaviracitaḥ aṣṭāṅgasaṅgrahaḥ induvyākhyāsahitaḥ*. Pune: Śrīmad Ātreya Prakāśanam, 1980.

BhS Sastri, Venkatasubramania and Raja Rajeswara Sarma, eds. *Bhelasaṁhitā*. New Delhi: Central Council for Research in Indian Medicine & Homeopathy, 1977.

CS Jādavaji Trikamji, ed. *The Carakasaṃhitā of Agniveśa, Revised by Charaka and Dṛidhabala, with the Āyurveda-Dīpikā Commentary of Cakrapāṇidatta*. New Delhi: Munshiram Manoharlal, 1941.

56 Cf. Kirfel, *Die fünf Elemente*.

SS Trikamji, Jādavji and Rām Nārāyaṇ, eds. *Suśrutasaṃhitā of Suśruta, with the Nibandhasaṅgraha Commentary of Śrī Dalhanāchārya and the Nyāyachandrikā Pañjikā of Śrī Gayadāsāchārya*. Chaukhamba Ayurvijnan Granthamala 42. Bombay: Nirṇaya Sāgar Press, 1938.

References

Angermeier, Vitus. "Regenzeiten, Feuchtgebiete, Körpersäfte: Das Wasser in der klassischen Indischen Medizin." Dissertation, Universität Wien, 2016. http://othes.univie.ac.at/45525/.

Bronkhorst, Johannes. *Greater Magadha: Studies in the Culture of Early India*. Leiden: Brill, 2007.

Chattopadhyaya, Debiprasad. *Science and Society in Ancient India*. Philosophical Currents 22. Amsterdam: B.R. Grüner, 1977.

Clark, Matthew. *The Tawny One: Soma, Haoma and Ayahuasca*. London: Muswell Hill Press, 2017.

Falk, Harry. "Soma I and II." *Bulletin of the School of Oriental and African Studies, University of London* 52 (1989): 77–90.

Falk, Harry. "Decent Drugs for Decent People: Further Thoughts on the Nature of Soma." *Orientalia Suecana* 51–52 (2003): 141–155.

Frauwallner, Erich. *Geschichte Der Indischen Philosophie*. Vol. 1. Salzburg: Otto Müller, 1953.

Houben, Jan E.M. "The Soma–Haoma Problem: Introductory Overview and Observations on the Discussion." *Electronic Journal of Vedic Studies* 9, no. 1a (2003).

Kirfel, Willibald. *Die Fünf Elemente, insbesondere Wasser und Feuer: Ihre Bedeutung für den Ursprung altindischer und altmediterraner Heilkunde*. Walldorf: Vorndran, 1951.

Lommel, H. "König Soma." *Numen* 2, no. 3 (1955): 196–205.

Lüders, Heinrich. *Varuṇa*, edited by Ludwig Alsdorf. Vol. 1. Göttingen: Vandenhoeck & Ruprecht, 1951.

Lüders, Heinrich. *Varuṇa*, edited by Ludwig Alsdorf. Vol. 2. Göttingen: Vandenhoeck & Ruprecht, 1959.

Meulenbeld, Gerrit Jan. *The Mādhavanidāna and Its Chief Commentary: Chapters 1–10*. Leiden: Brill, 1974.

Meulenbeld, Gerrit Jan. "Reflections on the Basic Concepts of Indian Pharmacology." In *Studies on Indian Medical History: Papers Presented at the International Workshop on the Study of Indian Medicine Held at the Wellcome Institute for the History of Medicine, 2–4 September 1985*, edited by Gerrit Jan Meulenbeld and Dominik Wujastyk, 1–17. Groningen: Forsten, 1987.

Meulenbeld, Gerrit Jan. *A History of Indian Medical Literature.* Vol. 1A. Groningen: Forsten, 1999.

Meulenbeld, Gerrit Jan. *A History of Indian Medical Literature.* Vol. 2A. Groningen: Forsten, 2000.

Oberlies, Thomas. *Die Religion des Ṛgveda.* Vol. 1: Das religiöse System des Ṛgveda. Publications of the De Nobili Research Library 27. Wien: Gerold, 1998.

Oberlies, Thomas. *Die Religion des Ṛgveda.* Vol. 2: Kompositionsanalyse der Soma-Hymnen des Ṛgveda. Publications of the De Nobili Research Library 27. Wien: Gerold, 1999.

Ray, Prafulla Chandra. *A History of Hindu Chemistry from the Earliest Times to the Middle of the Sixteenth Century, A.D.* London: Williams and Norgate, 1902.

Stuhrmann, Rainer. "Ṛgvedische Lichtaufnahmen: Soma botanisch, pharmakologisch, in den Augen der Kavis." *Electronic Journal of Vedic Studies* 13, no. 1 (2006): 1–93.

Wezler, Albrecht. "Epische Streifzüge / Roaming Through the Indian Epics I. Plaintive Cry of a *Kīṭa* Endangered by Traffic: How Are We to React to It?" In *Expanding and Merging Horizons: Contributions to South Asian and Cross-Cultural Studies in Commemoration of Wilhelm Halbfass*, edited by Karin Preisendanz, 641–652. Wien: Verlag der Österreichischen Akademie der Wissenschaften, 2007.

Wilden, Eva. *Der Kreislauf der Opfergaben im Veda.* Alt- und Neu-Indische Studien 51. Stuttgart: Steiner, 2000.

Wujastyk, Dominik. "Agni and Soma: A Universal Classification." *Studia Asiatica* 4–5 (2004): 347–369.

Zysk, Kenneth G. *Asceticism and Healing in Ancient India: Medicine in the Buddhist Monastery.* New York: Oxford University Press, 1991.

CHAPTER 4

Humoral Constitutions in the *Carakasaṃhitā*

Philipp Maas

The early-classical and classical Āyurvedic compendia of Caraka, Bhela, Suśruta, and Vāgbhaṭa contain mutually related and supplementary classification systems concerning the character or natural constitution of human beings that are respectively based on medical, psychological, and physical conceptions.[*,1] Of these, a specific medical classification system that is intimately related to the central Āyurvedic doctrine of the three pathogenetic substances wind (*vāta*), bile (*pitta*), and phlegm (*śleṣman*) (frequently called *doṣa* or humors) became "highly developed" in classical and medieval Āyurveda "and forms one of the most prominent parts of New Age ayurveda."[2] The present article deals with the earliest documented Āyurvedic doctrines of humoral constitutions as they appear in the *Carakasaṃhitā*, a Sanskrit work that can be roughly dated to the period of 100 BCE–200 CE for its older parts.[3] More precisely, the article draws attention to the fact that four different humoral classification systems are attested in the *Carakasaṃhitā*. These systems deviate from each other with regard to the acknowledged number of humoral categories and concerning their respective conceptions of the ideal state to be achieved (or re-established) by medical practice, i.e., health. In its final conclusion, the article argues that these deviations reflect different strategies of harmonizing the doctrine of

---

* Work on this research article was partly supported by the Austrian Science Fund (FWF) in the context of FWF projects P17300-G03 ("Philosophy and Medicine in Early Classical India I") and P19866-G15 ("Philosophy and Medicine in Early Classical India II"). An earlier version of the present article was presented at the 14th World Sanskrit Conference, Kyoto, in the panel "Physicians and Patients: Textual Representations in Pre-Modern South Asia" on 1 September 2009.

1 See Siegfried Lienhard, "Konstitution und Charakter nach den Lehren der altindischen Medizin," *Centaurus* 6 (1959): 82–93.

2 Dominik Wujastyk, *The Roots of Ayurveda* (London, New York, etc.: Penguin Books, 2003), 8. On New Age ayurveda, see Kenneth Gregory Zysk, "New Age Āyurveda or What Happens to Indian Medicine When It Comes to America," *Traditional South Asian Medicine* 6 (2001): 10–26.

3 See Gerrit Jan Meulenbeld, *A History of Indian Medical Literature*, vol. 1A (Groningen: Forsten, 1999), 114. On the complex redactorial history of the *Carakasaṃhitā*, see Philipp A. Maas, "On What Became of the *Carakasaṃhitā* after Dṛḍhabala's Revision," *eJournal of Indian Medicine* 3 (2010): 1–22.

 | DOI:10.1163/9789004438224_005

unchangeable humoral constitutions with the central Āyurvedic conception of health as resulting from a suitable amount of all three humors within the human body.

Carakasaṃhitā Vimānasthāna 8.95 introduces a humoral classification system of seven categories that is similar to the systems figuring in the Āyurvedic compendia of Suśruta, Bhela and Vāgbhaṭa.[4] Caraka's classification system appears in the context of the physician's initial examination of the patient, in which the physician should take into consideration ten factors (*bhāva*) in order to determine the patient's specific strength.[5] The first, and probably most important, factor is the patient's "humoral constitution" (*doṣaprakṛti*).[6] The following section of Vimānasthāna 8.96–100 provides a comprehensive account of this specific classification system.[7]

The initial section describes how the natural constitution of human beings is established by several factors that influence the origination of a body during conception: the sperm, the menstrual fluid, the season of conception, the womb, food and lifestyle of the mother, and the modification of the gross elements that make up the fetus' body. These factors are related to the production of one or two predominant humors in the fetal body, or they may lead to a suitable amount of humors, in which case neither a single humor nor the combination of two humors is predominant. Accordingly, each human being can be classified as possessing one out of seven possible humoral constitutions (see table 4.1 below). Once the above-mentioned factors have established the specific constitution during conception, this constitution remains unchanged for the whole of the human's life.

The immediately following passage, i.e., Vimānasthāna 96–98, comprehensively describes the determination of a specific constitution by means of a tripartite conclusion by analogy. In its first part, the characteristic qualities of the different humors are listed. Phlegm, for example, is said to be unctuous, smooth, soft, sweet, solid, viscid, slow, rigid, heavy, cold, smeary, and clear. Then every quality of each humor is related to a distinct bodily or mental feature of

4 See *Suśrutasaṃhitā* Śārīrasthāna 4.62–79, *Bhelasaṃhitā* Vimānasthāna 4.12–27 and *Aṣṭāṅgahṛdayasaṃhitā* Sūtrasthāna 1.10.

5 The remaining nine factors that are relevant for determining the patient's strength are [2] disease (*vikṛti*), [3] strong points of bodily elements (*sāra*), [4] compactness (*saṃhanana*), [5] measurement (*pramāṇa*), [6] habituation (*sātmya*), [7] character (*sattva*), [8] digestive power (*āhāraśakti*), [9] athletic power (*vyāyāmaśakti*), and [10] stage of life (*vayas*).

6 *Carakasaṃhitā* Vimānasthāna 8.95.

7 For a critical edition and translation of *Carakasaṃhitā* Vimānasthāna 95–100, see Appendix 1 below. The numbering of sections follows the numbering introduced by Jādvaji Trikamji in his 1941-edition of the *Carakasaṃhitā*.

TABLE 4.1 Seven humoral constitutions

Single humor constitutions	Double humor constitutions	Triple humor constitution
1. phlegmatic (*śleṣmala*)	4. phlegmatic-bilious (*saṃsṛṣṭa*)	7. suitable amount (*samadhātu*)
2. bilious (*pittala*)	5. phlegmatic-windy (*saṃsṛṣṭa*)	
3. windy (*vātala*)	6. bilious-windy (*saṃsṛṣṭa*)	

TABLE 4.2 Humoral constitutions and their characteristics, according to *Carakasaṃhitā* Vimānasthāna 8.96–89

Characteristics	Phlegmatic constitution	Bilious constitution	Windy constitution
Power	strong	of middling strength	mostly weak
Wealth	rich	wealthy with means of subsistence	little wealth / means of subsistence
Life expectancy	long-lived	of middling life span	short-lived
Education	learned	having general and practical knowledge	–
Other	energetic	–	having little offspring

the patient. Thus, for example, patients whose basic constitution is dominated by phlegm have clear eyes because phlegm is clear (*accha*). Besides the bodily features or qualities (*guṇa*) that are analogous to the qualities of the humors, each constitution dominated by a single humor is indicated by further typical qualities that concern physical strength, wealth, life span, education, and fertility. Phlegmatic persons are strong, rich, learned, energetic, and long-living, whereas bilious persons are of middling power, have an average life span, and possess general and practical knowledge, wealth and means of subsistence. Finally, windy persons are generally weak. They do not live for long, produce little offspring, have little means and little wealth. In this way, the three natural constitutions determined by a single humor classify humans hierarchically regarding their specific strength, wealth, life expectancy, education, and other factors (see table 4.2 above).

The correspondence between the three humors and the three degrees of strength, etc. is neat. However, the classification system contains four additional basic constitutions, for which the correspondence between the qualities of the humors and the qualities of the body is less clear, namely the three double humor constitutions and the single triple humor constitution. These four constitutions are characterized in Vimānasthāna 8.99–100:

> 99. Persons having mixed characteristics arise from a mixture [of humors]. 100. Persons having a suitable amount [of humors] are endowed with all qualities, ...[8]

The description of persons with constitutions dominated by multiple humors is succinct. The text just refers to "mixed characteristics" and to "all qualities" of the patients' body. What exactly are, however, these characteristics or qualities? Definitely neither strength of the body, nor life expectancy, nor the degree of prosperity can be meant, because the existence of one characteristic logically excludes the existence of its opposite. A mixture and an accumulation of the qualities "having much strength" and "being weak" are, for instance, impossible. The passage can therefore only refer to bodily features derived from the qualities of humors as, for example, the clearness of eyes in the case of a phlegmatic constitution. A mixture of features from different categories would then enable an inference to a mixed basic constitution. This interpretation, however, is also problematic, since even the lists of these bodily characteristics consist to a considerable degree in mutually exclusive pairs of opposites, such as having a body which is pleasant to the eye, smooth, and bright (*dṛṣṭisukhasukumārāvadātaśarīra*), which are characteristics of the phlegmatic constitution, whereas the lists of characteristic of a bilious constitution contains the opposite qualities. Moreover, the phlegmatic constitution is said not to start physical activities quickly, not to fall into despair quickly, and not to be diseased quickly, whereas the windy one is characterized by contrary qualities, etc.

A similar problem arises when a patient is characterised by a suitable amount of all three humors. How can a single person be regarded as having "all qualities" (*sarvaguṇa*), which would mean that this person possesses the above-mentioned, mutually exclusive features at the same time? The eleventh-century commentator Cakrapāṇidatta tried to solve this problem by taking the word *guṇa* not to mean just a general characteristic or quality but to refer exclusively to "good qualities."[9] Accordingly, patients with a constitution derived

8 *saṃsargāt saṃsṛṣṭalakṣaṇāḥ. sarvaguṇasamuditās tu samadhātavaḥ*, ... (see *Appendix*).

9 "Endowed with all qualities" means "connected with all good qualities that were mentioned concerning natural constitutions." *Ayurvedadīpikā* ad *Carakasaṃitā* Vimānasthāna 8.10: *sarvaguṇasamuditā iti sarvaprakṛtiṣūktapraśastaguṇayuktā* (p. 278a, l. 3f.).

from a suitable amount of humors would be characterised by "all good qualities." This interpretation, however, is unsatisfactory for two reasons: First, all inferences of humoral constitutions from the occurrence of bodily features are introduced with the word *evaṃguṇayogāt* "because [these persons] possess such qualities." Here the word *guṇa* clearly designates all listed characteristics, and not just "good qualities." Cakrapāṇidatta's interpretation of the word *guṇa* in the compound *sarvaguṇa* therefore implies that the author changed his terminology within a single, uniform context without indicating that he did so. This is highly unlikely from a pragmatic point of view. Secondly, several of the characteristics ascribed to the humors do not have a clear relational value, which makes it difficult to determine whether a certain characteristic is a good quality or not. For example, it may depend on socially determined values or on circumstances whether having soft hair of the beard is preferable to having rough hair.

In view of these difficulties, the references to combined humoral constitutions in Vimānasthāna 8.95 and 99 f. are problematic. The text neither provides satisfactory information on how the existence of a combined constitution can be determined, nor does it inform us about the influence of the different combined constitutions on the strength, life span, etc. of patients. For which purpose do the three categories of mixed humoral constitutions and the single one of suitable amounts of the humors exist at all?

As mentioned above, the passage Vimānasthāna 8.95–100 is not the only reference to doctrines of humoral constitutions in the *Carakasaṃhitā*. Caraka's compendium contains at least three further doctrines. The first of those occurs in Sūtrasthāna 7.39–40:

> From conception onward, some humans are seen to possess a suitable amount of bile, wind, and phlegm; some [others are seen] to possess superabundant wind, bile, or phlegm. Among these, those mentioned first are healthy, whereas those starting with "possessing superabundant wind" are always diseased, since the natural constitution of their body is taught to be based on a humor (*doṣa*).[10]

10 *samapittānilakaphāḥ kecid garbhādi mānavāḥ / dṛśyante vātalāḥ kecit pittalāḥ śleṣmalās tathā // teṣām anāturāḥ pūrve vātalādyāḥ sadāturāḥ / doṣānuśayitā hy eṣāṃ dehaprakṛtir ucyate //* (*Carakasaṃhitā* Sūtrasthāna 7.39–40, p. 52a, l. 17–20). For a different translation of this passage, see Hartmut Scharfe, "The Doctrine of the Three Humors in Traditional Indian Medicine and the Alleged Antiquity of Tamil Siddha Medicine," *Journal of the American Oriental Society* 119, no. 4 (1999): 618b.

This passage states that there are humans with a humoral constitution based on a suitable amount of all three humors, while the constitution of other humans is dominated by a single one. Accordingly, in contradistinction to the initially discussed doctrine of seven humoral constitutions, the present doctrine acknowledges the existence of just four human constitutions. Only humans with a constitution caused by a suitable amount of humors are said to be truly healthy, while the other three constitutions are regarded as always diseased (*sadātura*).

Two further doctrines concerning the natural constitution of humans are attested in *Carakasaṃhitā* Vimānasthāna 6.13. These are based on two partly contradictory concepts of health and disease. They also differ from each other with regard to their perception of the natural constitution of the body.

> In this regard, some say that no living beings with a suitable [amount of] wind, bile, and phlegm exist, because [all] men consume unsuitable food (i.e., food leading to an unsuitable amount of humors in the bodily elements), and therefore some men have a windy natural constitution, some have a bilious natural constitution, and some have a phlegmatic natural constitution. This, however, is not correct. For which reason? Because physicians consider (*icchanti*) a man with a suitable [amount of] wind, bile, and phlegm healthy, and because the basic constitution [of man] is health, and because health is the aim of the application of remedies (*bheṣajapravṛtti*)—this [application] is essentially medical practice (*ceṣṭārūpā*).[11]

11 *tatra kecid āhuḥ: "na samavātapittaśleṣmāṇo jantavaḥ santi, viṣamāhāropayogitvān manuṣyāṇām. tasmāc ca vātaprakṛtayaḥ kecit, kecit pittaprakṛtayaḥ, kecit punaḥ śleṣmaprakṛtayo bhavanti" iti. tac cānupapannam. kasmāt kāraṇāt? samavātapittaśleṣmāṇaṃ hy arogam icchanti bhiṣajaḥ, prakṛtiś* cārogyam, ārogyārthā ca bheṣajapravṛttiḥ—sā ceṣṭārūpā***. (*Carakasaṃhitā* Vimānasthāna 6.13, p. 255a, l. 41–p. 255b, l. 5 critically edited on the basis of the ten genealogically decisive manuscripts A^{d}, $Ap1^{d}$, $B1^{d}$, $B5^{d}$, $Ba1^{d}$, Ch^{d}, $Jp1^{d}$, Km^{d}, $P1^{ś}$, and $P4^{d}$, and the printed edition $Bo8^{E}$ (see appendix below); selected variant readings: * prakṛtiś] A^{d}, $Ap1^{d}$, $B1^{d}$, $B5^{d}$, $Ba1^{d}$, Ch^{d}, $Jp1^{d}$, Km^{d}, $P1^{ś}$, $P4^{d}$; yataḥ prakṛtiś $Bo8^{E}$. ** ceṣṭārūpā] $P1^{ś}$, $B1^{d}$; ceṣṭarūpa A^{d}, $Ap1^{d}$, $B5^{d}$, $Ba1^{d}$, Ch^{d}, $Jp1^{d}$, Km^{d}, $P4^{d}$, $Bo8^{E}$). The reading *ceṣṭārūpā* suggests itself on the basis of *Carakasaṃhitā* Vimānasthāna 8.77 which defines "activity" (*pravṛtti*) as the type of practice (*ceṣṭā*) aiming at health and on the basis of Vimānasthāna 8.129 which defines *pravṛtti* as the application of Āyurvedic remedies. For different translations of this passage based on the text as printed in $Bo8^{E}$, see Philipp A. Maas, "The Concepts of the Human Body and Disease in Classical Yoga and Āyurveda," *Wiener Zeitschrift für die Kunde Südasiens* 51 (2007/2008): 128 and Dominik Wujastyk "Post-Classical Indian Traditions of Medical Debate and Argumentation: The Rogārogavāda of Vīreśvara," *eJournal of Indian Medicine* 2 (2009): 72 f.

The just-quoted passage first refers to a doctrine of the humoral constitution according to which the constitution of every human being is determined by the predominance of one of the three humors. Accordingly, just three different natural constitutions exist, each one being determined by one particular humor. This view is, however, rejected, because the existence of natural constitutions dominated by a single humor contradicts the fundamental Āyurvedic doctrine concerning the natural state of the human body as healthy on account of a suitable amount of all three humors. Establishing a state in which a suitable amount of all three humors occurs within the human body is, after all, the very aim of medical practice. The text continues:

> Therefore, men with a suitable amount of wind, bile, and phlegm exist, but those with a windy natural constitution, those with a bilious natural constitution, or those with a phlegmatic natural constitution do not exist. Men are said to have a specific humoral constitution because of the superabundance of a specific humor. However, the state of having a natural constitution cannot arise when the humors are modified. Therefore, these are not natural constitutions. Windy, bilious, and phlegmatic men exist, but it should be understood that they are not in their natural constitution.[12]

Although this passage agrees in its conclusion with the passage from Sūtrasthāna 7.39–40 (quoted above) in that only persons with a suitable amount of humors are healthy, the passages differ on the point of how the term *prakṛti* is to be correctly understood. Sūtrasthāna 7.39–40 acknowledges the existence of four different natural constitutions, whereas Vimānasthāna 6.13 denies that different constitutions exist and affirms the existence of only one *prakṛti*, viz. the natural healthy condition of a suitable amount of humors.

Vimānasthāna 6.13 indicates that the theory of the existence of three different humoral constitutions was not readily accepted by the author (or by the redactors) of the *Carakasaṃhitā*, because it is in conflict with the fundamental Āyurvedic conception of health defined as a suitable amount of all three humors. Sūtrasthāna 7.39–40 and Vimānasthāna 6.13 show two different

---

12 *tasmāt santi samavātapittaśleṣmāṇaḥ, na khalu santi vātaprakṛtayaḥ pittaprakṛtayaḥ śleṣmaprakṛtayo vā. tasya tasya kila doṣasyādhikyāt sā sā doṣaprakṛtir ucyate manuṣyāṇām. na ca vikṛteṣu doṣeṣu prakṛtisthatvam upapadyate. tasmān naitāḥ prakṛtayaḥ santi. santi tu khalu vātalāḥ pittalāḥ śleṣmalāś ca, aprakṛtisthās tu te jñeyāḥ* (*Carakasaṃhitā* Vimānasthāna 6.13, p. 255b, l. 5–12).

strategies to solve this conflict. In the first passage, the existence of three constitutions caused by a single humor is admitted, but these constitutions are not accepted as truly healthy states. The state of being healthy is reserved for the fourth constitution, i.e., the one with a suitable amount of humors. The same holds good for Vimānasthāna 6, which, in addition, denies the existence of basic constitutions resulting from single humors.

A different strategy to deal with the theory of three basic constitutions is reflected in Vimānasthāna 8.95–100, where the assumption that three humoral constitutions exist appears as part of the concept of seven basic constitutions. In this more comprehensive doctrine, the constitutions dominated by a single humor retain their original relational values: a phlegmatic constitution is the best among the single-humor constitutions, whereas a bilious constitution is of a middle quality, and a windy constitution is of the lowest quality. However, even the phlegmatic constitution is of a lower quality than the constitution caused by a suitable amount of all three humors, which also surpasses any constitution determined by a mixture of two humors, the quality of which remains unspecified. This theory appears slightly modified in the *Aṣṭāṅgahṛdayasaṃhitā*, where Vāgbhaṭa assigned the lowest position within a triple hierarchy to constitutions established by a combination of two humors:

> These [three humors determine] separately three basic constitutions, viz. the lower, the medium, and the highest. The [constitution determined by a] suitable amount is the best of all; those originating from two humors are the poorest.[13]

As far as I can see, the passage *Carakasaṃhitā* Vimānasthāna 8.95–100 contains the only reference to the theory of seven basic constitutions in the entire work. Therefore, and in view of the fact that this reference is extremely brief and poorly integrated into its context, it is possible that the doctrine of a sevenfold humoral constitution was introduced into the work in the course of its complex transmission history. The relative simplicity of the style of writing in the passage of *Carakasaṃhitā* Vimānasthāna 8.95–124, which deals with the ten factors that the physician has to consider in order to determine the physical strength of the patient, may be taken as additional support of the hypothesis of the secondary nature of the sevenfold humoral doctrine in the

13 *taiś ca tisraḥ prakṛtayo hīnamadhyottamāḥ pṛthak / samadhātuḥ samastāsu śreṣṭhā nindyā dvidoṣajāḥ //* (*Aṣṭāṅgahṛdayasaṃhitā* Sūtrasthāna 1.10, p. 54).

Carakasaṃhitā. If this hypothesis should match historical facts, the insertion must have taken place at a very early stage in the transmission, because the pertinent text passage appears with some variation in all known manuscripts.[14] Possibly, it was the famous redactor Dṛḍhabala who inserted the passage into the *Carakasaṃhitā* in the course of his revision at some time during the period of 300–500 CE.[15]

14 For a critical edition of the passage of *Carakasaṃhitā* Vimānasthāna 8.95–100, see Appendix 1 below.

15 For Dṛḍhabala's date, see Meulenbeld, *A History*, 141.

Appendix. *Carakasaṃhitā* Vimānasthāna 8.95–100: Critically Edited Text and Translation

The critically edited text presented here is identical with the one in Karin Preisendanz, Philipp A. Maas, and Cristina Pecchia, *A Critical Edition of Carakasaṃhitā* Vimānasthāna *Chapter 8* (Vienna forthcoming). The present edition differs from the forthcoming one with regard to the documentation of variant readings in the apparatus. Whereas the present edition reports interesting variant readings only for a set of ten manuscripts and for the printed edition published in 1941 by Jādavjī Trikamjī Ācārya, the critical apparatus of the forthcoming edition documents interesting and acceptable readings from a set of 52 manuscripts and several printed editions.

The witnesses whose readings are presented here are the ten manuscripts that were identified by Philipp A. Maas as decisive for a reconstruction of a version of the text that resembles as closely as possible the version of the allegedly final redaction of the *Carakasaṃhitā* by Dṛḍhabala.[16] More specifically, the testimony of these witnesses allows in many cases for the reconstruction of the archetype A, or at least of the two hyparchetypes K and E (see the stemma in figure 4.1).

Besides the readings of these manuscripts, readings of the Trikamji-edition are also recorded in order to indicate the textual differences between the critically edited text and the text of this quasi-standard edition.

The critical apparatus records variant readings in lemmata that quote the critically edited text. These quotations, printed in bold and terminated by a closing square bracket, are preceded by the number of the line from which the critically edited text is extracted into the lemma. Next, all sigla of witnesses reading the critically edited text are listed. Following a semicolon, the (first) variant to the lemma text appears, which again precedes the list of sigla of witnesses that support this variant reading.

16 Philipp A. Maas, "Computer Aided Stemmatics: The Case of Fifty-Two Text Versions of *Carakasaṃhitā* Vimānasthāna 8.67–157," in *Text Genealogy, Textual Criticism and Editorial Technique*, eds. Jürgen Hanneder and Philipp A. Maas = *Wiener Zeitschrift für die Kunde Südasiens* 52–53 (2009–2010), 33 f. On a further stage of redactorial intervention that must have occurred after Dṛḍhabala's revision, see Philipp A. Maas, "On What Became of the *Carakasaṃhitā* after Dṛḍhabala's Revision," *eJournal of Indian Medicine* 3 (2010): 1–22.

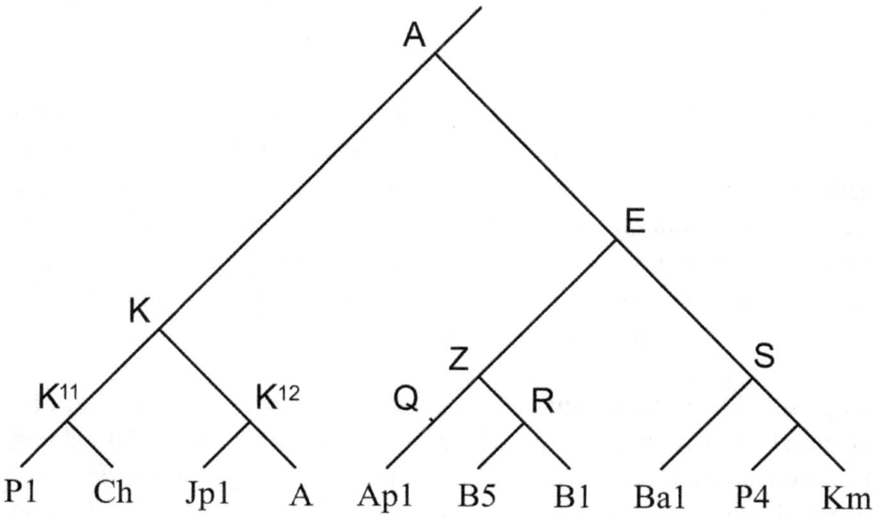

FIGURE 4.1 A hypothetical stemma of *Carakasaṃhitā* Vimānasthāna 8 for ten genealogically decisive manuscripts
Note: The stemma is identical with the rooted cladogram presented in Maas, *Computer Aided Stemmatics*, 95. On the differentiation of text-genealogical useful and less-useful manuscripts of the *Carakasaṃhitā*, see Cristina Pecchia, "Transmission-specific (In)utility, or Dealing with Contamination: Samples from the Textual Tradition of the *Carakasaṃhitā*," in *Text Genealogy, Textual Criticism and Editorial Technique* eds. Jürgen Hanneder and Philipp A. Maas = *Wiener Zeitschrift für die Kunde Südasiens* 52–53 (2009–2010), 121–159.

In the critical apparatus, the following abbreviations and symbols are used:

om.	omits	text missing from the respective witness(es)
(*vl*)	*varia lectio*	variant reading recorded in the printed edition Bo8^{E}
xy	wavy underline	the reconstruction of the archetypal reading is uncertain

Critical Edition and Translation of *Carakasaṃhitā* Vimānasthāna 8.95–100

95. tatrāmī prakṛtyādayo bhāvāḥ, tadyathā—śukraśoṇitaprakṛtiṃ kālagarbhāśayaprakṛtiṃ mātur āhāravihāraprakṛtiṃ mahābhūtavikāraprakṛtiṃ ca garbhaśarīram apekṣate. etāni hi yena yena doṣeṇādhikatamenaikenānekena vā samanubadhyante tena tena doṣeṇa garbho 'nubadhyate. tataḥ sā sā doṣaprakṛtir ucyate manuṣyāṇāṃ garbhādipravṛttā. tasmād vātalāḥ prakṛtyāḥ kecit, pittalāḥ kecit, śleṣmalāḥ kecit, kecit saṃsṛṣṭāḥ, kecit samadhātavaḥ prakṛtyā bhavanti. teṣāṃ hi lakṣaṇāni vyākhyāsyāmaḥ.

96. śleṣmā hi snigdhaślakṣṇamṛdumadhurasārasāndramandastimitaguruśītavijjalācchaḥ. tasya snehāc chleṣmalāḥ snigdhāṅgāḥ, ślakṣṇatvāc chlakṣṇāṅgāḥ, mṛdutvād dṛṣṭisukhasukumārāvadātaśarīrāḥ, mādhuryāt prabhūtaśukravyavāyāpatyāḥ, sāratvāt sārasaṃhatasthiraśarīrāḥ, sāndratvād upacitaparipūrṇasar-

1 **tatrāmī prakṛtyādayo bhāvāḥ**] $P1^{ś}$ Ch^{d} $Jp1^{d}$ A^{d}; tatremān prakṛtyādīn bhāvān anuvyākhyāsyāmaḥ $Ap1^{d}$ $B5^{d}$ $B1^{d}$ $Ba1^{d}$ $P4^{d}$ Km^{d}; tatra prakṛtyādīn bhāvān anuvyākhyāsyāmaḥ $Bo8^{E}$; tatreme prakṛtyādayo bhāvāḥ $Bo8^{E}$ (*vl*) ‖ **-prakṛtiṃ**[1]] $Bo8^{E}$ $P1^{ś}$ Ch^{d} $Jp1^{d}$ A^{d} $Ap1^{d}$; prakṛti $B5^{d}$ $B1^{d}$ $Ba1^{d}$ $P4^{d}$ Km^{d} 2 **-prakṛtiṃ mātur āhāra-**] $B1^{d}$; prakṛtim āturāhāra $Bo8^{E}$ $P1^{ś}$ Ch^{d} $Jp1^{d}$ A^{d} $Ap1^{d}$ $B5^{d}$ $Ba1^{d}$ $P4^{d}$ Km^{d} ‖ **-vihāra-**] $Bo8^{E}$ $P1^{ś}$ Ch^{d} $Jp1^{d}$ A^{d} $B5^{d}$ $B1^{d}$ $Ba1^{d}$ $P4^{d}$ Km^{d}; *om.* $Ap1^{d}$ ‖ **mahābhūtavikāraprakṛtiṃ**] $Bo8^{E}$ $P1^{ś}$ Ch^{d} $Jp1^{d}$ A^{d} $Ap1^{d}$ $Ba1^{d}$ $P4^{d}$ Km^{d}; *om.* $B5^{d}$ $B1^{d}$ ‖ **ca**] $Bo8^{E}$ $P1^{ś}$ Ch^{d} $Jp1^{d}$ A^{d}; *om.* $Ap1^{d}$ $B5^{d}$ $B1^{d}$ $Ba1^{d}$ $P4^{d}$ Km^{d} 3 **apekṣate**] $Bo8^{E}$ $P1^{ś}$ Ch^{d} $Jp1^{d}$ A^{d} $Ap1^{d}$ $Ba1^{d}$ $P4^{d}$ Km^{d}; avekṣeta $B5^{d}$ $B1^{d}$ ‖ **doṣeṇādhikatamenaikenānekena**] $P1^{ś}$ Ch^{d} $Jp1^{d}$ A^{d} $Ap1^{d}$ $B5^{d}$ $B1^{d}$ $Ba1^{d}$ $P4^{d}$ Km^{d}; doṣeṇādhikenaikenānekena $Bo8^{E}$; doṣeṇaikenādhikena samena $Bo8^{E}$ (*vl*) 4 **vā samanubadhyante**] $Bo8^{E}$ $P1^{ś}$ Ch^{d} $Jp1^{d}$ A^{d} $B5^{d}$ $B1^{d}$ $Ba1^{d}$ $P4^{d}$ Km^{d}; badhyante $Ap1^{d}$; vānubadhyante $Bo8^{E}$ (*vl*) ‖ **tataḥ**] $Bo8^{E}$ $P1^{ś}$ Ch^{d} $Jp1^{d}$ A^{d}; tasmāt $Ap1^{d}$ $B5^{d}$ $B1^{d}$; *om.* $Ba1^{d}$ $P4^{d}$ Km^{d} 5 **manuṣyāṇāṃ**] $Bo8^{E}$ $P1^{ś}$ Ch^{d} $Jp1^{d}$ A^{d} $Ap1^{d}$ $B5^{d}$ $B1^{d}$ $P4^{d}$ Km^{d}; mānuṣyāṇāṃ $Ba1^{d}$ ‖ **garbhādi-**] $Bo8^{E}$ $P1^{ś}$ Ch^{d} $Jp1^{d}$ A^{d} $Ap1^{d}$ $Ba1^{d}$ $P4^{d}$ Km^{d}; garbhāṇāṃ hi $B5^{d}$ $B1^{d}$ ‖ **vātalāḥ**] $P1^{ś}$ Ch^{d} $Jp1^{d}$ A^{d} $B5^{d}$ $B1^{d}$ $Ba1^{d}$ $P4^{d}$ Km^{d}; śleṣmalāḥ $Bo8^{E}$ $Ap1^{d}$ ‖ **prakṛtyāḥ**] $P1^{ś}$ Ch^{d} $Jp1^{d}$ A^{d}; prakṛtyā $Bo8^{E}$ $Ap1^{d}$ $B5^{d}$ $B1^{d}$; *om.* $Ba1^{d}$ $P4^{d}$ Km^{d} 6 **kecit śleṣmalāḥ**] $P1^{ś}$ Ch^{d} $Jp1^{d}$ A^{d} $Ba1^{d}$ $P4^{d}$ Km^{d}; *om.* $Ap1^{d}$ $B5^{d}$ $B1^{d}$ ‖ **śleṣmalāḥ**] $P1^{ś}$ Ch^{d} $Jp1^{d}$ A^{d} $Ap1^{d}$ $B5^{d}$ $B1^{d}$ $Ba1^{d}$ $P4^{d}$ Km^{d}; vātalāḥ $Bo8^{E}$ ‖ **kecit kecit**] $B5^{d}$ $B1^{d}$; kecit $Bo8^{E}$ $P1^{ś}$ Ch^{d} $Jp1^{d}$ A^{d} $Ap1^{d}$ $Ba1^{d}$ $P4^{d}$ Km^{d} ‖ **saṃsṛṣṭāḥ kecit**] $Bo8^{E}$ $P1^{ś}$ Ch^{d} $Jp1^{d}$ A^{d} $Ap1^{d}$ $B5^{d}$ $B1^{d}$; *om.* $Ba1^{d}$ $P4^{d}$ Km^{d} ‖ **sama-**] $Bo8^{E}$ $P1^{ś}$ Ch^{d} $Jp1^{d}$ A^{d} $Ap1^{d}$ $Ba1^{d}$ $P4^{d}$ Km^{d}; samā $B1^{d}$ 6–7 **prakṛtyā bhavanti**] Ch^{d} $Jp1^{d}$ A^{d} $B5^{d}$ $B1^{d}$ $Ba1^{d}$ $P4^{d}$ Km^{d}; kecit prakṛtyā bhavanti $P1^{ś}$; prakṛtyā bhavanti kecit $Ap1^{d}$; kecid bhavanti $Bo8^{E}$ 7 **hi**] $Bo8^{E}$ $P1^{ś}$ Ch^{d} $Jp1^{d}$ A^{d}; *om.* $Ap1^{d}$ $B5^{d}$ $B1^{d}$ $Ba1^{d}$ $P4^{d}$ Km^{d} ‖ **vyākhyāsyāmaḥ**] $Bo8^{E}$ $P1^{ś}$ Ch^{d} $Jp1^{d}$ A^{d} $Ap1^{d}$ $Ba1^{d}$ $P4^{d}$ Km^{d}; parīkṣet $B5^{d}$ $B1^{d}$ 8 **śleṣmā hi**] $Bo8^{E}$ $P1^{ś}$ Ch^{d} $Jp1^{d}$ A^{d} $Ap1^{d}$ $Ba1^{d}$ $P4^{d}$ Km^{d}; śleṣmādi $B5^{d}$ $B1^{d}$ 8–9 **-śīta-**] $Bo8^{E}$ $P1^{ś}$ Ch^{d} $Jp1^{d}$ A^{d} $Ap1^{d}$ $Ba1^{d}$ $P4^{d}$ Km^{d}; śītā $B5^{d}$ $B1^{d}$ 9 **-vijjalācchaḥ**] $Bo8^{E}$ $P1^{ś}$ Ch^{d} $Ap1^{d}$ $B5^{d}$ $B1^{d}$; bhijjalāchas $Jp1^{d}$ A^{d}; pichalāchaḥ $Ba1^{d}$ $P4^{d}$ Km^{d} ‖ **snigdhāṅgāḥ**] $Bo8^{E}$ $P1^{ś}$ Ch^{d} $Jp1^{d}$ A^{d} $Ap1^{d}$ $Ba1^{d}$ $P4^{d}$ Km^{d}; snigdhādrāḥ $B5^{d}$ $B1^{d}$ ‖ **ślakṣṇatvāc chlakṣṇāṅgāḥ**] $Bo8^{E}$ $P1^{ś}$ Ch^{d} $Jp1^{d}$ A^{d} $Ap1^{d}$ $B5^{d}$ $B1^{d}$; ślakṣṇatvān $Ba1^{d}$ $P4^{d}$ Km^{d} 10 **-sukumārāvadātaśarīrāḥ**] $P1^{ś}$ Ch^{d} $Jp1^{d}$ A^{d}; sukumārāvadātagātrāḥ $Bo8^{E}$ $Ap1^{d}$ $B5^{d}$ $B1^{d}$ $Ba1^{d}$ $P4^{d}$ Km^{d} 11 **sāndratvād**] $Bo8^{E}$ $P1^{ś}$ $Jp1^{d}$ A^{d} $Ap1^{d}$ $B5^{d}$ $B1^{d}$ $Ba1^{d}$ $P4^{d}$ Km^{d}; sārdratvād Ch^{d} 11–46.1 **-sarvagātrāḥ**] $P1^{ś}$ $Jp1^{d}$ A^{d} $Ap1^{d}$ $B5^{d}$ $B1^{d}$ $Ba1^{d}$ $P4^{d}$ Km^{d}; sarvāṅgāḥ $Bo8^{E}$ Ch^{d}

95. In this regard, these factors beginning with "natural constitution" are the following ones: The body of the embryo depends upon the nature of sperm and menses, upon the nature of the time [of conception], upon the nature of the womb, upon the nature of the food and lifestyle of the mother,[17] and upon the nature of the modifications of the gross elements. The fetus is related to exactly the same superabundant humor—either individually or in combination [with another one]—to which these [factors] are related. Accordingly, it is taught that humans have various humoral constitutions from conception onward. Therefore, some constitutions are windy, some are bilious, some are phlegmatic, some are mixed, and in some constitutions the humors occur [in a] suitable [amount]. We shall indeed describe their characteristics.

96. Phlegm is known to be unctuous, smooth, soft, sweet, solid, viscid, slow, rigid, heavy, cold, smeary, and clear. Because of its unctuousness, men with a phlegmatic constitution possess unctuous limbs. Because of its smoothness, their limbs are smooth. Because of its softness, their body is pleasant to behold, very tender and bright. Because of its sweetness, they have a lot of sperm, frequently have intercourse, and produce many children. Because of its solidness, their body is solid, compact, and steady. Because of its viscidness, their whole body is well-nourished and full. Because if its slowness, their behavior, food

17 On the reading *prakṛtiṃ mātur āhāra-* which is preferable to its alternative *prakṛtim āturāhāra*, although the latter was probably the reading of the archetype, see Philipp A. Maas, "On What to Do with a Stemma: Towards a Critical Edition of *Carakasaṃhitā* Vimānasthāna 8," in *Medical Texts and Manuscripts in Indian Cultural History*, eds. Dominik Wujastyk, Anthony Cerulli, and Karin Preisendanz (New Delhi: Manohar Publishers, 2013), 45–48.

vagātrāḥ, mandatvān mandaceṣṭāhāravyāhārāḥ, staimityād aśīghrārambhakṣobhavikārāḥ, gurutvāt sārādhiṣṭhitāvasthitagatayaḥ, śaityād alpakṣuttṛṣṇāsaṃtāpasvedadoṣāḥ, vijjalatvāt suśliṣṭasārabandhasaṃdhānāḥ, tathācchatvāt prasannadarśanānanāḥ prasannavarṇasvarāś ca. ta evaṃguṇayogāc chleṣmalā balavanto vasumanto vidyāvanta ojasvina āyuṣmantaś ca.

97. pittam uṣṇaṃ tīkṣṇaṃ dravaṃ visram amlaṃ kaṭukaṃ ca. tasyauṣṇyāt pittalā uṣṇāsahā asukhasukumārāvadātagātrāḥ prabhūtapipluvyaṅgapiṭakāḥ kṣutpipāsāvantaḥ kṣipravalīpalitakhālityadoṣāḥ prāyo mṛdvalpakapilaśmaśrulomakeśāḥ. taikṣṇyāt tīkṣṇaparākramās tīkṣṇāgnayaḥ prabhūtāśanapānāḥ kleśāsahiṣṇavo dandaśūkāḥ. dravatvāc chithilamṛdusandhibandhamāṃsāḥ prabhūtasṛṣṭasvedamūtrapurīṣāś ca. visratvāt pratatapūtivakṣaḥkakṣāsyaśiraḥśarīragandhāḥ. kaṭvamlatvād alpaśukravyavāyāpatyāḥ. ta evaṃguṇayogāt pittalā madhyabalā madhyāyuṣo jñānavijñānavittopakaraṇavantaś ca.

---

1 **-vyāhārāḥ**] *Bo8*^{E} *P1*^{ś} *Ch*^{d} *Jp1*^{d} *A*^{d} *Ap1*^{d} *Ba1*^{d} *P4*^{d} *Km*^{d}; vyāhārā *B5*^{d} *B1*^{d} 2 **-vikārāḥ**] *Bo8*^{E} *P1*^{ś} *Ch*^{d} *Jp1*^{d} *A*^{d} *Ap1*^{d} *Ba1*^{d} *P4*^{d} *Km*^{d}; vikārān *B5*^{d} *B1*^{d} ‖ **śaityād**] *Bo8*^{E} *P1*^{ś} *Ch*^{d} *Jp1*^{d} *A*^{d} *Ap1*^{d} *Ba1*^{d} *P4*^{d} *Km*^{d}; satyāc *B5*^{d} *B1*^{d} ‖ **alpa-**] *Bo8*^{E} *P1*^{ś} *Ch*^{d} *Jp1*^{d} *A*^{d} *Ap1*^{d} *Ba1*^{d} *P4*^{d} *Km*^{d}; cālpa *B5*^{d} *B1*^{d} 3 **-sāra-**] *Bo8*^{E} *P1*^{ś} *Ch*^{d} *Jp1*^{d} *A*^{d} *Ap1*^{d} *Ba1*^{d} *P4*^{d} *Km*^{d}; *om.* *B5*^{d} *B1*^{d} ‖ **-bandhasaṃdhānāḥ**] *P1*^{ś} *Ch*^{d} *Jp1*^{d} *A*^{d} *P4*^{d} *Km*^{d}; saṃdhibandhanāḥ *Bo8*^{E} *Ap1*^{d} *B5*^{d} *B1*^{d}; bandhasaṃdhibandhanāsaṃdhānās *Ba1*^{d} 4 **prasanna-**^{2}] *P1*^{ś} *Ch*^{d} *Jp1*^{d} *A*^{d}; prasannasnigdha *Bo8*^{E} *Ap1*^{d} *B5*^{d} *B1*^{d} *Ba1*^{d} *P4*^{d} *Km*^{d} ‖ **ca**] *P1*^{ś} *Ch*^{d} *Jp1*^{d} *A*^{d}; ca bhavanti *Bo8*^{E} *Ba1*^{d} *P4*^{d} *Km*^{d}; bhavanti *Ap1*^{d} *B5*^{d} *B1*^{d} ‖ **ta**] *Bo8*^{E} *P1*^{ś} *Ch*^{d} *Jp1*^{d} *A*^{d} *Ap1*^{d} *Ba1*^{d} *P4*^{d} *Km*^{d}; *om.* *B5*^{d} *B1*^{d} ‖ **-guṇa-**] *Bo8*^{E} *P1*^{ś} *Ch*^{d} *Ap1*^{d} *B5*^{d} *B1*^{d} *Ba1*^{d} *P4*^{d} *Km*^{d}; guṇā *Jp1*^{d} *A*^{d} 5 **balavanto vasumanto**] *Bo8*^{E} *P1*^{ś} *Ch*^{d} *Jp1*^{d} *A*^{d} *Ba1*^{d} *P4*^{d} *Km*^{d}; vasumanto *Ap1*^{d} *B5*^{d} *B1*^{d} ‖ **ojasvina**] *P1*^{ś} *Ch*^{d} *Jp1*^{d} *A*^{d} *Ap1*^{d} *B5*^{d} *B1*^{d}; ojasvinaś ca *Ba1*^{d} *P4*^{d} *Km*^{d}; ojasvinaḥ śāntā *Bo8*^{E} ‖ **ca**] *P1*^{ś} *Ch*^{d} *Jp1*^{d} *A*^{d}; ca bhavanti *Bo8*^{E} *Ap1*^{d} *Ba1*^{d} *P4*^{d} *Km*^{d}; ca bhavanti svapne ca śvetavasanabhavanasukumārajaladharapulinaramaṇīyāni rūpāṇi paśyantīti śleṣmaharāṇāṃ caiṣām upaśayo viparītānāṃ cānupaśaya iti śleṣmaprakṛtayaḥ *B5*^{d} *B1*^{d} 6 **dravaṃ**] *Bo8*^{E} *P1*^{ś} *Ch*^{d} *Jp1*^{d} *A*^{d} *B5*^{d} *B1*^{d} *Ba1*^{d} *P4*^{d} *Km*^{d}; drutaṃ *Ap1*^{d} ‖ **amlaṃ**] *Bo8*^{E} *Ap1*^{d} *B5*^{d} *Ba1*^{d} *P4*^{d} *Km*^{d}; amblaṃ *P1*^{ś} *Ch*^{d} *Jp1*^{d} *A*^{d} *B1*^{d} 7 **pittalā**] *em.*; pittalā bhavanty *Bo8*^{E} *Ap1*^{d} *B5*^{d} *B1*^{d} *Ba1*^{d} *P4*^{d} *Km*^{d}; *om.* *P1*^{ś} *Ch*^{d} *Jp1*^{d} *A*^{d} ‖ **uṣṇāsahā asukha-**] *em.*; uṣṇā'svāhāsukha *B5*^{d} *B1*^{d}; uṣṇāsahāḥ śuṣka *Ap1*^{d} *Ba1*^{d} *P4*^{d} *Km*^{d}; uṣṇasukha *P1*^{ś} *Ch*^{d} *Jp1*^{d} *A*^{d}; uṣṇāsahā uṣṇamukhāḥ *Bo8*^{E} ‖ **-piplu-**] *P1*^{ś} *Ch*^{d} *Jp1*^{d} *A*^{d}; pilla *Ba1*^{d} *P4*^{d} *Km*^{d}; viplu *Bo8*^{E} *Ap1*^{d} *B5*^{d} *B1*^{d} ‖ **-vyaṅga-**] *P1*^{ś} *Ch*^{d} *Jp1*^{d} *A*^{d}; vyaṅgatilaka *Ap1*^{d} *B5*^{d} *B1*^{d} *Ba1*^{d} *P4*^{d} *Km*^{d}; vyaṅgatila *Bo8*^{E} ‖ **-piṭakāḥ**] *P1*^{ś} *Ch*^{d} *Jp1*^{d} *A*^{d}; piḍakāḥ *Bo8*^{E} *Ap1*^{d} *B5*^{d} *B1*^{d} *Ba1*^{d} *P4*^{d} *Km*^{d} 8 **-khālitya-**] *Bo8*^{E} *P1*^{ś} *Ch*^{d} *Jp1*^{d} *A*^{d} *Ap1*^{d} *Ba1*^{d} *P4*^{d} *Km*^{d}; khālatya *B5*^{d} *B1*^{d} 9 **-keśāḥ**] *P1*^{ś} *Ch*^{d} *Jp1*^{d} *A*^{d} *Ap1*^{d} *B5*^{d} *B1*^{d} *Ba1*^{d} *P4*^{d} *Km*^{d}; keśāś ca *Bo8*^{E} 9–10 **taikṣṇyāt … 5 -māṃsāḥ**] *Bo8*^{E} *P1*^{ś} *Ch*^{d} *Jp1*^{d} *A*^{d} *Ap1*^{d} *Ba1*^{d} *P4*^{d} *Km*^{d}; *om.* *B5*^{d} *B1*^{d} 10 **dandaśūkāḥ**] *Bo8*^{E} *P1*^{ś} *Ch*^{d} *Jp1*^{d} *A*^{d} *Ba1*^{d} *P4*^{d} *Km*^{d}; dandasūkā iva *Ap1*^{d} ‖ **-bandha-**] *P1*^{ś} *Ch*^{d} *Jp1*^{d} *A*^{d} *Ap1*^{d} *B5*^{d} *B1*^{d} *Ba1*^{d} *P4*^{d} *Km*^{d}; *om.* *Bo8*^{E} 11 **ca**] *Bo8*^{E} *P1*^{ś} *Ch*^{d} *Jp1*^{d} *A*^{d} *Ba1*^{d} *P4*^{d} *Km*^{d}; *om.* *Ap1*^{d} *B5*^{d} *B1*^{d} ‖ **visratvāt**] *Bo8*^{E} *P1*^{ś} *Ch*^{d} *Jp1*^{d} *A*^{d} *Ap1*^{d} *B1*^{d} *Ba1*^{d} *P4*^{d} *Km*^{d}; viśrutatvāt *B5*^{d} ‖ **pratata-**] *P1*^{ś} *Ch*^{d} *Jp1*^{d} *A*^{d} *B5*^{d} *B1*^{d} *Ba1*^{d} *P4*^{d} *Km*^{d}; prabhūta *Bo8*^{E} *Ap1*^{d} ‖ **-pūti-**] *Bo8*^{E} *Ap1*^{d} *B5*^{d} *B1*^{d} *Ba1*^{d} *P4*^{d} *Km*^{d}; prati *P1*^{ś} *Ch*^{d} *Jp1*^{d} *A*^{d} ‖ **-kakṣāsya-**] *Bo8*^{E} *Ap1*^{d} *B5*^{d} *B1*^{d} *Ba1*^{d} *P4*^{d} *Km*^{d}; kakṣyāsya *P1*^{ś} *Ch*^{d} *Jp1*^{d} *A*^{d} ‖ **-śiraḥ-**] *Bo8*^{E} *P1*^{ś} *Jp1*^{d} *A*^{d} *Ap1*^{d} *B5*^{d} *B1*^{d} *Ba1*^{d} *P4*^{d} *Km*^{d}; *om.* *Ch*^{d} 12 **-amlatvād**] *Bo8*^{E} *A*^{d} *Ap1*^{d} *B5*^{d} *B1*^{d} *Ba1*^{d} *P4*^{d} *Km*^{d}; amblatvād *P1*^{ś} *Ch*^{d} *Jp1*^{d} 13 **madhyāyuṣo**] *P1*^{ś} *Ch*^{d} *Jp1*^{d} *A*^{d} *Ba1*^{d} *P4*^{d} *Km*^{d}; madhyāyu *B5*^{d} *B1*^{d}; madhyāyuṣo madhya *Bo8*^{E} *Ap1*^{d} ‖ **jñānavijñāna-**] *Bo8*^{E} *P1*^{ś} *Ch*^{d} *Jp1*^{d} *A*^{d} *B5*^{d} *B1*^{d}; jñāna *Ap1*^{d} *Ba1*^{d} *P4*^{d} *Km*^{d} ‖ **ca**] *P1*^{ś} *Ch*^{d} *Jp1*^{d} *A*^{d}; ca bhavanti *Bo8*^{E} *Ap1*^{d} *B5*^{d} *B1*^{d} *Ba1*^{d} *P4*^{d} *Km*^{d}

intake, and speech are slow. Because of its rigidness, they are not quick in their undertakings, are not quickly agitated, and not quickly diseased. Because of its heaviness, their motion is firm, regulated, and determined. Because of its coldness, they are not much affected by hunger and thirst, heat and sweat. Because of its smeariness, they have very firm and solid tendons and joint. And in the same way, because of its clearness, they have clear eyes and faces, and they have a clear complexion and a clear voice. Men with a phlegmatic constitution, because they are in this way endowed with qualities, are strong, rich, learned, energetic, and long-lived.

97. Bile is hot, sharp, liquid, stinking and sour and pungent. Because of its heat, men with a bilious constitution cannot bear heat; their body is neither pleasant, nor very tender, nor bright. They have many freckles, blemishes, and blisters, suffer from hunger and thirst, suffer from the fault of getting early [in their life] wrinkles, grey hair, and a bald head. They have mostly thin, little, or reddish hair of their beard, their body, and their head. Because of its sharpness, they are sharp in their enterprises, possess sharp bodily fires, eat and drink a lot, are unable to bear hardships, and they are mordacious. Because of its liquidity, they have loose and soft tendons, joints and muscle flesh, and they discharge a lot of sweat, urine, and feces. Because of its bad smell, they diffuse widely a stinking body smell from their breast, armpits, mouth, and head. Because it is sour and pungent, they have little sperm, rarely have intercourse, and have few children. Men with a bilious constitution, because they are in this way endowed with qualities, are of middling power and of a middling life span. They possess general and practical knowledge, wealth and means of subsistence.

98. vātas tu rūkṣalaghucalabahuśīghraśītaparuṣaviśadaḥ. tasyaraukṣyād vātalā rūkṣāpacitālpaśarīrāḥ pratatarūkṣakṣāmabhinnasaktajarjarasvarā jāgarūkāś ca. laghutvāl laghucapalagaticeṣṭāhāravyāhārāḥ. calatvād anavasthitasandhyasthibhrūhanvoṣṭhajihvāśiraḥskandhapāṇipādāḥ. bahutvād bahupralāpakaṇḍarāsirāpratānāḥ. śīghratvāc chīghrabhīmārambhakṣobhavikārāḥ śīghrottrāsarāgavirāgāḥ śrutagrāhiṇo 'lpasmṛtayaś ca. śaityāc chītāsahiṣṇavaḥ pratataśītakodvepakastambhāḥ. pāruṣyāt paruṣakeśaśmaśrunakhadaśanavadanapāṇipādāṅgāḥ. vaiśadyāt sphuṭitāṅgāvayavāḥ satatasandhiśabdagāminaś ca. ta evaṃguṇayogād vātalāḥ prāyeṇālpabalāś cālpāyuṣaś cālpāpatyāś cālpasādhanāś cādhanyāś ca.

99. saṃsargāt saṃsṛṣṭalakṣaṇāḥ.

100. sarvaguṇasamuditās tu samadhātavaḥ, ity evaṃ prakṛtitaḥ parīkṣeta.

---

1 **rūkṣalaghu-**] *Bo8*^{E} *P1*$^{\acute{s}}$ *Ch*^{d} *Jp1*^{d} *A*^{d} *B5*^{d} *B1*^{d} *Ba1*^{d} *P4*^{d} *Km*^{d}; laghurūkṣa *Ap1*^{d} ‖ **-cala-**] *Bo8*^{E} *P1*$^{\acute{s}}$ *Ch*^{d} *Jp1*^{d} *A*^{d} *Ap1*^{d}; capala *B5*^{d} *B1*^{d} *Ba1*^{d} *P4*^{d} *Km*^{d} ‖ **-viśadaḥ**] *Bo8*^{E} *Ap1*^{d} *B5*^{d} *B1*^{d} *Ba1*^{d} *P4*^{d} *Km*^{d}; viṣadaḥ *P1*$^{\acute{s}}$ *Ch*^{d} *Jp1*^{d} *A*^{d} 2 **-rūkṣa-**] *Bo8*^{E} *P1*$^{\acute{s}}$ *Ch*^{d} *Jp1*^{d} *A*^{d} *Ap1*^{d} *Ba1*^{d} *P4*^{d} *Km*^{d}; rū *B5*^{d} *B1*^{d} ‖ **-bhinna-**] *P1*$^{\acute{s}}$ *Ch*^{d} *Jp1*^{d} *A*^{d} *Ap1*^{d} *B5*^{d} *B1*^{d} *Ba1*^{d} *P4*^{d} *Km*^{d}; *om.* *Bo8*^{E} ‖ **-sakta-**] *P1*$^{\acute{s}}$ *Ch*^{d} *Jp1*^{d} *A*^{d}; sannasakta *Bo8*^{E} *B5*^{d} *B1*^{d} *Ba1*^{d} *P4*^{d} *Km*^{d}; sanna *Ap1*^{d} 3 **ca**] *P1*$^{\acute{s}}$ *Ch*^{d} *Jp1*^{d} *A*^{d}; ca bhavanti *Bo8*^{E} *Ap1*^{d} *B5*^{d} *B1*^{d} *Ba1*^{d} *P4*^{d} *Km*^{d} ‖ **-gati-**] *Bo8*^{E} *Ap1*^{d} *B5*^{d} *B1*^{d} *Ba1*^{d} *P4*^{d} *Km*^{d}; gata *P1*$^{\acute{s}}$ *Ch*^{d} *Jp1*^{d} *A*^{d} ‖ **calatvād**] *Bo8*^{E} *P1*$^{\acute{s}}$ *Jp1*^{d} *A*^{d} *Ap1*^{d} *B5*^{d} *B1*^{d}; capalatvād *Ch*^{d} *Ba1*^{d} *P4*^{d} *Km*^{d} ‖ **-sandhy-**] *Bo8*^{E} *Ch*^{d} *Jp1*^{d} *A*^{d} *Ap1*^{d} *B5*^{d} *B1*^{d} *Ba1*^{d} *P4*^{d} *Km*^{d}; ≈ sakthy *P1*$^{\acute{s}}$ 3–4 **-asthi-**] *P1*$^{\acute{s}}$ *Ch*^{d} *Jp1*^{d} *A*^{d} *B5*^{d} *B1*^{d} *Ba1*^{d} *P4*^{d} *Km*^{d}; akṣi *Bo8*^{E} *Ap1*^{d} 4 **-hanvoṣṭha-**] *Bo8*^{E} *P1*$^{\acute{s}}$ *Ch*^{d} *Jp1*^{d} *A*^{d} *Ap1*^{d} *Ba1*^{d} *P4*^{d} *Km*^{d}; hatvoṣṭha *B5*^{d} *B1*^{d} ‖ **-śiraḥ-**] *Bo8*^{E} *P1*$^{\acute{s}}$ *Ch*^{d} *Jp1*^{d} *A*^{d} *Ap1*^{d} *Ba1*^{d} *P4*^{d} *Km*^{d}; śira *B5*^{d} *B1*^{d} ‖ **-pāṇi-**] *Bo8*^{E} *P1*$^{\acute{s}}$ *Ch*^{d} *Jp1*^{d} *A*^{d} *Ap1*^{d} *Ba1*^{d} *P4*^{d} *Km*^{d}; rūpāṇi *B5*^{d} *B1*^{d} ‖ **bahu-**] *Bo8*^{E} *P1*$^{\acute{s}}$ *Ch*^{d} *Jp1*^{d} *A*^{d} *Ap1*^{d} *B5*^{d} *B1*^{d}; bahula *Ba1*^{d} *P4*^{d} *Km*^{d} 4–5 **-kaṇḍarā-**] *Bo8*^{E} *P1*$^{\acute{s}}$ *Ch*^{d} *Jp1*^{d} *A*^{d} *Ap1*^{d} *Ba1*^{d} *P4*^{d} *Km*^{d}; gaṃḍarā *B5*^{d} *B1*^{d} 5 **-sirā-**] *Bo8*^{E} *P1*$^{\acute{s}}$ *Ch*^{d} *Jp1*^{d} *A*^{d}; śirā *Ap1*^{d} *B5*^{d} *B1*^{d} *Ba1*^{d} *P4*^{d} *Km*^{d} ‖ **chīghrabhīmārambha-**] *P1*$^{\acute{s}}$ *Ch*^{d} *Jp1*^{d} *A*^{d} *Ap1*^{d} *Ba1*^{d} *P4*^{d} *Km*^{d}; chīghram ārambha *B5*^{d} *B1*^{d}; chrīghrasamārambha *Bo8*^{E} 5–6 **śīghrottrāsarāgavirāgāḥ**] *P1*$^{\acute{s}}$ *Ch*^{d} *Jp1*^{d} *A*^{d} *B5*^{d} *B1*^{d} *Ba1*^{d} *P4*^{d} *Km*^{d}; śīghratrāsarāgavirāgāḥ *Bo8*^{E}; *om.* *Ap1*^{d} 6 **'lpasmṛtayaś**] *Bo8*^{E} *Ap1*^{d} *B5*^{d} *B1*^{d} *Ba1*^{d} *P4*^{d} *Km*^{d}; 'smṛtayaś *P1*$^{\acute{s}}$ *Ch*^{d} *Jp1*^{d} *A*^{d} 7 **paruṣa-**] *Bo8*^{E} *P1*$^{\acute{s}}$ *Ch*^{d} *Jp1*^{d} *A*^{d} *B5*^{d} *B1*^{d} *Ba1*^{d} *P4*^{d} *Km*^{d}; paruṣasphuṭita *Ap1*^{d} ‖ **-śmaśru-**] *P1*$^{\acute{s}}$ *Ch*^{d} *Jp1*^{d} *A*^{d} *B5*^{d} *B1*^{d} *Ba1*^{d} *P4*^{d} *Km*^{d}; śmaśruloma *Ap1*^{d}; śmaśruroma *Bo8*^{E} ‖ **-daśanavadana-**] *Bo8*^{E} *P1*$^{\acute{s}}$ *Ch*^{d} *Jp1*^{d} *A*^{d} *B5*^{d} *B1*^{d}; vadanadaśana *Ba1*^{d} *P4*^{d} *Km*^{d}; daśana *Ap1*^{d} 8 **-pādāṅgāḥ**] *P1*$^{\acute{s}}$ *Ch*^{d} *Jp1*^{d} *A*^{d} *B5*^{d} *B1*^{d} *Ba1*^{d} *P4*^{d} *Km*^{d}; pādāḥ *Bo8*^{E} *Ap1*^{d} ‖ **vaiśadyāt**] *Bo8*^{E} *Ap1*^{d} *B5*^{d} *B1*^{d} *Ba1*^{d} *P4*^{d} *Km*^{d}; vaiṣadyāt *P1*$^{\acute{s}}$ *Ch*^{d} *Jp1*^{d} *A*^{d} ‖ **ca**] *P1*$^{\acute{s}}$ *Ch*^{d} *Jp1*^{d} *A*^{d} *Ap1*^{d} *B5*^{d} *B1*^{d}; ca bhavanti *Bo8*^{E} *Ba1*^{d} *P4*^{d} *Km*^{d} 9 **vātalāḥ**] *Bo8*^{E} *Ap1*^{d} *B5*^{d} *B1*^{d} *Ba1*^{d} *P4*^{d} *Km*^{d}; *om.* *P1*$^{\acute{s}}$ *Ch*^{d} *Jp1*^{d} *A*^{d} 9–10 **cālpāpatyāś cālpasādhanāś**] *Bo8*^{E} *Ap1*^{d} *B5*^{d} *B1*^{d}; cālpasādhanāś cālpāpatyāś *Ba1*^{d} *P4*^{d} *Km*^{d}; cālpasādhanāś *P1*$^{\acute{s}}$ *Ch*^{d} *Jp1*^{d} *A*^{d} 10 **cādhanyāś**] *P1*$^{\acute{s}}$ *Ch*^{d} *Jp1*^{d} *A*^{d} *B5*^{d} *B1*^{d} *Ba1*^{d} *P4*^{d} *Km*^{d}; cālpadhanāś *Bo8*^{E} *Ap1*^{d} ‖ **ca**] *P1*$^{\acute{s}}$ *Ch*^{d} *Jp1*^{d} *A*^{d}; ca bhavanti *Bo8*^{E} *Ap1*^{d} *B5*^{d} *B1*^{d} *Ba1*^{d} *P4*^{d} *Km*^{d} 11 **saṃsargāt saṃsṛṣṭa-**] *Bo8*^{E} *P1*$^{\acute{s}}$ *Ch*^{d} *Jp1*^{d} *A*^{d} *Ba1*^{d} *P4*^{d} *Km*^{d}; saṃsṛṣṭāt saṃsargā *Ap1*^{d}; saṃsṛṣṭāt saṃsṛṣṭa *B5*^{d} *B1*^{d} 12 **ity**] *Bo8*^{E} *P1*$^{\acute{s}}$ *Ch*^{d} *Jp1*^{d} *A*^{d}; bhavantīty *Ap1*^{d} *B5*^{d} *B1*^{d} *Ba1*^{d} *P4*^{d} *Km*^{d} ‖ **evaṃ**] *Bo8*^{E} *P1*$^{\acute{s}}$ *Ch*^{d} *Jp1*^{d} *A*^{d} *Ap1*^{d} *Ba1*^{d} *P4*^{d} *Km*^{d}; eva *B5*^{d} *B1*^{d} ‖ **parīkṣeta**] *Bo8*^{E} *P1*$^{\acute{s}}$ *Ch*^{d} *Jp1*^{d} *A*^{d} *Ap1*^{d} *Ba1*^{d} *P4*^{d} *Km*^{d}; parīkṣet *B5*^{d} *B1*^{d}

98. Wind, however, is dry, light, movable, manifold, quick, cold, rough, and transparent. Because of its dryness, men with a windy constitution have dry, thin, and short bodies; their voice is always dry, thin, interrupted, obstructed, and infirm, and they are wakeful. Because of its lightness, they walk, behave, eat, and speak unsteadily. Because of its moveability, their joints, bones, eyebrows, jaws, lips, tongue, head, shoulders, hands, and feet are unsteady. Because of its manifoldness, their chattering, their sinews, and the ramifications of their tubes are manifold. Because of its quickness, their undertakings, agitations, and diseases come quickly and fiercely. They quickly become frightened, attached, and detached. They understand what they hear and memorize little. Because of its coldness, they cannot bear the cold. They always feel cold, tremble, and are stiff. Because of its roughness, they have rough hair of the head, hair of the beard, nails, teeth, faces, hands, feet and limbs. Because of its clearness, the parts of their limbs are split, and they always produce sound with their joints when they move. These men with a windy constitution, because they are in this way endowed with qualities, are mostly weak, they do not live long, produce little offspring, and have little means and little wealth.

99. Men with mixed characteristics arise from a mixture [of humors].

100. Men with a suitable amount [of humors] are endowed with all qualities. In this way he should examine [the patient] according to his natural constitution.

References

Sigla of Witnesses for the Carakasaṃhitā

Superscripts	^{d} Devanāgarī; ^{ś} Śāradā; ^{E} Printed Edition
A^{d}	Alwar, Rajasthan Oriental Research Institute 2498
$Ap1^{d}$	Alipur, Bhogilal Leherchand Institute of Indology 5283
$B1^{d}$	Bikaner, Rajasthan Oriental Research Institute 1566
$B5^{d}$	Bikaner, Anup Sanskrit Library 3996
$Ba1^{d}$	Baroda, Oriental Institute 12489
$Bo8^{E}$	Text of the *Carakasaṃhitā* as printed in Trikamji 1941
Ch^{d}	Chandigarh, Lal Chand Research Library 2315
$Jp1^{d}$	Jaipur, Maharaja Sawai Man Singh II Museum 2068
Km^{d}	Kathmandu, Nepal-German Manuscript Preservation Project E-40 553
$P1^{ś}$	Pune, Bhandarkar Oriental Research Institute 555 of 1875–1876
$P4^{d}$	Pune, Anandashram 1546

Printed Sources

Aṣṭāṅgahṛdayasaṃhitā. See Vogel 1965.

Bhelasaṃhitā: Text with English Translation, Commentary, and Critical Notes, edited by Priya Vrat Sharma. Haridas Ayurveda Series 8. Varanasi: Chaukhama Visvabharati, 2000.

Carakasaṃhitā. For Vimānasthāna 8, see Preisendanz, Maas, and Pecchia, forthcoming; for other parts of the work, see Trikamji 1941.

Suśrutasaṃhitā, with the Nibandhasaṅgraha Commentary of Dalhanāchārya and the Nyāyacandrikā Pañjikā of Gayadāsāchārya on Nidānasthāna, edited by Jādavji Trikamji Āchārya from the beginning to the 9th *adhyāya* of Cikitsāsthāna and the rest by Nārāyaṇ Rām Āchārya. Chaukhamba Ayurvijnan Granthamala 42. Varanasi: Chaukhamba Surbharati Prakashan, 2003 [Bombay, 1933].

Lienhard, Siegfried. “Konstitution und Charakter nach den Lehren der altindischen Medizin.” *Centaurus* 6 (1959): 82–93.

Maas, Philipp A. “The Concepts of the Human Body and Disease in Classical Yoga and Āyurveda.” *Wiener Zeitschrift für die Kunde Südasiens* 51 (2007/2008): 125–162.

Maas, Philipp A. “On What Became of the *Carakasaṃhitā* after Dṛḍhabala’s Revision.” *eJournal of Indian Medicine* 3 (2010): 1–22.

Maas, Philipp A. “Computer Aided Stemmatics: The Case of Fifty-Two Text Versions of *Carakasaṃhitā* Vimānasthāna 8.67–157.” In *Text Genealogy, Textual Criticism and Editorial Technique*, edited by Jürgen Hanneder and Philipp A. Maas = *Wiener Zeitschrift für die Kunde Südasiens* 52–53 (2009–2010): 63–119.

Maas, Philipp A. "On What to Do with a Stemma: Towards a Critical Edition of *Carakasaṃhitā* Vimānasthāna 8." In *Medical Texts and Manuscripts in Indian Cultural History*, edited by Dominik Wujastyk, Anthony Cerulli, and Karin Preisendanz, 29–61. New Delhi: Manohar Publishers, 2013.
Meulenbeld, Gerrit Jan. *A History of Indian Medical Literature*. Vol. 1A. Groningen Oriental Studies 15. Groningen: Forsten, 1999.
Pecchia, Cristina. "Transmission-specific (In)utility or Dealing with Contamination: Samples from the Textual Tradition of the *Carakasaṃhitā*." In *Text Genealogy, Textual Criticism and Editorial Technique*, edited by Jürgen Hanneder and Philipp A. Maas = *Wiener Zeitschrift für die Kunde Südasiens* 52–53 (2009–2010): 121–159.
Preisendanz, Karin, Philipp A. Maas and Cristina Pecchia. *A Critical Edition of Carakasaṃhitā Vimānasthāna Chapter 8*. Vienna, forthcoming.
Scharfe, Hartmut. "The Doctrine of the Three Humors in Traditional Indian Medicine and the Alledged Antiquity of Tamil Siddha Medicine." *Journal of the American Oriental Society* 119, no. 4 (1999): 609–629.
Trikamji, J. (1941). *Caraka Saṃhitā by Agniveśa, Revised by Caraka and Dḍhabala, with the Āyurveda-Dīpikā Commentary of Cakrapāṇidatta*, edited by Jādavjī Trikamjī Ācārya. Krishnadas Ayurveda Series 66. Varanasi: Krishnadas Academy, 2000 [Bombay, 1941].
Vogel, Claus. *Vāgbhaṭa's Aṣṭāṅgahṛdayasaṃhitā: The First Seven Chapters of its Tibetan Version*. Abhandlungen für die Kunde des Morgenlandes 38, no. 2. Mainz: Deutsche Morgenländische Gesellschaft—Wiesbaden: Steiner (in Kommission), 1965.
Wujastyk, Dominik. *The Roots of Ayurveda: Selections from Sanskrit Medical Writings. Translated with an Introduction and Notes*. Third edition. London, New York, etc.: Penguin Books, 2003.
Wujastyk, Dominik. "Post-Classical Indian Traditions of Medical Debate and Argumentation: The Rogārogavāda of Vīreśvara." *eJournal of Indian Medicine* 2 (2009): 67–81.
Zysk, Kenneth Gregory. "New Age Āyurveda or What Happens to Indian Medicine When It Comes to America." *Traditional South Asian Medicine* 6 (2001): 10–26.

CHAPTER 5

Diachronic Migration of Ancient Indian Medical Literature: Divisions and Paratextual Elements in the *Carakasaṃhitā*

Cristina Pecchia

As a tribute to Professor Kenneth Zysk and a token of appreciation for the inspiring conversations we have had on the history of Indian medicine, I present a study that expands on a paper delivered at the *Asian Diversity in a Global Context* conference held in 2010 (November 11–13) at the University of Copenhagen.* The paper was part of the "Transmission of Sanskrit Medical Literature in India" panel convened by Professor Zysk. Inspired by the concept of his panel, in which the "migration of knowledge" was mentioned,[1] I have used the metaphor of migration to examine the transmission history of the *Carakasaṃhitā* (the oldest source of Āyurveda in Sanskrit)[2] from the point of view of its internal divisions—a subject also relevant to my ongoing exploration of the history of philology in South Asia.[3] This study, in particular, focuses on divisions within a thematic section of the *Carakasaṃhitā*. It has two main objectives, both of which are related to each other: 1) to examine how the

* Research for this article was conducted using the facilities of the Institute for the Cultural and Intellectual History of Asia (IKGA) at the Austrian Academy of Sciences, Vienna (Austria), which I gratefully acknowledge. The data concerning the *Carakasaṃhitā* derives from my work on a series of three projects on *Philosophy and Medicine in Early Classical India*, which the Austrian Science Fund (FWF) generously financed in the years 2006 to 2012 [Project Numbers P 17300–G03, P 19866–G15, and P 23330–G15]. I would also like to thank the editors for their helpful feedback and careful editing.

1 See https://asiandynamics.ku.dk/english/past-conferences/adi-conference-2010---asian-diversity/panels_and_workshops/panel04/ (accessed July 20, 2020).

2 Based on previous studies and his further examination of the available data, Jan Meulenbeld concluded that "the author called Caraka cannot have lived later than about A.D. 150–200 and not much earlier than about 100 B.C." (Jan Meulenbeld, *A History of Indian Medical History*, vol. 1A: Text (Groningen: Egbert Forsten, 1999), 114).

3 Philology and related expressions are used with reference to practices employed in "making sense of texts," according to the definition provided by Sheldon Pollock in "Future Philology? The Fate of a Soft Science in a Hard World" (*Critical Inquiry* 35 (Summer 2009): 949). For further elaborations starting from this definition, see Cristina Pecchia, "Introduction", in *Editors of Sanskrit Texts*, ed. C. Pecchia (Vienna, Austrian Academy of Sciences Press, forthcoming).

 | DOI:10.1163/9789004438224_006

divisions are signaled in the text itself and to analyze textual segments with a paratextual nature; 2) to reflect on the philological activity around the text as it is revealed by the fluctuation of textual divisions and paratextual elements during the migration of the text through many centuries and different cultural contexts. Some introductory remarks will clarify how this type of research is situated in the field of South Asian studies and in the larger framework of textual and cultural studies.

1 Texts and the Metaphor of "Migration"

It has been observed that "migrations of knowledge remain above all instances of *translation* in the 'narrow', non-metaphorical sense of that word; i.e., "the process of transforming a specific piece of one language (commonly a text of some sort) into another language"."[4] In fact, migrations of knowledge imply migrations of texts. The metaphor of migration has been applied to both texts and knowledge in William Sweet's collected volume *Migrating Texts and Traditions* (2012), which explores the migration of philosophical texts and traditions into other cultures. An analysis which explicitly adopts the term "migration" in connection with texts and their translations is offered by Andrew Pettegree in his study of how translations of the Spanish text *Amadis de Gaule* migrated through Europe in the sixteenth century.[5]

As shown by these studies, the metaphor of migration allows us to structure how we think about the presence and circulation of texts in an environment that is in various ways "foreign" to the texts themselves; that is, since the texts were produced in contexts culturally, geographically, and/or temporally distant from those into which they migrated. Based on these variables, we can speak in terms of synchronic and/or diachronic migrations respectively. Through this kind of framework, we can thus refer to the complex set of phenomena called reception and transmission of texts. If linguistic translation is one of the most visible operations which allow for texts to function in a foreign context, any act of physical reproduction of written texts through copying or printing can sim-

4 I. Amelung, J. Kurtz, and M. Lackner, "Introduction," in *New Terms for New Ideas: Western Knowledge and Lexical Change in Late Imperial China*, eds. I. Amelung, J. Kurtz, and M. Lackner (Leiden: Brill, 2001), 1; with a quotation from Scott L. Montgomery, *Science in Translation: Movements of Knowledge through Cultures and Time* (Chicago, London: The University of Chicago Press, 2000), 4.

5 Andrew Pettegree, "Translation and the Migration of Texts," in *Borders and Travellers in Early Modern Europe*, ed. Thomas Betteridge (Aldershot: Ashgate Variorum, 2007), 113–127.

ilarly engender their synchronic and diachronic migrations. The written word committed to a portable support can circulate anytime and anywhere, just as the written word committed to a fixed support (as in the case of inscriptions) can be read by newcomers, and thus travel with them through space and time.

Through the metaphor of migration certain phenomena can be considered as resulting from processes of adaptation to new contexts rather than from intentional or unintentional modifications. The present analysis of textual divisions as attested in the written transmission of the *Carakasaṃhitā* offers an example of such processes of adaption.

2 Divisions and Organization of the Text

Text divisions can be investigated at various levels of granularity, from a macro-level structure (which, using blanket terms from the European text and book cultures, corresponds to divisions such as books, chapters, or, more generally, sections) to a micro-level one. If, as noted by Marie-Paule Péry-Woodley, "to structure implies to segment, namely, to assemble and to divide at the same time,"[6] the study of the divisions of a text through its migration history implies dealing with changes in the view of its overall structure, too. To do so, one needs to identify text divisions, their shifting, and their (changing) markers, as well as a variety of conventions applied to the text in different times and places. This also implies formulating a hypothesis about the reasons and modalities of the acts of segmentation in each instantiation of the text, from its initial conception to the different phases of its reception—which requires investigating the uses of the text, interpretations of it, changes in mentality and taste, and approaches to textuality in particular times and places, with text segmentation revealing the agents' attitudes for a better reception of the text. In this connection, the study of text divisions is such that the process is as important as the goal, since the questions that arise on the way urge us to think from different perspectives on the text's history.

Useful conceptual tools can be found in the fields of literary criticism and linguistics, where divisions and organization of texts have become a special subject of reflection since the 1970s. In particular, they form part of the concept

6 Marie Paule Péry-Woodley, "Modes d' organisation et de signalisation dans des textes procéduraux," *Les discours procéduraux*, ed. Claudine Garcia-Debanc, in *Langages* 35, no. 141 (2001): 28, as translated in: Florence Bretelle-Establet and Stéphane Schmitt, "Introduction," in *Pieces and Parts in Scientific Texts*, eds. Florence Bretelle-Establet and Stéphane Schmitt (Cham: Springer, 2018), 9.

of paratext created by Gérard Genette and of the notions of textual segmentation and text architecture that have been developed in new trends of discourse analysis.[7] Péry-Woodley has, for example, elaborated on the concept of markers of text organization, distinguishing between visual, lexical, and syntactic markers.[8] As this requires considering a text as a sequence of linguistic signs as much as its different physical embodiments, such conceptual tools have been crucial in widening the scope of textual studies and go beyond content-centered investigations. However, since these tools have been developed based on a notion of text that mostly coincides with written texts and the printed word, their application to historical oriented research requires some adaptation. In the case of texts belonging to Sanskritic culture, one particularly has to take into account the oral performance of texts.[9]

3 Studies of Premodern South Asian Literature

The premodern literature of South Asia has only been sparsely investigated from the point of view of text division and organization. In a pioneering article published in 1957, "Les divisions dans les textes sanskrits," Louis Renou provided an overview of how major text divisions are designated in Vedic and classical literature. The richness of the materials he presented showed quite well the relevance of this kind of study—not only in an analytical approach to texts, but also from a historical point of view. Most interestingly, in his introductory remarks, Renou pointed out that text divisions can be even more fluctuating than content, that textual traditions differ from one another in this regard, and that certain works that formed a continuous unit were subsequently divided or segmented in a way quite different from the previous one:

7 Gérard Genette, *Seuils* (Paris: Editions du Seuil, 1987), translated into English in 1997 (see *References* below); Jacques Virbel, "Langage et métalangage dans le texte du point de vue de l'édition en informatique textuelle," *Cahiers de Grammaire* 10 (1985): 5–7. With a focus on the history of science and based on documents of the past from different parts of the world, these tools have been more recently applied in some of the papers comprising the 2015 volume *Texts, Textual Acts and the History of Science*, eds. Karine Chemla and Jacques Virbel; and, with a specific focus on texts from Asia, in the 2018 volume *Pieces and Parts in Scientific Texts*, eds. Bretelle-Establet and Schmitt.

8 See Péry-Woodley, "Modes d'organisation et de signalisation dans des textes procéduraux."

9 On oral performance of texts and its relation with manuscript cultures in South Asia, see the overview recently presented in Jan E.M. Houben and Saraju Rath, "Introduction: Manuscript Culture and Its Impact in 'India': Contours and Parameters," in *Aspects of Manuscript Culture in South India*, ed. Saraju Rath (Leiden: Brill, 2012), 1–53.

> Il apparait en effet à l'expérience que les divisions à l'intérieur des textes sont souvent flottantes—plus flottantes encore que la teneur même—, que les traditions ont varié et que, dans une partie des cas il a été noté secondairement un découpage pour certain œuvres qui primitivement formaient un ensemble continu ou avaient une fragmentation différente.[10]

Furthermore, with regard to Vedic literature, he identified two types of text division. The first division is based on the contents and what can be called "the internal development" of the work; the other is completely external and mechanical, clearly a more recent type of division because it is based on the Saṃhitā as a complete work:

> On sait qu'il existe deux divisions, l'une qui répond à la fois au contenu et à ce qu'on peut appeler le développement intrinsèque de l'ouvrage, l'autre tout extérieure et mécanique: celle-ci nécessairement plus récente, puisqu'elle repose sur la Saṃhitā achevée.[11]

The avenues of research outlined by Renou were taken up by a few scholars. Their contributions mostly focused on the history of texts notable for both their size and their circulation, as, for example, Dieter Schlingloff's "The Oldest Extant Parvan-List of the *Mahābhārata*" (1969), Oskar Von Hinüber's "Die Neun Aṅgas: Ein früher Versuch zur Einteilung buddhistischer Texte" (1994), and Asko Parpola's "On the Upagrantha(sūtra) of the Sāmaveda. 1. Contents and Textual Divisions" (1996). Furthermore, text structures were studied in connection with the oral performance and memorization of texts, which played a major role in determining the modalities of text segmentation in the early phases.[12]

The past decade has seen a renewed interest in the topic from within the field of South Asian studies. Lance Cousins and Anālayo, for example, have focused on the Nikāyas and Āgamas respectively, while Klaus Karttunen has offered another general overview of the names of text divisions, and Karin Preisendanz has analysed, among other texts, the *Mahābhārata*, the *Nyāyasūtra*, and the *Carakasaṃhitā*.[13] These studies have mostly focused on seman-

---

10 Louis Renou, "Les divisions dans les textes sanskrits," *Indo-Iranian Journal* 1 (1957): 1.

11 Renou, "Les divisions dans les textes sanskrits," 1–2.

12 See, for example, Mark Allon, *Style and Function: A Study of the Dominant Stylistic Features of the Prose Portions of Pāli Canonical Sutta Texts and their Mnemonic Function* (Tokyo: The International Institute for Buddhist Studies, 1997).

13 See *References* for details.

tic devices that indicate the macro-segmentation of texts. Birgit Kellner has instead examined the issues discussed by Indian commentators with regard to the order of chapters (*pariccheda*) in Dharmakīrti's *Pramāṇavārttika*, and has emphasized that they revolve around the question of how epistemology relates to the Buddha's doctrine.[14] Further, the present writer has analyzed how the structure of the *Pramāṇasiddhi* chapter of the *Pramāṇavārttika* was variously understood by interpreters in different times and regions, namely Indian commentators, Tibetan commentators and scholars, and contemporary scholars.[15]

All these cases clearly demonstrate the fluctuation of text divisions mentioned by Renou and the role of text structuring in the process of receiving a text into one's own cultural environment. Organizing a text was (and is) a matter of textual interpretation as much as a matter of the cultural representations and practices surrounding the text. What has come down to us are very likely successful interpretations with a long-lasting influence on the subsequent readings of the text. One reason for the influence of such interpretations is their use of various structuring devices to provide the text with a sense of clarity tailored to the habits or taste of the audience.[16] A most recent example is the success of the Tibetan analytical system of the *sa bcad* (synoptic subdivisions) among modern students of the *Pramāṇavārttika* from Europe, America, and Japan. The text divisions and organization of the *sa bcad* typically enhance clarity by means of structuring devices that include a systematic identification of the text segments. This clarity probably played an important role in the modern disregard for possible text structures revealed by the text itself or pointed out in the commentaries of the Sanskritic tradition.[17]

---

14 Birgit Kellner, "First Logic, then the Buddha? Remarks on the Chapter Sequence of Dharmakīrti's *Pramāṇavārttika*," *Hōrin* 11 (2004): 147–167. For an overview of the discussion, see Cristina Pecchia, *Dharmakīrti on the Cessation of Suffering* (Leiden: Brill, 2015), 43–45.

15 Pecchia, *Dharmakīrti on the Cessation of Suffering*, Chapter Four and Appendix C.

16 I use the term "clarity" following João Dionísio's remarks concerning tables of contents as "an auxiliary mechanism of reading that embodies an aspect of the rhetorical ideal of *claritas*" and his remarks concerning text division (*divisio textus*) as another auxiliary device connected with *claritas*; see João Dionísio, "Tables of Contents in Portuguese Late Medieval Manuscripts," in *The Book as Artefact: Text and Border*, issue 4 of *Variants*, eds. Anne Mette Hansen, Roger Lüdeke, Wolfgang Streit, Cristina Urchueguía, and Peter Shillingsburg (Amsterdam, New York: Rodopi, 2005), 94–95.

17 Pecchia, *Dharmakīrti on the Cessation of Suffering*, 63–65 and 73–74.

4 Text Divisions and Their Denomination: Paratext in a Historical Perspective

The markers of text divisions typically fall under what Genette termed "paratext." Limiting ourselves to written texts, the concept of paratext may prove problematic in connection with texts that circulated over a long period of time, typically as manuscript and printed books. On the one hand, in a text with a significant migration history, denominations of text divisions might appear only in some of the text instantiations. This makes it difficult to establish whether such phrases were part of the authorial text or not; how they were regarded by different audiences; and if, in their absence, an audience was nevertheless aware of specific text divisions. On the other hand, markers of text divisions do depend on strategies developed within different technologies of the book. Printed books often feature the use of visual and graphical features, such as indentation, new or blank lines, and different typefaces and dispositions on the page that mark off textual divisions and/or emphasize linguistic paratextual elements, such as headings, subheadings, or numeration. These kinds of markers are difficult to catch *prima facie* in manuscript books, because the visual environment determined by the *scriptio continua* may not indicate significant breaks or highlighting strategies (while it closely reflects the flow of sound in an oral textual performance). However, in instances of *scriptio continua*, as much as in cases of aerated script (in which, as observed by Paul Saenger, "space was present, but not consistently between every word"),[18] the flow of signs can display a visual organization of the text through a variety of strategies, from spaces to the application of colors, and from symbols to punctuation marks.[19] Linguistic forms of paratext may instead remain elusive in manuscripts, where they are rarely highlighted by visual means and therefore have to be detected through syntactical analysis. For this reason, as regard linguistic markers of text divisions, I prefer to speak of "liminal phrases," which set out segments within the text flow by expressly designating them verbally and/or numerically. If their function of paratext is not visually highlighted, liminal phrases can only be noted through attentive reading (as opposed to consultation); therefore, they can be distinguished from evident paratext and be considered "hidden" paratext.

18 Paul Saenger, *Space Between Words: The Origins of Silent Reading* (Stanford: Stanford University Press, 1997), 32.

19 The 2020 monograph *Dividing Texts* by Bidur Bhattarai (see *References* below) is to my knowledge the first attempt to describe systematically such visual strategies in manuscripts produced within a specific period in some regions of South Asia.

Since both text divisions and their markers (more, in particular, their denominations) most likely reflect the choice of an editor, rather than the intention of an author, their identification is quite relevant for the critical constitution of a text. Furthermore, since divisions and organization of a text are most likely modified through interventions that are global in scope rather than through occasional or unintentional changes, their fluctuation potentially indicates major phases of revision in the migration history of a text, together with interpretations and editorial strategies typical of a certain period. Given the absence of external sources of information, such as introductory remarks, editorial notes, or archive documents, the texts themselves can—and, indeed, must—function as sources for the study of editorial activities and interpretative attitudes over the centuries.

The following case study focuses on a short section of the *Carakasaṃhitā* known as the "*vādamārga* section." The section ranks far below larger divisions of the text's structure, such as *sthāna*s (books) and *adhyāya*s (chapters), which are uniformly repeated in the witnesses of the text, although with different orders of chapters within some books.[20] In fact, the "*vādamārga* section" is a good example of the fluctuation of text division markers, their different formats, and, more generally, of the problem of identifying the text's microstructure and its related markers. This is especially true of manuscripts and sometimes of early printed editions, as it is often difficult to identify paratexts and text divisions and to assess their origin, since explicit statements from the editors concerning the procedures and methods adopted are mostly absent.[21] However, even if the impact of the printer on early printed editions is imponderable, the layout and exterior appearance of the texts can be used as sources of information about the editors' views.

---

20 For a recent reconsideration of this issue, see Philipp Maas, "On What Became of the *Carakasaṃhitā* after Dṛḍhabala's Revision," *eJournal of Indian Medicine* 3 (2010): 1–22.

21 A notable exception is Rahula Sāṅkṛtyāyana, who in the introduction to his edition of Manorathanandin's *Pramāṇavārttikavṛtti* (p. iv) states: "I am responsible for supplying the headings and the sub-headings in the present work in which also I was helped by this commentary."

5 The "*Vādamārga* Section" of the *Carakasaṃhitā*, Vimānasthāna 8

5.1 *Divisions and Text Segment Denominations*

The eighth chapter of the Vimānasthāna of the *Carakasaṃhitā* (hereafter respectively abbreviated as Vimāna 8 and CaS) includes a section known among scholars as *vādamārga*, or "Course of Debate," in Satis Chandra Vidyabhushana's rendering from 1921.[22] It comprises a set of items that illustrate the procedures of a formal debate. Several scholars after Vidyabhushana have pointed out its peculiar character and importance in the history of debate in South Asia. A few studies have drawn attention to divergences in the arrangement of topical segments among the available witnesses,[23] which is a highly important aspect of the history of the textual transmission.[24] In Jādavji Trikamji's 1941 edition, which we take as our reference edition, the section extends from paragraph 27 to paragraph 66 of Vimāna 8. Here, the section begins with *imāni tu khalu padāni bhiṣagvādamārgajñānārtham adhigamyāni bhavanti* and concludes with *iti vādamārgapadāni yathoddeśam abhinirdiṣṭāni bhavanti*. The first paragraph lists all elements of the Course of Debate. Each of the subsequent paragraphs corresponds to a short text segment introduced by a liminal phrase which marks the beginning of a new topical division. By listing the liminal phrases in Trikamji's edition, the skeleton of the section on the Course of Debate becomes clear. This is represented in Table 5.1, where each element is sequentially numbered for the sake of convenience, and accompanied by Vidyabhushana's English translation[25] in round brackets:

TABLE 5.1 Liminal phrases in the *vādamārga* section[26]

1	*tatra vādo nāma* ... (debate)
2	*dravya-guṇa-karma-sāmānya-viśeṣa-samavāyāḥ* (substance, quality, action, generality, particularity, inherence) ... *ślokasthāne pūrvam uktāḥ*
3	*atha pratijñā—pratijñā nāma* ... (proposition)

22 Satis Chandra Vidyabhushana, *A History of Indian Logic* (Calcutta: Calcutta University, 1921), 31; the whole section is translated at pp. 31–35.

23 Ernst Prets has discussed the *vādamārga* section and provided a sketch of previous studies in his 2010 article, which is based on a paper delivered in 2003. Karin Preisendanz's contribution from 2013 (a revised version of a 2009 publication) has focused on historical and historiographical aspects of the text. See *References* for details.

24 This aspect is beyond the scope of our investigation; therefore, it is not reflected in the following analysis.

25 See n. 22.

26 The record of each item closely reproduces the text in Trikamji's edition.

TABLE 5.1 Liminal phrases in the *vādamārga* section (*cont.*)

4	*atha sthāpanā—sthāpanā nāma* ... (demonstration)
5	*atha pratiṣṭhāpanā—pratiṣṭhāpanā nāma* ... (counter-demonstration)
6	*atha hetuḥ—hetur nāma* ... (reason)
7	*atha dṛṣṭāntaḥ—dṛṣṭānto nāma* (example) ... *upanayo nigamanaṃ coktaṃ sthāpanāpratiṣṭhāpanāvyākhyāyām*
8	*athottaram—uttaraṃ nāma* ... (rejoinder)
9	*atha siddhāntaḥ—siddhānto nāma* ... (tenet or conclusion) *sa caturvidhaḥ*—[a] *sarvatantrasiddhāntaḥ*, [b] *pratitantrasiddhāntaḥ*, [c], *adhikaraṇasiddhāntaḥ*, [d] *abhyupagamasiddhāntaś ceti* *tatra* [a] *sarvatantrasiddhānto nāma* ... (a tenet or conclusion accepted by all schools) [b] *pratitantrasiddhānto nāma* ... (a tenet or conclusion accepted by a particular school) [c] *adhikaraṇasiddhānto nāma* ... (a tenet or conclusion accepted hypothetically) [d] *abhyupagamasiddhānto nāma* ... (a tenet or conclusion accepted by implication) *iti caturvidhaḥ siddhāntaḥ*
10	*atha śabdaḥ—śabdo nāma* ... (word) *sa caturvidhaḥ*—[a] *dṛṣṭārthaś ca*, [b] *adṛṣṭārthaś ca*, [c] *satyaś ca*, [d] *anṛtaś ceti.* *tatra* [a] *dṛṣṭārtho nāma* ... (that which refers to a matter which is seen) [b] *adṛṣṭārthaḥ punaḥ* ... (that which refers to a matter which is not seen) [c] *satyo nāma* ... (that which corresponds to what is real) ... [d] *anṛtaḥ* (that which does not correspond to what is real)
11	*atha pratyakṣam—pratyakṣaṃ nāma* ... (perception)
12	*athānumānam—anumānaṃ nāma* ... (inference)
13	*athaitihyam—aitihyaṃ nāma* ... (tradition)
14	*athaupamyam—aupamyaṃ nāma* ... (comparison)
15	*atha saṃśayaḥ—saṃśayo nāma* ... (doubt)
16	*atha prayojanam—prayojanaṃ nāma* ... (purpose)
17	*atha savyabhicāram—savyabhicāraṃ nāma* ... (uncertain)
18	*atha jijñāsā—jijñāsā nāma* ... (inquiry)
19	*atha vyavasāyaḥ—vyavasāyo nāma* ... (ascertainment)
20	*athārthaprāptiḥ—arthaprāptir nāma* ... (presumption)
21	*atha saṃbhavaḥ* ... (the originating cause)
22	*athānuyojyam—anuyojyaṃ nāma* ... (censurable)

TABLE 5.1 Liminal phrases in the *vādamārga* section (*cont.*)

23	*athānanuyojyam—ananuyojyaṃ nāma* ... (non-censurable)
24	*athānuyogaḥ—anuyogo nāma* ... (interrogation)
25	*atha pratyanuyogaḥ—pratyanuyogo nāma* ... (re-interrogation)
26	*atha vākyadoṣaḥ—vākyadoṣo nāma* (defect of speech) *yathā khalv asminn arthe* [a] *nyūnam,* [b] *adhikam,* [c] *anarthakam,* [d] *apārthakam,* [e] *viruddhaṃ ceti* ... *tatra nyūnam* ... [a] ... *nyūnaṃ* (inadequacy) [b] *athādhikam—adhikaṃ nāma* ... (redundancy) [c] *athānarthakam—anarthakaṃ nāma* ... (meaninglessness) [d] *athāpārthakam—apārthakaṃ nāma* ... (incoherence) [e] *atha viruddham—viruddhaṃ nāma* ... (contradiction) *iti vākyadoṣāḥ.*
27	*atha vākyapraśaṃsā—vākyapraśaṃsā nāma* ... (excellence of speech)
28	*atha cchalam—chalaṃ nāma* ... (quibble) *tad dvividham—*[a] *vākchalaṃ* [b] *sāmānyacchalaṃ ca* *tatra* [a] *vākchalaṃ nāma* ... (quibble in respect of a word) [b] *sāmānyacchalaṃ nāma* ... (quibble in respect of a generality)
29	*athāhetuḥ—ahetur nāma* ... (non-reason or fallacy) [a] *prakaraṇasamaḥ,* [b] *saṃśayasamaḥ,* [c] *varṇyasamaś ceti* *tatra* [a] *prakaraṇasamo nāmāhetur* ... (begging the question) [b] *saṃśayasamo nāmāhetur* ... (assumption based on doubt) [c] *varṇyasamo nāmāhetuḥ* ... (balancing the subject)
30	*athātītakālam—atītakālaṃ nāma* ... (mistimed)
31	*athopālambhaḥ—upālambho nāma* ... (attribution of censure)
32	*atha parihāraḥ—parihāro nāma* ... (avoidance of defect)
33	*atha pratijñāhāniḥ—pratijñāhānir nāma* ... (abandonment of a proposition)
34	*athābhyanujñā—abhyanujñā nāma* ... (admission)
35	*atha hetvantaram—hetvantaraṃ nāma* ... (shifting the reason)
36	*athārthāntaram—arthāntaraṃ nāma* ... (shifting the topic)
37	*atha nigrahasthānam—nigrahasthānaṃ nāma* ... (a point of defeat or an occasion for rebuke)

TABLE 5.2 Patterns of liminal phrases

A	*atha x—x nāma*
B	*x—x nāma*
C	*x nāma*
D	*x*

The witnesses of Vimāna 8—more than fifty manuscripts (see Appendix) and almost forty printed editions considering those published before 1950[27]—agree with Trikamji's edition on the number of divisions and their extent. However, the witnesses diverge significantly with regard to the liminal phrases at the beginning of each text segment and the sequential order of the text segments.

The available witnesses display four types of patterns with respect to the liminal phrases. Using *x* to denote the variable name of the item presented in a specific text segment, the four patterns of liminal phrases are as shown in table 5.2.

The main difference in these types of liminal phrases is that types A and B include an introductory element that can be considered a paratext, whereas types C and D do not. The reduplication of the variable name in patterns A and B shows that the liminal phrase was understood as comprising two parts, with the first (*atha x* or *x*) heading the topical segment and the second (*x nāma*) being the actual beginning of the description of the category. Liminal phrases of the types C and D simply mark the beginning of a new topical segment. It is to be noted that further subdivisions within a specific textual segment sometimes present an additional pattern, namely the presence of a liminal phrase (of type D) only at the end of the segment.

The very fact that the individual witnesses primarily follow one of these patterns shows that, as Genette observed with regard to paratext, "the discourse we are dealing with here is more "constrained" than many others."[28] Indeed, variation does not coincide with freedom in wording; in fact, it results from adopting or dropping a compositional element, which can be identified as paratext.

27 See Karin Preisendanz, Cristina Pecchia, and Vitus Angermeier, "Survey of Editions of the *Carakasaṃhitā* and their Sources" in: https://www.istb.univie.ac.at/caraka/Materials/120.

28 Genette, *Paratexts: Thresholds of Interpretation* (Cambridge: Cambridge University Press, 1997), 13.

5.2 *Liminal Phrases in the Printed Editions*

In the printed editions of the CaS, *atha x—x nāma* seems to be the most frequently occurring pattern for the liminal phrases in the *vādamārga* section. The pattern is found in Gangadhar Ray Kaviraj's edition of the CaS (Saidabad 1868)[29] and is repeated in most of the subsequent editions. The most notable exception to Gangadhar's version of the text appears in Jivananda Vidyasagar's printed edition of the CaS, published in Calcutta in 1877 (hereafter C1). Here, the regular pattern of the liminal phrases is *x nāma*, or, in a few cases, *x—x nāma* (items 4, 6, 8, 10[c], 11, 21, 26, 27, and 33). The repetition of *x* (e.g. *sthāpanā*) is not highlighted in C1 in any way. In fact, the first *x* element is sometimes printed together with the previous word and followed by a *daṇḍa*; therefore, it might initially be misinterpreted as marking the end of the previous sentence. Moreover, the layout of C1 does not allow the reader to easily identify syntactic limits or paratextual elements, because the printed text appears as a compact block where space is irregularly distributed between sequences of *akṣaras* (and not necessarily in accordance with semantic borders). Overall, a printed page in C1 is quite similar to the appearance of a manuscript folio written in what Saenger calls aerated script.

The C1 version of the *vādamārga* section was evidently adopted by Khemaraj Krishnadas in his edition of the CaS, published in Mumbai in 1898 (hereafter Bo2). The few deviations from C1 relate to items 4 and 8, both of which begin with the *atha x—x nāma* pattern, and items 6, 21, 26, 27, and 33, which follow the *x nāma* pattern.

A few editions display neither Gangadhar's version of the text, nor that of Jivananda Vidyasagar in C1 (or the revised form of it provided by Khemaraj in Bo2), but a combination of both. A mix of patterns was adopted by Jivananda himself in his second publication of the CaS (Calcutta 1896—hereafter C2). The text, though generally close to that of the Saidabad edition, partially retains the liminal phrase style of Jivananda's first edition of the CaS, namely C1, but it is "normalized" according to the pattern of the surrounding topical segments. So, *atha x—x nāma* appears throughout up to segment 26, where the *vākyadoṣas* are described. Exceptions to this style are the segments on *pratijñā, pratiṣṭhāpanā, uttara,* and *pratyanuyoga* (items 3, 5, 8, and 25), where the pattern is the unusual *atha x nāma*, and segment 12, on *anumāna*, which begins with *x nāma*. These cases do not seem to reflect any special design in presenting the organi-

29 Details about Gangadhar's edition are discussed in Cristina Pecchia, "Ayurveda, Philology and Print: On the First Printed Edition of the *Carakasaṃhitā* and its Context," in *Colonial Sciences and Indigenous Knowledge Systems in South Asia*, guest ed. Minakshi Menon, special issue of *South Asian History and Culture*, forthcoming.

zation of the section; in fact, they might be due to an inadvertent omission of the repeated element *x*. The liminal phrase changes to *x nāma* from segment 27 (on *vākyapraśaṃsā*, as in C1) to the end of the section. The general impression is that this part continues the style of the five subdivisions of segment 26 (although 26[b] begins with *atha ādhikyam / ādhikyaṃ nāma*), and therefore ranks below the larger division of the previous topical segments, which mostly begin with the pattern *atha x—x nāma*. The segments on *atītakāla* and *pratijñāhāni* (30 and 33) present the pattern *x—x nāma*, which might represent the editor's wish to mark the transition from a certain topic to another utterly unrelated topic.

Similarly to C2, a mixed style is adopted in both Shankar Daji Shastri Pade's editions of the CaS (Mumbai 1897–1898 and possibly 1903—B01 and B03), in Dattaram Narayan Caube's edition (Mumbai 1899–1900—B04), and Yashodanandan Sarkar's edition (Kolkata ^{2}1910–1911—C5). The three Mumbai editions and C5 agree with C2 in having the unusual pattern *atha x nāma* at the beginning of the segments on *pratijñā* and *pratyanuyoga* (3 and 25). In B01, sequences of *akṣaras* frequently appear at the center of the line as headings and subheadings. This is typically the case of the phrase *atha x* in the *vādamārga* section. Given its position on the page, this phrase is considered a paratext by the editor, while *x nāma* marks the beginning of the segment. In segment 33, the paratext only consists of the element *pratijñāhāniḥ*; in segment 30, *atītakālam* appears in one line together with *atītakālaṃ nāma* (which suggests a misprint). As a matter of fact, in the latter two segments, B01 reflects C2's *x—x nāma* pattern.

As noted above, liminal phrases may also appear only at the very end of a segment. This is the case in the segment on *nyūna* (which is part of *vākyadoṣa*—item 26[a]) and *anṛta* (which is part of *śabda*—item 10[d]). All printed editions share this particular textual order in both segments, but the above-mentioned printed editions, namely C1, C2, B01, B02, B04, and C5, begin the segment on *nyūna* with *tatra* or *atra*, while Trikamji's edition begins it with *tatra nyūnam*, and Gangadhar's printed edition neither has *tatra* nor *nyūnam*.

5.3 *Liminal Phrases in Manuscript Witnesses*

If printed books, in comparison with manuscripts, drastically innovated the *mise-en-page* of the text, editorial practices may not have undergone similarly drastic changes in the face of the new print technology.[30] In fact, one would rather expect a continuity of approach to textuality on the part of the literati

30 Pecchia, *Dharmakīrti on the Cessation of Suffering*, 83–84.

and texts specialists, and, to a certain extent, on the part of specialists of the book production.

The differences between liminal phrases that emerge in the collation of manuscript sources of the *vādamārga* section display certain interventions made in order to show the structure of the text as understood by their editors. Systematicness in the way text divisions are marked is one of the criteria that may have been applied, but other criteria may have been applied as well, such as, for example, the use of markers only for certain items depending on their thematic relevance or position in the list. Variation, once again, reflects interpretation and textual strategies in two key contexts—namely, use and community[31]—and points to nodal points in the history of the text migration; nodality resulting from major acts of interpretation, but possibly also from the accidents of fate which caused some sources to be preserved and others to be lost.

The variety of liminal phrases in the *vādamārga* section suggests that these phrases result from editorial activity similar to that found in the printed editions. Some manuscripts exhibit editorial interventions made either when they were copied from a previous exemplar, or at a different point in time, and possibly by a different agent. It is, in fact, impossible to say anything specific about how the editors interpreted any liminal phrases found in their exemplars. It is equally impossible to say whether, at the time of intervention, they sought to insert a paratext or amend Caraka's text, or a defective version of it. When considering the editorial process of the text as a whole, liminal phrases may have been augmented, changed, or eliminated, and other strategies for marking the text divisions may have been applied.

Even though contingency plays an important role in the texts transmitted from manuscript sources, the witnesses of the *vādamārga* section demonstrate the use of specific patterns of liminal phrases. Table 5.3 shows the patterns adopted by each of the four families into which the text witnesses can be grouped. The families are called K, Q, R, and S; they correspond to the four subarchetypes identified during the study of the sources for a critical edition of Vimāna 8 (see the Appendix for the list of witnesses and the families to which they belong).[32] In the table, colors are used to indicate which text is attested by which family of witnesses according to the following scheme:

---

31 See Gérard Colas, "Des rites de temple aux commentaires: les différentes transmissions du Pāramātmika," in *Écrire et transmettre en Inde classique*, eds. Gérard Colas and Gerdi Gerschheimer (Paris: École française d'Extrême-Orient, 2009), 221–239.

32 See Karin Preisendanz, Philipp A. Maas, and Cristina Pecchia, *A Critical Edition of Carakasaṃhitā Vimānasthāna Chapter 8*, forthcoming. For a stemma representing the rela-

atha x—x nāma	KQRS read *atha x—x nāma*
atha x—x nāma	Q reads *atha x—x nāma*; RSK read *x—x nāma*
atha x—x nāma	Q reads *atha x—x nāma*; RS read *x—x nāma*; K reads *x nāma*
atha x—x nāma	Q reads *atha x—x nāma*; R reads *x—x nāma*; K reads just *x*; S reads *x nāma*

When relevant witnesses clearly differ from the other witnesses with which they usually agree, their sigla are added to the list of family sigla.

The table puts the much more colorful record of the manuscript witnesses into a schematic frame. It shows that the four families followed three different patterns in introducing the topical segments of the section: the K manuscripts mainly use the *x nāma* pattern (type C), the Q manuscripts the *atha x—x nāma* pattern (type A), and both the R and S manuscripts the *x—x nāma* pattern (type B). As such, the liminal phrases in the K manuscripts differ from those of the Q, R, and S families in that they do not include a paratext; they do, however, include a paratext in eight out of thirty-six main categories and in a few sub-categories (highlighted in bold in Table 5.3) by repeating the variable name. In these cases, the K manuscripts give a pattern different from the pattern in the reference text, namely Trikamji's edition (for which, see Table 5.1); they are mostly followed by the printed edition C1 (see above). Moreover, a few manuscripts (notably B3 and C5) exhibit a strong fluctuation of patterns, which seemingly derives from a random use of different sources, hinting at a high degree of contamination.

As usual, the variation in the individual manuscripts can be due to accidental errors in the course of the textual transmission or due to the individual choices of copyists and editors. As can be imagined, the repetitive textual scheme of the *vādamārga* section made the text particularly prone to unintentional changes. Copyists might have unintentionally dropped the variable name in the liminal phrase or unintentionally reduplicated it. Such accidental changes can be easily identified because the resulting text diverges from the pattern generally presented by a specific textual witness in the rest of the section. However, changes that seem to be unintentional might also reflect textual usages which copyists or editors considered "normal" due to the forms of textuality they were familiar with. Therefore, except for evident unintentional errors,

tionships between manuscript sources, see Philipp A. Maas, "Computer Aided Stemmatics—The Case of Fifty-Two Text Versions of Carakasaṃhitā Vimānasthāna 8.67–157," in *Text Genealogy, Textual Criticism and Editorial Technique*, guest eds. Jürgen Hanneder and Philipp A. Maas, *Wiener Zeitschrift für die Kunde Südasiens* 52–53 (2009–2010), 65.

TABLE 5.3 Patterns adopted by K, Q, R, and S manuscript witnesses of Vimāna 8

1	*tatra vādo nāma …*
2	*dravyaguṇa° … ślokasthāne pūrvam uktāḥ*
3	*atha pratijñā—pratijñā nāma* … V2V3QRSK
4	*atha sthāpanā—sthāpanā nāma* … QKRS
5	*atha pratiṣṭhāpanā—pratiṣṭhāpanā nāma* … QRSK
6	*atha hetuḥ—hetur nāma* … QRSK
7	*atha dṛṣṭāntaḥ—dṛṣṭānto nāma* … QRSK
8	*athottaram—uttaraṃ nāma* … QKRS
9	*atha siddhāntaḥ—siddhānto nāma* … QRSK
	sa caturvidhaḥ—[a] *sarvatantrasiddhāntaḥ*, [b] *pratitantrasiddhāntaḥ*, [c] *adhikaraṇasiddhāntaḥ*, [d] *abhyupagamasiddhāntaś ceti. tatra*
	[a] *sarvatantrasiddhānto nāma* … KQRS
	[b] *pratitantrasiddhānto nāma* … KQRS
	[c] *adhikaraṇasiddhānto nāma* … KQRS
	[d] *abhyupagamasiddhānto nāma* … KQRS[33]
	iti caturvidhaḥ siddhāntaḥ.
10	*atha śabdaḥ—śabdo nāma* … QRSK
	sa caturvidhaḥ V2V3; *sa* KQR;[34] omitted in S
	[a] *dṛṣṭārthaś ca*, [b] *adṛṣṭārthaś ca*, [c] *satyaś ca*, [d] *anṛtaś ceti. tatra*
	[a] *dṛṣṭārtho nāma* … KQRS
	[b] *adṛṣṭārthaḥ punaḥ* … KQRS
	[c] *satyaḥ—satyo nāma* … QRSKS[35]Ap1V2V3
	[d] … *anṛtaḥ* QRSK
11	*atha pratyakṣam—pratyakṣaṃ nāma* … QKRS
12	*athānumānam—anumānaṃ nāma* … QRSK
13	*athaitihyam—aitihyaṃ nāma* … QRSK
14	*athaupamyam—aupamyaṃ nāma* … QRSK
15	*atha saṃśayaḥ—saṃśayo nāma* … QRSK
16	*atha prayojanam—prayojanaṃ nāma* … QB3CaL2RSK
17	*atha savyabhicāram—savyabhicāraṃ nāma* … QRSK

33 The S witnesses read *abhyupeta°* instead of *abhyupagama°*.

34 This is also the reading of the printed edition C1.

35 The siglum S is repeated twice because the S witnesses give two distinct liminal phrases. It is to be noted that the pattern *satyaḥ—satyo nāma* is not adopted in Trikamji's edition and other related editions, which instead have *satyo nāma*.

TABLE 5.3 Patterns adopted by K, Q, R, and S manuscript witnesses of Vimāna 8 (*cont.*)

18	*atha jijñāsā—jijñāsā nāma* ... QRSK
19	*atha vyavasāyaḥ—vyavasāyo nāma* ... QRSK
20	*athārthaprāptiḥ—arthaprāptir nāma* ... QRSK
21	*atha saṃbhavaḥ—saṃbhavo nāma* ... QKRS
22	*athānuyojyam—anuyojyaṃ nāma* ... QRSK
23	*athānanuyojyam—ananuyojyaṃ nāma* ... QRSK
24	*athānuyogaḥ—anuyogo nāma* ... QRSK
25	*atha pratyanuyogaḥ—pratyanuyogo nāma* ... V2V3(Q)[36]RSK
26	*atha vākyadoṣaḥ—vākyadoṣo nāma* QKRS
	yathā khalv asminn arthe [a] *nyūnam,* [b] *adhikam,* [c] *anarthakam,* [d] *apārthakam,* [e] *viruddhaṃ ceti. tatra* ...
	[a] ... *nyūnaṃ nyūnaṃ bhavati* KRSQS[37] ... *nyūnaṃ* KQRS
	[b] *athādhikam—adhikaṃ nāma* ... QRSK
	[c] *athānarthakam—anarthakaṃ nāma* ... QRSK
	[d] *athāpārthakam—apārthakaṃ nāma* ... QRSK
	[e] *atha viruddham—viruddhaṃ nāma* ... QSRK
	iti[38] *vākyadoṣāḥ.* KQRS
27	*atha vākyapraśaṃsā—vākyapraśaṃsā nāma* ... QKR[39]S
28	*atha cchalam—chalaṃ nāma* ... QRSK
	tad dvividham—[a] *vākchalaṃ* [b] *sāmānyacchalaṃ ca. tatra*[40]
	[a] *vākchalaṃ nāma* ... KQRS
	[b] *sāmānyacchalaṃ sāmānyacchalaṃ nāma* ... K[41]QRS
29	*athāhetuḥ—ahetur nāma* QRSK
	[a] *prakaraṇasamaḥ,* [b] *saṃśayasamaḥ,* [c] *varṇyasamaś ceti. tatra*
	[a] *prakaraṇasamo nāmāhetur* ... KQRS
	[b] *saṃśayasamo nāmāhetur* ... KQRS
	[c] *varṇyasamo nāmāhetuḥ* ... KQRS
30	*athātītakālam—atītakālaṃ nāma* ... QRSK

36 The relevant text is missing in most of the Q witnesses; therefore, it is not possible to state which version of the text they present.

37 The siglum S is repeated twice because the S witnesses give two distinct liminal phrases.

38 This is also the reading of the printed edition C1.

39 The R witnesses in fact read *atha vākyapraśaṃsā nāma,* which can be explained as a *saut du même au même.*

40 The word *tatra* is omitted by the S witnesses.

41 Most of the K witnesses reduplicate the *x* element, but P1 deletes the repeated *sāmānya-śchalaṃ.*

TABLE 5.3 Patterns adopted by K, Q, R, and S manuscript witnesses of Vimāna 8 (*cont.*)

31	*athopālambhaḥ—upālambho nāma* ... QRSK
32	*atha parihāraḥ—parihāro nāma* ... QRSK
33	*atha pratijñāhāniḥ—pratijñāhānir nāma* ... QKRS
34	*athābhyanujñā—abhyanujñā nāma* ... Q[42]RSK
35	*atha hetvantaram—hetvantaraṃ nāma* ... QRSK
36	*athārthāntaram—arthāntaraṃ nāma* ... QRSK
37	*atha nigrahasthānam—nigrahasthānaṃ nāma* ... QRSK

it is more probable that the variation we can observe in the available sources derives from editorial intervention at some point in the history of the textual transmission, and not from pure accident.

A first possible scenario is that the archetype itself presented an irregular distribution of paratext, possibly limited to segments with further subdivisions or to segments that were considered thematically distinct. Through processes of migration, such irregularity was either amplified (as in the K manuscripts), or normalized (as in the Q, R, S manuscripts). Another possible scenario is that the archetype included paratexts of types A or B, which were subsequently changed to types B or A respectively, or to type C. In this case, the K manuscripts should derive from a defective exemplar with liminal phrases of type C that an editor amended by using witness(es) with liminal phrases of type B. It cannot be excluded that liminal phrases of type B in K manuscripts are remnants of a version that was edited by collating a witness with liminal phrases of type C.

6 Conclusion

The patterns of liminal phrases in the *vādamārga* section of the CaS show different approaches to textual divisions and how to mark them. The three patterns that regularly appear display two main modalities of marking the transition to a new topical text segment; namely, either with or without a paratext functioning as a heading. The editors of the Q, R, and S sub-archetypes, in particular, chose to present the *vādamārga* section with divisions that were systematically marked through headings displaying two types of paratext—the

42 It is to be noted that except for V2 and V3, the other Q witnesses give different versions of a corrupt text which begins with *arthā°* instead of *athā°*.

more complex type A being perhaps later than the less complex type B. It is also apparent that a defective exemplar circulated, which was amended by the editor(s) of the K manuscripts by merging two styles of liminal phrases.

Any statement concerning the liminal phrases in the archetype is at present quite hypothetical, because the different patterns presented by the manuscript witnesses should be assessed based on considerations of the stylistic habits reflected in the text as a whole. At present, we are not able to form an opinion about these habits and, in view of this, about the distance between the text of the archetype and the choices of the editors and copyists in the migrational process of the CaS.

We can observe that the *vādamārga* section of the CaS largely circulated as a text with headings. It is this fact that requires an explanation, because such headings, or paratext, redundantly repeat the variable name *x* that appears at the beginning of each segment and clearly marks the transition from one segment to the next. The paratextual elements might reflect the intervention of editors or copyists-editors, but also, admittedly, of the author of the CaS, who wanted the audience of the text to have a clear outline of a content that was probably "difficult," or at least less familiar. Since the main interests and competencies of such an audience would have been related to medical science, providing the *vādamārga* section with paratexts indicating its divisions was in any case a way of enhancing clarity[43] and making the text more easily accessible for those studying it.

Appendix

Conspectus of Manuscript Witnesses of Vimānasthāna 8 of the Carakasaṃhitā, *Together with Their Sigla.*[44]

The sigla indicate where the relevant document is preserved and, in superscript, the script in which it is written (^{b} Bengali ^{d} Devanagari ^{k} Kannada ^{ś} Sharada). Superscript letters are only shown here in the Appendix.

The colors correspond to the families into which the witnesses are grouped: K Q R S

43 See fn. 16.

44 For a very general outline of the history of the manuscripts of the Vimānasthāna chapter of the CaS, see Cristina Pecchia, "Transmitting the *Carakasaṃhitā*: Notes for a History of the Tradition," *Indian Journal of History of Science* 44 (2009): 141–161.

A^{d}	Alwar, RORI 2498
Ab^{d}	Ahmedabad, B.J. Institute of Learning and Research 758
$Ap1^{d}$	Alipur, Bhogilal Leherchand Institute of Indology 5283
$Ap2^{d}$	Alipur, Bhogilal Leherchand Institute of Indology 5527
$B1^{d}$	Bikaner, RORI 1566
$B2^{d}$	Bikaner, Anup Sanskrit Library, 124
$B3^{d}$	Bikaner, Anup Sanskrit Library, 125
$B4^{d}$	Bikaner, Anup Sanskrit Library, 134
$B5^{d}$	Bikaner, Anup Sanskrit Library, 135
$B6^{d}$	Bikaner, Anup Sanskrit Library, 136
$Ba1^{d}$	Baroda, Oriental Institute 12489
$Ba2^{d}$	Baroda, Oriental Institute 25034
Bo^{d}	Bombay, Asiatic Society 172
$C1^{b}$	Calcutta, National Library RDS 101
$C2^{b}$	Calcutta, Library of Calcutta, Sanskrit College 23
$C3^{b}$	Calcutta, Library of Calcutta, Sanskrit College 24
$C4^{b}$	Calcutta, Asiatic Society G 4474/3
$C5^{b}$	Calcutta, Asiatic Society G 2503/1
$C6^{d}$	Calcutta, Asiatic Society G 4391
Ca^{b}	Cambridge, Trinity College Library R 15.85
Ch^{d}	Chandigarh, Lal Chand Research Library 2315
$Ib1^{d}$	Ilāhābad, G. Jha Kendriya Sanskrit Vidyapeetha 25398
$Ib2^{d}$	Ilāhābad, G. Jha Kendriya Sanskrit Vidyapeetha 8783/87
$Ib3^{d}$	Ilāhābad, G. Jha Kendriya Sanskrit Vidyapeetha 37089
$J1^{d}$	Jammu, Raghunath Temple Library 3266
$J2^{d}$	Jammu, Raghunath Temple Library 3209
$J3^{d}$	Jammu, Raghunath Temple Library 3330
$Jn1^{d}$	Jamnagar, Gujarat Ayurved University Library GAS 103
$Jn2^{d}$	Jamnagar, Gujarat Ayurved University Library GAS 118
$Jn3^{d}$	Jamnagar, Gujarat Ayurved University Library GAS 96/2
$Jp1^{d}$	Jaipur, MSMS Museum 2068
$Jp2^{d}$	Jaipur, MSMS Museum 2069
$Jp3^{d}$	Jaipur, MSMS Museum 2561
K^{d}	Koṭa, RORI 1563
Km^{d}	Kathmandu, NGMPP E-40553
$L1^{d}$	London, IOL Skt. ms. 335
$L2^{d}$	London, IOL Skt. ms. 881
$L3^{d}$	London, IOL Skt. ms. 1445b
M^{k}	Mysore, ORI 902
$P1^{ś}$	Pune, BORI 555 of 1875–1876

P2[d]	Pune, BORI 534 of 1892–1895
P3[d]	Pune, BORI 925 of 1891–1895
P4[d]	Pune, Ānandāśrama 1546
T1[d]	Tübingen, UB I.458
T2[d]	Tübingen, UB I.459
T3[d]	Tübingen, UB I.460 + I.474
U[d]	Udaipur, RORI 1474
V1[b]	Varanasi, Sarasvati Bhavan 44842
V2[b]	Varanasi, Sarasvati Bhavan 108824
V3[b]	Varanasi, Sarasvati Bhavan 108685
V4[d]	Varanasi, BHU C3688
V5a[d]	Varanasi, Sarasvati Bhavan 44870
V5b[d]	Varanasi, Sarasvati Bhavan 44870

References

Allon, Mark. *Style and Function: A Study of the Dominant Stylistic Features of the Prose Portions of Pāli Canonical Sutta Texts and their Mnemonic Function*. Tokyo: The International Institute for Buddhist Studies, 1997.

Amelung, I., J. Kurtz, and M. Lackner. "Introduction." In *New Terms for New Ideas: Western Knowledge and Lexical Change in Late Imperial China*, edited by M. Lackner, I. Amelung, and J. Kurtz, 1–13. Leiden: Brill, 2001.

Anālayo, Bhikkhu. "*Āgama* and *aṅga* in the Early Buddhist Oral Tradition." *Singaporean Journal of Buddhist Studies* 3 (2016): 9–37.

Bhattarai, Bidur. *Dividing Texts: Conventions of Visual Text-Organization in Nepalese and North Indian Manuscripts*. Berlin and Boston: DeGruyter, 2020.

Bretelle-Establet, Florence and Stéphane Schmitt. "Introduction." In *Pieces and Parts in Scientific Texts*, edited by Florence Bretelle-Establet and Stéphane Schmitt, 3–17. Cham: Springer, 2018.

Bretelle-Establet, Florence and Stéphane Schmitt, eds. *Pieces and Parts in Scientific Texts*. Cham: Springer, 2018.

CaS = *Carakasaṃhitā* with reference to: *Charakasaṃhitā by Agniveśa, revised by Charaka and Dṛidhabala, with the Āyurveda-Dīpikā commentary of Chakrapāṇidatta*, edited by Vaidya Jādavaji Trikamji Āchārya. Varanasi: Chaukhambha Prakashan, 2007 [Mumbai, 1941].

Chemla, Karine and Jacques Virbel, eds. *Texts, Textual Acts and the History of Science*. Cham: Springer, 2015.

Colas, Gérard. "Des rites de temple aux commentaires: les différentes transmissions du Pāramātmika." In *Écrire et transmettre en Inde classique*, edited by Gérard

Colas and Gerdi Gerschheimer, 221–239. Paris: École française d'Extrême-Orient, 2009.

Cousins, Lance S. "The Early Development of Buddhist Literature and Language in India." *Journal of the Oxford Centre for Buddhist Studies* 5 (2013): 89–135.

Dionísio, João. "Tables of Contents in Portuguese Late Medieval Manuscripts." In *The Book as Artefact: Text and Border*, issue 4 of *Variants*, edited by Anne Mette Hansen, Roger Lüdeke, Wolfgang Streit, Cristina Urchueguía, and Peter Shillingsburg, 89–109. Amsterdam, New York: Rodopi, 2005.

Genette, Gérard. *Paratexts: Thresholds of Interpretation*. Translated by Jane E. Lewin, Foreword by Richard Macksey. Cambridge: Cambridge University Press, 1997. Originally published in French as *Seuils*. Paris: Editions du Seuil, 1987.

Houben, Jan E.M. and Saraju Rath. "Introduction: Manuscript Culture and Its Impact in 'India': Contours and Parameters." In *Aspects of Manuscript Culture in South India*, edited by S. Rath, 1–53. Leiden: Brill, 2012.

Karttunen, Klaus. "Books, Songs, Adhyāyas, Sargas: Text Divisions in the Classical West and India." *Cracow Indological Studies* 13 (2011): 3–14.

Kellner, Birgit. "First Logic, then the Buddha? Remarks on the Chapter Sequence of Dharmakīrti's *Pramāṇavārttika*." *Hōrin* 11 (2004): 147–167.

Maas, Philipp A. "Computer Aided Stemmatics—The Case of Fifty-Two Text Versions of *Carakasaṃhitā* Vimānasthāna 8.67–157." In *Text Genealogy, Textual Criticism and Editorial Technique*, guest edited by Jürgen Hanneder and Philipp A. Maas. *Wiener Zeitschrift für die Kunde Südasiens* 52–53 (2009–2010): 63–119.

Maas, Philipp A. "On What Became of the *Carakasaṃhitā* after Dṛḍhabala's Revision." *eJournal of Indian Medicine* 3 (2010): 1–22.

Meulenbeld, G. Jan. *A History of Indian Medical History*. Vol. 1A: Text. Groningen: Egbert Forsten, 1999.

Montgomery, Scott L. *Science in Translation: Movements of Knowledge through Cultures and Time*. Chicago and London: The University of Chicago Press, 2000.

Parpola, Asko. "On the Upagrantha(sūtra) of the Sāmaveda: 1. Contents and Textual Divisions." In *Langue, style et structure dans le monde indien: centenaire de Louis Renou, actes du colloque international (Paris, 25–27 janvier, 1996)*, edited by N. Balbir and G.-J. Pinault, 91–121. Paris: Champion, 1996.

Pecchia, Cristina. "Transmitting the *Carakasaṃhitā*: Notes for a History of the Tradition." *Indian Journal of History of Science* 44 (2009): 141–161. Corrected and slightly revised version in *Medical Texts and Manuscripts in Indian Cultural History*, edited by Dominik Wujastyk, Anthony Cerulli, and Karin Preisendanz, 1–27. Delhi: Manohar Lal, 2013.

Pecchia, Cristina. *Dharmakīrti on the Cessation of Suffering: A Critical Edition with Translation and Comments of Manorathanandin's* Vṛtti *and Vibhūticandra's Glosses on* Pramāṇavārttika *II.190–216*. With the assistance of Philip Pierce. Leiden: Brill, 2015.

Pecchia, Cristina. "Introduction." In *Editors of Sanskrit Texts*, edited by C. Pecchia. Vienna: Austrian Academy of Sciences Press, forthcoming.

Pecchia, Cristina. "Ayurveda, Philology and Print: On the First Printed Edition of the *Carakasaṃhitā* and its Context." In *Colonial Sciences and Indigenous Knowledge Systems in South Asia*, guest edited by Minakshi Menon. Special issue of *South Asian History and Culture*, forthcoming.

Péry-Woodley, Marie-Paule. "Modes d'organisation et de signalisation dans des textes procéduraux." *Les discours procéduraux*, edited by Claudine Garcia-Debanc. *Langages* 35, no. 141 (2001): 28–46.

Pettegree, Andrew. "Translation and the Migration of Texts." In *Borders and Travellers in Early Modern Europe*, edited by Thomas Betteridge, 113–127. Aldershot: Ashgate Variorum, 2007.

Pollock, Sheldon. "Future Philology? The Fate of a Soft Science in a Hard World." *Critical Inquiry* 35 (Summer 2009): 931–961.

Preisendanz, Karin. "Logic Debate and Epistemology in Ancient Indian Medical Science: An Investigation into the History and Historiography of Indian Philosophy." In *Medical Texts and Manuscripts in Indian Cultural History*, edited by Dominik Wujastyk, Anthony Cerulli, and Karin Preisendanz, 63–139. New Delhi: Manohar Lal, 2013 [revised and updated version of the unauthorized, deprecated version published in *Indian Journal of History of Science* 44 (2009): 261–312].

Preisendanz, Karin. "Text Segmentation, Chapter Naming and the Transmission of Embedded Texts in South Asia, with Special Reference to the Medical and Philosophical Traditions as Exemplified by the *Carakasaṃhitā* and the *Nyāyasūtra*." In *Pieces and Parts in Scientific Texts*, edited by Florence Bretelle-Establet and Stéphane Schmitt, 159–220. Cham: Springer, 2018.

Preisendanz, Karin, Cristina Pecchia, and Vitus Angermeier. "Survey of Editions of the *Carakasaṃhitā* and their Sources" in https://www.istb.univie.ac.at/caraka/Materials/120 (last accessed 30-01-2019).

Preisendanz, Karin, Philipp A. Maas, and Cristina Pecchia. *A Critical Edition of Carakasaṃhitā Vimānasthāna Chapter 8*. Forthcoming.

Prets, Ernst. "On the Proof Passage of the *Carakasaṃhitā*: Editions, Manuscripts and Commentaries." In *Logic in Earliest Classical India: Papers of the 12th World Sanskrit Conference*. Vol. 10, no. 2, edited by Brendan S. Gillon et al., 67–85. Delhi: Motilal Banarsidass Publishers, 2010 [based on a paper delivered in 2003].

Renou, Louis. "Les divisions dans les textes sanskrits." *Indo-Iranian Journal* 1 (1957): 1–32 (= *Louis Renou: Choix d'études indiennes*. Vol. 2, collected by N. Balbir and G.-J. Pinault, 487–518. Paris: École française d'Extrême-Orient, 1997.)

Sāṅkṛtyāyana, Rahula, ed. *Dharmakīrti's Pramāṇavārttika with a Commentary by Manorathanandin*. Appendix to *Journal of the Bihar and Orissa Research Society* 24–26. Patna, 1938–1940.

Schlingloff, Dieter. "The Oldest Extant Parvan-List of the *Mahābhārata*." *Journal of the American Oriental Society* 89 (1969): 334–338.

Saenger, Paul. *Space Between Words: The Origins of Silent Reading*. Stanford: Stanford University Press, 1997.

Sweet, William, ed. *Migrating Texts and Traditions*. Ottawa: University of Ottawa Press, 2012.

Vidyabhushana, Satis Chandra. *A History of Indian Logic*. Calcutta: Calcutta University, 1921.

Virbel, Jacques. "Langage et métalangage dans le texte du point de vue de l'édition en informatique textuelle." *Cahiers de Grammaire* 10 (1985): 5–7.

Von Hinüber, Oskar. "Die Neun Aṅgas: Ein früher Versuch zur Einteilung buddhistischer Texte." *Wiener Zeitschrift für die Kunde Süd- und Ostasiens* 38 (1994): 121–135.

CHAPTER 6

A New Translation of *Carakasaṃhitā*, Vimānasthāna, Chapter 1, Based on the Vienna Critical Edition

Dominik Wujastyk

1 Medicine and the History of Indian Logic

This paper is written in affectionate dedication to my long-time friend and respected professional colleague Kenneth Zysk. Prof. Zysk has always said he enjoyed philology, and indeed he moved to a new continent in order to be even more philological. I hope he will enjoy what follows.

It has long been recognized that the earliest encyclopedias of medicine written in Sanskrit contained interesting philosophical and eristic material.[1] Since the medical treatises are so early, historians of philosophy have naturally raised the question of priority. Did professionalized philosophical discourse originate amongst physicians, to be taken up later by specialist philosophers? Did the physicians borrow from early philosophical thinkers or texts? Or were both communities, physician and philosophers, drawing upon a common earlier speculative literature? Amongst the many philosophically interesting topics in the medical treatises are the descriptions of the conduct of philosophical debates, together with quite formalized rules regarding the conduct of these debates. This tradition of debate became important and pervasive in the later history of philosophy in India.[2] The goal of the Philosophy and Medicine in

1 E.g., Rudolf Roth, 'Indische Medicin. Caraka,' *Zeitschrift der Deutschen Morgenländischen Gesellschaft* 26 (1872): 441–452, Satis Chandra Vidyabhusana, *A History of Indian Logic: Ancient, Mediaeval and Modern Schools*, Reprinted Delhi, Motilal Banarsidass, 1977 and often thereafter (Calcutta: Calcutta University, 1921), 25–35, Surendranath Dasgupta, *A History of Indian Philosophy*, Reprint of 1922 edition (Cambridge: Cambridge University Press, 1952–1961), v. 2, ch. 13, ISBN: 0-521-04778-1, accessed January 26, 2018, https://tinyurl.com/ratwdof "The Medical Schools."

2 E.g., Esther A. Solomon, *Indian Dialectics: Methods of Philosophical Discussion*, 2 vols., Sheth Bholabhai Jeshingbhai Institute of Learning and Research Research Series 70 (Ahmedabad: B.J. Institute of Learning / Research, Gujarat Vidya Sabha, 1976–1978), 74, accessed May 15, 2020, https://n2t.net/ark:/13960/t5jb4x70d; Sung-Yong Kang, *Die Debatte Im Alten Indien—Untersuchungen Zum Sambhāṣāvidhi Und Verwandten Themen in Der Carakasaṃhitā Vimā-*

 | DOI:10.1163/9789004438224_007

Early Classical India (PMECI) project at the University of Vienna was to explore such issues in the history of philosophy.[3]

Since many of these key philosophical topics occur particularly in a section of the *Compendium of Caraka* (Skt. *Carakasaṃhitā*) called the "Section on Specific Knowledge" (Vimānasthāna), and since the text had never been critically edited, it was decided early on in the PMECI project that a critical edition of the "Section on Specific Knowledge" should be undertaken so that scholarship on this work would be grounded on a reliable text. A survey of manuscripts and editions was conducted. Digital, film and photocopy manuscript surrogates were gathered in an archive at the University of Vienna and selected chapters were collated from the manuscripts. The stemmatic relationship between the fifty-two manuscripts that include chapter eight of the "Section on Specific Knowledge" was developed philologically by the members of the project over several years and also computationally and philologically by Maas (see Figures 6.1 and 6.2).[4] At the time of writing, a principled critical edition of the "Section on Specific Knowledge" is more than half-way completed.

The translation presented below of the first chapter (*adhyāya*) of the *Vimānasthāna* of Caraka's *Compendium* is based on the aforementioned unpublished critical edition of this part of the *Compendium* that was prepared by myself and other project participants in the context of the PMECI project.[5] The text of the Sanskrit edition was the result of many weeks of full-day meetings and discussions about every syllable of the text in the context of the manuscript readings, the stemmatic relationship of the manuscripts and the meaning of the text. The translation and annotation presented here are entirely my own.

The purpose of this particular contribution is not to explore "philosophy and medicine," the larger goal of the original Vienna project. This contribution offers a translation of an important chapter of this classic text in contemporary

---

nasthāna 8.15–28, Philosophica Indica Einsichten Ansichten 6 (Reinbek: Dr. Inge Wezler, Verlag für Orientalistische Fachpublikationen, 2003); Dominik Wujastyk, *Medical Dialectic in Late Precolonial India.* A Disquisition on Sickness and Health *by Vīreśvara* (in preparation).

3 Karin Preisendanz, 'Logic, Debate and Epistemology in Ancient Indian Medical Science: An Investigation into the History and Historiography of Indian Philosophy. Part I,' in *Medical Texts and Manuscripts in Indian Cultural History*, ed. Dominik Wujastyk, Anthony Cerulli, and Karin Preisendanz (Delhi: Manohar Publishers / Distributors, 2013), 63–139, ISBN: 978-9350980194.

4 Philipp André Maas, 'Computer Aided Stemmatics—the Case of Fifty-two Text Versions of Carakasaṃhitā Vimānasthāna 8.67–157,' special issue edited by Jürgen Hanneder and Philipp A. Maas, *Text Genealogy, Textual Criticism* and *Editorial Technique, Wiener Zeitschrift für die Kunde Südasiens* 52–53 (2010): 63–119. doi:10.1553/wzks2009-2010s63.

5 Project participants since 2000 have included Karin Preisendanz, Ernst Prets, Philipp Maas, Christina Pecchia, Vitus Angermeier and myself as well as the assistants who are listed on the project website.

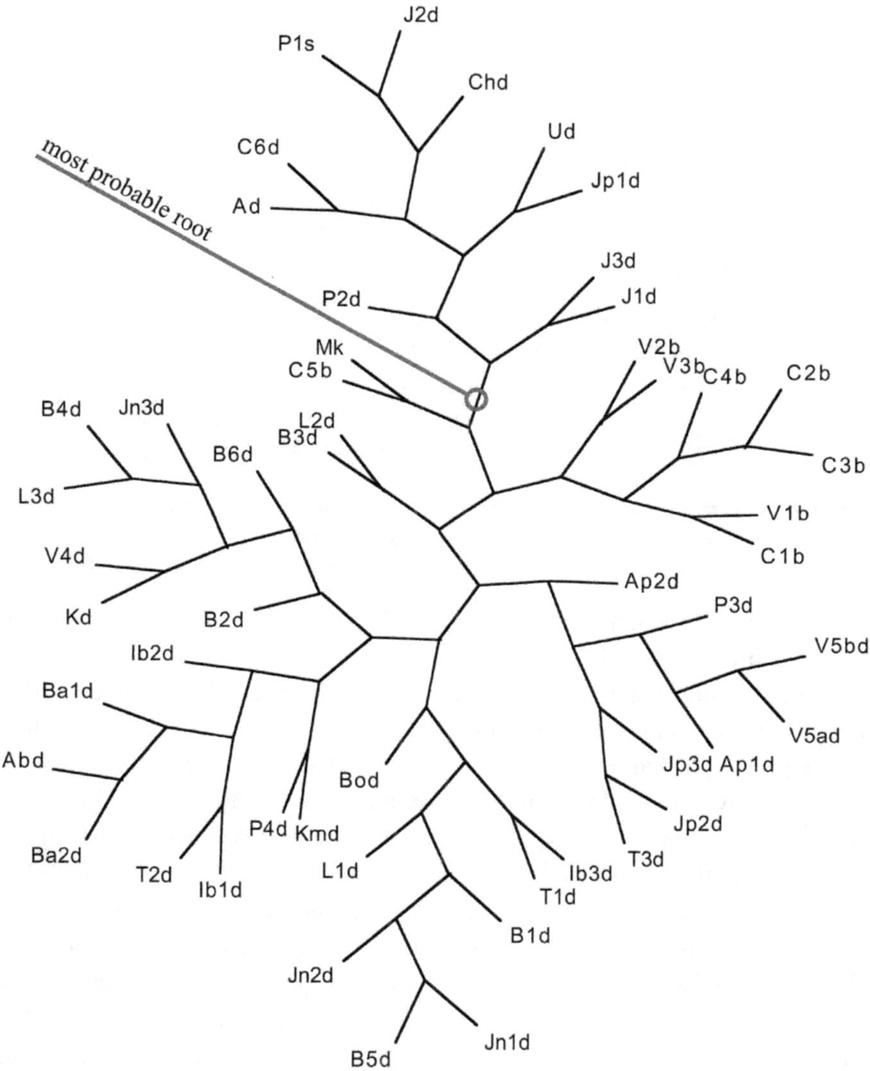

FIGURE 6.1 Unrooted cladogram reproduced from Maas, 'Computer Aided Stemmatics,' 74, showing the relationship between manuscripts after cladistic analysis and before the application of rooting and parsimony analysis. Sigla for the manuscripts are explained on the Caraka project website (Preisendanz et al., 'Caraka Project Website')

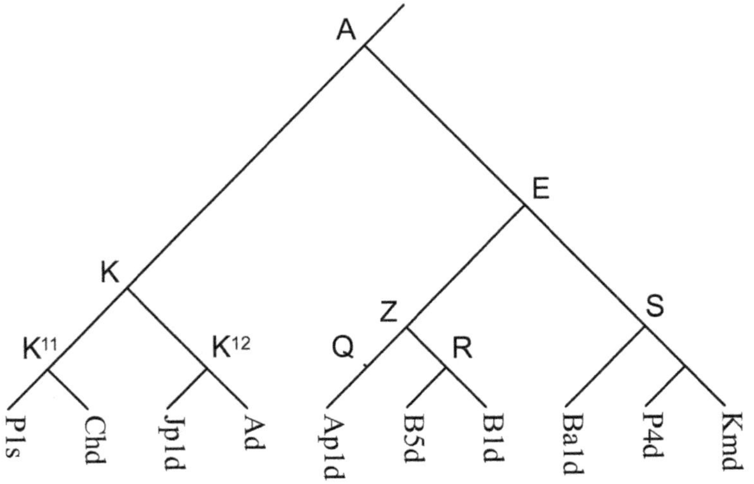

FIGURE 6.2 Stemma taken from Maas, 'Computer Aided Stemmatics,' 95, showing the most parsimonious cladogram of ten manuscript witnesses based on about one thousand genealogically significant varants

international English and as such it does offer carefully-considered and interpreted source material for future research.

But the principal goal of this paper is more technical and more modest. This paper invites you into the Caraka project's editorial workshop, to share some of the philological and hermeneutic intensity that has informed that project. It aims to give the reader an insight into how the Vienna critical edition differs from the nineteenth- and twentieth-century editions on which Indologists and medical historians usually rely. The publication of the full critical edition and apparatus of this chapter of the Vimānasthāna awaits the attention of the project members in Vienna. Meanwhile, I have designed the present paper with the purpose of answering the question so often put to me and others in the project, "how does the critically edited text differ from the commonly-available printed editions?" Therefore, much of the content of this paper is to be found in the footnotes, where these differences have been laid out as clearly as I know how. The attentive reader will see that the differences are small, but incremental. An immense amount of work has gone into the Vienna critical edition and it is inevitable that some evaluation should be made about whether the edition improves materially on the work of truly great editorial pioneers a century ago such as Vaidya Yādavaśarma Trivikramātmaja Ācārya. The present contribution will, I trust, provide serious historians and philologists with a quantitative and qualitative appreciation of how the accumulation of hundreds of small improvements and corrections results in an substantially clearer and more authoritative treatise.

2 Translation

All texts presented to readers as translations are heavy with their authors' philosophical presuppositions about the nature of language, textuality, and cultural history. These presuppositions too often remain unexamined; a translator should say something about them, and so should I. But this is not the place for an extended essay on what I understand and use from the academic field of translation theory. The following comments, therefore, are designed rather to orient the reader to the literature that has formed my views of what is important in producing a successful and valid translation. The literature on the theory of translation is very substantial.[6] Yet the field of South Asian translation studies remains insufficiently theorized.[7] Amongst eclectic studies that have most

6 Lawrence Venuti, ed., *The Translation Studies Reader*, 3rd ed. (London: Routledge, 2012), ISBN: 9780415613484; Mona Baker, ed., *Critical Readings in Translation Studies* (London: Routledge, 2010), ISBN: 9780415469548; Mona Baker and Gabriela Saldanha, eds., *Routledge Encyclopedia of Translation Studies*, 2nd ed. (London: Routledge, 2009), ISBN: 0-203-02911-9 provide entry points to the general field, while Enrica Garzilli, ed., *Translating, Translations, Translators from India to the West*, Harvard Oriental Series (Cambridge, MA: Harvard University, 1996), ISBN: 1888789026; Harish Trivedi Susan Bassnett, *Postcolonial Translation: Theory and Practice* (Routledge, 1998), ISBN: 0415147441; Niranjan Mohanty, 'Intranslability and the Translator's Task,' *Perspectives: Studies in Translation Theory and Practice* 4, no. 2 (1996): 163–171, doi:10.1080/0907676x.1996.9961284; Niranjan Mohanty, 'Translation: An Integration of Cultures,' *Perspectives: Studies in Translation Theory and Practice* 2, no. 2 (1994): 187–198, doi:10.1080/0907676x.1994.9961235 address issues specific to translation in South Asia. Further relevant studies include Daniel T. Rodgers, Bhavani Raman, and Helmut Reimitz, eds., *Cultures in Motion* (Princeton: Princeton University Press, 2013); Maria Tymoczko and Edwin Gentzler, *Translation and Power* (Amherst, Boston, etc.: University of Massachusetts Press, 2002), ISBN: 1558493581; Román Álvarez and M. Carmen-África Vidal, *Translation, Power, Subversion*, Topics in Translation 8 (Clevedon: Multilingual, 1996), ISBN: 9780585149806; Beatrice Fischer and Matilde Nisbeth Jensen, eds., *Translation and the Reconfiguration of Power Relations: Revisiting Role and Context of Translation and Interpreting*, Repräsentation—Transformation. Translating Across Cultures and Societies 7 (Zurich: LIT Verlag, 2012), ISBN: 9783643902832.

7 As was noted by Ramesh Krishnamurthy, 'Indian Tradition,' in *Routledge Encyclopedia of Translation Studies*, 2nd ed., ed. Mona Baker and Gabriela Saldanha (London: Routledge, 2009), 458, ISBN: 0-203-02911-9, in spite of the important reflections of Sundar Sarukkai, 'Translation As Method: Implications for History of Science,' *Indian Journal of History of Science* 51, no. 1 (2016): 105–117, doi:10.1163/9789004251410_014, the field needs further studies parallel to the mature reflections of Marwa S. Elshakry, 'Knowledge in Motion: The Cultural Politics of Modern Science Translations in Arabic,' *Isis* 99, no. 4 (2008): 701–730, accessed February 24, 2019, doi:10.1086/595767, https://www.jstor.org/stable/10.1086/595767, on Arabic scientific translation.

influenced my theory and practice of translation are Zimmermann[8] and Eco.[9] I have been deeply influenced by the arguments of Venuti,[10] although I disagree with the author's final thesis. However, I fully agree with Venuti's central idea—that different language communities have widely-varying expectations of the translations they receive. As he says,

> Fidelity cannot be construed as mere semantic equivalence: on the one hand, the foreign text is susceptible to many different interpretations, even at the level of the individual word; on the other hand, the translator's interpretive choices answer to a domestic cultural situation and so always exceed the foreign text. This does not mean that translation is forever banished to the realm of freedom or error, but that canons of accuracy are culturally specific and historically variable.[11]

This is important and, it seems to me, true. It demolishes the belief that a translation can be correct in any absolute sense, or indeed that one can ever provide translations that satisfy all audiences. The success of a translation is contingent on the reader's culture and mother tongue just as much as on the skills of the translator.

I have described some of my ideas about the theory of translation in the introductions to two of my previous books, *Metarules of Pāṇinian Grammar* and *Roots*. With many other translation theorists, I consider the act of translation to be, theoretically speaking, impossible. Languages, cultures and histories, are incommensurable.[12] In order to have a genuine experience of understanding the *Compendium of Caraka*, one must read it *and understand it* in its original language, without performing translation either mentally or in writing. For this, one has to bring a level of preparedness to the task, having acquired the source language as a child or through substantial later study. Only when

8 Francis Zimmermann, 'Terminological Problems in the Process of Editing and Translating Sanskrit Medical Texts,' in *Approaches to Traditional Chinese Medical Literature*, ed. Paul Unschuld (Dordrecht, London: Kluwer Academic, 1989), 141–151, ISBN: 1-55608-041-7, doi:10.1007/978-94-009-2701-8_13.

9 Umberto Eco, *Mouse or Rat? Translation As Negotiation* (London: Orion, 2004), ISBN: 9780297830016.

10 Lawrence Venuti, *The Translator's Invisibility: A History of Translation* (London, New York: Routledge, 1995).

11 Ibid., 37.

12 Maeve Olohan, 'History of Science and History of Translation: Disciplinary Commensurability?,' *The Translator* 20, no. 1 (2014): 9–25, doi:10.1080/13556509.2014.899091 discusses this dilemma in the context of the history of science.

one can read and understand without translating can one move on to translation. Translation is, in my view, a narrative account authored in the target language, of the mental states—especially sentences—experienced by the translator while reading and understanding in the original source language. These processes were eloquently theorized by George Steiner under the name "hermeneutic motion".[13] A particularly clear example of notational variation is the use of the table to present originally textual material, as in Table 6.1.[14]

The present translation is not aimed at the general public; far from it. It is a translation for historians of medicine and scholars of the Ayurvedic tradition who know Sanskrit and are closely interested in exactly what is being said in the *Compendium of Caraka*. It should not need to be argued, but it is my strongly-held view that an English translation should consist of English words throughout. Therefore, I have always used English and put Sanskrit technical terms in parentheses in the standard manner, not the other way around. An index of English and Sanskrit terms helps the reader in connecting ideas across different passages. It also helped me to improve consistency during the process of translation.

I hope this translation will be received by English readers as a fresh and engaging text that provokes reflection. Where there are difficulties in the underlying Sanskrit text, there are difficulties in the translation too. Parts of the text seem very well-organized and the ideas flow clearly; at other places the ideas seem to eddy and swirl and one suspects and intrusion of an editor who worked at an earlier stage of the text than can be recovered from the surviving manuscript transmission.

3 The Term *vimāna*

This part of the *Compendium of Caraka* is called, in manuscript colophons, "The Section (*sthāna*) on *Vimāna*." The meaning of the word *vimāna* in this context is not obvious. The common meaning of the word is "measurement".[15] But "mea-

13 George Steiner, *After Babel: Aspects of Language and Translation*, 3rd ed., Oxford paperbacks, first published in 1975 (Oxford: Oxford University Press, 1998), 296–303, ISBN: 0192880934.

14 This is an example of the process of "concept matching" or "symbolic rewriting" described by Sarukkai, 'Translation As Method,' 111 ff.

15 Vitus Angermeier, 'Über Die Auslöschung Von Land Und Leuten. Die Ursachen Von Massensterben Und Schutzmaßnahmen Gegen Diese Gemäß Carakasaṃhitā, Vi. 3: Leicht adaptierte und korrigierte Version der Diplomarbeit von 2007 (Betreuerin: O. Univ.-Prof. Dr. Karin C. Preisendanz)' (Diplomarbeit, University of Vienna, 2010), 16–17.

surement" does not appropriately describe this part of the *Compendium*, which includes chapters on an unusually wide range of topics, from flavours, parts of the stomach, special ways of recognizing diseases, tubes in the body, epidemics, rules of formal debate and the initiation of a medical student, but does not include any particular focus on quantitative information.[16] In the summary verses at the end of the present chapter, the "meaning of *vimāna*" is listed as one of the topics addressed in this chapter. That reference must be to the last part of section 3 of this chapter (p. 87 below), where the *Compendium* offers some remarks on the meaning of the term. Unfortunately, these remarks are not enough to definitively clarify what the *Compendium*'s author meant by the term.

The commentator Cakrapāṇidatta, in the eleventh century, already felt that the term required clarification. He explained,

> "*vimāna*" means that by which items such as humours and medicines are known, i.e., specifically understood. That is to say, the specific feature of things like the efficacy (*prabhāva*) of humours and of medicines.[17]

Cakrapāṇidatta was here telling the reader that he derived the word *vi-mā-na* from the Sanskrit root *mā-* (→*mīyate*), that he explains as meaning *jñāyate* "is known." He gestures to the meaning of the prefix *vi-* by his use of the adjective "specific feature" (*viśeṣa*). His understanding of *vimāna* is thus something like "knowledge, special knowledge, detailed knowledge."

This interpretation by Cakrapāṇidatta supports the argument made by Burrow[18] that there exists a distinct Sanskrit root *mā-* meaning approximately "ascertain" that appears in several words connected with cognition, especially *pra-mā-* and *anu-mā-*. Burrow's argument has met with both support[19] and criticism.[20] In my judgement, the critics of Burrow's hypothesis have not given

16 Dominik Wujastyk, *The Roots of Ayurveda: Selections from Sanskrit Medical Writings*, 3rd ed., Penguin Classics (London, New York, etc.: Penguin Group, 2003), 13, ISBN: 0-140-44824-1, cf. Angermeier, *idem.*

17 On Ca.vi.2: *viśeṣeṇa mīyate jñāyate doṣabheṣajādy aneneti vimānaṃ, doṣabheṣajādīnāṃ prabhāvādiviśeṣa ity arthaḥ.*

18 Thomas Burrow, 'Sanskrit MĀ- 'to Ascertain',' *Transactions of the Philological Society*, 1980, 134–140.

19 Eric P. Hamp, 'Ie *meha-,' *Münchener Studien zur Sprachwissenschaft* 43 (1984): 45, L. Isebaert, 'Quelques Considerations Sur Les Emprunts Iraniens En Tokharien. Le Traitement Des Diphtongues,' in *Studia Etymologica Indoeuropaea: Memoriae A.J. Van Windekens (1915–1989) Dicata*, ed. Lambert Isebaert, vol. 45, Orientalia Lovaniensia Analecta (Peeters Publishers, 1991), 145, http://tinyurl.com/o09fs8x.

20 Manfred Mayrhofer, *Etymologisches Wörterbuch des Altindoarischen*, 3 vols. (Heidelberg:

compelling reasons for rejecting it. They remained focussed on morphology while remaining too willing to derive the semantic field of "knowing" from that of "measuring." They have also not engaged with the extensive range of Burrow's semantic arguments and data-examples about particular Sanskrit usages from Vedic literature forward. Burrow's proposal retains merit and is at least supported from within the Sanskrit commentarial tradition by the present example. While Burrow's 1980 paper focussed on words unambiguously connected with the act of knowing, he only discussed the prefix-root combination *vi-mā* separately in the context of a quite different semantic field.[21]

It is clear that in the understanding of the commentarial tradition on the present usage, *vimāna* means "special knowing," "evaluation," or even "science" (if the English term were not so loaded) and can therefore be viewed as belonging to the word-cluster that includes *pramāṇa* "authority for correct knowledge" and *anumāna* "inferential knowledge."[22]

4 Typographical Conventions

Numerical footnotes provide references and discussion. Letter-footnotes provide text-critical readings where the Vienna critical edition (VCE) differs from the vulgate editions of Jīvānanda Vidyāsāgara Bhaṭṭācārya (1877) and Yādavaśarma Trivikrama Ācārya (1941), abbreviated C1 and B08 respectively. Such passages are marked in the text with a different colour of type.

Meulenbeld's *A History of Indian Medical Literature* offers a birds-eye view of the contents of this chapter, with substantive notes for further reading.[23]

Carl Winter, Universitätsverlag, 1986–2001), II, 342, ISBN: 3-533-03826-2., citing Jacques Duchesne-Guillemin, 'Post-script, Errata, Varia,' in *Languages and Cultures: Studies in Honor of Edgar C. Polomé*, ed. Mohammad Ali Jazayery and Werner Winter, vol. 36, Trends in Linguistics. Studies and Monographs (Amsterdam: Mouton de Gruyter, 1988), 125, ISBN: 3-11-010204-8. and Jean Haudry, 'Notes Sur Les Racines Indo-européennes *me-, *met-, *med- "mesurer",' *Études indo-européennes* 11 (1992): 54, ISSN: 0750–3547., who merely agrees with Duchesne-Guillemin.

21 Thomas Burrow, 'Sanskrit MĀ 'to Approach, Meet, Join; (trans.) to Bring To',' *Bulletin of the School of Oriental and African Studies* 44, no. 1 (1981): 89–92, ISSN: 0041–977X, http://www.jstor.org/stable/616297.

22 Cf. also with *nimāna* "price", used in Pāṇini 5.2.47, cited by Chlodwig H. Werba, *Verba Indoarica: Die Primären Und Sekundären Wurzeln Der Sanskrit-sprache*, vol. 1 (Vienna: Austrian Academy of Sciences Press, 1997), #331, which might perhaps be understood better as "value."

23 Gerrit Jan Meulenbeld, *A History of Indian Medical Literature*, 5 vols. (Groningen: E. Forsten, 1999–2002), IA.31, IB. 65 ff., ISBN: 9069801248 (hereafter cited as HIML).

Acknowledgements

This study is based on research done during the period 2010–2014 as part of the project "Philosophy and Medicine in Early Classical India" (PMECI) at the University of Vienna. The project was led by Prof. Dr Karin Preisendanz, and has been funded by research grants from the Austrian Science Fund (http://www.fwf.ac.at/, grant no. P23330-G15). The project's website offers much information about the history and aims of the project, and the reader is referred to that site for further details, including a bibliography of publications from 2000 onwards that have stemmed from the project.[24] Further research and completion of the contribution was conducted at the University of Alberta.

Abbreviations

C1 Jīvānanda Vidyāsāgara Bhaṭṭācārya, ed., *Carakasaṃhitā, Sūtra-nidāna-vimāna-śārīrendriya-cikitsita-kalpa-siddhisthānātmakā. Āyurvedīya Bhagavatā Ātreyeṇa Punarvasunā Upadiṣṭā Agniveśa-nāmadheyena Tatśiṣyeṇa Viracitā Carakābhidhena Ṛṣiṇā Pratisaṃskṛtā, Vi. E. Upādhidhāriṇā ŚrīJīvānanda Vidyāsāgara Bhaṭṭācāryyeṇa Bahūni Ādarśapustakāni Samālocya Saṃskṛtā Prakāśitā ca* (Kolkata: Sarasvati Press, 1877), accessed May 15, 2020, https://n2t.net/ark:/13960/t0sq9gf44 (hereafter cited as C1)

Bo8 Yādavaśarma Trivikrama Ācārya, ed., *Maharṣiṇā Punarvasunopadiṣṭā, Tacchiṣyeṇāgniveśena Praṇītā, Carakadṛḍhabalābhyāṃ Pratisaṃskṛtā Carakasaṃhitā, Śrīcakrapāṇidattaviracitayā Āyurvedadīpikāvyākhyayā Saṃvalitā*, 3rd ed. (Mumbayyāṃ: Nirnaya Sagara Press, 1941), accessed May 15, 2020, https://n2t.net/ark:/13960/t48q2f20n (hereafter cited as Bo8)

24 Preisendanz et al., 'Caraka Project Website.'

Translation of *Carakasaṃhitā, Vimānasthāna* 1

[1] Now I shall explain the special knowledge (*vimāna*) of the flavours (*rasa*),

[2] said Lord Ātreya.

[3] In this regard, the physician should enter into the particulars of the causes (*nimitta*), prodromes (*pūrvarūpa*), symptoms (*rūpa*), affinities (*upaśaya*), numbers (*saṅkhyā*), predominances (*prādhānya*), procedures (*vidhi*), combinations (*vikalpa*), forcefulnesses (*bala*) and durations (*kāla*) of diseases.

Next, he should attentively and exactly learn the concept (*māna*) of the humours (*doṣa*), drugs (*bheṣaja*), place (*deśa*), time (*kāla*), strength (*bala*), body, food, essence[a] affinities (*sātmya*),[25] mentality (*sattva*), temperament (*prakṛti*), and age (*vayas*).

Therapeutic actions (*kriyā*) depend upon knowing the concept (*māna*) of the humours, etc. For a physician who does not know the values of the humours, etc., is not capable of subjugating (*nigraha*) diseases. Therefore, O Agniveśa, I am teaching the chapter on special knowledge (*vimāna*) for the purpose of knowing the concept (*māna*) of the humours, etc.[b]

[Flavours, humours, substances, diseases and efficacies]

[4] In that context, I shall start by explaining the efficacies (*prabhāva*)[26] of flavours (*rasa*), substances (*dravya*), humours (*doṣa*), and diseases (*vikāra*).

[a] The critical text reads *śarīrāhārasāra*. C1 and Bo8 read *śarīrasārāhāra*, "body, essence, food."
[b] The critical text reads *doṣādimānajñānārtham*. One of the major witnesses to the transmission, MS B5, omits the word *-māna-* in this expression. While the archetype is very likely to have had *-māna-*, B5's reading is suggestive because it would mean that the word *jñāna* was being used as a direct gloss for the word *vimāna*.

25 See verse 20 below (p. 92) and Ca.vi.8.118 (Bo8, 279b): *sātmyaṃ nāma tad yat sātatyenopasevyamānam upaśete.* "Sātmya is defined as that which being used routinely, agrees with one."

26 The word *prabhāva* is here translated "efficacy." It expresses the idea of the effect, potency or power of an item, its likely outcome or power to produce an effect.

TABLE 6.1 Caraka's combinations in v. 6

	Wind	*Bile*	*Phlegm*
sweet	-	-	+
tangy	-	+	+
salty	-	+	+
bitter	+	+	-
spicy	+	-	-
acerbic	+	-	-

[Flavours]

There are six flavours: sweet (*madhura*), tangy (*amla*), salty (*lavaṇa*), bitter (*kaṭu*), spicy (*tikta*), and acerbic (*kaṣāya*). If they are treated[a] with respect, they make the body go. But wrongly used, they tend to irritate the humours.

[Humours and diseases]

[5] Furthermore, there are three humours: wind (*vāta*), bile (*pitta*), and phlegm (*śleṣman*). When they are in their natural state (*prakṛtibhūta*), they safeguard the body's welfare (*yogakṣema*),[b] but[c] if they go wrong they afflict the body with many different ailments (*vikāra*).

On the other hand, having become an alteration (*vikṛti*), they torment[d] the body with many different diseases (*vikāra*).

[6] In that connection, the flavours, taken in threes, generate each of the humours individually, and taken in threes they pacify them.[27]

For example, bitter (*kaṭu*), spicy (*tikta*) and acerbic (*kaṣāya*) generate wind, but sweet (*madhura*), tangy (*amla*) and salty (*lavaṇa*) pacify it.

[a] The critical text reads *upāsyamānāḥ*. C1 and Bo8 read *upayujyamānāḥ*, "they are used." [b] The critical text reads *śarīrayogakṣemakarā*. C1 and Bo8 read *śarīropakārakā*, "are assistants for the body." [c] The critical text reads *tu*. C1 and Bo8 read *tu khalu*, "but further." [d] The critical text reads *upatapanti*. C1 and Bo8 read *upatāpayanti*, "cause torment to."

27 On the combinatorics of humours and tastes in the *Carakasaṃhitā*, see Dominik Wujastyk, 'The Combinatorics of Tastes and Humours in Classical Indian Medicine and Mathematics,' *Journal of Indian Philosophy* 28 (2000): 480, doi:10.1023/a:1017514013759 that discusses a parallel passage in Ca.sū.26.14–22 (Bo8, 139–140).

Bitter, tangy (*amla*) and salty (*lavaṇa*) generate bile, but sweet (*madhura*), spicy (*tikta*) and acerbic (*kaṣāya*) pacify it.

Sweet, tangy (*amla*) and salty (*lavaṇa*) generate phlegm (*śleṣman*), but bitter (*kaṭu*), spicy (*tikta*) and acerbic (*kaṣāya*) pacify it.

[7] But when there is a conjunction (*sannipāta*) of flavours and humours, the flavours that have the same quality (*guṇa*) or predominantly the same quality as the respective humours cause them to grow. But those that are contrary[a], or predominantly of the contrary quality, pacify them, in regular use. This relationship is the reason why the flavours, not combined with one another, are taught as being six, and the humours as being three.

[8] For the extent of their optional combinations is infinite, because the different options are countless.

[Substances]

[9] In that context, then, when there are substances (*dravya*) with many flavours, and diseases that consist of many humours, then after examining the efficacy (*prabhāva*) of flavours and humours individually[b], one may gain certainty concerning what the efficacy (*prabhāva*) really is, as it applies to substances (*dravya*) and to diseases.

[10] But it is not like this in all cases. Because it is not possible to be certain of the truth about the efficacy of a combination by an inference based on the efficacy of the parts amongst the options that are assembled (*samaveta*) irregularly because of diseases (*vikṛti*), that are of various types, and that are in conflict with each other and with other options (*vikalpana*).

[11] When such a combination (*samudāya*) has been put together (*yukta*), then after understanding the actual truth about the efficacy of the combination (*samudāyaprabhāvatattva*) one may then be certain about the truth about the efficacy (*prabhāvatattva*) of the substances and illnesses.

---

[a] The critical text reads *viparītās tu*. C1 reads *vidārītaguṇās tu* "but those that have torn qualities." Bo8 reads *viparītaguṇā* "those that have contrary qualities." [b] The critical text reads *ekaikatvena*. Bo8 reads *ekaikaśyena*, "severally."

[12] Therefore, the truth (*tattva*) is taught[a] according to the efficacy of flavour (*rasa*), the efficacy of substance (*dravya*), the efficacy of humour (*doṣa*), and the efficacy of illness (*vikāra*).

[Efficacy]

[13] In that connection, the efficacy of flavour[b] has just been taught. Now we shall teach the efficacy of substance (*dravyaprabhāva*).

Sesame oil, clarified butter, honey[c]: these are the substances (*dravya*)[d] used to calm down wind, bile and phelgm.

[14] In that connection, when sesame oil is used routinely, it calms[e] wind, because it is imbued with oiliness, heat and heaviness. Because wind has the opposite qualities[f], being imbued with sereness (*raukṣya*), coolness and lightness.[28] For when there is a conjunction of opposing qualities, the lesser one is always completely overwhelmed[g]. That is why sesame oil, used routinely, always conquers wind. In just the same way, too, ghee conquers bile, because of its sweetness, coolness and sluggishness. For bile is non-sweet, hot and sharp. And honey, phlegm[h] by the fact of being sere, sharp, and astringent. For phlegm is oily, sluggish and sweet[i].

And any other substance whatsoever that is used routinely will conquer wind[j], bile and phlegm, if it is qualitatively contrary[k] to them.

[15] There are three substances that, more than any other substances, one should not over-use. They are long pepper, alkali, and salt.

---

[a] The critical text reads *upadiśyate*. C1 and Bo8 read *upadekṣyāmaḥ*, "we shall teach." [b] The critical text reads *rasaprabhāva*. C1 reads *rasadravyadoṣavikāraprabhāva*, "efficacy of flavour, substance, humour and illness." [c] The critical text reads *tailaṃ sarpir madhv iti*. C1 and Bo8 read *tailasarpirmadhūni*, "sesame oil, clarified butter and honey." [d] The critical text reads *dravyāṇi*. C1 and Bo8 read *dravyāṇi bhavanti*, "they are the substances (*dravya*)." [e] The critical text reads *śamayati*. C1 and Bo8 read *jayati*, "defeats." [f] The critical text reads *viruddhaguṇaḥ*. C1 and Bo8 read *viruddhaguṇo bhavati*, "it is such as has opposite qualities." [g] The critical text reads *avajīyate 'yantam*. C1 reads *avacīyate*, "gathered." [h] The critical text reads *śleṣmānam*. C1 and Bo8 read *śleṣmānaṃ jayati*, "conquers phlegm." [i] The critical text reads *snigdhamandamadhuraḥ*. C1 and Bo8 read *snigdho mando madhuraḥ*, "oily, sluggish and sweet." [j] The critical text reads *vāta-*. C1 and Bo8 read *evaṃ vāta-*, "similarly wind." [k] The critical text reads *viruddham*. C1 and Bo8 read *viparītam*, "opposite."

---

28 "Sereness" translates *raukṣya*, meaning "dryness" in the particular sense of "not oily." *Raukṣya* may be contrasted with words like *śuṣka* and *śoṣa*, which mean "dry, parched" as in "not wet with water." The English word "dry" does not make this distinction, so I have used "sere" for oil-dry and "dry" for water-dry.

[16] For peppers, being[a] bitter (*kaṭuka*), having sweet post-digestive flavour, and being heavy, are not overly oily (*snigdha*) or hot, and are moistening (*prakledin*). They are a desirable medicine and cause good and bad results rapidly.

They are beneficial from the outset because their excellence (*sādguṇya*) manifests at the same time (*sama*) they are used. They are associated with a build-up of humours. Used all the time, they upset (*utkleśayanti*) the phlegm because of being heavy and moistening; the bile because of heat; and they do not contribute to soothing the wind because of being low in oiliness and heat[b]. They are, in fact, catalysts (*yogavāhin*), so one should not over-use peppers.[29]

[17] Alkali is endowed with hotness, sharpness and lightness. Initially it wets, and subsequently it dries. It is used[c] for the purpose of maturing (*pacana*), burning (*dahana*), and loosening (*bhedana*). If over-used, it harms the hair, eyes, heart and masculinity. Those villages (*grāma*), towns (*nagara*), cities (*nigama*), and regions (*janapada*) where it is used all the time suffer from blindness, impotence, baldness, and grey hair, and have a tearing away of the heart. For example, easterners and the Chinese (*cīna*). For that reason, alkali should not be used too much.

[18] Salt is endowed with severe heat (*auṣṇya*) and sharpness (*taikṣṇya*)[d]. It is not too heavy, not too oily, it is moistening, it is capable of causing something to fall away, and it makes food and drink enjoyable. It is beneficial at the very outset, because its excellence[e] manifests at the same time as it is used. It is followed

[a] The critical text reads *satyaḥ*. C1 reads *sadyaḥ*, "suddenly." [b] The critical text reads *auṣṇya*. C1 and Bo8 read *uṣṇa*, "hot." [c] The critical text reads *prayujyate*. C1 and Bo8 read *upayujyate*, "it is employed." [d] The critical text reads *bhṛśauṣṇyataikṣṇya-*. C1 and Bo8 read *punar auṣṇyataikṣya-*, "heat and sharpness, however." [e] The critical text reads *-sādguṇyāt*. C1 reads *-ṣāḍguṇyād*, "fact of having six qualities."

29 The concept of *yogavāhin*, a catalyst (in a non-technical sense) or an item which intensifies the delivery of a drug, is rare in early ayurvedic literature, occurring only four times in the *Carakasaṃhitā* (Ca.sū.27.249, vi.1.16, śā.6.4, ci.3.38. The single occurrence in the *Suśrutasaṃhitā*, sū.45.142, is identical to Ca.sū.27.249.) The term gains frequency in the *Aṣṭāṅgahṛdayasaṃhitā* and in alchemical literature. See Oliver Hellwig, *Wörterbuch Der Mittelalterlichen Indischen Alchemie* (Groningen: Barkhuis & University of Groningen, University Library, 2009), 460, ISBN: 9789077922620. Cakrapāṇidatta (fl. 1075) characterizes it thus, while describing honey as the prime example of a catalyst: "Honey is as it is because it is generated from flowers that have various flavours, potencies, and so forth. Because of that, it has various latent powers. So whatever it is joined to, be it an emetic, a tonic, a aphrodisiac, or something else efficacious, it performs that same substance (*dravya*)'s action. I.e., because its substance (*dravya*), which is a mimesis of the same thing, has its power activated." (on Ca.sū.27.249 (Bo8, 167)).

by an accumulation of the humours. It is used[a] for the purpose of enhancing tastiness (*rocana*), for maturing (*pācana*), moistening (*upakledana*), and for making things fall away (*visraṃsana*).

If it is used too much, it brings about lethargy (*glāni*)[b], slackness (*śaithilya*), and weakness (*daurbalya*), in the body.

Those villages (*grāma*), towns (*nagara*), cities (*nigama*), and regions (*janapada*) where it is used all the time are extremely exhausted (*glasnu*), and they[c] have slack (*śithila*) flesh and blood. They are unable to bear suffering (*parikleśa*). For example, people from Balkh, Saurāṣṭra[d], Sindh and Sauvīra. They even consume salt with milk.[e]

In those regions of the earth that are excessively saline (*ūṣara*), plants (*oṣadhi*), bushes (*vīrudh*), fruit trees (*vanaspati*) and flowering trees (*vānaspatya*) will not grow, or will have little lustre (*tejas*), because it is damaged by salt.[30]

For that reason, salt should not be used too much. Because even people who have a great natural affinity (*sātmya*) for salt go bald (*khālitya*), grey, completely hairless (*indralupta*), and have premature wrinkles.[f]

[19] Therefore, where they are concerned, it is best for someone to distance themselves gradually from a natural affinity (*sātmya*) to them. Because even if there is a natural affinity, a gradual disengagement[g] will cause little or no harm (*doṣa*).

[Affinity]

[20] Natural affinity (*sātmya*) means whatever is appropriate out of familarity[h] to the self.[31] For the meaning of affinity is the same as the meaning of appropri-

---

[a] The critical text reads *prayujyate*. C1 and Bo8 read *upayujyate*, "it is used." [b] C1 omits this word. [c] The critical text reads *te'pi*. C1 and Bo8 read *te*, "they." [d] The critical text reads *saurāṣṭra*. C1 and Bo8 read *saurāṣṭrika*, "from Saurāṣṭra." [e] The critical text reads *te'pi aśnanti payasāpi lavaṇam*. C1 reads *te hi payasāpi saha lavaṇam aśnanti* "for they eat salt even together with milk." Bo8 reads *te hi payasāpi sadā lavaṇam aśnanti* "for they always eat salt even with milk." [f] The critical text reads *khālityapālityendraluptāni*. C1 reads *khālityendriyaluptapālityāni* "bald, wrinkled and completely hairless." Bo8 reads *khālityapālityāni* "bald and wrinkled." [g] The critical text reads *upanirvartyamāna*. C1 and Bo8 read *upanivartyamāna*, "repetition." [h] The critical text reads *okād upaśete*. C1 Bo8 read *upaśete* "it is appropriate".

---

30 Ca.sū.1.71–72 describes these four types of medicinal plant: *oṣadhi* is a plant that ends after fruiting, *vīrudh* is a plant that branches out, *vanaspati* is a tree with fruit, and *vānaspatya* is a tree with fruit and flowers.

31 *okād*, "from home, out of familarity" is read in the Kashmir transmission and seems to be original, but is omitted in most manuscripts from the rest of the transmission.

ateness. It is divided into three kinds, superior, inferior, and average. And it is of seven kinds, according to the flavours (*rasa*) taken one by one and all together. In that context, the use of all the flavours is superior, and a single flavour is inferior. And standing in the middle between superior and inferior is the average. One should encourage the the inferior and the average types of affinity towards the superior type of affinity, but only in step by step manner. Even if one has successfully achieved the affinity characterised by all the flavours, one should adhere to using only good things, after considering all[a] the foundations for the special rules of eating.

[Factors involved in eating]

[21] In that connection, these are the eight foundations for the special rules of eating. For example, constitution (*prakṛti*), processing (*karaṇa*), combination (*saṃyoga*), volume (*rāśi*), location (*deśa*), time (*kāla*), standard for consumption (*upayogasaṃsthā*), and consumer (*upayoktṛ*).[b]

[Constitution]

[22.1] In that context, constitution (*prakṛti*) is said to be that which is inherent (*svabhāva*). It is the inherent conjunction with the qualities of heaviness etc., that are present in the substances of foods and medicines, for example, of black gram (*māṣa*) and green gram (*mudga*) or of pork[c] and venison.[32]

[Processing]

[22.2] Processing, to continue, is the preparation (*saṃskāra*) of inherent (*svābhāvika*) substances. Preparation is said to be the activation of qualities[d]. Those qualities are activated by water, fire, proximity, purity, stirring, place, time storage (*vāsa*)[e], and steeping (*bhāvanā*)[f], and by means of time, intensity (*prakarṣa*), and the cooking pot (*bhājana*), etc.

[a] The critical text reads *sarvāṇy*. Bo8 reads *prakṛtyādy upayoktraṣṭamāni sarvāṇy*, "all, the eight things from constitution to consumer." [b] C1 and Bo8 add *bhavanti*, "they are." [c] The critical text reads *sūkara*. Bo8 reads *śūkara*, "pork." [d] The critical text reads *guṇādhāna*. Bo8 reads *guṇāntarādhāna*, "activation of different qualities." [e] The critical text reads *vāsa*. Bo8 reads *vāsana*, "being covered." [f] The critical text reads *bhāvanā*. C1 reads *bhavana*, "being."

32 The commentator Cakrapāṇidatta notes that black gram is heavy, while green gram is light, pork is heavy and venison is light.

[Combination]

[22.3] Combination, though, is melding together two or more substances. It is something special that arises that is not the substances individually[a]. Thus, the combination of honey and ghee, or of honey, fish and milk.[33]

[Volume]

[22.4] Volume, though, is the total and aggregated accumulation[b], for the purpose of ascertaining[c] the result of right or wrong measure.

In that context, the "total accumulation (*sarvagraha*)" is taking the measure of all the food that is eaten as a single unit. The "aggregated accumulation (*parigraha*)," on the other hand, is taking the measure of foods and substances taken individually[d]. "Total accumulation" is said to mean "taking everything," "aggregated accumulation" is "taking in from each and every one."

[Location]

[22.5] Location, moreover, means place. And it means that the production and application of substances[e] have an affinity arising from location (*deśasātmya*)[f].

[Time]

[22.6] Time, then, is perpetual (*nityaga*) and it is circumstantial (*āvasthika*). Thus, the circumstantial refers to the illness (*vikāra*), and the perpetual refers to the affinity arising from the season (*ṛtusātmya*).

[a] The critical text reads *yan naikaśo dravyāṇi*. C1 and Bo8 read *yaṃ punar naikaikaśo dravyāṇy ārabhante*, "that is not, however, the substances arising one by one." [b] The critical text reads *sarvaparigraha*. C1 and Bo8 read *sarvagrahaparigraha*, "the total accumulation and the aggregated accumulation." [c] The critical text reads *viniścayārtha*. C1 reads *viniścayārthaḥ prakṛtaḥ*, "put forward for the purpose of ascertaining." [d] The critical text reads *ekaikaśyena*. C1 reads *ekaikaśena*, "[misprint]." [e] The critical text reads *dravyotpattipracārau*. C1 and Bo8 read *dravyāṇām utpattipracārau*, "the production and application of substances." [f] The critical text reads *deśasātmya*. C1 reads *deśasāmya*, "identity arising from location."

33 The *Carakasaṃhitā* notes elsewhere that these particular combinations are toxic because of a clash of conflicting potencies (Ca.sū.26.82–84, Bo8, 149–150).

[Standard for consumption]

[22.7] The standard for consumption (*upayogasaṃsthā*), moreover, is the constraint (*niyama*) on consumption. It has reference to the character of digestion.

[The consumer]

[22.8] The consumer (*upayoktṛ*), moreover, is the person who consumes that food. What one depends upon becomes the acquired affinity (*okasātmya*).[34]

These are the eight[a] foundations for the special rules of eating.

[23] Their special properties are that they have healthy and unhealthy results that help and hinder one another. One should seek to understand them. And after understanding them, one should strive only for what is healthy (*hita*). One should not, out of delusion or lack of attention, partake of a favourite consumable or anything else that is not healthy, or that results in later discomfort.

[Modes of eating]

[24] In that context, this regulation about the rules of consumption is for the healthy and for some of the sick, who are eating the healthiest items (*hitatama*) on time and according to constitution (*prakṛti*).

After properly examining oneself, one should eat food that is

- hot,
- oily,
- in appropriate measure,
- after digestion,
- that has no conflict of potency (*vīrya*),
- in a selected location,
- with selected utensils,
- not too quickly,

---

[a] The critical text reads *aṣṭau*. C1 reads ... *bhavanti* "they are ...." Bo8 reads ... *vyakhyatāni bhavanti* "these eight ... have been explained."

---

34 Caraka defined "home-affinity (*okasātmya*)" as food that starts to suit someone as a result of regular usage (Ca.sū.6.49, (Bo8, 48 and note 3): *upaśete yad aucityād okasātmyaṃ tad ucyate*). As in English, one speaks of an "acquired taste." The commentator Cakrapāṇidatta took the notion a step further, saying that even something that is unhealthy for one may, through regular usage, cause no illness. He gave the example of a snake, which is not harmed by its own poison, even though this poison is permanently present.

- not too slowly,
- without chattering,
- without laughing[a],
- and with attention[b].

[25] We shall elaborate on its excellence (*sādguṇya*)[c].[35]

[Hot]

[25.1] One should eat something hot (*uṣṇa*). For one relishes something hot being eaten. And once eaten, it promotes the fire in the belly[d]. It is digested quickly, it smooths out wind, and it dries out[e] phlegm. That is why one should eat something hot.

[Oily]

[25.2] One should eat something oily (*snigdha*). For one relishes something oily being eaten. And once eaten, it promotes the fire in the belly[f]. It is digested quickly, it smooths out wind, it makes the development of the body firm[g], it leads to an increase of strength[h], and it produces clarity of the voice. That is why one should eat something oily.

[Measure]

[25.3] One should eat in measure. For something eaten in measure, not causing harm[i] to wind, bile and phlegm, uniquely extends life itself[j]. It goes around the intestines comfortably, it does not impair the heat (*ūṣman*), and it is digested without trouble. That is why one should eat in measure.

---

[a] The critical text reads *ahasan*. C1 reads *sahasam*, "with laughter." [b] The critical text reads *tanmanā*. C1 reads *stambhanā*, "becoming rigid." [c] The critical text reads *sādguṇya*. C1 reads *ṣāḍguṇya*, "fact of having six qualities." [d] The critical text reads *agnim udaryam*. Bo8 reads *cāgnim audaryam* "and [once eaten] the fire associated with the belly", C1 reads *cāgnim udīryyam* "and [once eaten] the fire is to be raised". [e] The critical text reads *pariśoṣayati*. Bo8 reads *parihrāsayati*, "it shortens." [f] The critical text reads *agnim udaryam*. Same variants as in footnote d. [g] The critical text reads *dṛḍhīkaroti śarīropacayam*. Bo8 reads *śarīram upacinoti dṛḍhīkarotīndriyāṇi*, "it builds up the body, it makes the senses firm." [h] The critical text reads *balavṛddhim*. C1 and Bo8 read *balābhivṛddhim*, "an augmentation of strength." [i] The critical text reads *apratipīḍayad*. Bo8 reads *apīḍayad* "not hurting," C1 reads *aprapīḍayad* "not hurting." [j] The critical text reads *āyurevābhivardhayati*. C1 and Bo8 read *āyureva vivardhayati*, "particularly extends life itself."

---

35 I.e., the elements that constitute the special excellence of food.

[Completed digestion]

[25.4] One should eat when digestion is complete. Because, when digestion has not happened, the person eating (*bhuñjāna*)[a], mingling the unconverted (*apariṇata*) chyle (*rasa*) of the earlier meal with the chyle of the later meal, immediately irritates all the humours.

But, when digestion of the food is complete, with the humours being in their proper locations, with the digestive fire (*agni*) boosted, when the desire to eat has arisen, when the openings of the tubes (*srotas*) are cleansed and when the spittle (*udgāra*) is clean, when the heart is purified, when wind is where it should be (*vātānulomya*), and when the urges of urine and feces have been released, when the digestion has taken place, the kind of food that has been eaten uniquely enhances the vitality (*āyus*), while not corrupting the body tissues (*dhātu*) of the entire body.

For that reason, one should eat when digestion is complete.

[No conflict of potency]

[25.5] One should eat things that do not have conflicting potencies (*viruddhavīrya*). Because someone eating things without conflicting potencies is not afflicted (*upasṛj-*) by diseases that arise from food with conflicting potencies. That is why one should eat things that do not have conflicting potencies.

[Selected location]

[25.6] One should eat[b] in a choice location. Because, someone eating (*bhuñjāna*) in a choice location does not incur the mental conflict due to the feelings that bring about mental conflict that arise out of being in an undesired location. And it is the same with the choice utensils (*upakaraṇa*). So one should eat[c] in a choice location.

---

[a] The critical text reads *bhuñjānaḥ*. Bo8 reads *bhuñjānasyābhyavahṛtam āhārajātaṃ*, that includes the phrase "the kind of food that has been eaten" that does not construe here, being duplicated through *saut du même au même* from later in the paragraph [b] The critical text reads *aśnīyāt*. Bo8 reads *iṣṭasarvopakaraṇaṃ cāśnīyāt*, "and one should eat the choice utensils." [c] The critical text reads *aśnīyāt*. C1 and Bo8 read *tatheṣṭasarvopakaraṇaṃ cāśnīyāt*, "one should thus eat the choice utensils."

[Unhurriedly]

[25.7] One should not eat in a hurry. Because, a person eating hurriedly has regurgitation (*utsnehana*)[a],[36] and sinking down (*avasādana*), and the food will not be settled. The appreciation (*upalabdhi*) of the faults and excellences (*sād-guṇya*) of what should be eaten is restricted[b].[37]

[Not too slowly]

[25.8] One should not eat very slowly. Because, a person eating (*bhuñjāna*) very slowly does not achieve satisfaction (*tṛpti*), he eats too much. His food gets cold, and it is digested unevenly. Therefore, one should not eat very slowly.

[Without chatter, with attention]

[25.9] One should eat without chattering, without laughing, and with concentration (*tanmanā*). A person eating (*bhuñjāna*) who is chattering, laughing or has his mind on something else will have those self-same problems (*doṣa*) as the person eating too fast. Therefore, one should eat without chattering, without laughing and with concentration.

[a] The critical text reads *bhuñjānasyotsnehanam*. C1 reads *bhuñjānasya tat snehanam*, "the person eating has that oily thing." [b] The critical text reads *niyatā*. C1 and Bo8 read *ca na niyatā*, "and is not definite."

36 The meaning given for this word is conjectural, and based on context. The prefix *ud-* means "upwards" and the word *snehana* means "being oily, smearing." The word is glossed by Cakrapāṇidatta as *unmārgagamana*, which is normally interpreted as "going the wrong way," but could also mean "going by the upward path." All later dictionaries appear to repeat or paraphrase the first meaning, without thought or reference to further literature. The *ud-* prefix is probably meant to contrast with the *ava* "downwards" of the next word.

37 Many manuscripts read *niyatā* with a negation in this sentence: *ca na niyatā* or *cāniyatā* "not sure, not certain." The meaning would then be that the hasty eater cannot discriminate about the qualities of his food. But the earliest reconstructible reading of this sentence is that of the K[ashmir] branch, which omits the negation. To make sense of this, I take the meaning of "*niyatā*" as "restricted, controlled, suppressed," for which there is copious lexicographical precedent. The overall meaning of the sentence is similar in either case.

[After self-examination]

[25.10] After examining oneself, one should eat food properly. Knowing, "this would be suitable for me and[a] this would not be suitable," his own self[b] gains an affinity for itself (*ātmasātmya*). Therefore, one should eat after examining oneself[c].

[Summary verses]

[26] There are verses about this:[d]

The man who knows flavours, substances, humours and illnesses, as far as efficacy is concerned, who knows place and time, and the body, that man is indeed[e] a physican.[38]

[27–28] All the following has been clarified in "The Evaluation of Taste:"
- The meaning (*artha*) of special knowledge (*vimāna*),
- Taste (*rasa*), substance (*dravya*), humour (*doṣa*) and disease (*roga*), as far as efficacy (*prabhāva*) is concerned,
- substances (*dravya*) that should not be over-used and also
- the threefold natural affinity (*sātmya*),
- the eight foundations of consumption, and also
- the excellence of food.

Here ends the Special Knowledge about Flavours.[f]

[a] The critical text reads *ca*. Bo8 and C1 omit this. [b] The critical text reads *asyātmanaḥ*. C1 reads the irregular *asyātmena*. [c] The critical text reads *ātmānam*. C1 reads *ātmātmānam*, "the self of one's self." [d] The critical text reads *tatra ślokāḥ*. Bo8 reads *bhavati cātra* "and on this there is." C1 omits this, but reads *tatra ślokau* between 26ab and 26cd. Bo8 reads *tatra ślokau* before verses 27–28. [e] The critical text reads *sa no*. C1 reads *mano*, "the mind." [f] The critical text reads *iti rasavimānaṃ samāptam*. C1 and Bo8 read *agniveśakṛte tantre carakapratisaṃskṛte vimānasthāne rasavimānaṃ nāma prathamo'dhyāyaḥ*, "Here ends the first chapter, entitled Special Knowledge about Flavours, in the Section on Special Knowledge in the treatise created by Agniveśa that was Re-edited by Caraka."

38 "That man is indeed" translates *sa no*. The expression is odd, though probably in the archetype. The commentator Cakrapāṇidatta reads it as sandhi for *sa naḥ*, the oblique case of the plural first person pronoun: "he is our [physician]." However, it can more plausibly be construed as *sa nā u* "that man, indeed [is a physician]." The argument against this is that the enclitic *u* is archaic, known mainly from Vedic and Brāhmaṇa literature, though not entirely unknown in classical Sanskrit. But the idea is supported by one lesser branch of the manuscript transmission that reads *sa nā* "that man."

References

Ācārya, Yādavaśarma Trivikrama, ed. *Maharṣiṇā Punarvasunopadiṣṭā, Tacchiṣyeṇāgniveśena Praṇītā, Carakadṛḍhabalābhyāṃ Pratisaṃskṛtā Carakasaṃhitā, Śrīcakrapāṇidattaviracitayā Āyurvedadīpikāvyākhyayā Saṃvalitā.* 3rd ed. Mumbayyāṃ: Nirnaya Sagara Press, 1941. Accessed May 15, 2020. https://n2t.net/ark:/13960/t48q2f20n.

Álvarez, Román, and M. Carmen-África Vidal. *Translation, Power, Subversion.* Topics in Translation 8. Clevedon: Multilingual, 1996. ISBN: 9780585149806.

Angermeier, Vitus. 'Über Die Auslöschung Von Land Und Leuten. Die Ursachen Von Massensterben Und Schutzmaßnahmen Gegen Diese Gemäß Carakasaṃhitā, Vi. 3: Leicht adaptierte und korrigierte Version der Diplomarbeit von 2007 (Betreuerin: O. Univ.-Prof. Dr. Karin C. Preisendanz).' Diplomarbeit, University of Vienna, 2010.

Baker, Mona, ed. *Critical Readings in Translation Studies.* London: Routledge, 2010. ISBN: 9780415469548.

Baker, Mona, and Gabriela Saldanha, eds. *Routledge Encyclopedia of Translation Studies.* 2nd ed. London: Routledge, 2009. ISBN: 0-203-02911-9.

Bhaṭṭācārya, Jīvānanda Vidyāsāgara, ed. *Carakasaṃhitā, Sūtra-nidāna-vimāna-śārīrendriya-cikitsita-kalpa-siddhisthānātmakā. Āyurvedīya Bhagavatā Ātreyeṇa Punarvasunā Upadiṣṭā Agniveśa-nāmadheyena Tatśiṣyeṇa Viracitā Carakābhidhena Ṛṣiṇā Pratisaṃskṛtā, Vi. E. Upādhidhāriṇā Śrījīvānanda Vidyāsāgara Bhaṭṭācāryyeṇa Bahūni Ādarśapustakāni Samālocya Saṃskṛtā Prakāśitā ca.* Kolkata: Sarasvati Press, 1877. Accessed May 15, 2020. https://n2t.net/ark:/13960/t0sq9gf44.

Burrow, Thomas. 'Sanskrit MĀ 'to Approach, Meet, Join; (trans.) to Bring To'.' *Bulletin of the School of Oriental and African Studies* 44, no. 1 (1981): 85–104. ISSN: 0041–977X. http://www.jstor.org/stable/616297.

Burrow, Thomas. 'Sanskrit MĀ- 'to Ascertain'.' *Transactions of the Philological Society,* 1980, 134–140.

Dasgupta, Surendranath. *A History of Indian Philosophy.* Reprint of 1922 edition. Cambridge: Cambridge University Press, 1952–1961. ISBN: 0-521-04778-1, accessed January 26, 2018. https://tinyurl.com/ratwdof.

Duchesne-Guillemin, Jacques. 'Post-script, Errata, Varia.' In *Languages and Cultures: Studies in Honor of Edgar C. Polomé,* edited by Mohammad Ali Jazayery and Werner Winter, 36:123–131. Trends in Linguistics. Studies and Monographs. Amsterdam: Mouton de Gruyter, 1988. ISBN: 3-11-010204-8.

Eco, Umberto. *Mouse or Rat? Translation As Negotiation.* London: Orion, 2004. ISBN: 9780297830016.

Elshakry, Marwa S. 'Knowledge in Motion: The Cultural Politics of Modern Science Translations in Arabic.' *Isis* 99, no. 4 (2008): 701–730. Accessed February 24, 2019. doi:10.1086/595767. https://www.jstor.org/stable/10.1086/595767.

Fischer, Beatrice, and Matilde Nisbeth Jensen, eds. *Translation and the Reconfigura-*

tion of Power Relations: Revisiting Role and Context of Translation and Interpreting. Repräsentation—Transformation. Translating Across Cultures and Societies 7. Zurich: LIT Verlag, 2012. ISBN: 9783643902832.

Garzilli, Enrica, ed. *Translating, Translations, Translators from India to the West*. Harvard Oriental Series. Cambridge, MA: Harvard University, 1996. ISBN: 1888789026.

Hamp, Eric P. 'Ie *meha-.' *Münchener Studien zur Sprachwissenschaft* 43 (1984): 45–6.

Haudry, Jean. 'Notes Sur Les Racines Indo-européennes *me-, *met-, *med- "mesurer".' *Études indo-européennes* 11 (1992): 43–55. ISSN: 0750–3547.

Hellwig, Oliver. *Wörterbuch Der Mittelalterlichen Indischen Alchemie*. Groningen: Barkhuis & University of Groningen, University Library, 2009. ISBN: 9789077922620.

Isebaert, L. 'Quelques Considerations Sur Les Emprunts Iraniens En Tokharien. Le Traitement Des Diphtongues.' In *Studia Etymologica Indoeuropaea: Memoriae A.J. Van Windekens (1915–1989) Dicata*, edited by Lambert Isebaert, 45:141–150. Orientalia Lovaniensia Analecta. Peeters Publishers, 1991. http://tinyurl.com/o09fs8x.

Kang, Sung-Yong. *Die Debatte Im Alten Indien—Untersuchungen Zum Sambhāṣāvidhi Und Verwandten Themen in Der Carakasaṃhitā Vimānasthāna 8.15–28*. Philosophica Indica Einsichten Ansichten 6. Reinbek: Dr. Inge Wezler, Verlag für Orientalistische Fachpublikationen, 2003.

Krishnamurthy, Ramesh. 'Indian Tradition.' In *Routledge Encyclopedia of Translation Studies*, 2nd ed., edited by Mona Baker and Gabriela Saldanha, 449–458. London: Routledge, 2009. ISBN: 0-203-02911-9.

Maas, Philipp André. 'Computer Aided Stemmatics—the Case of Fifty-two Text Versions of Carakasaṃhitā Vimānasthāna 8.67–157.' Special issue edited by Jürgen Hanneder and Philipp A. Maas, *Text Genealogy, Textual Criticism and Editorial Technique, Wiener Zeitschrift für die Kunde Südasiens* 52–53 (2010): 63–119. doi:10.1553/wzks2009–2010s63.

Mayrhofer, Manfred. *Etymologisches Wörterbuch des Altindoarischen*. 3 vols. Heidelberg: Carl Winter, Universitätsverlag, 1986–2001. ISBN: 3-533-03826-2.

Meulenbeld, Gerrit Jan. *A History of Indian Medical Literature*. 5 vols. Groningen: E. Forsten, 1999–2002. ISBN: 9069801248.

Mohanty, Niranjan. 'Intransla[ta]bility and the Translator's Task.' *Perspectives: Studies in Translation Theory and Prạctice* 4, no. 2 (1996): 163–171. doi:10.1080/0907676x.1996.9961284.

Mohanty, Niranjan. 'Translation: An Integration of Cultures.' *Perspectives: Studies in Translation Theory and Practice* 2, no. 2 (1994): 187–198. doi:10.1080/0907676x.1994.9961235.

Olohan, Maeve. 'History of Science and History of Translation: Disciplinary Commensurability?' *The Translator* 20, no. 1 (2014): 9–25. doi:10.1080/13556509.2014.899091.

Preisendanz, Karin. 'Logic, Debate and Epistemology in Ancient Indian Medical Science: An Investigation into the History and Historiography of Indian Philosophy.

Part 1.' In *Medical Texts and Manuscripts in Indian Cultural History*, edited by Dominik Wujastyk, Anthony Cerulli, and Karin Preisendanz, 63–139. Delhi: Manohar Publishers / Distributors, 2013. ISBN: 978–9350980194.

Preisendanz, Karin, et al. 'Caraka Project Website.' 2002–. Accessed August 11, 2016. http://www.istb.univie.ac.at/caraka.

Rodgers, Daniel T., Bhavani Raman, and Helmut Reimitz, eds. *Cultures in Motion*. Princeton: Princeton University Press, 2013.

Roth, Rudolf. 'Indische Medicin. Caraka.' *Zeitschrift der Deutschen Morgenländischen Gesellschaft* 26 (1872): 441–452.

Sarukkai, Sundar. 'Translation As Method: Implications for History of Science.' *Indian Journal of History of Science* 51, no. 1 (2016): 105–117. doi:10.1163/9789004251410_014.

Solomon, Esther A. *Indian Dialectics: Methods of Philosophical Discussion*. 2 vols. Sheth Bholabhai Jeshingbhai Institute of Learning and Research Research Series, 70, 74. Ahmedabad: B.J. Institute of Learning / Research, Gujarat Vidya Sabha, 1976–1978. Accessed May 15, 2020. https://n2t.net/ark:/13960/t5jb4x70d.

Steiner, George. *After Babel: Aspects of Language and Translation*. 3rd ed. Oxford paperbacks. First published in 1975. Oxford: Oxford University Press, 1998. ISBN: 0192880 934.

Susan Bassnett, Harish Trivedi. *Postcolonial Translation: Theory and Practice*. Routledge, 1998. ISBN: 0415147441.

Tymoczko, Maria, and Edwin Gentzler. *Translation and Power*. Amherst, Boston, etc.: University of Massachusetts Press, 2002. ISBN: 1558493581.

Venuti, Lawrence, ed. *The Translation Studies Reader*. 3rd ed. London: Routledge, 2012. ISBN: 9780415613484.

Venuti, Lawrence. *The Translator's Invisibility: A History of Translation*. London, New York: Routledge, 1995.

Vidyabhusana, Satis Chandra. *A History of Indian Logic: Ancient, Mediaeval and Modern Schools*. Reprinted Delhi, Motilal Banarsidass, 1977 and often thereafter. Calcutta: Calcutta University, 1921.

Werba, Chlodwig H. *Verba Indoarica: Die Primären Und Sekundären Wurzeln Der Sanskrit-sprache*. Vol. 1. Vienna: Austrian Academy of Sciences Press, 1997.

Wujastyk, Dominik. *Medical Dialectic in Late Precolonial India*. A Disquisition on Sickness and Health *by Vīreśvara*. In preparation.

Wujastyk, Dominik. *Metarules of Pāṇinian Grammar: The Vyāḍīyaparibhāṣāvṛtti*. 2nd ed. Critical Studies on Pāṇinian System 1. Delhi: Motilal Banarsidass, 2017. ISBN: 9788120839823.

Wujastyk, Dominik. 'The Combinatorics of Tastes and Humours in Classical Indian Medicine and Mathematics.' *Journal of Indian Philosophy* 28 (2000): 479–495. doi: 10.1023/a:1017514013759.

Wujastyk, Dominik. *The Roots of Ayurveda: Selections from Sanskrit Medical Writings*.

3rd ed. Penguin Classics. London, New York, etc.: Penguin Group, 2003. ISBN: 0-140-44824-1.

Zimmermann, Francis. 'Terminological Problems in the Process of Editing and Translating Sanskrit Medical Texts.' In *Approaches to Traditional Chinese Medical Literature*, edited by Paul Unschuld, 141–151. Dordrecht, London: Kluwer Academic, 1989. ISBN: 1-55608-041-7. doi:10.1007/978-94-009-2701-8_13.

Lexical Index

Sort order is by Latin alphabet, ignoring accents.

CHAPTER 7

On the Textual History of the *Suśrutasaṃhitā*, (2): An Anonymous Commentary and Its Identified Citations

Andrey Klebanov

1 Introduction

This paper continues a series of introductory studies of unpublished material related to the textual history of the *Suśrutasaṃhitā* (SS).[*,1] It examines a hitherto unknown commentary on the SS, which has so far only been accessible to me in a single modern transcript (see the description below). Neither the author nor the title of the work is stated anywhere within the actual text, so the label *Suśrutavyākhyā* (a commentary on the *Suśruta*[*saṃhitā*]), which is given to the manuscript in the catalogs, was likely intended as a description of its content. For the sake of convenience, however, in the present article I refer to the text with the title **Suśrutavyākhyā* and, more often, the abbreviation *SVyā, but I mark these with an asterisk to signal their provisory character. The current paper is my first public engagement with this arguably significant piece of Āyurveda's textual history. Therefore, I first provide a brief general introduction to the text and then turn to the main subject, an overview of identified quotations from medical literature. The focus of this introductory study was influenced by Prof. Zysk's 2010 paper,[2] in which he demonstrated how a survey of citations can provide insight into the working techniques and intellectual background of a given author. Unlike the aforementioned study, however, my paper concentrates on quotations from works other than the SS. Further articles, which highlight the *SVyā's unique readings and interpretations of the SS,

* Preparation of this article was financially supported by JSPS KAKENHI grant number 17K17835.

1 The first study of what I wishfully call here a "series" was submitted more than five years ago for publication in a special issue of *eJournal of Indian Medicine*, which has so far, unfortunately, not come into being. I still hope, however, that it will appear prior to the release of the current volume so that I can allow myself occasional references to this forthcoming study.

2 Kenneth Zysk, "Citations in Jajjaṭa's *Nirantarapadavyākhyā*," *eJournal of Indian Medicine* 3, no. 2 (2010): 69–99.

 | DOI:10.1163/9789004438224_008

discuss its relationship with other commentaries, and present a diplomatic edition of selected chapters, are currently in preparation.

2 Description of the Manuscript

2.1 *General Information*

The examined manuscript (MS) belongs to the collection of the Government Oriental Manuscripts Library in Chennai (GOML) and is reported under the title *Suśrutavyākhyā* in at least two specialized catalogues.[3] In both cases, the MS is referred to by the number R 3422, which is also found in the alphabetical index of manuscripts in the GOML.[4] The same identifier appears on the first page of the codex, whereas the back cover contains two additional labels, Sa 1923 and T-1–175, the latter likely being the shelf mark assigned to the MS at one of its former locations.

2.2 *Physical Description, Format, and Layout*

The format and layout of R 3422 are modeled on a printed book rather than on a palm-leaf MS. It is written on ruled paper in portrait orientation and bound by Smyth sewing with a simple, paper soft cover. The ruling layout consists of 20 widely spaced feint lines as well as two vertical lines, which delimit the left and the right margins; the head and foot margins are unruled. The exact size of the used paper is unknown.[5] The MS is written in modern Devanāgarī script below the ruled lines. An average text line consists of ca. 45–50 *akṣaras*. The pages are continuously numbered with modern numerals (1 to 217) in the upper outer margin of each page (i.e., on each folio side).

Each chapter of the text is headed with // *śrīḥ* // and, in the majority of cases, with an indication of the chapter number, such as, for example, // *dvitīyo*

3 See (1) K.V. Sharma, *Science Texts in Sanskrit in the Manuscript Repositories of Kerala and Tamilnadu* (New Delhi: Rashtriya Sanskrit Sansthan, 2002), 191 and (2) Rama B. Rao and Vinod K. Bhatnagar, *Sanskrit Medical Manuscripts In India* (New Delhi: Central Council for Research in Ayurveda & Siddha, 2005), 191.

4 See nr. 24161 in Subrahmanya P.P. Sastri, *An Alphabetical Index of Sanskrit Manuscripts in the Goverment Oriental Manuscripts Library, Madras*, vol. 2 (Madras: Government of Madras), 907.

5 Among other things, the back cover of the codex contains the figure "105/8 x 93/8", which could indicate the size of the MS in inches (i.e., 333.375 × 295.275 mm). These dimensions do not seem to correspond to any standard page format, and, given that I missed noting the measurements when I inspected and photographed the MS myself in August 2009, I cannot be sure whether my interpretation is correct or not.

'dhyāyaḥ // on page 28. Similarly, each chapter is concluded with a repetitive formula, such as // *iti dvitīyo 'dhyāyaḥ samāptaḥ* // on page 32. All of these captions are invariably written in the middle of individual lines and sometimes (especially in the beginning of the MS) in slightly bigger characters. While the prose passages are written in—characteristic for Indic manuscripts—*scriptio continua*, it is noteworthy that the verses, whenever recognized by the scribe, follow the layout common for printed Sanskrit books: each half-verse (or, rarely, a *pāda* of a long meter) is indented and occupies an individual line.

2.3 *Scribal Colophon and Annotations*

At the end of the Sanskrit text on the last page of the MS (p. 217), we find the following colophon written in cursive handwriting (in English):

> *Transcribed from a MS 7 V.N. Narayanan Mambudiri (!) Vaidyamatam Tiruthala post. in the year 1920–1921.*
> *31-3-21*

This colophon is followed by three subsequent additions made, probably, by three different hands:

> *P.S. 16/5/21*
> *CC 17/5/21*
> *Compared with original. A.M. 18.9.21.*

The above date explains why the MS was not mentioned in the twenty-third volume of the Descriptive Catalogue of the GOML, the one focusing on medical literature, which was published in 1918—that is, prior to the completion of the transcript.[6] Dr. Madhu K. Parameswaran[7] pointed out to me that the owner of the original MS was likely the celebrated physician Ashtavaidyan Vaidyamadham Valiya Narayanan Namboodiri (fl. 1882–1959).[8]

A further notable feature of the MS is the figures indicating what appear to be the folio numbers of its template. They are invariably placed in the left-hand margin and begin with the number 2, which is found several lines below the

6 Kuppuswami Śastri, *A Descriptive Catalogue of the Sanskrit Manuscripts in the Government Oriental Manuscripts Library, Madras*, vol. 23 (Madras: Government of Madras, 1918).

7 Madhu Parameswaran, e-mail message to author, February 27, 2019.

8 For a brief and largely legendary life sketch of the scholar, see S.K. Rajesh and G.R. Kumar, "Vaidyamadham Cheriya Narayanan Namboodiri," *Journal of Ayurveda and Integrative Medicine* 1, no. 2 (April 2010): 136–138.

opening of the text on page 1, and end with 223 (clearly an error for 123) on page 215 around the conclusion of SS Sū 45. Based on these marks, we can estimate that a single folio side of the original MS roughly corresponded to a single page of the transcript in terms of the amount of text it contained. Given the large format of the latter (ca. 900–1000 *akṣaras* per page), it appears likely that it was copied from a palm-leaf MS.

2.4 *Extent of the MS and Special Features*

Although the MS is complete as a codicological unit, the text it contains is preserved only partly and fragmentarily. It begins somewhere in the commentary on, presumably, the first sentence of SS Sū 1 (the initial part of the text is extremely lacunar) and ends abruptly with the introduction of a quote by Bhoja in the commentary on SS Sū 46.53.[9]

The transcript exhibits a great number of lacunas of varying lengths. Longer omissions (or other kinds of corruptions) in the template are typically marked with the word *granthapātaḥ* (~ text damage) written diagonally over several lines. In this way, for example, on page 4, the copyist indicated the omission of the entire folio 6. Shorter omissions and other corruptions of various kinds are signaled by one of three methods: Many times omissions etc. are indicated by (1) one or a series of ellipses or, alternatively, (2) with the help of one or several long dashes. It may be noted that, according to my current observations, the number of ellipses or dashes does not necessarily reflect the length of the concerned textual corruption. In other instances, the copyist marked lost portions of the text by (3) merely leaving blank space, the length of which was probably comparable to the one occupied by the unrestorable *akṣaras*. This method in particular is often employed in the case of verses, where the number of corrupted syllables can be estimated based on the meter, and where the specific layout allows one to indicate graphically which portions and of what length are missing.

In transcribing the text from the MS, in the present article I utilize ellipses "…" in order to represent (1) and (2) and underscores "_" for (3).

---

9 All references to passages of the SS in this article follow Yadavjī Trikamjī and Nārāyaṇa Rāmā Ācārya 'Kāvyatīrtha', *Suśruta Saṃhitā of Suśruta with the Nibandhasaṃgraha Commentary of Śrī Ḍalhaṇācārya & Nyāyacandrikā Pañjikā of Śrī Gayadāsācārya on Nidānasthāna* (Varanasi: Chowkhamba Krishnadas Academy, 2004 [1938]) unless otherwise stated.

3 General Characteristics of the Text

The *SVyā is composed in a relatively straightforward variety of Śāstric commentarial Sanskrit and exhibits many stylistic and structural features characteristic of this literary genre.[10] In accordance with these, the *SVyā only scarcely provides pure glosses of individual words and usually does not deal with every sentence of the SS. Rather, it concentrates on selected terms or subjects. The length and level of elaboration vary depending on the individual chapters. So, for example, Sū 10, 19, 28, 20, 30–34, 38, 42, and 44 each receive not more than a single page of commentary, while, to give another (extreme) example, eight pages of commentary on Sū 11 essentially deal with a single word (*kṣārodaka*) found in Sū 11.11.

In keeping with the common structure of a commentary on a scientific text, the major part of the *SVyā is "concerned with discussing the contents and the implication of the text."[11] On many occasions, the commentary makes use of a debate-like dialectic approach, which routinely utilizes complex nested arguments. On other occasions, however, the *SVyā is eclectic and simply lists various alternative interpretations without providing any judgment about their relative value (in so doing, the text is reminiscent of a commentary on belles lettres). Anticipating that the latter approach may appear unsatisfying to readers of a medical work, at the conclusion of an elaborate discussion of Sū 1.28,[12] the *SVyā introduces a brief, but certainly curious, discussion, where it questions its own methodology. I would like to quote this passage in full in order to exemplify the style of the text (p. 26, ll. 4 ff.):

> *atrāha—ekaṃ vākyam ekam evārtham abhidhatta iti evam ekasya sūtrasya bahavo 'rthā*[13] *na vaktavyā vaktavya eka eveti* |
> *ucyate—kartṛmukhāgṛhītatvāt sūtrasyānekārthatvāc ca bahavo 'rthāḥ parikalpyante vaktṛbhiḥ* | *tatra punaḥ parīkṣakair yo 'rtho yuktarūpaḥ sa gṛhītavya iti* | *uktaṃ ca—*
> *ekapade 'pi ca bahvyo*[14] *vyākhyās tāsām atīva yā yuktā* |

---

10 For an overview of the various linguistic, stylistic, and structural peculiarities of Śāśtric (that is, scholastic) commentaries in Sanskrit, see Gary Tubb and Emery Boose, *Scholastic Sanskrit. A Manual for Students* (New York: Columbia University, 2007), 173–175.

11 Tubb and Boose, *Scholastic Sanskrit*, 173.

12 Note that in Sū 1, the *SVyā follows an order of sentences that is distinctly different from the one found in Trikamjī and 'Kāvyatīrtha', *Suśruta Saṃhitā*. The commentary on Sū 1.28, for example, comes after the explanation of Sū 1.38.

13 °*rthā*] conj., °*sthā*° MS.

14 *bahvyo*] conj., *bāhyo* MS.

tām adhitiṣṭhed dhīraḥ kṛtsnāṃ[15] *kṛtvāvyatikrāntaḥ* ||
api ca loke hi na sarve tulyamatayo bhavanti, yathā kasmaicid amlaṃ rocate kasmaicit tiktaṃ rocate, tadvad artheṣv api pravṛttir iti kṛtvaikasya bahūnām arthānāṃ vacanam iti |

The argument can be summarized as follows: An objection is raised that, just as a single sentence expresses a single meaning, so also, when explaining a *sūtra*, one should restrict oneself to a single interpretation instead of giving multiple interpretations. The *SVyā disapproves of this comparison, saying that, on the one hand, the commentators do not learn the *sūtras* directly from their original composers and that, on the other hand, *sūtras* do after all have multiple meanings, so it is legitimate for commentators to propose several different interpretations. It is, the *SVyā continues, up to a discriminating reader to choose the most reasonable interpretation. To substantiate this, an anonymous *āryā* verse is quoted: "Even for a single word there are many commentaries (that is, interpretations). [Therefore,] an intelligent [student should choose] one commentary [that she considers] the most appropriate of all and master it, having completed it without deviations."[16] The *SVyā continues with a "psychological" argument that is further supported by a mundane example. Interpretations of *sūtras*, so the *SVyā claims, may differ just because people vary in their mental predispositions. In fact, just as some people are fond of sour things, while others favor the bitter ones, so also certain people are attracted to one kind of interpretation, while others will choose a different kind.[17]

Furthermore, it must be highlighted that the text of the SS, which is followed by the *SVyā, is distinctly different from all other versions of the treatise that have been available to us to date. It is noteworthy that the *SVyā often pro-

---

15 *kṛtsnāṃ*] conj., *kṛtsnā°* MS.

16 *Kṛtsnām* is my tentative conjecture of what is most likely a corruption of *kṛtsnā*. At the moment, I understand the clause *kṛtsnāṃ kṛtvā* to mean something like "having completed [the study of that single commentary]". Admittedly, I am not entirely sure if this conjecture and its translation are right. *Avyatikrānta* (nondeviant), although I translated it as an adverb, is, of course, an adjective describing *dhīra*. The basic structure of a more literal translation would be, therefore, "a *dhīra*, being nondeviant, should complete and master."

17 Among other things, it is worth noting that in the above discussion the author of the *SVyā calls sentences of the SS *sūtras*. From the available material, it is impossible to judge whether she applies this appellation only to sentences of SS Sū or throughout the entire SS. On another occasion, I have shown, however, that the *SVyā uses both terms, *Ślokasthāna* and *Sūtrasthāna*, interchangeably when referring to SS Sū. See Andrey Klebanov, "On the Textual History of the *Suśrutasaṃhitā* (1): A Study of Three Nepalese Manuscripts," *eJournal of Indian Medicine* (forthcoming): § 3.1.3.

vides elaborate arguments for the correctness of the accepted text of the SS and hardly ever neutrally reports variant readings (i.e., *pāṭhāntara*s commonly found in other commentaries on medical *saṃhitās*). My working hypothesis—further reasons for which I hope to demonstrate in a separate article—is that the author of the *SVyā must have purposefully altered the root text in accordance with the proposed interpretations of it. Here are a few simple examples of such modifications:

(1) The SS Sū 15, which is commonly known as *Doṣadhātumala**kṣayavṛd-dhi**vijñānīya* (On the knowledge about the decrease and increase of humors, bodily elements and waste materials),[18] is called *Doṣadhātu-mala**kṣaya**vijñānīya* (… about the decrease of …) in the *SVyā. The latter title is more fitting, and thus used by the commentary, because (1) the chapter deals mainly with the decrease of elements and (2) because it signals that the decrease of elements causes particularly great distress.[19]

(2) Further emphasizing the prominence of the decrease of elements, the *SVyā omits the entire section (Sū 15.4–6) that deals with their general features and reinterprets the clause "their characteristics" (*eteṣāṃ lakṣaṇam*) from the preceding introductory sentence in Sū 15.3b as "their characteristics, when they are diminished" (*lakṣaṇam eteṣāṃ kṣīṇānām*).[20]

(3) Similarly, the actual description of the increase of elements (Sū 15.14–17) is removed from its position immediately following the account of their decrease and placed after the treatment of *ojas* and before the introduction of a new topic (corpulent and lean patients) in Sū 15.32.

Among the structural originalities of the SS commented upon in the *SVyā, one may note, furthermore, the order of the last sixteen (from Sū 30 onward) chapters of the *Sūtrasthāna*. With reference to the sequence of chapters accepted by Ḍalhaṇa,[21] the *SVyā arranges them in the following way: Sū 30, 33, 32, 31, 34,

18 Note that the Nepalese MSS of the SS transmit another variation in the chapter's title and omit the final °*vijñānīya*. See Andrey Klebanov, *The *Nepalese Version of the* Suśrutasaṃhitā *and Its Interrelation with Buddhism and the Buddhists*, (University of Hamburg: M.A. thesis, 2011), 86.

19 P. 34: *doṣadhātumalakṣayavṛddhivijñānīyam iti avacanam, atra bāhulyād doṣādikṣayavijñānasya, kṣayasya mahātyayapradarśanārthatvāc ca* /

20 Sū 15.3b in Trikamjī and 'Kāvyatīrtha', *Suśruta Saṃhitā*, 67: *tasmād eteṣāṃ lakṣaṇam ucyamānam upadhārayasva* /. Here, the *SVyā most likely accepts a slightly different reading, where the words *eteṣāṃ* and *lakṣaṇam* are inverted. The available portion of the commentary reads (p. 34, l. 10): *tasmāl lakṣaṇam eteṣām iti* / *atra kṣīṇānām iti vākyaśeṣaḥ* / *yasmād doṣadhātumalamūlaṃ śarīram, tasmāl lakṣaṇam eteṣāṃ kṣīṇānām ucyamānam upadhārayasva* /

21 Trikamjī and 'Kāvyatīrtha', *Suśruta Saṃhitā*, 136ff.

35, 38, 39, 37, 40, 41, 42, 43, 44, 36, 45, and 46. Since the commentary on each chapter begins with an indication of the logical connection between the current and the preceding chapter (*saṃgati*), there cannot be any doubt about the incidental displacement of chapters.

4 Identified Citations from Medical Literature[22]

In the remaining pages, I will concentrate on another remarkable and historically significant feature of the *SVyā: its unusual, copious amount of literary quotations. These originate from a wide spectrum of medical works, and about half of them include an identified source. It is not exactly clear yet whether the author of the *SVyā followed any definite pattern when deciding to identify a quotation or leave it unidentified. It is likely, however, that many anonymous quotes come from texts that were well known to the commentator. On pp. 76–7, for example, in the commentary to Sū 14.7, we come across the following unidentified quote from the medical treatise by Bhoja, a text abundantly cited throughout the *SVyā:

> *uktaṃ hi tantrāntare—*
> *sirāsnāyurajastanyatvaco* (!) *gativivarjitāḥ* /
> *dhātubhya*___________ *dhātavaḥ* //[23]

This verse was quoted and attributed to Bhoja by Gayadāsa, Cakrapāṇidatta, and Śivadāsasena.[24] Both earlier commentaries must have at least had independent access to the work, so its authorship can hardly be contested.

Similarly, at the end of the commentary to Sū 1.35 on pp. 24–25 we find:

> *dviruktilakṣaṇaṃ hy etat—*
> *gadyokto yaḥ punaḥ ślokair arthaḥ samanuvarṇyate* /
> *_vyaktivyavasāyārthaṃ dviruktiḥ sā na garhitā* //

22 Note that the current survey intends to primarily demonstrate the width of the quoted material rather than its depth. Therefore, in the following section, the majority of Sanskrit citations are given without an English translation.

23 For a translation and discussion of the verse, see Das, *The Origin of the Life*, 320 ff., which quotes it in whole: *sirāsnāyurajaḥstanyatvaco gativivarjitāḥ* / *dhātubhyaś copajāyante tasmāt ta upadhātavaḥ* //

24 See Das, *The Origin of the Life*, 320 for the exact references.

This verse is found with minor variants in CS Ni 2.41,[25] a text which, at other times, is quoted with an explicit attribution.

Further noteworthy is the density of citations, and what this amounts to is that there are only a very few pages in the whole MS on which I have not yet detected a quotation. So far, I was able to find citations of the following medical authors and works (ordered roughly according to the frequency of their appearance; the texts mentioned last are unknown to me):

1. Bhoja
2. Bhāluki (*Bhālukīya*)
3. Kṣārapāṇi
4. Puṣkalavāta (or *Pauṣkalavātiya*)
5. Vaitaraṇa
6. Viśvāmitra (or *Viśvāmitrīya*)
7. Kāśyapa (or *Kāśyapīya*)
8. Caraka
9. Hiraṇyākṣa (or *Hiraṇyākṣīya*)
10. Kapilabala
11. Nimi
12. *Vaidehīya*
13. Śālihotra
14. Jātukarṇa
15. Parāśara
16. Pālakāpya
17. Vṛṇda's *Siddhayoga* (quoted anonymously)
18. Siddha (?)
19. *Lakṣaṇaśāstra* (?)
20. *Pāryovida* (?!?)

1) Bhoja[26]

The authority quoted the most often is without a doubt, Bhoja. Based on other extant quotations from his work, we can infer that it was closely linked with the

25 Trikamjī, *Carakasaṃhitā of Agniveśa*, 204 prints the following text: *gadyokto yaḥ punaḥ ślokair arthaḥ samanugīyate / tadvyaktivyavasāyārthaṃ dviruktaṃ tan na garhyate //*. The same verse is repeated at the end of a collection of *paribhāṣās* appended to the Ka of the *Aṣṭāṅgasaṃgraha*. See Vaidya Ananta Dāmodara Āṭhavale, *Aṣṭāṅgasaṅgrahaḥ: Induvyākhyāsahitaḥ*, (Pune: Śrīmad Ātreya Prakāśaam 'Naṃdanaṃdana,' 1980), 614–618.

26 See Jan Gerrit Meulenbeld, *A History of Indian Medical Literature* (Groningen: Egbert Forsten, 1999), 1a, 690–691.

SS.[27] Gayadāsa, to give but one example, who extensively quotes Bhoja's treatise, calls upon its authority as a *samānatantra* or a *tulyatantra vis-à-vis* the SS in order to justify an emendation in SS Ni 6 14–18.[28] Also noteworthy is that all three old Nepalese manuscripts of the SS[29] include Bhoja's name in the list of students who approach Kāśīrāja Divodāsa in Sū 1.3 to ask him for instruction in medical science (Sū 1.4).[30]

In the *SVyā, Bhoja's citations are identified with the name of the author, either in the instrumental *bhojena* (e.g., *bhojenāpy uktam* on p. 14), or, more commonly, in the locative *bhoje* (e.g., *uktaṃ hi bhoje* twice on p. 52), which should, perhaps, be understood as a contraction of something like **Bhoja* [*viracitagranthe*].[31] All the citations from Bhoja that I have been able to locate in the *vyākhyā* so far are metrical and add up to well over 100 verses. The absolute majority of verses are composed in the *pathyā* variety of *anuṣṭubh*, with occasional use of regular *vipulā*s, and only three verses are written in *upajāti*.[32] Bhoja's quotations are found in nearly every chapter of the commentary and, in this way, cover a broad variety of topics. In line with the overall polemic

---

27 With reference to the *Nibandhasaṃgraha* on SS Sū 27.9, Meulenbeld, *History*, 1a, 691 and 1b, 760, n. 109 remarks that "Bhoja is almost always quoted with respect, but on one occasion Ḍalhaṇa rejects his view." The final statement is not exactly right, because what Ḍalhaṇa actually criticizes here is not Bhoja's view but the fact that some scholars incorporate his verse into the SS. See Trikamjī and 'Kāvyatīrtha', *Suśruta Saṃhitā*, 126: *"parśukāntarapatitāni ca" ity asyāgre, "anuttuṇḍitaśalyāni* [...] *chedānubandhataḥ"— iti bhojasaṃhitoktaṃ kecit pāṭhaṃ paṭhanti, sa ca pramādapāṭhaḥ, nibandheṣv adarśanāt.* On another occasion, Meulenbeld, *History*, 1a, 691 observes that "Śrīkaṇṭhadatta remarks that Suśruta follows Bhoja." The reference is to the *Madhukośa* on *Mādhavanidāna* 55.1. Here, however, Śrīkaṇṭhadatta states merely that several types of *kṣudraroga*, which are specified by Bhoja, are not explicitly mentioned, but, given that disorders are innumerable, are accepted by Suśruta as well. See Upādhyāya and Śāstri, *Mādhavanidānam*, 193: *bhoje tu mūṣikākārṇamuṣkakośaphalārśaḥprabhṛtayo 'dhikavikārāḥ paṭhyante, te ca suśrute vikārasyānantyād anumatā eva.*

28 See Klebanov, "On the textual history (1)", § 3.2.2 for further details.

29 See Klebanov, "On the textual history (1)" for a detailed description of the MSS in question.

30 Ḍalhaṇa, whose text of the SS does not include Bhoja, explains that he is after all referred to by the word *prabhṛti-* at the end of the compound.

31 This is a common linguistic phenomenon, which is often seen in reference to the titles of medical treatises. I have briefly discussed its grammatical justification in Klebanov (forthcoming, n. 14).

32 By way of example, Meulenbeld, *History*, 1b, 759, n. 79 records three quotations from Bhoja's text that are composed in prose. To these one may add, for instance, another quote in Ḍalhaṇa's NS on SS Sū 27.12. However, in view of the fact that the overwhelming majority of Bhoja's citations are versified, one may wonder about the exact position of the few prose passages within the original text.

style of the *SVyā, its author and her many imagined opponents (*pūrvapakṣins*) often appeal to Bhoja's authority in order to prove an opinion, establish the correctness of a certain reading, or merely to elaborate upon Suśruta's explanations.

2) Bhāluki / *Bhālukīya*[33]

This is another important authority closely associated with the transmission of the SS and often quoted by its commentators. In case of the current *vyākhyā*, too, though not as copious as Bhoja, Bhāluki's views are regularly reviewed in connection with various topics. For the attribution, the commentary uses either the name of the author (e.g., *bhālukināpy uktam* on p. 150, or *āha bhālukiḥ* twice on p. 33), or refers to the actual work as *Bhālukīya* (e.g., *bhālukīye coktam* on p. 85).[34] The majority of identified quotes are verses composed in *anuṣṭubh* (about 50 verses), with a few exceptions in *upajāti* (four verses so far). There are, however, also several prose passages.

Apart from the main bulk of technical subjects, which need to be treated in a separate study, several verses attributed to Bhāluki address more general topics. So, for example, on page 33 in the commentary to Sū 3.1, we find a verse exalting the gift of teaching and the persona of a teacher:

> *sarveṣāṃ dānānām adhyayanasaṃpradānaṃ prakṛṣṭam, tasmād āha Bhālukiḥ—*
> *vasudaṃ bhogadaṃ śrīdaṃ kāmadaṃ dhanadhānyadam /*
> *prāṇadaṃ sarvadaṃ prāhuḥ vidyādaṃ varadaṃ gurum //*

Other references offer insight into the compositional features of Bhāluki's treatise. For example, a verse, found soon after the one just quoted, introduces the structure of the text, which is, as we can see, notably parallel to that of Vāgbhaṭa's works:

> *āha bhālukiḥ—*
> *ślokāḥ pūrvam ihādhyāyāḥ śārīrāṇi tataḥ param /*
> *nidānāni cikitsāś ca kalpaś caiṣa kramaḥ smṛtaḥ //*

33 Cf. Meulenbeld, *History*, 1a, 689–690.

34 On account of the locative form, we understand that the nominal *bhālukīye*, which grammatically speaking is an adjective, must refer to a nonfeminine noun, in this case, perhaps, the neuter *tantra*.

In light of this citation, it appears surprising that both Niścalakāra and Śivadāsasena, authors of the *Ratnaprabhā* (RPbhā) and the *Tattvacandrikā* (TaCa) commentaries on Cakrapāṇi's *Cakradatta* (CD) respectively, refer to the *Siddhisthāna* of the *Bhālukitantra*.[35] One may wonder whether one of the sections of the *Bhālukīya* was also known under the title *Siddhisthāna*, or, whether Niścalakāra and Śivadāsasena may not have had access to the whole text of Bhāluki's treatise.

Further noteworthy is an indirect reference to Bhāluki and others (*bhālukiprabhṛti*) on pages 28–29, according to which these scholars disapproved of the statement *atha khalu vatsa suśruta* at the beginning of Sū 2 and all subsequent chapters. They argued that the teacher (Divodāsa? Suśruta?) was fond of brevity (just as the founders of other Śāstras) and it was unnecessary for him to repeat *atha khalu vatsa suśruta* at the beginning of each chapter to indicate that every chapter was taught in reply to Suśruta's request. Instead, it was sufficient for him to state an even shorter *atha khalu* only once (perhaps at the beginning of the text?) to show that, in fact, the whole treatise was propounded in response to an initial request:

> *bhālukiprabhṛtayas(!) tu pratyadhyāyaṃ pṛṣṭaprativacananyāyena mahatārambheṇa sampratipatteḥ, ayaṃ khalv ācāryo lāghavikaḥ sann "atha khalu"-śabdenaiva pṛṣṭaprativacananyāyenedaṃ pravṛttam iti jñāpitavān.*

This reference suggests either that Bhāluki directly voiced an opinion about the wording of the SS or, given that in this case we would expect the *SVyā to quote the concerned passage, that Bhāluki pointed at it indirectly by accepting a certain reading in his own text. The latter option, which is more likely in view of what we know about *Bhālukīya* so far, implies that, in the *SVyā author's opinion, this text did not have any statement such as **atha khalu vatsa bhāluke* at the beginning of its chapters.[36]

---

35 For the RPbhā on CD 1.157–158, see Sharma, *Cakradatta-Ratnaprabhā*, 72: *siddhivihitaṃ = bhālukitantroktasiddhisthānoktam ity arhtaḥ*. For the *Tattvacandrikā* on CD 1.134, see Vidyābhūṣaṇa and Vidyāratna, *Cakradatta*, 38: *asya yogasya bhālukitantroktatvāt, bhālukitantrasyaiva siddhisthāne jñeyam.*

36 We know (cf. Meulenbeld, *History*, 1a, 689) that *Bhālukīya* contained addresses to Bhāluki in the running text of the work (similar to what we find in SS Sū 3.33, 46.3, 45.445, etc.). For this, see RPbhā on *Cakradatta* 1.150 (Sharma, *Cakradatta-Ratnaprabhā*, 69): *yad āha bhālukiḥ—bhāluke śuṇe* (!) *yathārthyaṃ sannipātaviśeṣaṇam* [...].

3) Kṣārapāṇi[37]

Among the representatives of the *kāyacikitsā* school of medicine, it is Kṣārapāṇi's[38] work that is quoted and identified in the *SVyā the most frequently. The majority of quotes are verses in *anuṣṭubh*, but there are also several prose passages. Kṣārapāṇi's views are referred to at a number of rather diverse occasions and often seem quite unique.

On pages 4–5, in the commentary to Sū 1.8, the *SVyā appeals to Kṣārapāṇi in order to support the view that *rasāyana* and *vājīkaraṇa* are opposites, just as *bṛṃhaṇa* (roborant therapy) and *lekhana* (attenuating therapy) are:

> [...] *tadanantaraṃ rasāyanatantram* [...] *rasāyanavyājīkaraṇayor* (!) *lekhanabṛṃhaṇapratidvandvasambandhāt, uktaṃ ca kṣārapāṇinā—*
> *abhīkṣṇaṃ doṣaharaṇaṃ mārdvikaṃ madhusaṃskṛtam* /
> *yathānnam āsavās tīkṣṇā rasāyanavidhiś ca yaḥ* //
> *vyāyāmaś copavāsaś ca karśane vidhir iṣyate*[39] /
> *bṛṃhaṇaṃ tu kṛśasyeṣṭam vyājīkarmavidhir*[40] *yathā* //

Further noteworthy is a prose quotation found in the commentary to Sū 14.7. The same excerpt is also quoted and attributed to Kṣārapāṇi in the *Kalyāṇakāraka* (KK).[41] It treats the production of the *upadhātus* and, as such, has been extensively analyzed in Das (*The origin of the life*, 316 ff.) and Klebanov (The **Nepalese* Suśrutasaṃhitā, 162–164). The *SVyā (p. 76) reads as follows (highlighting is mine):

> *evaṃ hy āha kṣārapāṇiḥ*
> *svena ca prāṇatejasā rasībhūtvā dvaidhībhavati stanyaṃ śoṇitaṃ ca, śoṇitaṃ bhūtvā dvaidhībhavati rajas-***tvaṅ**[42] *māṃsaṃ ca, māṃsaṃ*

37 Meulenbeld, *History*, 1a, 168–169.

38 According to CS Sū 1.31, Kṣārapāṇi studied alongside Agniveśa, Bhela, Jatūkarṇa, Parāśara, and Hārīta.

39 *iṣyate*] conj., *icyate* MS. The conjecture *iṣyate*, I think, is preferable to *ucyate*, because (1) a corruption of the consonant cluster "*ṣy*" to "*cy*" is easier to explain than of the syllable "*ri*" to "*ru*", and because (2) *iṣyate* is weakly supported by *iṣṭam* in the next *pāda*.

40 °*vidhir*] conj., °*pidhīr* MS.

41 Meulenbeld, *History*, 2a, 151–155 provisionally dates the text to the ninth century. As I was not able to procure a copy of this work, in the following I rely on the citation in Das, *The origin of the life*, 317, n. 1095.

42 *rajastvaṅ*] *SVyā, *rajo* KK.

> *bhūtvā dvaidhībhavati sirā-**tvaṅ**[43] medaś ca, medo bhūtvā dvaidhībhavati **vasā**-snāyv[44] asthi ca, asthi bhūtvā dvaidhībhavati vasā majjā ca, majjā bhūtvā dvaidhībhavati majjā caiva śukraṃ ca, śukrād garbhaḥ sambhavati.*

Several things are notable about this quote. On the one hand, it provides some evidence against Das's proposal that the passage may be a late fabrication.[45] On the other hand, it demonstrates that Kṣārapāṇi's text (or at least the text from which this quote comes) has undergone some redaction aimed at updating the theory it contains. In fact, even taken in their own right, the three highlighted words seem disturbing, because in each case they add an additional third entity to a list of two products of a given *dhātu*. The only way to resolve this obvious contradiction is to construe them as members of rather unintelligible compounds (i.e., *rajastvac, sirātvac, vasāsnāyu*). In view of the readings preserved in the KK, however, the secondary character of the highlighted words becomes fairly obvious. This interpolation can be explained as an attempt to include the "missing" element *tvac* (see Bhoja's verse in note 23) into Kṣārapāṇi's account. The reading *vāsasnāyu* (instead of *snāyu*) may have occurred as a result of an attempt to make the whole sentence appear stylistically more cohesive.

Further quotes from Kṣārapāṇi provide, for example: a comprehensive list of sixteen therapeutic actions, all of which are based merely on the six tastes (on p. 21, on Sū 1.28); a discussion of countermeasures to be taken in case of *purīṣakṣaya* (decrease of the secondary element feces) and *mūtrakṣaya* (decrease of urine, see p. 85, Sū 15.13); recommendations to reject a patient who lost his senses (p. 87, Sū 15.31); a comparison of obese and lean patients (pp. 88–9, Sū 15.35); and descriptions of the qualities of various types of meats (several quotes in the *SVyā on Sū 20), and many more.

4,5) Puṣkalavāta/*Pauṣkalavātiya*[46] and Vaitaraṇa[47]

Two further authorities with close links to the SS are Puṣkalavāta (quoted in the *SVyā by his name and with reference to his work, *Pauṣkalavātiya*) and Vaitaraṇa. Both of them are mentioned in all known versions of SS Sū 1.3 as Suśruta's fellow students. In addition to this, Puṣkalavāta is referred to in Sū 4.9,

---

43 *sirātvaṅ*] *SVyā, *sirā* KK.

44 *vasāsnāyv*] *SVyā, *snāyv* KK.

45 Das, *The origin of the life*, 317–318.

46 Meulenbeld, *History*, 1a, 371.

47 Ibid.

a verse that enumerates four basic treatises on *śalya* (surgery).[48] These, Cakrapāṇi, Ḍalhaṇa, and the author of the *SVyā agree, should be consulted in case of any doubt concerning surgery. It is noteworthy that, commenting on this verse, the *SVyā cites Vaitaraṇa:

> *vaitaraṇenāpy uktam—*
> *yas tv ācāryair abhimataḥ sūtrajñaḥ karmakovidaḥ* /[49]
> *sa pṛṣṭas tatra tattvaṃ ca _ _ sūtraṃ prabhāṣate* //
> *aupadhenavam aurabhaṃ pauṣkalāvatasauśrute* /[50]
> *kulāny etāni catvāri jñānānāṃ saṃbhavas tv iha* //

Among other notable quotes by Vaitaraṇa, one may highlight, for example, a long passage encompassing seven *anuṣṭubh* verses and a short sentence in prose found in the commentary to Sū 18.13–14 on page 107 (incidentally, just a few lines above, the same discussion refers to Puṣkalāvata). This citation is noteworthy because three of its verses are also quoted in Cakrapāṇi's *Bhānumatī* (BhM) in the commentary on the same sentence of SS. Among many other things, this co-occurrence is notable from the text-historical point of view. Both versions contain several significant differences, which I will demonstrate with a single example. Consider a single half-verse quoted in the *SVyā, which, just as the previous quote, contains a *na-vipulā*:

> *rātrau tu pīlaṇam(!) ṛte nālepanam avacārayet* /

The even *pāda* is hypermetrical and can be easily restored to *nālepam avacārayet*. Compare this to the parallel line in the BhM:[51]

> *rātrau tu śītalam muktvā pralepam avacārayet* /

Note, for example, how the version in the BhM avoids the *na-vipulā* by replacing *ṛte* with *muktvā*. As for the *SVyā's *pīlaṇam*, note that Trikamjī and Bhisha-

---

48 SS Sū 4.9 (Trikamjī and 'Kāvyatīrtha', *Suśruta Saṃhitā*, 18): *aupadhenavam aurabhraṃ sauśrutaṃ pauṣkalāvatam / śeṣāṇāṃ śalyatantrāṇāṃ mūlāny etāni nirdiśet* //. Note that the *SVyā thinks the author of *Aupadhenava* is Upadhenava (and not Aupadhenava). See, e.g., p. 37, line 15: *upadhenavasyedam śāstram aupadhenavam.*

49 Note the regular *na-vipulā* in *pāda* A.

50 Note an occurrence of what Steiner, "Die Lehre der Anuṣṭubh," 248 calls "alte (vedische) Anuṣṭubh" in *pāda* A.

51 Yadavjī Trikamjī and Nandikishor Bhishagacharya, *Suśruta-Saṃhitā (Sūtra Sthān), with Bhānumatī Commentary by Cakrapāṇidatta* (Varanasi: Krishnadas Academy, 2001), 139.

gacharya[52] report an alternative reading to *śītalam*—namely, *pīḍaṇam*. Given that *l* and *ḍ* are commonly confused with each other in Sanskrit, one can safely accept *pīḍaṇam* as the intended readings of the MS. However, a seemingly minor orthographic difference between *nālepam* and *pralepam* makes the meanings of the two versions quite irreconcilable with each other. As far as the *SVyā is concerned, it is likely that the preserved reading is, in fact, the one intended by the author, because the current group of verses is introduced in the following way (p. 107, l. 5):

> *kasyāṃcid avasthāyāṃ rātrāv ālepanaṃ pratiṣiddham, tad yathā* …

All in all, quotations from the works by Puṣkalavāta and Vaitaraṇa are far less frequent when compared to the works of Bhoja, Bhāluki, or Kṣarapāṇi. Unlike citations from *Viśvāmitrīya* (see the next section), however, they cover a far broader span of topics. Vaitaraṇa's quotes are, as we saw above, mainly in verse and consider, for example, some fundamental aspects of Āyurveda (*puruṣa* as a substratum of treatment in Sū 1.22 and the relationship between living beings, food, tastes, and substances in Sū 1.28), authoritative treatises on surgery (in Sū 4.9, see above), the seasons *hemanta* and *śiśira* (in Sū 6), as well as the application of *lepa* at night according to the condition of the swellings and the particular ingredients of the paste (in Sū 18.13–14, see above).

Excerpts from Puṣkalavāta's treatise, on the other hand, seem to be more frequently in prose. His opinion is recollected, for example, in connection with various basic concepts of Āyurveda, such as the *svābhāvika* diseases in Sū 1.24[53] or the characteristics of a healthy person in Sū 15.33. He also talks, for example, about the seasons *varṣā* and *grīṣma* (in Sū 6), a specific kind of *lepa* to be applied at night (in Sū 18.13–14), the qualities of *kauśeya* as a material for bandaging (in Sū 18.15), the qualities of honey (in Sū 15.43), and the five *ubhayatobhāgadravyas* (medical substances which evacuate from both sides; in Sū 39.5).

52 Ibid.

53 Although Pauṣkalāvata's quote aims primarily to authorize the view that *svābhāvika* diseases occur without a cause, we also learn that he unconventionally distinguished only three groups of diseases (omitting *mānasa*)—namely, *svābhāvya*, *āgantuka*, and *śārīra*. See p. 18: *pauṣkalāpatenāpy*(!) *uktaṃ—trividhā vyādhayaḥ, svābhāvyāgantuśārīrāḥ. tatra jarāmṛtyū svābhāvikāv acintyau, svabhāvasyāhetukatvād iti.*

6) Viśvāmitra/*Viśvāmitrīya*[54]

So far, I have been able to spot just over ten verses attributed to Viśvāmitra (or the work called *Viśvāmitrīya*) in the *SVyā. It is, perhaps, noteworthy that ten of these verses are found in the commentary to Sū 11, a chapter on caustics, *kṣāra*. In the commentary to Sū 11.4, for example, the *SVyā quotes Viśvāmitra's etymological explanation of the term *kṣāra* (p. 61):

> *uktaṃ ca viśvāmitrīye—*
> *guṇān kṣarati rogeṣu vyādhīn kṣārayatīti vā /*
> *akṣate kṣatakṛttvāc ca tasmāt kṣāraḥ smṛta iti //*

Later, in the commentary to 11.5, *SVyā mentions Viśvāmitra's opinion (which is at odds with Suśruta's) that *kṣāra* is of both an *āgneya* (fiery) and *saumya* (watery) nature (p. 62):

> *uktaṃ ca viśvāmitrīye—*
> *paicchilyāc chauklyavikledāt saumyaḥ śaityāt tathaiva ca /*
> *pācanādibhir āgneyo dvividhaḥ kṣāra ucyate //*
> *agniprabala eva syāt sāmānye saty api dvayoḥ /*
> *agniṣomātmakas tasmāt kṣāra ity eva nirṇayaḥ //*

The remaining seven verses quoted in the same chapter all deal with the preparation of *kṣārodaka* (alkali water) in Sū 11.11. Among these, one of two verses on plane *muṣkaka* (p. 63), as well as another two verses on the exact measurement of employed substances (p. 65), are also cited in the BhM (the verse on *muṣkaka* is repeated in the RPbhā on CD 5.142–146). Note that the rejected reading reported in this edition of the BhM[55] is the one that corresponds to the *SVyā's quote. Another half-verse on *kṣārodaka* that is attributed to Viśvāmitra in the *SVyā (p. 65) is cited (with the same attribution) along with its second half in Candraṭa's commentary on Tīsaṭa's *Cikitsākalikā* 161–164.[56]

Outside of Sū 11, we find Viśvāmitra's definition of the *samyagdagdha* type of cauterization (which is essentially the same as the *māṃsadagdha* in SS Sū 12.8) quoted in Sū 12.13 (pp. 72–3):

---

54 Meulenbeld, *History*, 1a, 179: "Viśvāmitra is one of the members of the assembly of sages described in" CS Sū 1.8–15.

55 Trikamjī and Bhishagacharya, *Suśruta-Saṃhitā with Bhānumatī*, 84, n. 1.

56 On Tīsaṭa see Meulenbeld, *History*, 2a, 148–151, on Candraṭa ibid. (122–125).

viśvāmitrīye 'py uktam—
kapotavarṇaṃ paruṣaṃ śuṣkaṃ saṅkucitavraṇam /[57]
samyagdagdhaṃ vijānīyād alpaśvayathuvedanam //

Finally, we find another verse by Viśvāmitra quoted in the commentary to Sū 2.3, which states that Āyurveda, given its status as an *upaveda*, is particularly important for Brahmins to study, although it should be practiced by members of all three higher *varṇas* (p. 29):

uktaṃ ca viśvāmitrīye—
idaṃ vedopavedatvāt śāśtraṃ varṇais tribhiḥ śubhaiḥ /
adhyetavyaṃ tv apatitaiḥ brāhmaṇais tu viśeṣataḥ //

7) Kāśyapa (*Kāśyapīya*)

Several verses in the *SVyā are attributed to Kāśyapa and his work, the *Kāśyapīya*. There are, for example, two verses on the purpose and method of ear piercing (both on Sū 16.4, pp. 95–6),[58] as well as four verses related to wholesome and unwholesome food items in the commentary to Sū 20 (pp. 122, 123, 124, 127).

8) Caraka

Verses attributed to Caraka are rather scarce in the *SVyā, but, as mentioned above, it is likely that more verses from the CS were quoted without an identification. The degree of the reproduced text of the CS's deviation from its vulgate edition[59] varies from case to case. In the commentary to SS Sū 14 on page 83, for example, the *SVyā quotes a single block of two verses that correspond to CS Sū 6.41 and 44. Apart from the omission of the two intermediate verses, the reading of the *SVyā differs just in a single instance of *pāda* B in 6.44: *śoṇitasya ca mokṣaṇam.*[60] On page 136, however, in the commentary to Sū 21, the *SVyā quotes a distinctly divergent version of CS Ci 15.9–11:[61]

---

57 Note, a *bha-vipulā*.

58 *Kāśyapasaṃhitā* Sū 21, which deals with the piercing, is almost entirely lost.

59 Yādavjī Trikamjī, *Carakasaṃhitā of Agniveśa, Elaborated by Caraka and Dṛḍhabala, with the Āyurvedadīpikā Commentary by Cakrapāṇidatta* (Varanasi: Chaukhamba, 2008 [1941]).

60 Trikamjī, *Carakasaṃhitā of Agniveśa*, 48 instead reads *vireko raktamokṣaṇam.*

61 Trikamjī, *Carakasaṃhitā of Agniveśa*, 512 reads CS Ci 15.9–11 as follows: *annasya bhuktamātrasya ṣaḍrasasya prapākataḥ / madhurādyāt kapho bhāvāt phenabhūta udīryate // paraṃ tu pacyamānasya vidagdhasyāmlabhāvataḥ / āśayāc cyavamānasya pittam accham*

uktaṃ hi carake—
āhārasyāttamātrasya pacyamānasya tejasā |
madhurādyaiḥ kapho bhāvaiḥ phenabhūta udīryate ||
ataḥ paraṃ vidagdhasya amlībhūtasya pacyataḥ |
āśayāc cālyamānasya pitta … … …
punar agninā |[62]
paripiṇḍitapakvasya vāyus syāt kaṭupākinaḥ ||

Furthermore, in the commentary to Sū 15.39 (p. 90), the *SVyā quotes a slightly altered version of CS Sū 17.62 and attributes it to Āgniveśya. I have not found this identifier anywhere else in the *SVyā, so I cannot confirm if it always refers to the CS or not. The reading of the quote is as follows:

uktaṃ cāgniveśyena—
doṣāḥ pravṛddhāḥ svaliṅgaṃ darśayanti pṛthak pṛthak |
tyajanti hīnāḥ svaṃ karma samā svaṃ karma kurvate ||[63]

9) Hiraṇyākṣa (*Hiraṅyākṣīya*)[64]

So far, I was only able to find the following three identified quotes from this work. A half-verse emphasizing *rasa* as the root and means to nurture other bodily elements (*dhātu*) in the *SVyā to Sū 14.10 on page 78:

hiraṇyākṣīye coktam—
dhātūnāṃ śoṇitādīnāṃ sa mūlaṃ vardhano rasaḥ[65]

In the commentary to Sū 21.10 on page 135, Hiraṇyākṣa summarizes the activities of *agni* as being, perhaps, incited by the *udānavāta* (cf. note 67). It is

---

udīryate || pakvāśayaṃ tu prāptasya śoṣyamāṇasya vahninā | paripiṇḍitapakvasya vāyuḥ syāt kaṭubhāvataḥ ||

62 This remainder of a *pāda* and the following half-verse is not recognized by the scribe as a part of the quote.

63 Trikamjī, *Carakasaṃhitā of Agniveśa*, 102 prints CS Sū 17.62 as follows: *doṣāḥ pravṛddhāḥ svaṃ liṅgaṃ darśayanti yathābalam | kṣīṇā jahati liṅgaṃ svaṃ samāḥ svaṃ karma kurvate ||*

64 Meulenbeld, *History*, 1a, 159–160. Hiraṇyākṣa is another member of the assembly at the beginning of the CS.

65 I understand *vardhana* here as a verbal noun in the meaning of an instrument (formation with affix *LyuṬ* in accordance with *Aṣṭādhyāyī* 3.3.117: *karaṇādhikaraṇayoś ca*). *Vardhana* and *mūla* both qualify the main word, *rasa*.

noteworthy that Hiraṇyākṣa speaks of *agni* and not of *pitta* here and, possibly, uses these terms interchangeably (as proposed in SS Sū 21.9):[66]

uktaṃ ca hiraṇyākṣīye—
sa pacaty agnir āhāraṃ vivinakti tridhā ca tam |
rasaṃ doṣaṃ malaṃ cāntaḥ_ _ pānasamīritaḥ ||[67]

Finally, Hiraṇyākṣa is quoted once again in the same chapter in the commentary to Sū 21.28. While the actual sentence in the SS deals with the third *kriyākāla* (period of medical intervention)[68]—namely, *prasara* (spread)—the quoted passage refers, perhaps, to the second stage (*prakopa*, aggravation) and explains how *prasara* in Sū 21.28 can happen in combinations (p. 138):

hiraṇyākṣeṇāpy uktam—
avaśyaṃ vardhamānas tu vidāhaḥ pittaraktayoḥ |
kupitaḥ kupayaty āśu pra_ _ kaphamārutau ||

10) Kapilabala[69]

The first quote attributed to Kapilabala is found in the commentary to Sū 40.14–5 (according to the order of chapters in the *SVyā, it is chapter 39) on page 191:

uktaṃ ca kapilabale—
rasavīryavipākānāṃ sāmye karmaviśeṣaṇam |
dravyasvabhāvaṃ jānīyāt śṛṇu cātra nidarśanam ||
kaṭukaḥ_ _ _ _ _ _ _ _ citrako mataḥ |
kaṭupākarasā dantī svabhāvāt sā virecanī ||
maṇīnāṃ dhāraṇīyānāṃ karma coktam prabhāvataḥ ||

---

66 For a comprehensive study of the relationship between *agni* and *pitta* in the SS, see Jan Meulenbeld, "Some Neglected Aspects of Ayurveda II—The Suśrutasaṃhitā." *Traditional South Asian Medicine* 8: 16–31 (2008).

67 Given that Hiraṇyākṣa's text is quoted to provide further grounds for the opinion of some (*eke*) who hold that *agni* digests (*pacati*) are instigated by the *udāna* type of bodily wind, it is likely that *pānasamīritaḥ* at the end of the verse is a corruption of [*u*]*dānasamīritaḥ*.

68 This translation of the term is adopted from Priya Vrat Sharma, *Suśruta-Saṃhitā, with English Translation of the Text and Ḍalhaṇa's Commentary along with Critical Notes*, vol. 1 (Varanasi: Chaukhambha Visvabharati, 2001).

69 Meulenbeld, *History*, 1a, 659.

It is curious for a number of reasons. First, all three of the quoted verses closely resemble CS Sū 26.67–68 and 70ab.[70] This observation indicates that although the secondary literature generally accepts that "Kapilabala is [...] credited with a śalyatantra,"[71] his text must have also had some affinity with the CS. Furthermore, more relevant for the current study, is a short introduction to these verses furnished in the *SVyā (pp. 190–191):

> *iha citrakaḥ kaṭuko rasataḥ pākataś ca ... coṣṇaḥ,*[72] *sa virecayati / virecanaṃ tāvat pṛthvyapāṃ guṇodrekād bhavati / citrake punā rasapākaviryāṇi laghubhūtajāny eva dṛśyante / ataḥ svabhāvena karmakaro bhavatīti mantavyam / uktaṃ ca kapilabale—*
> *rasavīryavipākānāṃ* [...]

The analysis of the verses presented in the extant portion of the commentary is rather striking. Although rather complex and profound at first sight, it is not only contradictory to the understanding of the described phenomena argued for in the CS and some other medical text,[73] but it also seems to go against some basic pharmacological facts and, what is more, to disturb the inner logic of the verse. I conclude, therefore, that either the transmitted text of the *SVyā is seriously corrupt here, or, what seems more likely, that the author of the *SVyā held a rather unique set of views.

The situation can be summarized as follows: The quoted verses by Kapilabala say that, in cases when different substances (*dravya*) have the same taste (*rasa*), digestive alteration (*vipāka*), and potency (*vīrya*), one should consider their specific properties (*svabhāva*) in order to find out their individual therapeutic effects. An example of two substances, *citraka*[74] and *dantī*,[75] is given. Both have exactly the same *rasa*, *vīrya* and *vipāka*, but *dantī* is purgative and

70 CS Sū 26.67–70 (Trikamjī, *Carakasaṃhitā of Agniveśa*, 148): *rasavīryavipākānāṃ sāmānyaṃ yatra lakṣyate / viśeṣaḥ karmaṇāṃ caiva prabhāvas tasya sa smṛtaḥ // 67 // kaṭukaḥ kaṭukaḥ pāke vīryoṣṇaś citrako mataḥ / tadvad dantī prabhāvāt tu virecayati mānavam // 68 //* [...] *maṇīnāṃ dhāraṇīyānāṃ karma yad vividhātmakam / tat prabhāvākṛtaṃ teṣāṃ prabhāvo 'cintya ucyate // 70 //*

71 Meulenbeld, *History*, 1b, 723, n. 46 lists three historians of Āyurveda.

72 The dots in the transcription could, perhaps, be filled out with something like *vīryataś*.

73 A similar statement is, in fact, found in *Aṣṭāṅgasaṃgraha* Sū 17.52cd–53ab (= *Aṣṭāṅgahṛdaya* Sū 9.26) and in the BhM on SS Sū 40.11.

74 *Citraka* is generally identified as *Plumbago zeylanica* L. See, for example, Jan Meulenbeld, *The Mādhavanidādana and its chief commentary* (Leiden: E.J. Brill, 1974), 553.

75 *Dantī* is *Baliospermum solanifolium* (J. Burm.) Suresh. See, e.g., Meulenbeld, *The Mādhavanidāna*, 561.

citraka is not. Note that according to SS Sū 45.124, *dantī* is, indeed, *adhobhāgadoṣahara* (that is, removes the *doṣa*s through the lower parts of the body),[76] while nothing like that is mentioned in case of *citraka* in SS Sū 46.239ab. Now, the *SVyā seems to quote this passage in order to illustrate a situation, in which the *svabhāva* of a certain *dravya* is decisive for its therapeutic effect, which could not be otherwise explained based on taste, etc. The *SVyā observes that *citraka* has a pungent *rasa*, a pungent *vipāka*, and a hot *vīrya* (this word is lost, but we can easily supply it based on the context as well as the parallel in the CS). While this statement is in perfect agreement with the general understanding, what follows next takes a rather unexpected turn. According to the commentary, it is *citraka* that acts as a purgative,[77] which is by itself a phenomenon that seems incompatible with what the *rasa*, *vipāka* and *vīrya*, of the drug suggest. Purging, according to the *SVyā, is a therapeutic effect related to an excess of the elements earth and water in a given drug (this is, indeed, in agreement with SS Sū 41.5).[78] However, all the mentioned properties of *citraka*—that is, pungent taste and *vipāka*, as well as its hot potency—point toward a predominance of the other three elements (fire, wind, and space), because these properties can only be produced from those three elements. Here, the *SVyā refers to several basic principles of Suśruta's pharmacology: the division of elements into heavy (earth and water) and light (fire, wind, and space) in SS Sū 40.10,[79] the association of the pungent taste with the predominance of fire and wind in SS Sū 42.3,[80] and the association of the pungent *vipāka* with light elements in SS Sū 40.10,[81] as well as the association of the hot potency with the element of fire in SS Sū 41.11.[82] Consequently, the commentary concludes, it must be the *svabhāva* that is responsible for the particular therapeutic effect—that is, purging.

---

76 Cf. Ḍalhaṇa (Trikamjī and 'Kāvyatīrtha', *Suśruta Saṃhitā*, 206): *adhobhāgadoṣaharā iti virecanena doṣaharā iti.*

77 The masculine pronoun *saḥ* certainly cannot refer to the feminine word *dantī*, which could have been lost somewhere in the omission.

78 SS Sū 41.6 (Trikamjī and 'Kāvyatīrtha', *Suśruta Saṃhitā*, 182): *tatra virecanadravyāṇi pṛthivyambuguṇabhūyiṣṭhāni* [...].

79 SS Sū 40.10 (Trikamjī and 'Kāvyatīrtha', *Suśruta Saṃhitā*, 179): *tatra pṛthivyaptejovāyvākāśānāṃ dvaividhyaṃ bhavati, guṇasādharmyād gurutā laghutā ca, pṛthivyāpaś ca gurvyaḥ śeṣāṇi laghūni* [...].

80 SS Sū 42.3 (Trikamjī and 'Kāvyatīrtha', *Suśruta Saṃhitā*, 184): [...] *vāyvagniguṇabāhulyāt kaṭukaḥ* [...].

81 SS Sū 40.10 (Trikamjī and 'Kāvyatīrtha', *Suśruta Saṃhitā*, 179): *āgame hi dvividha eva pāko madhuraḥ kaṭukaś ca / tayor madhurākhyo guruḥ kaṭukākhyo laghuriti / tatra pṛthivyaptejovāyvākāśānāṃ* [...].

82 SS Sū 41.11 (Trikamjī and 'Kāvyatīrtha', *Suśruta Saṃhitā*, 183): *tatra ya ime guṇā vīryasaṃjñakāḥ śītoṣṇasnigdharūkṣamṛdutīkṣṇapicchilaviśadāḥ, teṣāṃ tīkṣṇoṣṇāv āgneyau* [...].

As we can see, the *SVyā puts forward a sound, and rather ingenious, analysis, which is, however, based on a rather dubious proposition (that *citraka* and not *dantī* has laxative properties)!

The second quote attributed to Kapilabala is found in the commentary on Sū 42.8. It is a single *anuṣṭubh* verse that postulates a direct relation between tastes and *doṣas* and summarizes how the latter can be manipulated by means of the former. In addition, this quote possesses a further text-historical significance, because a distinctly different version of the same (or possibly a closely related) verse is quoted and attributed to Kapilabala in the *Aṣṭāṅgasaṃgraha* (AS) Sū 20.17.[83] The reading preserved in the MS of the *SVyā (p. 203) is as follows:

> *kapilabalenāpy uktam:*
> *kaṭvāmlalavaṇaṃ*(!)[84] *pittaṃ, svāduḥ śleṣmānilaḥ kaṭuḥ* /
> *teṣāṃ tulyarasair vṛddhir viparītair viparyayaḥ* //

11,12) Nimi and/or *Vaidehīya*[85]

Nimi is credited with two *anuṣṭubhs*. One of them appears in the commentary to Sū 1.28 (p. 21) and is quoted in order to endorse Kṣārapāṇi's idea that the whole variety of therapeutic actions is based entirely on the six tastes of the therapeutic substances (see above):

> *nimināpy uktam:*
> *rasamūlā cikitseyaṃ catuṣpādā*[86] *ṣaḍāśrayā*[87] /
> *bhiṣaguttamasampannā prāpnoti mahatīṃ kriyām* //

A further verse attributed to Nimi is found in the commentary to Sū 40 (39), 14–15, immediately after the long quote by Kapilabala discussed above. It seems to refer to a view, according to which a *dravya* (presumably on account of its *svabhāva*) can be associated with specific actions—namely, *pūraṇa* (filling), *dhāraṇa* (supporting), etc.—in the same way as *rasa*, *pāka*, and *vīrya* are asso-

83 AS Sū 20.17 (Āṭhavale, *Aṣṭāṅgasaṅgrahaḥ. Induvyākhyāsahitaḥ*, 151): *kapilabalas tv eṣāṃ svalakṣaṇāni rasato nirdideśa: kaṭvamlalavaṇaṃ pittaṃ svādvamlalavaṇaḥ kaphaḥ / kaṣāyatiktakaṭuko vāyur dṛṣṭo 'numānataḥ* //

84 The long *ā* in *kaṭvāmla°* must be a simple mistake for a short *a*.

85 Meulenbeld, *History*, 1a, 171–173.

86 The four quarters (*pādas*) of treatment are enumerated, e.g., in SS Sū 34.15 (but also in CS Sū 9.3, *Aṣṭāṅgahṛdayasaṃhitā* Sū 1.27, or *Kāśyapasaṃhitā* Sū 26.2) as physician, patient, drug, and attendant.

87 *Ṣaḍāśrayā* (*cikitsā* based on the six) refers perhaps to the six tastes.

ciated with theirs. I must confess, however, that neither the meaning of the verse nor the reference to *pūraṇa* as a *karma* of *dravya* is clear to me:

api ca, dravyaṃ pūraṇadhāraṇādi karma karoti /
yathā—odana[88]*stambhaprabhṛtīni*(!) */ uktaṃ ca niminā—*
pūraṇārthaṃ tu tad dravyam āsvādārthas tu yo rasaḥ /
doṣadhātumalān pāko vīryaṃ karmaṇi cocyate // iti //

Another extended quote, comprised of seven *anuṣṭubh* verses and listing the qualities of milk, is quoted in the commentary on Sū 45.9 and attributed to *Vaidehīya*, which may (or may not) be the work composed by Nimi, the kind of Videha (pp. 213–214):

uktaṃ vaidehīye—
gokṣīraṃ madhuraṃ śītaṃ vipāke madhuraṃ guru /
vātapittaharaṃ snigdhaṃ ___ kaphavardhanam /
guruvīryoṣṇam ājaṃ tu tridoṣaśamanaṃ payaḥ /
madhuraṃ sakaṣāyaṃ ca vipāke madhuraṃ tathā /
pāke rase ca madhuraṃ kaṣāyānurasaṃ guru /
pittamārutayoḥ pathyam uṣṇavīryaṃ ca māhiṣam /
aurabhram uṣṇaṃ madhuraṃ gurupittāpahaṃ smṛtam /
___ madhuraṃ snigdhaṃ śleṣmaṇaś ca vivardhanam /
madhuraṃ rasapākābhyāṃ nārīstanyaṃ viśeṣataḥ /
cakṣuṣyaṃ śītavīryaṃ ca tridoṣaśamanaṃ smṛtam /
hastinyās tu payo vṛṣyaṃ bṛṃhaṇaṃ prāṇavardhanam /
pāke rase ca madhuraṃ cakṣuṣyaṃ dīpanaṃ __ /
uṣṭrakṣīraṃ salavaṇaṃ madhuraṃ rasapākataḥ /
arśoghnadīpanīyaṃ ca balyam indriyavardhanam / iti /

13) Śālihotra[89]

Śālihotra is the first among the authorities mentioned so far whose works, according to Meulenbeld's assessment,[90] "cannot be dated accurately, but [...] may be from or somewhat earlier than the tenth century." In contrast to the topics that are usually dealt with in the quotes ascribed to this author,[91] in the

88 *Odana*°, I believe, is a corruption for *hlādana*°. See the previous note.
89 Meulenbeld, *History*, 2a, 575–576.
90 Meulenbeld, *History*, 2a, 576.
91 See notes 757–771 in Meulenbeld, *History*, 2b, 609.

*SVyā Śālihotra is quoted twice in the commentary to Sū 6 on topics vaguely related to astronomy. See, for example, page 47:

uktaṃ ca śālihotreṇa—
śuklapakṣe pratipade śiśire cottarāyaṇe /
māghamāse śravi+su[92]yugaṃ samparivartate //

14) Jātukarṇa[93]

The single quote attributed to Jātukarṇa is unfortunately only fragmentary. It is found in the commentary to Sū 21.9 and describes a debate among medical sages concerning the question of what entity is in charge of the process of digestion. In this debate, Ātreya pronounces the final opinion (it is the fiery heat situated in three abodes), while Bhadraśaunaka also makes an appearance, but his position is lost (pp. 133–134):

jātukarṇenāpy uktam—
tatra praśna āhārasya jantoḥ kaḥ pacatīti / vātaḥ pacatīty eke, kaphaḥ pacatīty eke, anupāna ity eke / nety ātreyaḥ / yato vātaḥ paced [*ced*] *anya eva bahuvātās teṣāṃ kṣipratāṃ pacyate* (?) */ tathā pittabhūyiṣṭhānām anupānaśīlānām iti / sa punar āhāram asya jantoḥ kaḥ pacatīti bhadraśaunako … … … treya uvāca / śarīre taijasa uṣmā sa triṣu sthāneṣv āhāraṃ pacati, śleṣmapittavātāśayeṣu / iti*

15) Parāśara[94]

A single verse in the commentary to Sū 15.36 (p. 89) is attributed to Parāśara. It summarizes three reasons for the decrease of bodily elements (lack of its own properties, an increase of the opposite ones, and mutual disturbance):

uktaṃ ca parāśareṇa—
svaguṇānām alābhena pratyanīko_ _yeṇa[95] vā /
kṣīyante dhātavaḥ puṃsām anyonyābhigraheṇa vā //

92 I cannot decipher the consonant cluster between *vi* and *su*.
93 Meulenbeld, *History*, 1a, 161–162.
94 Meulenbeld, *History*, 1a, 173–175.
95 Here, one expects something like *pratyanīkocchrayeṇa*.

16) Pālakāpya[96]

A single verse by Pālakāpya is quoted (as an intermediate position) in the commentary to Sū 24.5 during the discussion of the origin of the *mānasa* type of *ādibālapravṛtta* (hereditary) diseases (p. 155):

api coktaṃ pālakapāye—
kruddho _ _ prahṛṣto vā durbalo balavān api /
yadbhāvo maithunaṃ kuryāt tadbhāvaṃ labhate sutam // iti //[97]

17) *Siddhayoga* by Vṛndamādhava (quoted anonymously)[98]

Given that, with the exception of Śālihotra, the quoted treatises cannot be dated with any degree of certainty, the discovery of a citation originating from a work that can be dated, at least to a certain period of time, is rather significant. The concerned verse is found at the beginning of the commentary on Sū 23 (p. 145) that deals with the title of the chapter, *kṛtyākṛtyavijñānīya* (a chapter on the knowledge of *kṛtya* and *akṛtya* diseases). The *SVyā offers different interpretations of the words *kṛtya* and *akṛtya*. According to one of them, *kṛtya* = *sādhya* (curable) and *akṛtya* = *asādhya* (incurable), both of which can be subdivided into two subtypes. To illustrate this opinion, the *SVyā quotes the following verse:

uktaṃ ca—
kṛcchropāyaḥ sukhopāyo dvividhaḥ sādhya ucyate /
asādhyo dvividhaḥ prokto yāpyo yaś cāpratikriyaḥ // iti

This verse is identical to *Siddhayoga* 1.11[99] by Vṛnda, who "lived later than Mādhava, the author of the Nidāna, and earlier than Cakrapaṇidatta, probably in the period 800–950."[100]

Apart from these, the *SVyā identifies several sources that, so far, I have not been able to identify:

---

96 Meulenbeld, *History*, 2a, 570–574.

97 Note that the *SVyā quotes a similar verse by Bhoja as well (o. 154–155): *uktaṃ hi bhojena—yadācāro yadākāro yadbuddhir maithunaṃ caret / tadācāraṃ tadākāraṃ tadbuddhiṃ janayet sutam //*

98 Meulenbeld, *History*, 2a, 78–85.

99 Mahādev Āpṭe, *Siddhayogaḥ* (*Vṛndamādhavāparanāma*), *Śrīkanṭadattaviracitavyākhyā-Kusumāvalyākhyaṭīkāsametaḥ* (Pune: Ānandāśrama, 1894), 4.

100 Meulenbeld, *History*, 2a, 82.

18) Siddha (does not seem to be in the *Siddhayoga*) in Sū 11 on page 67:

siddhenāpy uktam—
yāḥ kriyāḥ pratiṣedhavyāḥ, tāḥ kriyāḥ kurvato bhṛśam |
tadvyādhivṛddhir maraṇam anyā vā vyapado dhruvā ||

19) *Lakṣaṇaśāśtra*, at Sū 35.3 on page 176:

uktaṃ ca lakṣaṇaśāstre—
yasya snātānuliptasya pūrvam ūrdhvaṃ viśuṣyati |
sa dīrghāyur iti jñeyo bṛhatsrotāś ca mānavaḥ ||[101]

20) *Pāryovida* (?!?), at Sū 13, 78:

evam āha pāryovidaḥ (?)—
yathā hi puṃsāṃ śukrāntā rasādyāḥ sapta dhātavaḥ |
tathā hi dhātavaḥ saptaḥ nārīṇām iti niścayaḥ ||

5 Conclusion

The main objective of this article has been to introduce a newly discovered commentary on the SS that was hitherto unknown to the scholarly community. Given the considerable length of the text, the fact that it has not been printed so far, and the circumstance that it is only available in an extremely lacunar and partly corrupt MS (itself, I believe, an early modern transcript of a palm-leaf MS), I chose a methodology that: first, allowed me to have a look at the whole text; secondly, was not entirely dependent on an accurate reconstruction of the original wording; and thirdly, and most importantly, highlighted a unique feature of the text that, in my opinion, makes its historical importance directly perceivable—namely, the extraordinarily large amount of citations from other medical treatises, most of which have been lost.

To begin with, the abundance of these citations provides important value on its own. It allows us to access and evaluate the actual views of many ancient authorities, a large portion of whom have remained unknown to us to date. It

101 Note, this verse is exactly the opposite of the CS in 12.5: *yasya snātānuliptasya pūrvaṃ śuṣyaty uro bhṛśam | ārdreṣu sarvagātreṣu so 'rdhamāsaṃ na jīvati ||*

is, therefore, particularly gratifying that the *SVyā assigns the majority of its citations to particular authors and texts, and thus they can be studied in their own right.

The citations furnish us, furthermore, with a window into the study room of a learned medieval *vaidya*. They allow us to form an opinion about a group of "canonical" works—that is, medical treatises that were considered important and authoritative and whose opinions could have been used to prove a point or settle a debate. A close study of these citations may help us to access further biographical data about the author of the *SVyā. On the basis of what we have seen so far, we can conclude, for example, that the commentator must have had separate access to a number of ancient medical treatises (such as *Bhālukīya* or *Vaitāraṇa*) that were scarcely quoted by other important medieval scholars (e.g., Niścalakāra and Cakrapāṇi). In this connection, I would also like to highlight the second quote from Kapilabala's treatise, which reads differently from the vulgate version of the *Aṣṭāṅgahṛdayasaṃhitā*. Based on the above survey, we also learn that the *SVyā must have been composed after the *Siddhayoga* (800–950). Further attention to the citations may provide us with clues about the author's provenance, her philosophical or religious affiliations, etc.

Finally, and potentially most importantly, in the discussion of several citations, as well as in the general introduction at the beginning of my article, I hope to have sufficiently emphasized the *SVyā's most unorthodox and often ingenious interpretations and, in fact, modifications of the SS as well as other cited works. As a matter of fact, the majority of the citations mentioned in this article are found amid lively and complex discussions on various aspects of Āyurvedic doctrine and practice. A thorough study of the *SVyā will, therefore, undoubtedly provide a fascinating insight into the world of early medieval Āyurveda.

References

Āpṭe, Mahādev Cimaṇājī. *Siddhayogaḥ (Vṛndamādhavāparanāma), Śrīkanṭadattaviracitavyākhyā-Kusumāvalyākhyaṭīkāsametaḥ*. Pune: Ānandāśrama, 1894.

Āṭhavale, Vaidya Ananta Dāmodara. *Aṣṭāṅgasaṅgrahaḥ: Induvyākhyāsahitaḥ*. Pune: Śrīmad Ātreya Prakāśaam "Naṃdanaṃdana," 1980.

Das, Rahul Peter. *The Origin of the Life of A Human Being: Conception and Female According to Ancient Indian Medical and Sexological Literature*. Delhi: Motilal Banarsidass Publishers Private Limited, 2003.

Klebanov, Andrey. "The *Nepalese version of the *Suśrutasaṃhitā* and its Interrelation with Buddhism and the Buddhists." M.A. thesis, Hamburg University, 2011.

Klebanov, Andrey. Forthcoming. "On the Textual History of the *Suśrutasaṃhitā* (1): A Study of Three Nepalese Manuscripts." *eJournal of Indian Medicine*.

Meulenbeld, Gerrit Jan. *The Mādhavanidāna and Its Chief Commentary: Chapters 1–10. Introduction, Translation and Notes*. Leiden: E.J. Brill, 1974.

Meulenbeld, Gerrit Jan. *A History of Indian Medical Literature*. Vols. 1a, 1b, 2a, 2b, 3. Groningen: Egbert Forsten, 1999–2002.

Meulenbeld, Gerrit Jan. "Some Neglected Aspects of Ayurveda II—The Suśrutasaṃhitā." *Traditional South Asian Medicine* 8 (2008): 16–31.

Rajesh, S.K. and G.R. Kumar. "Vaidyamadham Cheriya Narayanan Namboodiri." *Journal of Ayurveda and Integrative Medicine* 1, no. 2 (2010): 136–138. https://doi.org/10.4103/0975-9476.65093.

Rao, B. Rama and Vinod Kumar Bhatnagar. *Sanskrit Medical Manuscripts In India*. New Delhi: Central Council for Research in Ayurveda & Siddha, Department of AYUSH, Ministry of Health and Family Welfare, Govt. of India, 2005.

Sastri, P.P. Subrahmaṇya. *An Alphabetical Index of Sanskrit Manuscripts in the Goverment Oriental Manuscripts Library, Madras*. Vol. 2. Madras: Government of Madras, 1940.

Śastri, S. Kuppuswami. *A Descriptive Catalogue of the Sanskrit Manuscripts in the Government Oriental Manuscripts Library, Madras*. Vol. 13: Medicine. Madras: Government of Madras, 1918.

Sharma, K.V. *Science Texts in Sanskrit in the Manuscript Repositories of Kerala and Tamilnadu*. New Delhi: Rashtriya Sanskrit Sansthan, 2002.

Sharma, Priya Vrat. *Cakradatta-Ratnaprabhā: The Cakradatta (Cikitsā-Saṅgraha) of Cakrapāṇidatta with the Commentary Ratnaprabhā by Mahāmahopadhyaya Śrī Niścala Kara*. Jaipur: Swami Jayaramdas Ramprakash Trust, 1993.

Sharma, Priya Vrat. *Suśruta-Saṃhitā, with English Translation of the Text and Ḍalhaṇa's Commentary along with Critical Notes*. Vol. 1. Haridas Ayurveda Series. Varanasi: Chaukhambha Visvabharati, 2001.

Sharma, Priya Vrat. *Cikitsākalikā of Tīsaṭācārya, Containing Sanskrit Commentary of His Son Candraṭa*. Varanasi: Chaukhambha Surbharati Prakashan, 2014.

Steiner, Roland. "Die Lehre Der Anuṣṭubh Bei Den Indischen Metrikern." In *Suhṛllekhāḥ. Festgabe Für Helmut Eimer*, edited by Michael Hahn, Jens-Uwe Hartmann, and Roland Steiner, 227–247. Swisttal-Odendorf: Indica et Tibetica Verlag, 1996.

Trikamjī Ācārya, Vaidya Yādavjī. *Carakasaṃhitā of Agniveśa, Elaborated by Caraka and Ḍṛḍhabala, with the Āyurvedadīpikā Commentary by Cakrapāṇidatta*. The Chaukhamba Ayurvijnan Granthamala 34. Varanasi: Chaukhamba, 2008 [1941].

Trikamjī Ācārya, Vaidya Yādavjī and Nārāyaṇa Rāma Ācārya 'Kāvyatīrtha'. *Suśruta Saṃhitā of Suśruta, with the Nibandhasaṃgraha Commentary of Śrī Ḍalhaṇācārya & Nyāyacandrikā Pañjikā of Śrī Gayadāsācārya on Nidānasthāna*. Krishnads Ayurveda Series 51. Varanasi: Chowkhamba Krishnadas Academy, 2004 [1938].

Trikamjī Ācārya, Vaidya Yadavjī and Nandikishor Sharma Bhishagacharya. *Suśruta-Saṃhitā (Sūtra Sthān), with Bhānumatī Commentary by Cakrapāṇidatta*. Krishnadas Ayurveda Series 74. Varanasi: Krishnadas Academy, 2001.

Tubb, Gary Alan and Emery R. Boose. *Scholastic Sanskrit. A Manual for Students*. Treasure of the Indic Sciences. New York: The American Institute of Buddhist Studies at Columbia University, 2007.

Upādhyāya, Yadunandana and Sudaraśana Śāstri. *Mādhavanidānam, Mahāmahopādhyāya-Śrīvijayarakṣita-Śrīkaṇṭhdattābhyāṃ Viracitayā "Madhukośa"-Vyākhyayā Vibhūṣitam*. Vol. 2. Second edition. Varanasi: The Chaukhamba Sanskrit Series Office, 1961.

Vidyābhūṣaṇa, Āśubodha and Nityabodha Vidyāratna. *Cakradatta (Cikitsāsaṃgrahagranthaḥ), Śrīcakrapāṇinā Viracitaḥ. Śrīśivadāsasenaviracitayā Tattvacandrikāsamākhyayā Vyākhyayā Samalaṅkṛtaḥ*. Calcutta: Vācaspatyayantram, 1914.

Zysk, Kenneth. "Citations in Jajjaṭa's *Nirantarapadavyākhyā*." *eJournal of Indian Medicine* 3 (2010): 69–99.

CHAPTER 8

On Attaining Special Powers through *Rasāyana* Therapies in Sanskrit Medical Literature

Dagmar Wujastyk

In this article, I will examine the association of a particular branch of Āyurveda called *rasāyana* (vitalization therapy) with the development or attainment of extraordinary capabilities or special powers (*siddhi*). Special powers or abilities are often described in Indic literature as the necessary or incidental products of spiritual advancement: For example, a specific set of special powers (*siddhi*) forms an integral part of the yogic path of the *Pātañjalayogaśāstra*.[1] But what role do special powers play in medicine? Are they also part of a spiritual path, or do they have a different function in medicine? In the following, I will examine how the generation of special powers is described in Sanskrit medical literature in the context of *rasāyana* and will explore what this tells us about the Sanskrit medical authors' understanding of medicine.[2]

1 See Knut A. Jacobsen, *Yoga Powers: Extraordinary Capacities Attained Through Meditation and Concentration* (Leiden: Brill, 2011) on the general theme of yoga powers in Indian thought; Corrado Pensa, "On the Purification Concept in Indian Tradition, with Special Regard to Yoga," *East and West* 19, no. 1–2 (1969): 194–228 on the role of *siddhi*s in Patañjali's work and in Buddhism; Philipp A. Maas, "On the Meaning of *rasāyana* in Classical Yoga and Āyurveda," *History of Science in South Asia Special Issue "Rejuvenation, Longevity, Immortality in South Asian Medical, Alchemical and Yoga Traditions"* 5, no. 2 (2017): 66–84 on the generation of yogic powers through the use of *rasāyana* in the *Pātañjalayogaśāstra* and its commentaries; and James Mallinson and Mark Singleton, *Roots of Yoga* (London: Penguin Random House UK, 2017), 359–394 on yogic powers in medieval yoga literature.

2 The following Sanskrit medical works were consulted for this article: the *Carakasaṃhitā* (early centuries CE), *Suśrutasaṃhitā* (early centuries CE), *Aṣṭāṅgahṛdayasaṃhitā* (c. seventh century), *Aṣṭāṅgasaṃgraha* (c. seventh century CE), *Siddhayoga* (c. tenth century CE), *Cakradatta* (c. eleventh century CE), *Vaṅgasenasaṃhitā*, (c. eleventh/twelfth century CE), *Śārṅgadharasaṃhitā* (c. thirteenth/fourteenth century CE), *Bhāvaprakāśa* (sixteenth century CE), *Yogaratnākara* (eighteenth century CE), and *Bhaiṣajyaratnāvalī* (eighteenth/nineteenth century CE). Apart from the *Śārṅgadharasaṃhitā*, all these works contain separate chapters on *rasāyana*.

DOI:10.1163/9789004438224_009

1 Background

The Indian medical tradition of Āyurveda is transmitted to us through a series of compilations, most of which are written in Sanskrit. The earliest of these, the *Carakasaṃhitā* and *Suśrutasaṃhitā* date to the early centuries of the Common Era.[3] These works represent the first attempts at the systematization and codification of medical knowledge as a discipline in its own right, defining medicine as a professional occupation with a circumscribed body of knowledge. As the titles suggest,[4] these works are collections, or assemblages, and as such, they unite in themselves a range of approaches to their subjects. At the same time, they are not miscellanies, but (multiply) redacted works that reflect which subjects, medical theories and practices were considered valid or at least relevant by the compilers and redactors. Later medical works build upon the early ones, ensuring continuity through an adaptive reuse of the older materials. The innovation of the later compilers and authors lies in restructuring and freshly interpreting older materials, introducing new ideas, and removing and/or replacing parts.

The early Sanskrit medical works propose a division of medical knowledge into eight subject areas, each representing medical specializations. In the *Carakasaṃhitā* (Sūtrasthāna 30.28), these are listed as: 1. general medicine (*kāyacikitsā*); 2. the surgical treatment of body parts above the shoulders (*śālākya*); 3. the removal of foreign bodies, surgery (*śalyāpahartṛka*); 4. toxicology (*viṣagaravairodhikapraśamana*); 5. the treatment of possession by supernatural beings (*bhūtavidyā*); 6. pediatrics (*komārabhṛtyaka*); 7. vitalization therapy (*rasāyana*); and 8. aphrodisiacs and treatments relating to virility (*vājīkaraṇa*). The *Suśrutasaṃhitā* records a similar list, albeit in a different order.[5] While

3 Both works have complicated transmission histories with several historical layers. For a survey of the proposed dates of the different redactions of the *Suśrutasaṃhitā*, see Jan Gerrit Meulenbeld, *A History of Indian Medical Literature* (Groningen: Egbert Forsten, 1999–2002), IA, 342–352. The *Carakasaṃhitā* is presented as a revised version of an ancient work on medicine (the *Agniveśatantra*), the date of which is uncertain. The version of the *Carakasaṃhitā* that has been transmitted to us is a redacted version that was edited and added to by a scholar called Dṛḍhabala in about the fifth century. See Meulenbeld, *A History of Indian Medical Literature*, IA: 105–115 for a detailed discussion of the issues surrounding the dating of the *Carakasaṃhitā*. Also see the article by Philipp A. Maas, "On What Became of the *Carakasaṃhitā* After Dṛḍhabala's Revision," *eJournal of Indian Medicine* 3, no. 1 (2010): 1–22.

4 "*Saṃhitā*" means "collection."

5 See *Suśrutasaṃhitā* Sūtrasthāna 1.7. The list reads: 1. surgery (*śalya*); 2. surgery above the shoulders (*śalākya*); 3. general medicine (*kāyacikitsā*); 4. treatment of posession (*bhūtavidyā*); 5. pediatrics (*kaumārabhṛtya*); 6. toxicology (*agadatantra*); 7. vitalization therapy (*rasāyana*); and 8. virility therapy (*vājīkaraṇatantra*).

both these works note that a physician should have proficiency in all areas of medicine, their descriptions of the different medical subjects display marked differences in methodological approaches, which may reflect older divisions among medical practitioners into groups with distinct specializations. For example, while general medicine was predicated on humoral theory and associated physiological processes, these initially played no role in the context of toxicology, or vitalization and virility medicine.[6]

2 What Is *Rasāyana*?

The branch of Āyurveda called *rasāyana*, which I have rendered as "vitalization therapy" above, is dedicated to the prolongation of life, rejuvenation, anti-ageing and the preservation or promotion of health and well-being.[7] The term *rasāyana* also denotes the therapies that together constitute this branch of medicine and the medical substances and formulations used in these therapies. Many Sanskrit medical works dedicate whole sections to the topic of *rasāyana*. Often, these sections are juxtaposed or, especially in later works, merged with sections on virility therapy (*vṛṣya*/*vājīkaraṇa*).[8] *Rasāyana* formulations and references to substances that function as *rasāyana*s are also found interspersed throughout Sanskrit medical works.

The medical treatises offer some definitions of *rasāyana*, or rather, of what is meant to be achieved through *rasāyana*. For example, the *Carakasaṃhitā*

6 This changed: For example, humoral theory was integrated into vitalization therapy (*rasāyana*) from the c. seventh-century medical work *Aṣṭāṅgahṛdayasaṃhitā* onward. I would like to thank Prof. Francis Zimmermann for this important observation, made at the Ayuryog workshop in October 2017 (http://ayuryog.org/event/ayuryog-workshop-rejuvenation-longevity-immortality-perspectives-rasāyana-kāyakalpa-and-bcud). However, the humors were still only of marginal importance in the *Aṣṭāṅgahṛdayasaṃhitā*'s description of vitalization therapy and remained a fairly minor element in vitalization therapy in general. The exception to this rule is the eleventh-/twelfth-century *Vaṅgasenasaṃhitā*, which describes an array of tonics specifically formulated for treating disorders of the humors.

7 The term *rasāyana* is somewhat elusive, due to the polyvalent meanings of its elements, in particular of *rasa*. See Dagmar Wujastyk, "Acts of Improvement." *History of Science in South Asia Special Issue "Rejuvenation, Longevity, Immortality in South Asian Medical, Alchemical and Yoga Traditions"* 5, no. 2 (2017): 1–2 and especially Maas, "On the Meaning of *rasāyana*," 75–79, which discusses the term at length and concludes that the different interpretations of the term by the Sanskrit medical authors show a certain level of uncertainty concerning its meaning. I therefore leave it untranslated in this article, unless it is clearly used in the sense of "tonic."

8 See Kenneth G. Zysk, "Potency Therapy in Classical Indian Medicine," *Asian Medicine* 1, no. 1 (2005): 101–118 on potency (= virility) therapy in classical Indian medicine.

defines *rasāyana* as therapies that promote strength and vigor in the healthy.[9] Its more detailed definition reads as follows:

> Through *rasāyana*, a man gains longevity, memory, mental power, health, youthful vigor, a great radiance, complexion, and voice, an extremely strong body and keen senses, mastery of speech, respect, and beauty.[10]

One of the *Carakasaṃhitā*'s listed effects, "mastery of speech" (*vāksiddhi*), points to the aspect of *rasāyana* we will examine further here, namely the attainment of special powers. As explained elsewhere, "mastery of speech" could be understood to refer to excellent oratorical skills, or even just to a strong or pleasant voice.[11] Regaining or attaining a strong voice is certainly documented as one of the functions of *rasāyana*: a number of formulations are indicated for strengthening the voice and combating hoarseness (*svarabheda*).[12] However, the function of curing hoarseness seems already covered by "*svaraudāryam*"—"greatness of voice." The attainment of "mastery of speech" in the sense of oratorical skills or eloquence is perhaps less easily explained in terms of overcoming a pathology, unless mastering speech is understood to mean counteracting stammering or other problems with coordinating speech.[13] Since the term *vāksiddhi* to my knowledge only occurs twice in the *Carakasaṃhitā* (both times without explanation), it is impossible to know what

---

9 See *Carakasaṃhitā* Cikitsāsthāna 1.1. 1–8. In the edition of the *Carakasaṃhitā* used here, the *rasāyana* chapter is located in the section on therapeutics (Cikitsāsthāna) and it is divided into four subchapters (*pāda*): 1. *abhayāmalakīyo rasāyanapāda* ("the *rasāyana* section dedicated to the chebulic and emblic myrobalans"); 2. *prāṇakāmīyo rasāyanapāda* ("the *rasāyana* section dedicated to the desire for vital breath"); 3. *karapracitīyo rasāyanapāda* ("the *rasāyana* section dedicated to hand-plucked (emblic myrobalan fruits)"); and 4. *āyurvedasamutthānaīyo rasāyanapāda* ("the *rasāyana* section dedicated to the uplift of the science of life").

10 *Carakasaṃhitā* Cikitsāsthāna 1.1.7–8ab: *dīrgham āyuḥ smṛtiṃ medhām ārogyaṃ taruṇaṃ vayaḥ / prabhāvarṇasvaraudāryaṃ dehendriyabalaṃ param //7// vāksiddhiṃ praṇatiṃ kāntiṃ labhate nā rasāyanāt /* This passage recurs in a number of later medical texts, albeit with some variations. See Wujastyk, "Acts of Improvement," 5–7.

11 Dagmar Wujastyk, "On Perfecting the Body: *rasāyana* in Sanskrit Medical Literature," AION—*sezione filologico-letteraria* 37 (2015): 57. Maas, "On the Meaning of *rasāyana*," 74.

12 On hoarseness, see, for example, *Carakasaṃhitā* Cikitsāsthāna 1.1.71, 1.2.3, 1.3.27. The strengthening of the voice is often mentioned in conjunction with the increase of strength (*bala*), digestion (*agni*) and complexion (*varṇa*).

13 Stammering is typically associated with a dysfunction of the humor wind and is not explicitly addressed in *rasāyana* chapters. On stammering as a wind- (*vāta*) induced disease, see, for example, *Suśrutasaṃhitā* Cikitsāsthāna 1.85.

exactly was meant. However, in the eleventh century, the commentator Cakrapāṇidatta explained *vāksiddhi* as a special power through which "whatever is said, inevitably arises."[14] According to this interpretation, someone in possession of *vāksiddhi* would automatically speak the truth since any statement they made would consequently become true. In other words, not only would they speak the truth, they would produce truth, creating reality through the act of speaking.[15]

The idea that humans can gain access to special powers, usually through spiritual practices, is found in many sources in Indian literature and is also discussed in early medical literature. While *rasāyana* chapters in medical works describe the attainment of special powers through the intake of medicine, the *Carakasaṃhitā* also gives a short list of special powers attained through yogic practice in Śārīrasthāna 1.140–141:

> Entering someone's mind, knowledge of objects, free action, vision, hearing, recollection, beauty, and invisibility at will: these are called the eightfold lordly powers that yogins have. All that arises from meditation with a purified mind.[16]

Notably, *vāksiddhi* is not among this list, although one could perhaps make an argument that "free action" may describe a similar phenomenon. Vision, hearing, recollection, and beauty are central aims of *rasāyana* treatments, while invisibility does not feature as an effect of *rasāyana*. Thus, while there is some overlap, the extraordinary capabilities described in its *rasāyana* chapter do not fully correspond to the group of eight powers attained through yogic practice outlined in the *Carakasaṃhitā*'s Śārīrasthāna.

A complex picture of the variety of goals pursued with *rasāyana* treatment emerges when we look at the various *rasāyana* sections of the medical and alchemical works and examine what effects are attributed to their formula-

---

14 *vāksiddhiḥ yad ucyate tad avaśyaṃ bhavatīty arthaḥ*—Cakrapāṇidatta on *Carakasaṃhitā* Cikitsāsthāna 1.1.7–8.

15 See the discussion of this passage in Maas, "On the Meaning of *rasāyana*," 72–75. Also see Lloyd W. Pflueger, "Holding On and Letting Go: The In and Out of Power in Classical Yoga," in *Yoga Powers: Extraordinary Capacities Attained Through Meditation and Concentration*, ed. Knut Jacobsen (Leiden: Brill, 2011), 251 on the concept of the conformation of one's words with reality.

16 Dominik Wujastyk, "The Path to Liberation through Yogic Mindfulness in Early Ayurveda," in *Yoga in Practice*, ed. David Gordon White (Princeton: Princeton University Press, 2011), 40. This list of eight differs from the more common list of eight special powers of the *Pātañjalayogaśāstra*.

tions. The described effects of *rasāyana* medicines and procedures can be broadly divided into five, partly overlapping, thematic groups:[17]

1. Health: Medicines and procedures used for the prevention or cure of disease, and for the establishment of overall well-being, including the proper functioning of the body and senses.
2. Ageing: Medicines and procedures attributed with effecting rejuvenation, anti-ageing, or the increase of lifespan.
3. Cognitive power: Medicines and procedures that restore, establish or enhance memory power, memorization ability and intelligence.
4. Virility: Medicines and procedures which restore or enhance sexual stamina, bodily strength and fertility.
5. Special powers/extraordinary capacities/perfections (discussed below).

Rasāyana formulations most often unite a range of benefits from several, and in some cases all, thematic groups. However, there are many examples of *rasāyana* tonics that have a specific focus or a bias toward one group, such as the "*medhyarasāyana*," medicines for the improvement of cognitive faculties. It is also not always possible to neatly discern one group from the other. Health and ageing could both be considered the overarching category, since the occurrence of certain diseases, or the loss of cognitive or sensory powers or problems concerning sexual stamina may be understood in terms of health problems, but also in terms of ageing, i.e., health problems caused or exacerbated by the ageing process. These are also the dominant terms in the definitions of *rasāyana* in later medical literature, ambiguously presented in a compound (*jarāvyādhi*) that allows to understand them as the separate categories of "ageing and diseases," or as a combined category of "diseases caused by ageing."[18]

However, the group of special powers does not in all cases fit either category of health or ageing. I would like to suggest a division into two kinds of special powers. The first kind consists of powers unconnected to a medical framework of a patient's bodily and mental health and well-being, or the absence or pres-

17 See Dagmar Wujastyk, "Acts of Improvement" for a more detailed discussion of the applications of *rasāyana*. Also see Oliver Hellwig, "Rasāyana und die āyurvedische Krankheitskunde," *Traditional South Asian Medicine* 8 (2008): 32–64, which discusses the aims of *rasāyana* of treating "self-arisen" (*svabhāva*) diseases in the *Suśrutasaṃhitā*.

18 See *Siddhayoga* 69 (*rasāyanādhikāra*): *yaj jarāvyādhividhvaṃsi bheṣajam tad rasāyanam* / ("a *rasāyana* is a remedy that removes ageing and disease (or: that removes diseases [caused by] ageing)"). The same phrase appears in *Cakradatta* 66 (*rasāyanādhikāra*) 1; *Vaṅgasenasaṃhitā rasāyanādhikāra* 371; and *Bhaiṣajyaratnāvalī Uttarakhaṇḍa* 73 (*rasāyanaprakaraṇam*), 1. *Śārṅgadharasaṃhitā* 1.4.13 gives a similar definition: *rasāyanaṃ ca taj jñeyaṃ yaj jarāvyādhināśanaṃ* // ("*rasāyana* is known as that which removes disease caused by ageing").

ence of disease.[19] Rather, their function seems to be to manipulate the world in favor of the person using the *rasāyana*, in particular influencing the person's social standing and general fortune and therefore affecting not only the consumer, but also other persons or the environment. *Rasāyana* preparations associated with these kinds of powers address human vulnerability to malign forces (both human and non-human) and the evil intent of others.

The second kind of special powers does not so much stand outside a medical framework of health and well-being, but rather extends it by redefining what is possible in terms of perfecting the body and its functions. This kind of *rasāyana* addresses a wider spectrum of human corporeality and functioning.

3 Good Luck and Protection from Malign Forces

A number of *rasāyana* medicines are attributed with affording the consumer fame, success, wealth, charm, valor, good fortune, eloquence, or success in debates; securing the favor of the king or other powerful persons; subduing the will of others; preventing misfortune; protecting from the evil intentions and actions of others (in particular from witchcraft and damaging mantras), or from possession by evil spirits or bad dreams; and removing wrongdoing (*pāpman, pāpa*). We meet with these kinds of effects throughout the examined *rasāyana* chapters, though the frequency of their occurrence differs from work to work.

The *Suśrutasaṃhitā*'s presentation of *rasāyana* in Cikitsāsthāna 28 provides a number of recipes that are specially dedicated to this kind of effect:

> Gold, lotus seeds, honey, parched grain and beautyberry (*priyaṅgu*), taken with cow's milk, should ward off bad luck.[20]

> A decoction of the petals of the blue lotus boiled with cow's milk together with gold and with sesame is thought to destroy bad luck.[21]

---

19 The objective of Āyurveda is defined as freeing the diseased of disorders and protecting the health of the healthy in *Suśrutasaṃhitā* Sūtrasthāna 1.14. This is echoed in the beginning of the *Carakasaṃhitā*'s *rasāyana* chapter (Cikitsāsthāna 1.1.4), which states that healing (*bheṣaja*) is twofold, dividing it into what promotes strength in the healthy, and what eradicates disorders in the diseased.

20 *suvarṇaṃ padmabījāni madhu lājāḥ priyaṅgavaḥ / gavyena payasā pītam alakṣmīṃ pratiṣedhayet //* 14

21 *nīlotpaladalakvātho gavyena payasā śṛtaḥ / sasuvarṇas tilaiḥ sārdham alakṣmīnāśanaḥ smṛtaḥ //* 15

> Wild asparagus (*śatāvarī*) butter used properly every day
> together with honey and gold will place the king under one's power.[22]

> A compound of sandalwood, *mohanikā*, licorice, pyrites, honey
> and gold should be taken by one who desires good fortune.[23]

This kind of special effect is somewhat less developed in the *Carakasaṃhitā*. We find examples in two recipes:

> A man who takes an iron tonic for a year will become
> intelligent, famous, a master of speech,
> one who retains what he has heard, and very wealthy.[24]

> This powerful *aindra* (waterhyssop?) tonic is extremely energizing
> and neither enchantments, nor bad luck, poison or disease can withstand it.[25]

Similarly, such effects also play only a minor role in the *Aṣṭāṅgahṛdaysaṃhitā*, where they occur in only two (out of about sixty-seven) recipes in the *rasāyana* chapter. The first, a multiple-herb ghee called *brāhmyādighṛta* found in Uttarasthāna 39.50–53 promises the user a multitude of benefits, among them prosperity and a bright and beautiful appearance, as well as the cure of problems caused by spells. The latter are specified as "*atharvamantra*," i.e., spells

22 *śatāvarīghṛtaṃ samyagupayuktaṃ dine dine / sakṣaudraṃ sasuvarṇaṃ ca narendraṃ sthāpayed vaśe* // 21

23 *gocandanā mohanikā madhukaṃ mākṣikaṃ madhu / suvarṇam iti saṃyogaḥ peyaḥ saubhāgyam icchatā* // 22

I was not able to identify the plant "*mohanikā*" and the occurrence of this name in the *Suśrutasaṃhitā* was the only one I was able to find. The name translates as "the confusing".

24 *dhīmān yaśasvī vāksiddhiḥ śrutadhārī mahādhanaḥ / bhavet samāṃ prayuñjāno naro lauharasāyanam* // *Carakasaṃhitā* Cikitsāsthāna 1.3.22. The full recipe is found in verses 15–23.

25 *paramojaskaraṃ caitat siddham aindraṃ rasāyanaṃ / nainat prasahate kṛtyā nālakṣmīr na viṣaṃ na ruk* // *Carakasaṃhitā* Cikitsāsthāna 1.3.28. The full recipe is found in verses 24–29. Regarding *ojaskara*, "bestowing *ojas*," as "energizing": *Ojas* is a concept connected to the digestive process in Āyurvedic literature. It is the end product of the transformation process of one bodily element (*dhātu*) into the next. It is sometimes interpreted as something akin to life force, hence my rendering as "energizing". On the varying meanings of *ojas* in different Indic literatures over time, see Jan Gerrit Meulenbeld, "The Woes of *ojas* in the Modern World," in *Modern and Global Ayurveda: Pluralism and Paradigms*, eds. Dagmar Wujastyk and Frederick M. Smith (Albany: SUNY Press, 2008), 157–175.

associated with the *Atharvaveda*, the fourth of the Vedic liturgical scriptures. Presumably, this is a reference to mantras categorized in the *Atharvaveda* as "*aṅgirāmantra*" or "*āngiras*," i.e., mantras for harming others (*abhicāra*).[26] The second recipe is found in *Aṣṭāṅgahṛdaysaṃhitā* Uttarasthāna 39.169–173. It details the preparation of a medicated ghee called "*nārasiṃha*" ("associated with Narasiṃha," i.e., the man-lion incarnation of the god Viṣṇu), and lists becoming wealthy (*śrīmat*) and devoid of wrongdoing (*nirdhūtapāpman*) among the medicine's many effects.

The *Aṣṭāṅgasaṃgraha*'s *rasāyana* chapter, which presents about 183 recipes, is much longer than the *Aṣṭāṅgahṛdayasaṃhitā*'s. Correspondingly, it contains more recipes for the attainment of special powers, though the majority of its recipes are focused on eradicating disease, supporting cognitive power and procuring longevity. The chapter includes variations of a number of the recipes from the *Suśrutasaṃhitā* presented above: A recipe of gold, lotus seeds, honey, grains and beautyberry against bad luck;[27] a compound of gold, wood-apple, sweet flag, and clarified butter for prosperity (and beauty, health, longevity, and nourishment);[28] a preparation of wild asparagus and clarified butter with gold and honey to subjugate the king.[29] The *Aṣṭāṅgasaṃgraha* also reiterates recipes associated with special powers from the *Aṣṭāṅgahṛdayasaṃhitā*: the recipe for *brāhmyādighṛta*,[30] and the one for *nārasiṃhaghṛta*.[31] It further contains a description of the characteristics of bakuchi (*bākucī*) as calming, bestowing wealth, and dispelling poisons, evil spirits, misfortune, and wrongdoing.[32] A recipe for sweet flag ghee promises success in all activities;[33] one for elephant creeper root ghee protection from sickness, possession by spirits, wrongdoing, and bad luck.[34] A recipe for *śivā guṭikā* ("auspicious pill") is described in much detail. Based on rock resin or bitumen (*śilājatu*), this pill is presented as an effective wide spectrum medicine (forty-seven disease groups

26 On this classification of mantras for harming others in the *Atharvaveda* and its ancillary texts, see Teun Goudriaan, *Māyā Divine and Human* (Delhi, Varanasi, Patna: Motilal Banarsidass, 1978): 221.

27 *Aṣṭāṅgasaṃgraha* Uttarasthāna 49.57.

28 *Aṣṭāṅgasaṃgraha* Uttarasthāna 49.60.

29 *Aṣṭāṅgasaṃgraha* Uttarasthāna 49.61.

30 *Aṣṭāṅgasaṃgraha* Uttarasthāna 49.69–72 and *Aṣṭāṅgahṛdayasaṃhitā* Uttarasthāna 39.50–53.

31 *Aṣṭāṅgasaṃgraha* Uttarasthāna 49.394–398 and *Aṣṭāṅgahṛdayasaṃhitā* Uttarasthāna 39.169–173.

32 *Aṣṭāṅgasaṃgraha* Uttarasthāna 49.155 a: *iti śamanī dhanyeyaṃ viṣagarabhūtopasargapāpmaghnī* |

33 *Aṣṭāṅgasaṃgraha* Uttarasthāna 49.159: *sarvārtheṣu siddhimān* (...) *bhavati*.

34 *Aṣṭāṅgasaṃgraha* Uttarasthāna 49.343–346.

are listed), and an antidote to all sorts of poisons (animal, herbal, and mineral). It is also attributed with protecting the consumer from malevolent spells, bad dreams, evil spirits, evil, and misfortune, as well as with bestowing wealth, beauty, fame, and good fortune. Finally, it is also meant to procure the favor of the king; and to make a person win debates.[35] This formula remained popular in the following centuries: it reappears a number of times in later works in *rasāyana* chapters.[36]

The Bower Manuscript's *rasāyana* chapter mostly describes formulations with medicinal functions, i.e., for general health and against specific disease groups. However, its version of the widely adopted "increasing long pepper" (*pippalīvardhamāna*) recipe is ascribed with effecting good fortune (*lakṣmī*), a growth of luck (*śrī*) and general success (*siddhi*).[37]

The *Siddhayoga*'s *rasāyana* chapter contains only one recipe for a medicine with extra-medical special effects, namely a *rasāyana* based on puncture vine (*gokṣura*) that is supposed to make the consumer powerful and fortunate (as well as beautiful, long-lived, and sexually active).[38] In the *Cakradatta*, extra-medical special effects are noticeably absent, the one exception being the inclusion of a recipe for *śivā guṭikā* (spelled "*guḍikā*" in this version), which lists the same effects as the older version from the *Aṣṭāṅgasaṃgraha*.[39]

The *Vaṅgasenasaṃhitā*'s *rasāyana* chapter is structured very differently than those of all the other works examined here. It includes materials that seem to be interpolations rather than integral parts of the subject of *rasāyana*. For example, the chapter begins with recipes for various kinds of vinegars and medicated fats not found in any other *rasāyana* chapter. It also contains a discussion of *marmans*, i.e., vital points in the body,[40] a description of a method of treatment with a head-pouch (*śirovasti*)[41] and lists of diseases associated with particular humoral substances.[42] The other parts have a strong focus on the medical effects of its *rasāyana* formulations: about half of its recipes are for medicines aimed at specific disease groups. Otherwise, the focus is on longevity and

35 *Aṣṭāṅgasaṃgraha* Uttarasthāna 49.309–325.

36 Recipes for *śivā guṭikā* are found in *Cakradatta* 66. 172–193, *Vaṅgasenasaṃhitā rasāyanādhikāra* 14–30, and *Bhaiṣajyaratnāvalī Uttarakhaṇḍa* 73.151–175.

37 See Bower Manuscript 2.6.4 and 22.

38 *Siddhayoga* 69.48–49.

39 See *Cakradatta* 66. 172–193.

40 *Vaṅgasenasaṃhitā rasāyanādhikāra* 332–350.

41 *Vaṅgasenasaṃhitā rasāyanādhikāra* 331.

42 *Vaṅgasenasaṃhitā rasāyanādhikāra* 351–360 diseases associated with the humor wind (*vāta*), 361–365 diseases associated with the humor bile (*pitta*), 366–369 diseases associated with the humor phlegm (*kapha*).

rejuvenation and the enhancement of cognitive powers. Nevertheless, some recipes do list special "good luck" powers among their effects. The *Vaṅgasenasaṃhitā* includes the recipe for *śivā guṭikā*, again with the same expected effects of protecting the consumer from malevolent spells, bad dreams, evil spirits, evil, and misfortune; as well as with bestowing wealth, beauty, fame, and good fortune.[43] Its flagship recipe, a formulation based on mica (*abhraka*),[44] offers a multitude of benefits, including becoming brave (*śūra*) and a good orator (*vāgmin*), and enjoying the fruits of the aims of human life (*trivargaphalabhājana*),[45] i.e., experiencing success in all spheres of life.

Neither the *Bhāvaprakāśa*, nor the *Yogaratnākara* include such special effects in their descriptions of the effects of *rasāyana*. Finally, the *Bhaiṣajyaratnāvalī* reintroduces the theme by including a recipe for *śivā guḍikā* (151–175) with the same effects as described in the older works. It also describes a recipe for a pill called *śrīsiddha* (44–53), which it attributes with letting the consumer attain the eight special powers (*siddhi*).[46] These are presumably the classic eight yoga powers detailed in the *Pātañjalayogaśāstra*, though the author may have also had alternative lists of *siddhi*s in mind, such as those described in the *Carakasaṃhitā*.

4 Extended Capacities, Bodily Perfection

The other kind of special effects of *rasāyana* medicines and therapies aim at extending human capacities and attaining a god-like perfection of the body, mind, and senses. This includes the attainment of extreme longevity, superhuman strength and beauty, complete immunity of the body, and extraordinary eyesight.

A passage from the *Suśrutasaṃhitā*, which describes the use of Soma,[47] may serve as an example of these kinds of special effects of *rasāyana* medicines:

---

43 *Vaṅgasenasaṃhitā rasāyanādhikāra* 15–30.

44 *Vaṅgasenasaṃhitā rasāyanādhikāra* 136–166.

45 The three aims in human life (*trivarga*) are *artha*, *kāma* and *dharma*, i.e., wealth; physical pleasure and love; and virtue.

46 See *Bhaiṣajyaratnāvalī Uttarakhaṇḍa* 73.151–175, and 44–53, respectively.

47 Soma is frequently mentioned in the *Ṛgveda* (one of the ancient Indian liturgical texts), with many hymns praising its energizing qualities. It is not clear, however, in what relation the Soma of the medical works stands to that of Vedic literature. On the identification of the Soma plant, see Harry Falk, "Soma I and II," *Bulletin of the School of Oriental and African Studies* 52 (1989): 77–90. On Soma in the *Suśrutasaṃhitā*'s *rasāyana* chapter, see Dagmar Wujastyk, "On Perfecting the Body," 62–63.

> The visionary man who makes use of the king of plants, Soma, wears a new body for ten thousand years. Neither fire, nor water, neither poison, blade nor projectile are powerful enough to take his life. He gains the strength of a thousand well-bred, sixty-year-old, rutting elephants. If he wants to go to the lands of northern legend, to the milky ocean, or even to the abode of the king of the gods, nothing can stand in his way. He is as beautiful as the god of love, as attractive as the second moon. He is radiant and brings joy to the hearts of all creatures. He truly knows all sacred knowledge, with all its branches and sub-branches. He moves like a god through the whole world, with infallible power.[48]

This potion promises all: extreme longevity; immunity, indeed indestructibility, of the body; superhuman strength (including virile power) and beauty; a transcendence of bodily constraints in the ability to go wherever one wishes; and a complete grasp of all sacred knowledge. In other words, the intake of Soma lets its consumer attain a god-like state.

Compare this with the more moderate effects of another *rasāyana* recipe from the *Suśrutasaṃhitā*:

> One whose impurities have been removed should enter the hut,[49] and should consume a lump of sweetflag about the size of an emblic myrobalan after it has been sanctified by oblations together with milk. Once this has been digested, milk, clarified butter, and boiled rice serve as food. Having done this for twelve days, his hearing becomes unimpeded, after using it for twice the amount of time, he becomes someone who has good memory, after using it for thrice the amount of time, it lets him retain what he hears, using it for four times twelve days, he transcends all sins, attains eagle[50] vision and a lifespan of one hundred years.[51]

Barring the transcendence of all sins and perhaps the attainment of eagle vision (unless this simply means very good vision and not literally vision like that of the divine Garuḍa), the effects of this *rasāyana* substance fall well

48 Translation of *Suśrutasaṃhitā* Cikitsāsthāna 29, 14–19 by Dominik Wujastyk, *The Roots of Ayurveda* (London: Penguin Books, 2003), 130.

49 On the use of special buildings as part of *rasāyana* therapy, see Arion Roşu, "Considerations sur une technique du *rasāyana* āyurvedique," *Indo-Iranian Journal* 17 (1975): 1–29 and 395 and Dagmar Wujastyk, "On Perfecting the Body."

50 *Tārkṣya* is a synonym of the legendary bird Garuḍa, the mount of the god Viṣṇu, who was attributed with extremely good vision.

51 *Suśrutasaṃhitā* Cikitsāsthāna 28.7.

within the range of ordinary expectations of medicine, i.e., a general improvement of the senses (in this case hearing and eyesight) and cognitive power (memory and memorization ability) and a good life expectation.

I would like to suggest that these two recipes and their effects describe different points on a spectrum of the human condition that spans a range of conditions: Bad health, deficient cognitive and sensory function, and early death on the one end; good health, satisfactory cognitive and sensory function, and normal life expectancy in the middle; perfect health, outstanding cognitive and sensory powers, and extreme longevity or even immortality on the far end of the spectrum.[52] The second recipe is mostly aimed at achieving the middle conditions, the first the conditions at the far end of the spectrum.

The perfection of the body to a god-like state and the extreme extension of lifespan are central themes for these kinds of special powers at the far end of the spectrum.

5 Bodily Perfection: When the Body Becomes like a Stone

Many *rasāyana*s are described as effecting either a return to youthfulness or a halting of the ageing process. Youthfulness is associated with bodily strength, proper functioning of the senses (especially eyesight), and beauty. The more superlative descriptions promise a bodily perfection in which the body is beautiful and strong, but also immune to deterioration and impervious to damage.

The stability of the body, described with terms such as *sthira* ("firm") or *dṛḍha* ("hard"), or by likening the body to hard materials, is frequently mentioned. For example, *Carakasaṃhitā* Cikitsāsthāna 1.1.77 asserts that the body will become firm as a stone (*gātram aśmavat sthirībhavati*), while later medical and also alchemical literature refers to a "diamond body" or "adamantine body" (*vajrakāya*).[53] This is a concept well-known from Mahāyāna scriptures, where it is associated with the extreme longevity or even absolute permanence, freedom from sickness, and purity of a Buddha's body.[54] In medical and alchemical

52 I would like to thank Dominik Wujastyk for suggesting the idea of a spectrum of the human condition to me.

53 For example, in *Vaṅgasenasaṃhitā rasāyanādhikāra* 100 (full recipe: 93–100). For instances of the term in Sanskrit alchemical literature, see *Rasaratnākara Rasendrakhaṇḍa* 4. 19 and 86 and *Rasaratnākara Rasāyanakhaṇḍa* 2.11, 2.42, 3.214, 4.14 and 4.47, and in *Rasārṇava* 18.185.

54 On the changing concepts of the *vajrakāya* in Buddhist literature, see Michael Radich, "Immortal Buddhas and Their Indestructible Embodiments," *Journal of the International Association of Buddhist Studies* 34. no. 1–2 (2011(2012)): 227–290.

literature, it is used for its imagery of hardness, the idea of both impenetrability and unbreakableness. We encounter this idea early on, as seen in the example of the *Suśrutasaṃhitā*'s description of the effects of Soma, namely that neither fire, nor water, poison, or any weapons can damage the transformed body; and in an iron preparation in the *Aṣṭāṅgasaṃgraha* that lets the consumer "overcome all troubles and become unassailable by disease, ageing, death, weapons, fire, water, and poison."[55]

The idea of perfecting the body is, however, also connected with that of rebuilding the body. And this necessitates its prior disintegration. The fullest expression of this process is again found in the *Suśrutasaṃhitā*'s description of *rasāyana* therapy with Soma: a long process, in which the body progressively disintegrates, shedding impurities and old layers, to then begin its renewal and gradual transformation into god-like perfection.[56] The *Carakasaṃhitā* describes comparable effects for a treatment which involves sealing the patient's body in a covered wooden barrel post-*rasāyana*-intake. The body would gradually dissolve and then regenerate after six months, having become similar to the gods in appearance, knowledge and powers.[57]

The theme of disintegration and regeneration is also present in another context, namely in the *rasāyana* treatment of *kuṣṭha*, i.e., a group of skin diseases and particularly a disease with the symptoms of what now might be diagnosed as leprosy.[58] As mentioned, many *rasāyana* formulations are described as cures to various diseases, especially serious ones considered particularly difficult or even impossible to treat. *Kuṣṭha* is one of them, and indeed probably the most widely mentioned disease in *rasāyana* chapters. Its symptoms, the falling off of skin and extremities, parallel the body's disintegration described in the *Suśrutasaṃhitā*'s and *Carakasaṃhitā*'s extreme *rasāyana* therapies. The medical texts promise a complete cure of *kuṣṭha* through *rasāyana*, but more than that, they describe a reconstitution of the body reminiscent of the regeneration of

55 *Aṣṭāṅgasaṃgraha* Uttarasthāna 49. 53.

56 See *Suśrutasaṃhita* Cikitsāsthāna 29.10–12. See also the translation of the passage in Dominik Wujastyk, *Roots*, 126–129 and a discussion of it in Roşu, "Considerations," 13–15.

57 See *Carakasamhitā* Cikitsāsthāna 1.4.7. The theme of shedding layers of the body for a non-ageing and long-lived body is also found in Sanskrit alchemical literature, as for example in *Rasaratnākara rasāyanakhaṇḍa* 4.59, which reads: *jīved brahmadinaṃ yāvat sarpavat kañcukaṃ tjayet* // ("He would live a day of Brahma while shedding his layers like a snake").

58 *Kuṣṭha* can describe a group of diseases characterized by symptoms primarily showing on the skin. Here, the symptoms described seem to point to leprosy more specifically. However, in modern medical terms, leprosy is a long-term infection by particular bacteria. Āyurvedic works do not have a concept of infection by bacteria and offer a different model of how the different kinds of *kuṣṭha* develop.

the disintegrated body after Soma or barrel therapy, with the nose, ears, fingers and indeed any limbs that have fallen off growing back again.[59] The cure of *kuṣṭha* is already a bold claim; the *rasāyana*'s effect of giving someone the power to regenerate the body, however, is extraordinary. It confirms the image of *rasāyana* medicines and therapy as effective in otherwise hopeless cases, a cure that can cheat death and overcome the ravages of time and disease.

6 Longevity

Suśrutasaṃhitā Sūtrasthāna 35.7–11 describes a lifespan of seventy years as a medium life expectation and twenty-five years as a short lifespan.[60] A long life would necessarily go beyond a medium life expectation of seventy years, though we are not told by how much. The promise of reaching one hundred years in a healthy condition occurs frequently in *rasāyana* chapters and can perhaps be understood as a natural, or normal limit. However, some formulations are credited with providing even longer spans of life. In the *Suśrutasaṃhitā*'s *rasāyana* chapter, we find recipes that are described as prolonging life up to 300, 500, 800, 2000, and 10000 years. Such claims are found throughout the examined medical works. However, in the *Carakasaṃhitā*'s narrative, such extended lifespans are usually associated with a deep past and a particular demographic, namely Vedic sages. These are presented as the original recipients of Āyurvedic knowledge and the first consumers of *rasāyana* medicines.[61] Rather than stating that a particular *rasāyana* will let someone live for several centuries, the treatise narrates how a formulation affected sages in the past: Sage Cyavana became young and virile again, the Vaikhānasa and Bālakhilya sages attained an unlimited lifespan.[62] As prototypical consumers of *rasāyanas*, the sages also set the standard for future recipients of *rasāyana*

59 See, for example, *Aṣṭāṅgahṛdayasaṃhitā* 39.109–110, which describes the application of a *somarājī* seed medicine for growing new limbs in the case of *kuṣṭha*.

60 See Christèle Barois, "Stretching Out Life, Maintaining the Body: Part 1—Vayas in Medical Literature," *History of Science in South Asia Special Issue "Rejuvenation, Longevity, Immortality in South Asian Medical, Alchemical and Yoga Traditions"* 5, no. 2 (2017): 37–65 on different presentations of normal lifespan in Sanskrit medical literature.

61 For example, see *Carakasaṃhitā* Cikitsāsthāna 1.1.75, which states that the sages lived for countless hundreds of years (*anekavarṣaśatajīvano*) after taking *rasāyana* medicines, or Cikitsāsthāna 1.1.54 and Cikitsāsthāna 1.2.21 which relates that sages attained an unlimited lifespan (*amitāyus*) through *rasāyana*.

62 See the recipe for *cyavanaprāśa* in *Carakasaṃhitā* Cikitsāsthāna 1.1.62–74 and for *brahmarasāyana* in Cikitsāsthāna 1.1. 41–57.

therapy, who, according to *Carakasaṃhitā* Cikitsāsthāna 1.4.30–35, need to be self-restrained, of impeccable moral conduct, and intent on spiritual practices. Those who do not fulfil these requirements would not be able to benefit from treatment. This general idea also appears in the *Suśrutasaṃhitā* (Cikitsāsthāna 28.32, 30.4 and 30.28–29), where the crucial ingredients of Soma or a divine herb can only be found and used by worthy persons. Nevertheless, the treatises do allow for the possibility that anyone following certain behaviors can reap the same benefits of *rasāyana* treatment the sages purportedly enjoyed. The *Carakasaṃhitā* (Cikitsāsthāna 1.1.78–80) states:

> The application of *rasāyana* was to the great sages of old as nectar was to the gods and ambrosia to the serpent-deities.
>
> In old times, those occupied with *rasāyana* experienced neither ageing nor weakness, illness or death for thousands of years.
>
> One who undertakes *rasāyana* according to rule does not merely attain a long lifespan in this world. He follows the auspicious path pursued by the divine sages and attains the eternal Absolute.[63]

The last verse, which is repeated in several later medical works,[64] seems to promise a kind of immortality, depending on how we interpret *brahma tathaiti cākṣayam*, here rendered as "attains the eternal Absolute." Cakrapāṇidatta's eleventh-century commentary on *Carakasaṃhitā* Cikitsāsthāna 1.1.78–80 glosses *brahma* with *mokṣa*, i.e., liberation from the cycle of rebirths.[65] This suggests immortality outside an embodied existence.

But what about embodied immortality? Many medical works allow for longevity beyond the paradigmatic one hundred years as an outcome of *rasāyana* therapy,[66] and some formulations are described as increasing lifespan accord-

---

63 *yathāmarāṇām amṛtaṃ yathā bhogavatāṃ sudhā /*
tathābhavan maharṣīṇāṃ rasāyanavidhiḥ purā // 78
na jarāṃ na ca daurbalyaṃ nāturyaṃ nidhanaṃ na ca /
jagmur varṣasahasrāṇi rasāyanaparāḥ purā // 79
na kevalaṃ dīrgham ihāyur aśnute rasāyanaṃ yo vidhivan niṣevate /
gatiṃ sa devarṣiniṣevitāṃ śubhāṃ prapadyate brahma tathaiti cākṣayam // 80

64 See *Siddhayoga* 69.72, *Vaṅgasenasaṃhitā rasāyanādhikāra* 510, *Yogaratnākara* 76.37, and *Bhāvaprakāśa Uttarakhaṇḍa* 2.15.

65 For the text of Cakrapāṇidatta's commentary of the *Carakasaṃhitā*, see the edition of the *Carakasaṃhitā* by Trikamji Āchārya.

66 For example, in the *Aṣṭāṅgahṛdayasaṃhitā*, one recipe is for 200 years of life, two recipes

ing to how long they are taken. For example, the *Suśrutasaṃhitā* claims that each month of intake of a false black pepper (*viḍaṅga*) compound increases lifespan by a hundred years.[67] The same concept is found in the *Carakasamhitā* (Cikitsāsthāna 1.2.13) where the number of emblic myrobalan fruits eaten equals the number of years added to one's life. Theoretically, that should mean that lifespan could be extended indefinitely through a continuous intake of *rasāyanas*.

Still, it should be noted that promises of bodily immortality are unusual and occur only in a few instances, though we find the idea of keeping death at bay or preventing an untimely death a number of times. Sometimes, it is not entirely clear whether formulations are meant to bring about a rescue from death at a particular time, or whether the effect is meant to be permanent. The *Aṣṭāṅgasaṃgraha* gives a recipe for an iron preparation through which the consumer would "overcome all troubles and become unassailable by disease, ageing, death, weapons, fire, water, and poison."[68] And some four hundred years later, the *Vaṅgasenasaṃhitā*, which shows influence from alchemical literature in a number of its *rasāyana* formulations, describes a multi-herb compound as providing the consumer with a body that is free from ageing, disease and death (*jarārujāmṛtyuvimuktadeha*).[69] A further recipe for a compound containing iron and guggul uses the phrase *yāti amaratvam*—"he reaches immortality"—which seems fairly unambiguous.[70] Note, however, Olivelle on Thieme's 1968 study of the related term *amṛta* and the verbal root *mṛ* in Vedic literature: "The term *amṛta* does not always mean immortal in the sense we usually attach to it; it often means vitality or vital energy (Thieme's Lebenskraft)—it is a full and prosperous life and all things that sustain and promote such a life, including food, drink, cattle, and medicine. *Amṛta* can thus indicate both life/immortal-

---

for 300 years, one recipe for 1000 years, one recipe for 500 years, and one recipe for a double life span. In the *Siddhayoga*, we find recipes promising 500 and 1000 years of life.

67 See the recipe in *Suśrutasaṃhitā* Cikitsāsthāna 27.7.

68 *tataḥ sarvopatāpān jitvā vyādhijarāmaraṇaśastrāgnitoyaviṣāṇām agamyaḥ* (…) //53// The full recipe is found in *Aṣṭāṅgasaṃgraha* Uttarasthāna 49. 53–54.

69 *Vaṅgasenasaṃhitā rasāyanādhikāra* 396. The same recipe is found in *Yogaratnākara* 76.13–14 and in *Bhāvaprakāśa Uttarakhaṇḍa* 2.11–12. See also *Vaṅgasenasaṃhitā rasāyanādhikāra* 116, where a mixture of sulphur, the three myrobalans and false daisy (*bhṛṅgarāja*) is attributed with removing death and ageing (*mṛtyujarāpaha*); or *rasāyanādhikāra* 119, which describes a sulphur and iron preparation as letting someone live to the end of the moon and sun (*jīved ācandratārakam*).

70 See variants of the recipe in *Vaṅgasenasaṃhitā rasāyanādhikara* 423, *Yogaratnākara* 76.27 and *Bhāvaprakāśa Uttarakhaṇḍa* 2.14.

ity, as well as instruments that sustain life and ward off death."[71] Perhaps we can apply similar meanings to the related terms *mṛtyu* and *amara*. Correspondingly, the "body that is free from death" (… *mṛtyuvimuktadeha*), in the example above, then might be understood as a body full of vitality, rather than one that is free from death forever, while to attain *amaratvam* might be understood in terms of a gain in vitality instead of immortality.

In the early medical works up to the *Vaṅgasenasaṃhitā*, bodily immortality is typically not presented as an outcome of medical intervention. The author of the thirteenth/fourteenth-century medical treatise *Śārṅgadharasaṃhitā* (1.5 .50cd-52ab) states this explicitly:

> The conjunction of body and breath is called life.
> Their separation through time is called death by the wise.
>
> No living being born anywhere on earth is immortal.
> Hence, death is inevitable. However, one should ward off disease.[72]

Importantly, the author gives a reason why health should be pursued:

> Given that the body is an instrument for accomplishing righteousness, wealth, pleasure, and liberation, a man who understands the consequences of past actions should protect the body from diseases.[73]

Thus, the body is used for the pursuit of this-worldly aims but is also a crucial means for the attainment of religious merit and the final goal of liberation from the cycle of rebirths.[74]

---

71 Patrick Olivelle, "Amṛtā: Women and Indian Technologies of Immortality." *Journal of Indian Philosophy* 25 (1997): 428.

72 *śarīraprāṇāyor eva saṃyogād āyur ucyate* // 50cd
kāḷena tadviyogāc ca pañcatvaṃ kathyate budhaiḥ /
na jantuḥ kaścid amaraḥ pṛthvyāṃ jāyate kvacit // 51
ato mṛtyur avāryaḥ syāt kiṃtu rogān nivārayet /

73 *dharmārthakāmamokṣaṇāṃ śarīraṃ sādhanaṃ yataḥ* / *ato rugbhyas tanuṃ rakṣen naraḥ karmavipākavid* //53//

74 See also *Carakasaṃhitā* Sūtrasthāna 1. 6, where disease is described as an impediment to religious observances, such as austerities (*tapas*), abstinence (*upavāsa*), study (*adhyayana*), celibacy (*brahmacarya*), and vows (*vrata*), as well as to lifespan. See also *Carakasaṃhitā* Sūtrasthāna 1. 15, which states that freedom from disease is the root cause of righteousness, wealth, pleasure, and liberation—*dharmārthakāmamokṣāṇāṃ ārogyaṃ mūlam uttaman*. Notably, none of the *rasāyanas* or other medicines in the *Śāṛṅgadharasṃhitā* are described as especially extending lifespan, so that the suggestion seems to be that the attainment of liberation (*mokṣa*) is possible within a regular lifespan.

7 Concluding Reflections

Special powers as effects of *rasāyana* medicines are found throughout the medical treatises. They are an integral, though minor part of *rasāyana*.

Many of the special effects of *rasāyana* medicines concern averting misfortune and malevolent acts. *Rasāyanas* here function as shields against attacks, rendering them ineffectual. Notably, the possibility of (counter-)attacks is missing: the purpose of *rasāyanas* is defensive, not offensive. The other function of these *rasāyanas* is to bring about positive effects, often directed at improving social standing, such as fame, wealth, and success in one's endeavors. Here, *rasāyanas* may not only affect the consumers, but also their social environment: the powers bestowed on the consumers determine society's reaction to them. The first function of these *rasāyanas* points to an essentially hostile environment in which all sorts of calamities may befall a person, many of them caused by the malign actions of others. The second seems to reflect an optimism about the possibility to nevertheless live a happy life in favorable conditions. These effects of *rasāyana* medicines are reminiscent of some of the effects attributed to mantras and rituals of the *Atharvaveda* and its ancillary texts. The difference is that many of the *Atharvaveda*'s spells are dedicated to damaging one's enemy—these kinds of spells are alluded to in the *Aṣṭāṅgahṛdayasaṃhitā* and *Aṣṭāṅgasaṃgraha*, as we have seen, but only as something the *rasāyana* will guard against, not as something a *rasāyana* would effect. It should also be mentioned that while mantras are sometimes cited in *rasāyana* chapters, typically to accompany a *rasāyana*'s intake, they are not specifically associated with special powers for good luck or protection from malign forces.[75]

75 The exceptions to this are *Suśrutasaṃhitā* Cikitsāsthāna 28.25, which states that the Gāyatrī mantra should be spoken during the making of the formulations that confer such special powers; and Cikitsāsthāna 30.27, which prescribes the incantation of a mantra during the harvest of divine herbs (*divyauṣadhi*). Otherwise, mantras occur in *Suśrutasaṃhitā* Cikitsāsthāna 28.25, *Aṣṭāṅgahṛdayasaṃhitā* Uttarasthāna 39.89, and *Aṣṭāṅgasaṃgraha* Uttarasthāna 49.83 in the context of chaulmoogra oil (*tuvaraka*), an oil used against skin diseases (*kuṣṭha*); in *Aṣṭāṅgasaṃgraha* Uttarasthāna 49. 262–263 in the context of taking guggul against diseases caused by the humors wind and phlegm; in the *Siddhayoga* 69.31 for the intake of flame of the forest (*hastikarṇa*) and 69.41 for taking marking nut (*bhallātaka*) for general good health and enhancing cognitive power; in the *Cakradatta* 66.100–106 in the context of taking the iron tonic called *amṛtasāralauha* for a beautiful, stable, and healthy body, and in 66.19–21 for the intake of flame of the forest (*hastikarṇa*) for longevity; in the *Vaṅgasenasaṃhitā*'s *rasāyana* chapter, verse 120, for the use of a sulphur and iron compound which would confer divine sight and longevity on the consumer. Mantras are not found in the *rasāyana* chapters of the *Carakasaṃhitā*, the *Bhāvaprakāśa*, or the *Yogaratnākara*.

The other kind of effects of *rasāyana*s are more easily integrated into a medical framework of physical health as an extension of normal human corporeality and functioning. Here, bad eyesight becomes good vision and then divine vision; a breaking voice becomes smooth and powerful, turning into an instrument for eloquence and oration skills, which in turn culminate in "mastery of speech" (*vāksiddhi*); an untimely death is averted and replaced by the appropriate life-span of a healthy human being, or extended to an extraordinary length, or expanded indefinitely. The ultimate promise of *rasāyana* is the perfection of the body that will allow a sustained pursuit of both worldly pleasures and transcendent goals.

It is tempting for a modern reader to explain the inclusion of special powers as an effect of *rasāyana* as hyperbole: exaggerations perhaps used to claim an elevated position for *rasāyana* within medicine. Accordingly, 1000 years of lifespan would just mean "a long life;" divine sight 20/20 vision; an adamantine body good immunity; and the cure of leprosy and regeneration of lost limbs an alleviation of symptoms. This is certainly how modern proponents of Āyurveda tend to frame the workings of *rasāyana* medicines. That is, special powers or extreme effects are simply left unmentioned, played down or dismissed. We can see this mindset at work in translations, but also in the careful wording and emphasis on health benefits in research papers and other publications on *rasāyana* formulations.[76] This is not surprising, given that modern practitioners of Āyurveda have long been trained to work within the modern biomedical idiom.

Examining the texts as historical sources, however, we can note that the special powers, while not especially given prominence, form part and parcel of the medical writers' representations of *rasāyana*. Summarizing research findings on yogic powers in different Indic literatures, Jacobsen notes that "(t)he traditions assumed that the powers could be explained rationally, neither considered unreal, nor denied as irrational."[77] This is also the impression I gain from the medical texts, though there is a certain ambiguity in some expressions

76 An online search for the term *rasāyana* brings up many research publications on specific *rasāyana* formulations or single drugs. These are all focused on their health benefits and do not mention special powers as possible outcomes of *rasāyana* therapy. See also the representations of *rasāyana* in Chittabrata Palit and Nupur Dasgupta, *Ancient Indian System of Rasayana Suvarnatantra. A Treatise on Alchemy/Rasayana* (Delhi: Kalpaz Publications, 2009), S. Mahdihassan, *Indian Alchemy or Rasayana in The Light of Ascetiscism and Geriatrics* (Delhi: Motilal Banarsidass, 1991), Ram Karan Sharma, *Rejuvenative Health Care in Āyurveda* (Delhi: Pratibha Prakashan, 1996), and Ajay Kumar Sharma, *Elements of rasāyana Therapy in Ayurveda* (Delhi: Sri Satguru Books, 2005).

77 Jacobsen, *Yoga Powers*, 15.

that allow for either a more mundane or a supramundane interpretation. The boundaries between normal and extraordinary are fluid and are not marked in the texts: the cure of cough can be juxtaposed with the extension of life for eight hundred years. Where do we draw the line? Do we accept claims of efficacy in Āyurvedic medicines? And if so, what is the limit to this acceptance? How do we distinguish between reasonable and unreasonable claims, between normal and extraordinary? The medical writers do set certain limits, as we have seen, for example, in the descriptions of normal lifespan. These limits are then, however, stretched in the *rasāyana* chapters. This may reflect the juxtaposition of materials collected from different sources. After all, the medical works are collections, and the doctrine of *rasāyana* may have been based on a different set of rules than other branches of medicine.

In his pioneering book, *Asceticism and Healing in Ancient India*, Kenneth Zysk argued that the early medical works display "empirico-rational" and "magico-religious" elements that reflect the different cultural milieus in which medicine was developed. According to Zysk, the empirical and rational theories of medicine (especially humoral theory) originated in ascetic milieus, while the magico-religious elements were part of a "brahmanic veneer" applied while adapting medical theory and practice to brahmanic culture. The division between "empirico-rational" (as indicative of modern ideas of science) and "magico-religious" was critiqued by Engler on the grounds that "Ayurveda did not manifest standard criteria of "science" (e.g., materialism, empirical observation, experimentation, falsification, quantification, or a developed conception of proof) and that Vedic aspects of the classical texts are too central to be considered inauthentic or marginal."[78] More recently, Slouber (2016) has shown how certain religious and magical elements found in the texts follow a systematic logic and should therefore not be categorized as irrational. Nevertheless, the difference in approaches in the medical texts is fairly clear and indeed is explicitly referenced in the source texts themselves. For example, the *Carakasaṃhitā* divides therapy into three categories: 1. that which depends on the sacred (*daivavyapāśraya*); 2. that which depends on reasoning (*yuktivyapāśraya*); and 3. that which triumphs through good character (*sattvāvajaya*). The first category includes "the use of mantras, herbs, jewels, good-luck ceremonies, food offerings to the gods, presents to the gods, fire oblations, pledges, penances, fasts, benedictions, prostrations, and pilgrimage."[79] The sec-

78 Steven Engler, "'Science' vs. 'Religion' in Classical Ayurveda," *Numen* 50 (2003): 416.

79 *Carakasaṃhitā* Sūtrasthāna 11.54. Translation by Dominik Wujastyk, *Roots*, 33.

ond employs diet and medicines. For the third, the mind is turned away from detrimental thought.[80] *Rasāyana* encompasses all three approaches.

I have avoided the use of the words "supernatural" or "magical" to describe the effects of *rasāyana* medicines, since the way in which medical writers incorporate special powers points to their seeing them as part of how the world functions, not as a disruption of any natural laws. Even the words "special" or "extraordinary" must be used with some caution, since they depend on a specific understanding of what is normal, or common. In his study on yoga powers in *Śaiva* tantric traditions, Somadeva Vasudeva suggests the use of "supramundane abilities" for the word *siddhi* to reflect the idea that these abilities are not common for ordinary humans, though they are for accomplished yogins.[81] Perhaps we can similarly understand the special effects of *rasāyana* as something that is not commonly available to all, but can be attained under special circumstances by certain kinds of persons, letting them achieve the "normal" state of higher beings, such as seers, sages, and gods. In the world view of the medical authors, these effects of *rasāyana*s are then only special or extraordinary in that they require special or extraordinary effort.

Acknowledgements

This article was made possible through funding from the European Union's Horizon 2020 research and innovation program under grant agreement No. 639363.

References

Primary Sources

[*Aṣṭāṅgahṛdayasaṃhitā*:] *Vāgbhaṭa's Aṣṭāṅga Hṛdayam: Text, English Translation, Notes, Appendix and Indices*, translated by K.R. Srikantha Murthy. 3 vols. Fourth edition. Varanasi: Krishnadas Academy, 1999–2000.

80 The *Suśrutasaṃhitā* offers a tripartite classification of diseases in Sūtrasthāna 24.4. One of these is the category of "*ādhidaivika*", explained in 24.7 as caused by offending gods, curses, black magic (*atharvaṇa*) and misfortune. However, the passage offers no corresponding therapeutic approach here.

81 Somadeva Vasudeva, "Powers and Identities: Yoga Powers and the Tantric Śaiva Traditions," in *Yoga Powers: Extraordinary Capacities Attained Through Meditation and Concentration*, ed. Knut Jacobsen (Leiden: Brill, 2011), 265.

[*Aṣṭāṅgasaṃgraha*:] *Aṣṭāṅga Saṃgraha of Vāgbhaṭa: Text, English Translation, Notes, Indices etc.*, translated by K.R. Srikantha Murthy. 3 vols. Varanasi: Chaukhamba Orientalia, 1995–1997.

Bhaiṣajyaratnāvalī of Shri Govinda Dasji, edited and enlarged by Brahmashankar Mishra, commented upon by Ambikadatta Shāstrī, English translation by Kanjiv Lochan, translation technically reviewed by Anand K. Choudhary. 3 vols. Varanasi: Chaukhambha Sanskrit Bhawan, 2006.

Bhāvaprakāśa of Bhāvamiśra: Text, English Translation, Notes, Appendeces[*sic*] *and Index*, translated by K.R. Srikantha Murthy. 2 vols. Varanasi: Krishnadas Academy, 1998–2000.

[Bower Manuscript:] *Nāvanītakam or (The Bower Manuscript): Critically Edited with Various Readings and Restorations for the First Time from the Editio Princeps of the Late D. Hoernle by Kavirāj Balwant Singh Mohan, Vaidya-Vacaspati*. Lahore: Mehar Chand Lachman Das, 1925.

Carakasaṃhitā: Agniveśa's Treatise Refined and Annotated by Caraka and Redacted by Dṛḍhabala, Text with English Translation, edited and translated by Priyavat Sharma. 4 vols. Eighth edition. Varanasi: Chaukhamba Orientalia, 2003.

[*Carakasaṃhitā*:] *The Charakasaṃhitā of Agniveśa: Revised by Charaka and Dṛidhabala, with the Āyurveda-Dīpikā Commentary of Chakrapāṇidatta*, edited by Vaidya Jādavaji Trikamji Āchārya. Fifth edition. Delhi: Munshiram Manoharlal, 1992.

Cakradatta, Text with English Translation: A Treatise on Principles and Practices of Ayurvedic Medicine. Varanasi, Delhi: Chaukhamba Orientalia, 1994.

Vaṅgasena Saṃhitā or Cikitsāsāra Saṃgraha of Vaṅgasena: Text with English Translation, Notes, Historical Introduction, Comments, Index and Appendices by Nirmal Saxena. 2 vols. Varanasi: Chowkhamba Sanskrit Series Office, 2004.

[*Rasaratnākara* Rasendrakhaṇḍa:] *Śrīḥ siddhanityanāthapraṇītaḥ rasaratnākaraḥ* / (*samastarasagranthānāṃ śirobhūṣaṇam*). *Māthuravaiśyā 'yurvedoddhāraka śālagrāmakṛta bhāṣāṭīkāvibhūṣitaḥ / sa ca Kṣemarāja Śrīkṛṣṇādāsaśraṃṣṭhinā*. Mumbai: Śrīveṅkaṭeśvara Steam Press, 1966.

[*Rasaratnākara* Rasāyanakhaṇḍa:] *Rasāyanakhanda: Fourth Part of Rasaratnākara of Śrī Nitya Nātha Siddha*, edited by Vaidya Jādavjī Tricumjī Āchārya. Benares: Vidya Vilas Press, 1939.

Rasārṇava, accessed at Oliver Hellwig's online Digital Corpus of Sanskrit (DCS) at http://www.sanskrit-linguistics.org/dcs/.

[*Śārṅgadharasaṃhitā*:] *Śārṅgadhar-Saṃhitā (A Treatise on Āyurveda) by Śārṅgadhara*, translated by Āyurveda-Vidwān Prof. K.R. Srikantha Murthy. Jaikrishnadas Ayurveda Series 58. Varanasi, Delhi: Chaukhamba Orientalia, 1984.

[*Siddhayoga*:] *The First Treatise of Āyurveda on Treatment: Vṛndamādhava or Siddha Yoga*, edited and translated by Premvati Tewari and Asha Kumari. 2 vols. Varanasi: Chaukhambha Visvabharati.

[*Suśrutasaṃhitā*:] *Suśruta-saṃhitā with English Translation of Text and Ḍalhaṇa's Commentary Along with Critical Notes*, edited and translated by Priya Vrat Sharma. 3 vols. Varanasi: Chaukhambha Visvabharati, 1999–2001.

[*Yogaratnākara*:] *A Complete Treatise of Āyurveda: Yogaratnākara*, edited and translated by Asha Kumari and Premvati Tewari. 2 vols. Varanasi: Chaukhambha Visvabharati, 2010.

Secondary Sources

Barois, Christèle. "Stretching Out Life, Maintaining the Body: Part I—Vayas in Medical Literature." *History of Science in South Asia Special Issue "Rejuvenation, Longevity, Immortality in South Asian Medical, Alchemical and Yoga Traditions"* 5, no. 2 (2017): 37–65. https://doi.org/10.18732/hssa.v5i2.31.

Dasgupta, Nupur. "Rasa, Rasayana, Rasatantra: Exploring Concepts and Practices." In *Ancient Indian System of Rasayana Suvarnatantra: A Treatise on Alchemy/Rasayana*, edited by Chittabrata Palit and Nupur Dasgupta, 9–61. Delhi: Kalpaz Publications, 2009.

Engler, Steven. "'Science' vs. 'Religion' in Classical Ayurveda." *Numen* 50 (2003): 416–463.

Falk, Harry. "Soma I and II." *Bulletin of the School of Oriental and African Studies* 52 (1989): 77–90.

Filliozat, Jean. *The Classical Doctrine of Indian Medicine: Its Origins and its Greek Parallels*. First English edition, translated by D.R. Chanana. Delhi: Munshiram Manoharlal, 1964.

Fenner, Edward Todd. *Rasayana Siddhi: Medicine and Alchemy in the Buddhist Tantras*. Ph.D. thesis, University of Wisconsin-Madison, 1979.

Goudriaan, Teun. *Māyā Divine and Human*. Delhi, Varanasi, Patna: Motilal Banarsidass, 1978.

Hellwig, Oliver. "Rasāyana und die āyurvedische Krankheitskunde." *Traditional South Asian Medicine* 8 (2008): 32–64.

Jacobsen, Knut A. *Yoga Powers: Extraordinary Capacities Attained Through Meditation and Concentration*. Leiden: Brill, 2011.

Maas, Philipp A. "On What Became of the *Carakasaṃhitā* After Dṛḍhabala's Revision." *eJournal of Indian Medicine* 3, no. 1 (2010): 1–22.

Maas, Philipp A. "On the Meaning of *rasāyana* in Classical Yoga and Āyurveda." *History of Science in South Asia Special Issue "Rejuvenation, Longevity, Immortality in South Asian Medical, Alchemical and Yoga Traditions"* 5, no. 2 (2017): 66–84. https://doi.org/10.18732/hssa.v5i2.32.

Mahdihassan, S. *Indian Alchemy or Rasayana in The Light of Ascetiscism and Geriatrics*. Delhi: Motilal Banarsidass, 1991.

Mallinson, James and Mark Singleton. *Roots of Yoga*. London: Penguin Random House UK, 2017.

Meulenbeld, J.G. *A History of Indian Medical Literature*. 5 vols. Groningen: Egbert Forsten, 1999–2002.

Meulenbeld, J.G. "The Woes of *ojas* in the Modern World." In *Modern and Global Ayurveda: Pluralism and Paradigms*, edited by Dagmar Wujastyk and Frederick M. Smith, 157–175. Albany: SUNY Press, 2008.

Olivelle, Patrick. "Amṛtā. Women and Indian Technologies of Immortality." *Journal of Indian Philosophy* 25 (1997): 427–449.

Palit, Chittabrata and Nupur Dasgupta, eds. *Ancient Indian System of Rasayana Suvarnatantra. A Treatise on Alchemy/Rasayana*. Delhi: Kalpaz Publications, 2009.

Pensa, Corrado. "On the Purification Concept in Indian Tradition, with Special Regard to Yoga." *East and West* 19, nos. 1–2 (1969): 194–228.

Pflueger, Lloyd W. "Holding On and Letting Go: The In and Out of Power in Classical Yoga." In *Yoga Powers: Extraordinary Capacities Attained Through Meditation and Concentration*, edited by Knut A. Jacobsen, 241–264. Leiden: Brill, 2011.

Radich, Michael. "Immortal Buddhas and Their Indestructible Embodiments." *Journal of the International Association of Buddhist Studies* 34, nos. 1–2 (2011 (2012)): 227–290.

Roşu, Arion. "Considerations sur une technique du *rasāyana* āyurvedique." *Indo-Iranian Journal* 17 (1975): 1–29 and 395.

Sharma, Ajay Kumar. *Elements of rasāyana Therapy in Ayurveda*. Delhi: Sri Satguru Books, 2005.

Sharma, P.V. *Āyurveda kā vaijñānika itihāsa*. Vārānāsi: Caukhambā Oriyanṭāliyā, 1975.

Sharma, Ram Karan. *Rejuvenative Health Care in Āyurveda*. Delhi: Pratibha Prakashan, 1996.

Slouber, Michael. *Early Tantric Medicine: Snakebite, Mantras, and Healing in the Garuda Tantras*. Oxford: Oxford University Press, 2016.

Vasudeva, Somadeva. "Powers and Identities: Yoga Powers and the Tantric Śaiva Traditions." In *Yoga Powers: Extraordinary Capacities Attained Through Meditation and Concentration*, edited by Knut A. Jacobsen, 265–302. Leiden: Brill, 2011.

Thieme, Paul. "Ambrosia." In *Indogermanische Dichtersprache*, edited by Rüdiger Schmitt. Darmstadt: Wissenschaftliche Buchgesellschaft, 1968.

Wujastyk, Dagmar. "Acts of Improvement." *History of Science in South Asia Special Issue "Rejuvenation, Longevity, Immortality in South Asian Medical, Alchemical and Yoga Traditions"* 5, no. 2 (2017): 1–36. https://doi.org/10.18732/hssa.v5i2.26.

Wujastyk, Dagmar. "On Perfecting the Body: *rasāyana* in Sanskrit Medical Literature." *AION—sezione filologico-letteraria* 37 (2015): 55–80.

Wujastyk, Dominik. "The Path to Liberation through Yogic Mindfulness in Early Ayurveda." In *Yoga in Practice*, edited by David Gordon White, 31–42. Princeton: Princeton University Press, 2011.

Wujastyk, Dominik. *The Roots of Ayurveda*. London: Penguin Books, 2003.

Zysk, Kenneth G. "Potency Therapy in Classical Indian Medicine." *Asian Medicine* 1, no. 1 (2005): 101–118.

Zysk, Kenneth G. *Asceticism and Healing in Ancient India: Medicine in the Buddhist Monastery*. Delhi: Motilal Banarsidass, 2000 [1991].

CHAPTER 9

Karman, Prāyaścitta, and Disease: A Reading of the *Karmavipāka* Digression in Mādhavācārya's *Parāśaramādhavīya*

Mikael Aktor

The subject of this chapter—the relationship between penance and notions of *karman*—is a good example of the multiplicity of semantic domains involved in key Dharmaśāstric notions. *Karman* combines the moral and natural domains. Natural events, like disease or accidents, are explained in terms of morals as the outcome of the person's former demeritorious acts. Like witchcraft for the great British social anthropologist E.E. Evans-Pritchard's Azande informants, *karman* explains unfortunate events and supplies the missing link between people's lives and what happens to them.

Prāyaścitta is generally articulated as an individual's soteriological management, but at the same time, it is clear enough from the texts that it is also—or primarily—a matter of social control and public reputation. *Prāyaścitta* is both individual and subjective as well as social and objective. It is presented in the Dharmaśāstric texts as a powerful means of avoiding unlucky results after death for immoral acts done while alive. But this was not always so. In the early formulations of doctrines of *karman*, as we know them from the early Upaniṣads and the Buddhist Pāli canon, the overall understanding is that *karman* is not something that can be avoided or destroyed, it can only be exhausted. The possibility of annihilating bad *karman* by the performance of penance is the special perspective of Dharmaśāstra.

The well-known doctrine of the five fires and the two paths leading from the cremation fire, as it is presented in *Bṛhadāraṇyakopaniṣad* 6.2 and *Chāndogyopaniṣad* 5.3–10, locates rebirth within a ritualistic world view from which the Upaniṣadic voices dissociate themselves. The villagers who circle through the five exegetical stages of the rebirth cycle (heaven, air, earth, man and woman) are those people who do not know the esoteric interpretations of the sacrificial fire. The early doctrine of *karman* and rebirth was, therefore, on a soteriological level, a projection of the social outlook of renouncers who downgraded ritual as a means of fulfilling worldly desires and family life as the social framework and biological root of these desires. The idea of performing penances in order to improve one's life conditions in future births was irrelevant to these renounc-

 | DOI:10.1163/9789004438224_010

ers, who expected to stay in the world of *brahman* forever. A hint at another means of overcoming bad *karman* is, however, indicated in the *Chāndogya* version. Those who know the secret of the five fires are not affected by the sin of associating with grievous sinners, but this is not necessarily an expression of the *karman*-destroying potential of that knowledge, but perhaps more an expression of the general potential of esoteric Upaniṣadic knowledge to transcend opposite social categories, among other things.

The Theravāda view of *karman*, with its monastic focus, is even more uncompromising. For the severest offenses, the *pārājikas*, no rehabilitation is possible. These offenses are tantamount to excommunication, metaphorically a suicide. The Vinaya notion of *pācittiya*—the Pāli equivalent of Sanskrit *prāyaścitta*—did not involve performing penitential rituals, only confession, and this confession has no bearing on the personal *karman* created by the transgression of a moral law, but only on the offense of breaking a social rule; it guarantees the integrity of the *saṅgha* that was reestablished each *uposatha* day at *pātimokkha*. The texts distinguish between, on the one hand, a moral transgression (*pāpa kamma*) belonging to the category of universal offenses (*loka vajjā*), and, on the other hand, a social or institutional transgression (*āpatti*) belonging to the category of conventional offenses (*sammuti vajjā*).[1] And actually, the authority of the *saṅgha* was only effective in relation to these conventional offenses; it cannot bend the universal law of *kamma*, as there *is* no institution with such a power. Mohan Wijayaratna summarizes the idea in terms of the offense of killing an animal:

> [P]romising through their confession to take better care of living creatures in the future, [monks] were exonerated from the institutional offense of killing an animal, but not from the universal offense, for which they would still suffer retribution in accordance with the law of *kamma*.[2]

The Upaniṣadic and the Theravāda views share a common skepticism toward the efficacy of ritual actions. Instead, they emphasize the absolute importance of mental activities, like desires, esoteric knowledge, or emotions. Thus, in the Buddhist case, the only effective means of hindering the fatal karmic effects of killing living creatures is to cultivate kindness and to root out the hatred that manifests as violent actions.

---

1 Mohan Wijayaratna, *Buddhist Monastic Life According to the Texts of the Theravāda Tradition* (Cambridge: Cambridge University Press, 1990), 148–149.

2 Wijayaratna, *Buddhist Monastic Life*, 150.

Therefore, the primary difference between these early ideas of *karman* and the Dharmaśāstric perspective is the positive attitude to rituals that is axiomatic for the latter tradition. Rituals are attributed with extraordinary powers for securing success and procuring change. And the introduction, development, and promotion of penitential rituals in the Dharmaśāstric tradition is a logical consequence of that ritualistic thinking. Through penance, it is possible to change and even eradicate the universal causes that the early renouncer traditions regarded as necessary and unavoidable.

The earliest comprehensive and systematic account of penance in a Dharmaśāstric work is that of Gautama. But according to P.V. Kane,[3] this early text also contains a remnant of the rejection of penance articulated in the early renouncer traditions. This is expressed in the *pūrvapakṣa* view that penance is not required, because, in principle, "an act can never be wiped out" (GDhS 19.5). This, however, is refuted, and the subsequent *sūtras* describe the general elements of penance, as well as the more elaborate penances, in detail. Chapter 20 of the same text describes the rules for excommunicating sinners who refuse to undergo the prescribed penance. Comparing this ritual, the *ghaṭasphoṭa*, to the uncompromising rules of *pārājika* offenses in the Vinaya, we notice a difference: symmetrical and parallel to the *ghaṭasphoṭa* ritual, we find a readmission ritual that is to be performed when the sinner has observed a penance (GDhS 20.10 ff.). A similar readmission was not conceived of by the monastic rules of the *saṅgha*, which did not accept the transformative effects of penitential rituals on *karman*.

1 The Effects of Sin and the Scope of Penance

It is not my intention in this chapter to present the historical development of penance as it unfolded through the following generations of Dharmaśāstric works.[4] Rather, I will make a jump to a time when the institution and rules of penance had developed fully. For the rest of the paper, I want to focus on the mid-fourteenth-century, probably South Indian, commentary on the *Parāśarasmṛti* (PS), the *Parāśaramādhavīya* (PM) attributed to Mādhavācārya,

3 Pandurang Vaman Kane, *History of Dharmaśastra: Ancient and Mediæval Religious Law*, 5 vols. (Pune: Bhandarkar Oriental Research Institute, 1962–1975), 5.1589.

4 For a general overview of penance, see Wilhelm Gampert, *Die Sühnezeremonien in der altindischen Rechtslieteratur* (Prague: Oriental Institute, 1939) and recently, David Brick, "Penance: *prāyaścitta*," in *Hindu Law: A New History of Dharmaśāstra*, eds. Patrick Olivelle and Donald R. Davis, Jr. (Oxford: Oxford University Press, 2018), 313–324.

better known for his later career as a Vedānta monk by the monastic name of Vidyāraṇya, and the works composed during that period. There is still much debate about the certainty of this identity and the extent of his authorship,[5] but that would require a full chapter of its own.

In his *Mādhavīya*, Mādhavācārya follows in the footsteps of *bhāṣyakāras* such as Bhāruci, Medhātithi, and Vijñāneśvara: He does not settle for a mere gloss or a compendium of *smṛti* verses but often contributes lengthy discussions of the subject at hand. One of these discussions is a continuation of Gautama's presentation of the opposite views on the scope of penance in terms of its effects on *karman*. More precisely, this matter had developed over time into a specific discussion of the question about whether or not penance has any effect in situations of deliberate and intentional offenses. According to Mādhava:

> The capacity of sin is double; it is the cause of hell and it is a hindrance to worldly transactions. Hence, the capacity of penance, which removes that evil, should also be distinguished as double. It wards off hell, and it makes social transactions possible. With respect to this argument, it is the case that, among the sages who proclaim there is no penance in the case of intentional sins, what is meant is that hell cannot be prevented in that case, whereas among those who proclaim penance is effective even in this case [intentional sins], what is meant is that this option is only with respect to the capacity to make social transactions possible. This is evidently the conclusion put forward by Yājñavalkya: "The sin which is committed in ignorance vanishes by penances. But if it is committed intentionally, penance, according to the texts, only makes the sinner eligible for others to deal with economically."[6]

The succeeding discussion presents various objections but eventually returns to the basic distinction of the *pūrvapakṣa*, which is only modified with two further additions: the penance for a deliberate sin is double the amount of that

5 Mikael Aktor, *Ritualisation and Segregation: The Untouchability Complex in Indian Dharma Literature with Special Reference to Parāśarasmṛti and Parāśaramādhavīya* (Turin: Corpus Iuris Sanscrititicum, 2008), 11–17.

6 *atra kecin nirṇayam āhuḥ—dvividhā hi pāpasya śaktiḥ / narakotpādikā vyavahāravirodhikā cety atas tannivartakasya prāyaścittasyāpi śaktir dvividhā bhidyate / narakanivārikā vyavahārajananī ceti / tatra prāyaścittābhāvavādināṃ munīnāṃ narakanivāraṇābhāvo 'bhipretaḥ / sadbhāvavādināṃ tu vyavahārajananī śaktir abhipretā / ayaṃ ca nirṇayo yājñavalkyena* [YDhŚ 3.226] *vipaṣṭam abhihitaḥ—prāyaścittair apaity eno yad ajñānakṛtaṃ bhavet / kāmato vyavahāryas tu vacanād iha jāyate //* (PM 2,1, p. 201).

which is observed for an unintentional sin, and for the grievous sins, the *mahāpātaka*s, only a penance that involves the sinner dying from the observance will ward off the unpleasant stay in hell.

The distinction between some effects that exclude the sinner from economic transactions and other effects that drop the sinner into hell when he dies is a parallel to the Buddhist distinction between conventional and universal offenses—the former are dealt with by the *saṅgha*, the latter are not possible to expiate at all. If the parallel is more than coincidental, it indicates that the economic isolation introduced in the Yājñavalkya verse quoted by Mādhava, unlike the stay in hell, is not really a matter of karmic causality but only of social integrity. It also indicates that, in the case of intentional sin, what is repaired by penance in the first place is not the karmic damage but only the social damage. In particular, it prevents other good people, who engage in economic transactions with the sinner, from committing an offense by transacting with such a man. Only in the case of the *mahāpātaka*s is it explicitly spelled out that a complete repair of the karmic damage is in fact possible, but this will involve the sinner undergoing one of the *prāyaścitta*s that end with the sinner's own death. On the other hand, judged from the almost ubiquitous distinction between unintentional and intentional acts—even in cases where it becomes quite absurd, having sex with a cow for instance—there seems little doubt that the category of "unintentional" sin must have created space for considerable flexibility in the interpretation of the situation and the stipulation of the penance.

Apart from economic isolation and hell, the other three general effects of sin recorded in the Dharmaśāstric works prior to the medieval commentaries are birth in a low caste or in lower life, congenital physical disabilities, and disease. It is rare, however, that accidents or catastrophes causing bodily damage are mentioned in a context that involves personal *karman*; as of right now, I can only think of the case that opens chapter eight of our text, that of a cow that dies accidentally, for instance by being strangled by its tether. This is considered a karmic calamity, which strikes the owner who will therefore have to undergo penance. Mādhava's discussion about intentional and unintentional sin that I have just summarized is precisely part of his commentary on that situation.

2 *Karmavipāka*: Definitions

Mādhava's discussion of *karmavipāka*, the ripening of *karman*, is inserted, as the last part of a long digression on penance, between verse eighty and the concluding two verses of the twelfth and last chapter of *Parāśarasmṛti*, the chapter on *prāyaścitta*. In Islāmapurkar's edition (PM), this *karmavipāka* chapter goes from pages 205 to 278. The largest part of the chapter, however, is a compendium of *karmavipāka* texts from other sources, but the remaining prose discussion still offers some interesting points.

The structure is clear, and I will simply go through the argument while presenting its central points.[7] First, we are presented with the necessary definitions. In principle, when we talk about the ripening of *karman*, we should count both good and bad actions and results, Mādhava points out, but in the context of the present chapter, which is *prāyaścitta*, what comes to mind is primarily the frightening results of bad acts. *Karmavipāka* is analyzed as the ripening or maturing of *karman*, and it is defined as *phaladaśā*, that is, the condition, state, or period of karmic results. But it may very well be that the word *daśā* here, apart from its general abstract meaning—condition, state, or period—is also supposed to produce associations with astrology, where the term has a more specific meaning, namely, as a planetary period in the horoscope. It is only during these periods that the otherwise latent *saṃcitakarman*—the stock of previously accumulated actions that still await fruition—can be foreseen by the astrologer. Mādhava had composed a large *Kālanirṇaya* shortly before he wrote his *Parāśaramādhavīya*, so astrological associations should be considered.

After this short definition, Mādhava refers to *Yogasūtra* 2.13, which defines the results of *karman* in terms of *jāti*, *āyus*, and *bhoga*. Mādhava exemplifies these with textual illustrations that more precisely characterize these three as: 1) life-form—*jāti*, that is, from worm to human, including low or high human birth; 2) life-length—*āyus*, that is, in the sense that very short lives, like five or ten years, are seen as the effects of *karman* from previous lives; and finally 3) consumption—*bhoga*, in the form of experiencing the manifestation of the results of *karman*. This experience is exemplified by verses about going to hell because of the effects of sin.

7 *yady api loke karmaśabdaḥ sukṛtaduṣkṛtayoḥ sādhāraṇaḥ tathāpy atra prāyaścittasya prakṛtatvāt tannivartyatayā duṣkṛtasyaiva buddhisthatvāt tatraiva paryavasyati | viśeṣena pacanaṃ vipākaḥ phaladaśā | karmaṇāṃ vipākaḥ karmavipākaḥ | sa ca vipākas trividho janma cāyur bhogaś ceti | etat sarvaṃ patañjalir yogaśāstre sūtrayāmāsa—sati mūle tadvipāko jātyāyurbhogāḥ* [*Yogasūtra* 2.13] | *iti* | *asya sūtrasyāyam arthaḥ—avidyāsmitārāgadveṣābhiniveśātmakāḥ pañca kleśāḥ karmaṇāṃ mūlam* | *kleśamūlaḥ karmāśayaḥ* | *iti sūtritatvāt* |

[…] *tatra janma kṛmyādi* | *patitaṃ yājayitvā tu krimiyonau prajāyate* [*Yogasūtra* 2.12] | *ityādismaraṇāt* | *āyuḥ pañcadaśavarṣādikam* | *tatra jīvati varṣāṇi daśa vā pañca vā dvijaḥ* | *ityabhidhānāt* | *bhogo yamayātanādiḥ* | *yamaloke tu te ghorāṃ labhante pariyātanām* | *ityādismaraṇāt* | (PM 2,2., pp. 205–206).

3 *Karmavipāka* as *Arthavāda*

After these introductory definitions, Mādhava further characterizes *karmavipāka* in terms of a textual-ritual analysis, in which he categorizes *karmavipāka* texts as *arthavādas*.[8] This should not, I think, be read as an ontological statement meaning that the descriptions of *karmavipāka* have no actual root in reality—that they are *merely* a kind of textual stunt in order to frighten people into performing the required penances. They *are* indeed frightening, he says, and they *are* employed as such to make people act properly and observe penance when they have not done so, but I do not think we should read the ontological "merely" into his statements. This would compromise his overall argument, which is that penances do work, they are in fact capable of inhibiting the ripening of *karman*, and they are also capable of repairing or even stopping the already manifesting *prārabdhakarman*. Rather, Mādhava wants to make the precise status and function of these *karmavipāka* texts *as* texts clear. Kane mentions Mādhava's argument that is preceded by the *Mitākṣara* and a few other texts.[9] He paraphrases these texts as meaning that the *karmavipāka* texts are not to be taken literally. But, I think this must be one of Kane's frequent apologetic statements. He wants to assure the shocked reader (Western I suppose, or with a modern outlook at least) that these absurd lists of weird consequences for stealing grain, eating meat, etc. are not meant as ontological facts but just as literary devices. But this is a simplification. The detailed chapters on penance in Dharmaśāstric works, including the *karmavipāka* descriptions, are not to be read naïvely as only being attempts to control the ill-behaving masses or as a means of retaining a social reputation of Brahmin purity. Although these texts are often composed for kings, whose duty it was to guarantee that penances were enforced, they were primarily didactic, with a view to the authors' own class, the Brahmins. In fact, most of the examples of sins involve the Brahmins as sinners. When a sin is unknown to the public, secret penances are recommended. Why would any Brahmin want to undergo secret penances for sins that are unnoticed by his surroundings if he did care about his own *karman* from a personal soteriological perspective? This does not support the idea that the Brahmin authors of our texts were aware of themselves as deliberately producing fictional narratives merely to persuade their clients to observe penance.

---

8 *prāyaścittārthavādatvena tadupayogāt* [tad = *karmavipāka*] / *arthavādas tu vihite pravartakatvena pratiṣiddhān nivartakatvena ca vidhipratiṣedhaikavākyatāṃ bhajate—iti nyāyavidāṃ maryādā / atra ca prājāpatyādyanuṣṭhānarūpasya prāyaścittasya kleśātmakatvād rāgataḥ pravṛttir na sambhavatīti tatra pravṛttisidhaye prarocanā kartavyā* / (PM 2,2, pp. 206–207).

9 Kane, *History of Dharmaśāstra*, 4.177.

Rather, *karmavipāka* supplies *vidhipratiṣedhaikavākyatā*—explicit evidence of single injunctions and prohibitions. But it does so in different ways according to whether we talk of an injunction or a prohibition. This is the argument developed on the following pages of the text.[10] When texts describe the results of uplifting rituals, like the *agniṣṭoma*, this is by means of favorable results like heaven, etc., but when they describe the results of forbidden acts, this must of course be by means of the unfavorable results of decay and disease. On the other hand, this piling up of the frightening consequences of bad acts can be seen as both an *arthavāda* in support of a prohibition against these sinful acts and one in support of the injunction to perform a penance when they have nevertheless been committed.

This distinction is further linked to the two types of *karman*, the *saṃcita* and the *saṃceṣyamāṇa*, which are the earlier accumulation of *karman* from previous lives and the future accumulation that will be produced in the present life. When *karmavipāka* is seen as an *arthavāda* in support of a prohibition, this is with respect to the *saṃceṣyamāṇa*, the future accumulation; when it is seen as being in support of the injunction to perform penances, it is related to the *saṃcita*, the past accumulation of karman. That is to say, in the former case, *saṃceṣyamāṇa*, the effect of the *arthavāda* is to make a not-yet-sinning man desist from committing sin, whereas in the latter case, *saṃcita*, its effect is to make the man who is already a sinner engage in the prescribed penance. The conclusion of this section is therefore that the *arthavāda* element of the *karmavipāka* descriptions should not be understood one-dimensionally but must be understood in this double sense, and it will therefore be examined in the text as "both consisting of decay and disease [in case of sin] as well as the ces-

---

10 *nanu—vihiteṣv agniṣṭomādiṣu svargādivipākopavarṇanena prarocayituṃ śakyate / svargādīnām anukūlaphalatvāt / kṣayarogādīnāṃ tu pratikūlaphalatvena tadupavarṇanaṃ na prarocanāyopayuktam /*

viṣama upanyāsaḥ / tatra hi svargādisādhane prarocanam utpādanīyam / na tv atra kṣayarogādisādhane prarocanam utpādyate / kiṃ tarhi kṣayarogādinivartake prāyaścitte / tannivṛttiś cānukūlaṃ phalam /

nanu—evam api kṣayarogādinivṛttirūpa eva prāyaścittātmakasya karmaṇo vipākaḥ prarocakatvena vaktavyaḥ / na tu brahmahatyādivipākaḥ / tasyāprarocakatvāt /

maivam / dvividhā hi brahmahatyādayaḥ / saṃcitā saṃceṣyamāṇāś ca / tatra saṃceṣyamāṇeṣu pratiṣedhaviṣayeṣu brahmahatyādiṣu kṣayarogādyupanyāsaḥ brāhmaṇāvagoraṇādau śatayātanādivat tannivartakārthavādatvenānveti / saṃciteṣu pratikūlaṃ phalaṃ śrutvā tata udvijamānas tannivṛttyupāyam ākāṅkṣati / ata udvegaprayojakatayā prāyaścitte 'pi pravartakatvaṃ bhavatīti pravartakanivartakadvividhārthavādatvenāyaṃ karmavipākopanyāsa upayujyate /

[…] *ataḥ kṣayarogādirūpas tannivṛttirūpaś ca vipāko 'tra nirūpyate /* (PM 2,2, pp. 207–208).

sation of the same [in case of penance]": *ataḥ kṣayarogādirūpas tannivṛttirūpaś ca vipāko 'tra nirūpyate.*

At this place, the text starts to pile up textual examples of *karmavipāka* from known sources, mainly Purāṇas and Dharmaśāstras. This is by far the largest part of the whole chapter. In Islāmapurkars edition it starts at page 208 and ends at page 272. The quotes are ordered thematically: texts that describe existences in various types of hell according to specific sins; other texts that describe different kinds of low births in lower life, low castes, or even as ghosts or demonic beings; and theoretical categorizations, like the division between different scales of results according to the *karman* being bodily, verbal, or mental, or according to its categorization in terms of the three *guṇas*. Special sections are reserved for the different kinds of birth that are the result of violent actions, stealing various products, drinking what is not to be drunk, eating what is not to be eaten, having sex with persons that should not be approached for that purpose, etc. An interesting quote from the *Yājñavalkyadharmaśāstra* (that is, YDhŚ 3.129–136) identifies the Self with the primal *puruṣa* and then wonders how this God can be born in human beings of low birth ("*pāpayoniṣu*") (YDhŚ 3.129–130). It pushes the identification further, as the Self-*puruṣa* is spoken of as the *jīva* in the body in v. 131, and from this point, *karmavipāka* is explained by saying that the manifold stages of the *jīva* are responsible for the manifold variations in birth from lower to higher:

> For some, the ripening of action occurs after death, for some here in this world, or it occurs here and in the next world for some; the state [of the *jīva*] is causing this.[11]

Having piled up all this evidence, the text concludes by stressing that, however many examples are given, there will always be other cases that can only be inferred by indication or examples, because, ultimately, *karman* is endless.[12]

---

11 *vipākaḥ karmaṇo pretya keṣāṃcid iha jāyate / iha cāmutra vai keṣāṃ bhāvas tatra proyojanam //* (YDhŚ 3.133).

12 *tad evaṃ kānicic chāstrāṇy udāhṛtya kiyān api karmavipāko nirūpitaḥ / anayaiva diśā vipākāntarāṇy unneyāni / nahi niḥśeṣeṇa kenacid udāhartuṃ śakyante /* [...] */ aśaktiś ca karmānantyāt / yady apy atipātakādīni prakīrṇakāntāni pāpāni daśavidhatvena viṣṇunā saṃkalitāni tathāpy atipātakādijātyupādhāv eva tatsaṃkalanam / pāpavyaktīnāṃ tu svarūpato deśakālakartravasthādibhedāc ca neyattā / yadā karmaṇām eveyattā nāsti tadā tadvipākānām iyattā dūrāpāstā / ekaikasya karmaṇo bahuvidhavipākavat tv asmaraṇāt / udāhṛtāś ca bahuvipākābhidhāyikāḥ śvasūkaretyādismṛtayaḥ / tathāpi teṣāṃ niḥśeṣakathanaṃ manasāpi cintayitum aśakyam / ato diṅmātraṃ pradarśitam /* (PM 2,2, pp. 272–273).

Even detailed categorizations, like the list in *Viṣṇusmṛti* of various types of sin, ranging from the very severe *atipātaka*s to the miscellaneous *prakīrṇaka*s, are incomplete when it comes to the concrete manifestations of *karman*. And, as the possibilities of different sinful acts are endless, so also is their ripening in the form of concrete results.

4 Can Penance Stop Even the *Prārabdhakarman*?

So far, the text has examined *karmavipāka* in relation to the not yet manifesting *saṃcita-* and the future *saṃceṣyamāṇakarman*. Generally, penance is relevant specifically in relation to the *saṃcitakarman* that has accumulated but not yet ripened. Penance can restore purity by expiating actually committed acts, and the instructions in the law books are mostly used for settling the exact penances to be observed in cases of such specific acts. For the rest of the text, however, it takes up the more unsettled and controversial, but nonetheless much more pressing, issue of whether penances are also effective in relation to the part of the *saṃcita* that has already started to manifest itself with and in the present body, that is, the *prārabdhakarman*. It does not require much imagination to understand why this *karman* calls for specific attention. It is when disease or sorrow strikes that the question about how to remove the effects of the former sinful acts, which have caused these sufferings, becomes really acute. But generally, when the law books mention the many different diseases that will be the results of certain acts, it is to motivate people to observe the relevant penance when they have committed these acts; it is not to let them wait until these acts ripen as disease or other unfortunate conditions, which may be far away in some future life. Other texts, however, are open to the possibility that disease can be cured by certain ritual acts, like *homa* or *japa*, and Mādhavācārya mentions some of these texts in the discussion that follows. But generally, such rituals, like *śānti*-rituals, *vrata*s, and *tīrthayātra*s, are not mentioned as cures of already manifesting diseases. Rather, the scenario is about certain evil omens (*nimitta*), typically a dream, and the rituals are performed in order to avoid the unlucky, but not fully known, results that such omens may anticipate.

Nevertheless, Mādhava now quotes a saying about how to cure a disease, which mentions medicine along with gifts, fire offerings, *japa*, and worship as means to this end.[13] Except for medicine, all these are classical elements of penance according to, for instance, Gautama 19.11 and Manu 12.228, which may

---

13 *niṣedhā ācārakāṇḍe prapañcitāḥ / prāyaścittavidhayaś ca prāyaścittakāṇḍe / tāni ca prāyaścittāni saṃcitaviṣayāṇi / prārabdhe tu vicāraṇīyam / kiṃ prāyaścittam asty uta neti /*

be one reason why Mādhava connects this quote with the question of whether penance is also effective in relation to *prārabdhakarman*. He starts with a reservation. Unlike the case of the not yet unfolding *saṃcita*, which is unproblematic, this question is to be examined carefully: *prārabdhe tu vicāraṇīyam*. Is penance effective in that case or is it not? It is, according to some, who put forward the saying just mentioned. This verse structures the rest of Mādhava's discussion. It reads:

> A sin committed in a previous life gives pain in the form of disease [in the present life]. Its alleviation is obtained by medicine, gifts, soft recitation, fire offerings, worship, and the like.[14]

The verse is mentioned by Mādhava as an anonymous "saying" (*vacana*), and none of the editors have been able to identify it. I have found a clear parallel, however, in *Śātātapasmṛti* 1.4–10.[15] But there is a minor difference. Where Mādhava's *vacana* speaks of *auṣadha*, medical drugs, *Śātātapa* in verse 4 uses the word *upakrama*, which generally means a practical means or remedy, but which also can have the specific meaning of medicine. Kane, who paraphrases the *Śātātapa* verses, therefore reads *upakrama* together with the methods mentioned in the second half-verse: "the remedies of the recitation of Vedic texts, worship of gods, *homa*s and gifts." Mādhava, in contrast, clearly reads the parallel *auṣadha* in his version of the saying as a method of its own, that is, as medicine apart from *dāna*, *japa*, *homa*, etc. This is evident from his further discussion of the matter.

---

tatra tāvad astīti kecid āhuḥ / udāharanti ca vacanam—pūrvajanmakṛtaṃ pāpaṃ vyādhirūpeṇa bādhate / tacchāntir auṣadhair dānair japahomārcanādibhiḥ // iti / (PM 2,2, pp. 273–274).

14 See the verse at the end of the text quoted in fn. 13.

15 *prāyaścittavihīnānāṃ mahāpātakināṃ nṛṇām / narakānte bhavej janma cihnāṅkitaśarīriṇām* // 1.1
pratijanma bhavet teṣāṃ cihnaṃ tatpāpasūcakam / prāyaścitte kṛte yāti paścāt tāpavatāṃ punaḥ // 1.2
mahāpātakajaṃ cihnaṃ saptajanmani jāyate / upapāpodbhavaṃ pañca trīṇi pāpasamudbhavam // 1.3
duṣkarmajā nṛṇāṃ rogā yānti copakramaiḥ śamam / jāpyaiḥ surārcanair homair dānais teṣāṃ śamo bhavet // 1.4
pūrvajanmakṛtaṃ pāpaṃ narakasya parikṣaye / bhādate vyādhirūpeṇa tasya japyādibhiḥ śamaḥ // 1.5 (*Śātātapasmṛti* 1.1–5 in Āpte, *Smṛtīnāṃ Samuccayaḥ*, 396).
Kane's paraphrase [translation rather] of this text, v. 1.4: "[T]he diseases arising from sinful deeds subside by adopting *the remedies of the recitation of Vedic texts, worship of gods, homas and gifts*." (Kane, *History of Dharmaśāstra*, 4.173).

But before this, he first mentions the opposite view: it is not possible to remove the *karman* that has already started to work. This view is immediately refuted by again referring to the quoted saying, which now is regarded as Śāstric evidence.[16] This, after all, is the only means of eventually deciding whether penance is or is not valid in the present context.

5 An Objection from Vedānta

There is, however, another objection to the idea of stopping the *prārabdhakarman*. This comes from Vedānta, and Mādhava now presents this argument.[17] Just as an arrow that has already been shot at its target cannot be reverted, so the result that has already started to work cannot be undone. Quoting *Brahmasūtra* 4.1.15, the reason for this is explained: The liberating knowledge of the Self blocks all *saṃcitakarman* that has not already started to work, it hinders the *saṃceṣyamāṇa* from ever ripening, but it cannot stop the *prārabdhakarman*, which has to run out during the length of the life of the body. In fact, the life of the body presupposes some *prārabdhakarman*, and the idea of living on when all *prārabdhakarman* is gone is illogical. This view is textually substantiated further by quoting *Chāndogyopaniṣad* 6.14.2, "for me there is delay only as long as I am not freed, but then I will arrive"[18]—the same text that Śaṅkara quotes in his commentary to the same *sūtra*. This is interpreted by Mādhava as meaning that only when a person is freed of the body will he also be freed of all *prārabdhakarman*.

This would perhaps seem to be a powerful argument, but in the overall context, which is to recommend the observance of penance even for stopping some types of *prārabdhakarman*, the preliminary character of the argument is nev-

16 […] / *athavā prāyaścittasadasadbhāvaḥ śāstraikasamadhigamyaḥ* / *śāstraṃ cet phalasyāpi prāyaścittaṃ brūte kiṃ tava chidyate* / *udāhṛtaṃ ca śāstram—tac chāntir auṣadhaiḥ* / *ityādi* / *tasmād ārabdhaphalasyāpy asty eva prāyaścittam* / (PM 2,2, p. 274).

17 *yathā muktasyeṣor lakṣyavedho 'pratisamādheyaḥ tathā prārabdhaphalaṃ na pratisamāḍhātuṃ śakyate* [Cf. Śaṅkara's Bhagavadgītābhāṣya on BhG 13.23] / *etad evābhipretya bhagavān bādarāyaṇaḥ sūtrayāmāsa—anārabdhakārya eva tu pūrve tadavadheḥ* [Brahmasūtra 4.1.15] / *iti* / *asyāyam arthaḥ—ye sukṛtaduṣkṛte tattvajñānanivartyatvena pūrvayor adhikaraṇayoḥ prakṛte te anārabdhakārya eva* [More or less a quote from Śaṅkara's commentary to BS 4.1.15] / *na tv ārabdhaphalayoḥ sukṛtaduṣkṛtayor nivṛttiḥ* / *kutaḥ* / *tadavadheḥ* / *tasya tāvad eva ciram* [ChU 6.14.2] / *iti śrutāv utpannajñānasyāpi mokṣam prati śarīrapātāvadhiśravaṇād iti* / (PM 2,2, p. 275)

18 On the difficulties of translating these sentences, see Patrick Olivelle, *The Early Upaniṣads: Annotated Text and Translation* (Oxford: Oxford University Press, 1998), 562–563.

ertheless evident. In fact, Mādhava could have refuted it more explicitly by referring to the specific context of the Vedānta argument. He does not do that and simply moves on to another argument. The context of *Brahmasūtra* 4.1.15 is knowledge (i.e., a *nivṛtti* activity), not *pravṛtti* activities like penances, fire offerings, or gifts that have the positive capacity of creating changes in a person's life conditions. Apart from this, and unlike the idea in *Brahmasūtra* 4.1.15, there is no question of destroying *all prārabdhakarman* in the overall context of Mādhava's discussion. That, as previously mentioned, would be absurd. The issue is only about destroying a part of that *karman*, namely what manifests as disease in the present moment.

6 Penitential vs. Medical Means of Curing Disease Caused by *Prārabdhakarman*

The next argument, which leads to Mādhava's conclusion, compares the working of medicine to that of rituals like *dāna*, *japa*, and *homa*.[19] The text does not seem fully reliable here. There are many instances of variant and conflicting readings, and therefore, the translation is uncertain. It starts by referring once more to the quoted saying about medicine, gifts, etc. being means of alleviating the *prārabdhakarman* that has manifested as disease. It rhetorically asks what its *gati* is, that is, what is meant by it, or perhaps what is its reason, purpose, or context. This is answered by saying that penance is not mentioned at all in this saying (or in the Vedānta argument—the reference of the *atra* is not quite clear)—because it is not found in works on penance, only in works on medical practice (*cikitsā*). And as these are different, so are the natures of the means being employed in each context, i.e. penance and medical practice. Medicine works in a visible, uncontroversial way as a means of stopping a disease, whereas gifts, etc. work in invisible ways. In this sense, the invisible penitential potential of gifts is not like medicine. More precisely, gifts produce some particularly meritorious and powerful *apūrva*, or extraordinary potency, which is not to be perceived or inferred from the visible act of presenting someone

19 *evaṃ tarhi tacchāntir ityādeḥ kā gatir iti cet / ucyate / nahy atra prāyaścittam abhidhīyate / tatprakaraṇe pāṭhābhāvāt / cikitsāprakaraṇe hīdaṃ paṭhitam / tathā ca cikitsā yathā roganivṛtter dṛṣṭa upāyas tathā dānādir apy adṛṣṭa upāyaḥ / tathā sati cikitsāvad dānāder nātra prāyaścittatvam / kiṃ tu roganivartakaṃ kiṃcit prabalaṃ sukṛtāpūrvam utpādayanti dānādayaḥ / tac ca sukṛtaṃ pravṛttāny api phalāni pratibadhya svaphalaṃ prayacchati / pratibaddhaś ca phalaśeṣaḥ svapnādau janmāntare vopabhujyate /* (PM 2,2, pp. 275–276).

with a certain visible object, but which nevertheless possesses the capability of stopping a disease. This *apūrva* can subdue the negative *prārabdhakarman* and yield its own positive results. A virtual fight between good and evil *karmans* within the same person seems to be going on. The rest of the subdued evil *prārabdhakarman* will then just be experienced in dreams or perhaps in some other future life.

This is supported by epic and Purāṇic evidence of similarly powerful acts, like blessings and curses, that have the capacity to transform people's life forms, for instance turning a man into a python, as with Nahuṣa in the *Mahābhārata*.[20] Since such extraordinary transformations are possible, how much more possible are they in cases of trivial diseases, like loose nails for instance. In this manner, the author again—like earlier with the anonymous saying—relies heavily on what is presented as traditionally accepted texts for his main proof. He has demonstrated, however, that there are conflicting views on the matter, but he asks his reader to side with texts that support a positive view of the possibility of affecting even the *prārabdhakarman* through penance.

As further evidence, he refers to texts that support this view in relation to exactly those methods that have been mentioned in the anonymous saying, that is, medicine, gifts, *japa*, *homa*, and worship. For medicine, he simply refers to Āyurveda, and for gifts to the *Brahmāṇḍapurāṇa*. He is more detailed with *japa*, where he relies mainly on examples given in the *Ṛgvidhāna*, and similarly with *homa*, where he quotes the *Taittirīyasaṃhitā*. Finally, for the idea that worship can also be an effective means of stopping a disease, he refers in general to Śaiva and Vaiṣṇava Āgamas and the Yogaśāstras.

Thus, having weighed textual evidence that expresses a positive view against other texts that express a negative view on the issue, he arrives at the final conclusion: people ought to have confidence in the capacity of penance, even in relation to *karman* whose unwished results are already manifesting.[21] He uses

20 *mahābhārate nahuṣasyendrapadaprepsor agastyaśāpenājagaratvaṃ prāptam ity abhihitam / yadā jātasya dehasyāpi prabalena nimittenānyathābhāvo 'sti tadā kim u vaktavyaṃ kaunakhyāder anyathābhāva iti / evaṃ ca sati rogaśāntyādipratipādakāni sarvāṇi śāstrāṇi prārabdhabhogavirodhenānugṛhītāni bhaviṣyanti /* (PM 2,2, p. 278).

21 *ayaṃ ca vyādhipratīkāranyāyaḥ sarveṣv api aniṣṭeṣu prārabdhaphaleṣu karmasu yojanīyaḥ / yadi japādīnāṃ prāyaścittatvam aṅgīkṛtya prārabdhakarmanivṛttir abhyupeyate yadi vā tadanaṅgīkāreṇa phalamātrapratibandhaḥ sarvathāpi vihite japādau visrambheṇa pravṛttiḥ sidhyati /*

yadā bhogaikanivartyatvaśaṅkāyukteṣv api prārabdhphaleṣv īdṛśapratīkāravisrambhas tadānārabdhaphaleṣu karmasu prāyaścittapravṛttivisrambho 'stīti kim u vaktavyam / tasmād yathokteṣu prāyaścitteṣu niḥsaṃdigdhatayaiva pravartitavyam ity aśeṣam atimaṅgalam // (PM 2,2, p. 286).

the word *visrambha* or *viśrambha*, confidence, perhaps to express that the issue is not one that can be settled with certainty but remains to some extent a matter of personal inclination and faith. This impression is strengthened by his final remarks, which mention the view that negative *prārabdhakarman* can only be overcome by experiencing the suffering in which it has manifested. He ends by stating somewhat ambiguously that if we can have confidence in the capacity of penance to prevent the already manifesting *prārabdhakarman*, even though we may doubt that, in fact, this *karman* can be overcome without experiencing its results, then we can have even more confidence in performing penance in relation to the *karman* whose results have not yet started to work. This is in line with the classical Dharmaśāstras, which, as he explained at the start of the discussion, are generally only focused on preventing the *saṃceṣyamāṇa* of future sinful acts or on the benefit of observing penance when such acts have actually been committed in the present life and have been stored as fresh *saṃcita*, but not on preventing the *prārabdhakarman*. Therefore, in any case, *prārabdha* or not, we should certainly go on observing penances since they are given in the texts. This brings complete happiness. And here ends his discussion on *karmavipāka*.

7 A Case: The *Brahmakūrca* Penance[22]

A further discussion of the relation between the visible and invisible causes of diseases and the visible and invisible means of curing them is offered in the context of the powerful *brahmakūrca* penance. This penance is prescribed in the text as a cure for drinking water, curd, clarified butter, or milk from vessels belonging to people who are *abhojyānna*, that is, whose food should not be eaten (PS 2.11.25). It consists of a fast for one day and drinking *pañcagavya* (the five products of the cow: butter, curd, milk, dung, and urine) prepared with Vedic mantras the next day.[23] It is a prominent penance in the *mūla* text (PS), which spends fifteen verses (2.11.26–40) on its instructions. Not only are the five ingredients collected from specially selected animals (29–30), but during every stage of the entire procedure of collecting and preparing the ingredients, Vedic mantras are recited. When all the ingredients are ready, a portion is offered in the fire (35–36), and what remains is the portion that should be drunk together

---

22 This section is taken from Aktor, *Ritualisation and Segregation*, 211–213, with some changes.

23 Kane, *History of Dharmaśāstra*, 4.146–147.

with a decoction of *kuśa* grass. At this point, the *mūla* text offers a description of the effects of this elixir. It is presented here with Mādhava's introduction and subsequent remarks:[24]

> It might be objected that it is only because the negative [moral] result of, for instance, killing a cow is invisible, that it is reasonable that it can be averted by the extraordinary [and invisible] power [*apūrva*] of the merit generated by penance. But the negative result generated by eating forbidden food[25] is not only invisible; some is visible, since that food is transformed into skin, bones, etc. Hence, it does not seem reasonable that it [i.e., also the visible physical transformation] can be averted by the observance. Therefore [in order to reject this objection], he says:
>
> > "Whatever sin there may reside in the body of an embodied being, lying in the skin and bones, *brahmakūrca* burns it all like a kindled fire burns up the fuel."
>
> Just as articles of forbidden food are transformed into skin, bones, etc., in the same way do cow urine and the other ingredients also transform themselves into the same. Having been digested they destroy, like fire destroys the firewood, these disagreeable transformations of the forbidden food also by invisible means and not only by means of the visible. Thus, the negative result of eating what is forbidden is stopped by *brahmakūrca*.

24 *nanu—govadhādiṣu yaḥ pratyavāyaḥ sa kevalādṛṣṭarūpatvāt tasya prāyaścittajanyena sukṛtāpūrveṇa nivṛttir yuiyate / abhakṣyabhakṣaṇajanyas tu pratyavāyo na kevalam adṛṣṭarūpaḥ / kiṃ tu dṛṣṭarūpo 'pi / tasyāhārasya tvagasthyādirūpeṇa pariṇatatvāt / ato na tasya vratena nivṛttir yujyate—ity ata āha—yat tvagasthigataṃ pāpaṃ dehe tiṣṭhati dehinām / brahmakūrco dahet sarvaṃ pradīptāgnir ivendhanam // iti /* PS 2.11.37c–38b / *abhakṣyāṇi yathā tvagasthyādirūpeṇa pariṇatāni tathā gomūtrādīny api tena rūpeṇa pariṇamanti / pariṇamya cāgniḥ kāṣṭānīva svavirodhyabhakṣyapariṇāmān adṛṣṭamukhenāpi vināśayanti / na tu kevalaṃ dṛṣṭenaiva mukhena / tasmād brahmakūrcenābhakṣyabhakṣaṇanivṛttir upapadyate /* PM 2.11.37c–38b with introduction.

25 Although the context of these Smṛti-verses is the subject of eating food belonging to people whose food should not be eaten, that is, food which is *abhojya* (PS 2.11.25a), Mādhava here and in the following uses the word *abhakṣya*. Generally, *abhakṣya* refers to products that are forbidden as food stuff, for instance garlic, whereas *abhojya* refers to articles of food that come from people whose food should not be eaten, that is, unfit food. See Patrick Olivelle, "Abhakṣya and Abhojya: An Exploration in Dietary Language," in *Language, Texts, and Society: Explorations in Ancient Indian Culture and Religion* (Florence: Firenze University Press, 2005), 352–353. But either Mādhava does not follow this distinction here, or he ignores the context in a case such as this where the focus is less on the sin and more on the cure.

Mādhava is provoked into this discussion by the formulation of the Smṛti text, which stresses that sins affect the physical body of the sinner. As the overall context is the eating of unlawful food, the suggestion is understood in connection with the physical transformations of food through digestion.

What, then, is the relation between these visible transformations of food and the invisible quality of the sin of transgressing a Smṛti rule? The question has an additional layer of complexity because even these invisible qualities produce visible physical changes in the body, for instance as disease, if the sin is not expiated. It would have been interesting if Mādhava had addressed these questions in more depth. Here he seems somewhat cursory. The objection of the anonymous opponent is that since the force of penance (that is, the *apūrva* of the merit (*sukṛta*) of performing the penance) is invisible, it should only be able to inhibit the still invisible effects of sin before they ripen as disease and other visible changes in the body. On the other hand, Mādhava seems to mix up the immediately visible bodily transformations involved in consuming and digesting food and *pañcagavya* with the long-term, visible karmic results of unexpiated sin and the power of the *brahmakūrca* penance to cure these visible changes. What he suggests is that, although the *pañcagavya* consists of ingredients that are eaten and digested—just like unlawful food—and therefore are also transformed into skin, bones, etc., it is the invisible *apūrva* of the rite and the Vedic mantras involved that overcome the long-term, negative bodily effects of the sin of eating unlawful food.

Compared to the ancient Vedic and Buddhist texts on the possibility of averting negative *karman*, the discussions of *karmavipāka* in Mādhava's work have moved far away. Not only is it possible to prevent the negative *karman* that has not yet started to work but even parts of the *prārabdhakarman* that manifest in people's present life in the form of disease can be averted. By being open to the possibility of using penance also in case of present disease, Mādhava's discussion is an example of a late development. As such, it is an example showing that the process of Dharmaśāstra is not as static and conservative as it is often presented, but rather—within its own idiom—there is an ongoing dynamic process of adaptations and change.

References

Aktor, Mikael. *Ritualisation and Segregation: The Untouchability Complex in Indian Dharma Literature with Special Reference to Parāśarasmṛti and Parāśaramādhavīya.* Turin: Corpus Iuris Sanscrititicum, 2008.

Āpṭe, Hari Nārāyaṇa. *Smṛtīnāṃ Samuccayaḥ* [Ānandāśrama Sanskrit Series, 48]. Pune: Ānandāśrama, 1905.

Brick, David. "Penance: *prāyaścitta*." In *Hindu Law: A New History of Dharmaśāstra*, edited by Patrick Olivelle and Donald R. Davis, Jr., 313–324. Oxford: Oxford University Press, 2018.

Gampert, Wilhelm. *Die Sühnezeremonien in der altindischen Rechtslieteratur*. Prague: Oriental Institute, 1939.

GDhS. *Gautamadharmasūtra*. In *Dharmasūtras: The Law Codes of Āpastamba, Gautama, Baudhāyana, and Vasiṣṭha*, edited by Patrick Olivelle, 116–189. Delhi: Motilal Banarsidass, 2000.

Kane, Pandurang Vaman. *History of Dharmaśastra: Ancient and Mediæval Religious Law*. 5 vols. Pune: Bhandarkar Oriental Research Institute, 1962–1975.

Olivelle, Patrick. *The Early Upaniṣads: Annotated Text and Translation*. Oxford: Oxford University Press, 1998.

Olivelle, Patrick. "Abhakṣya and Abhojya: An Exploration in Dietary Language." In *Language, Texts, and Society: Explorations in Ancient Indian Culture and Religion*, 351–366. Florence: Firenze University Press, 2005.

PM. *Parāśaramādhavīya*. In *Parāśaradharmasaṃhitā or Parāśarasmṛti with the Commentary of Sāyaṇa Mādhavācārya*, edited by Vāman Śāstri Islāmapurkar. 6 vols. Mumbai: The Department of Public instruction, 1893–1919.

PS. *Parāśarasmṛti*. Included in PM.

Wijayaratna, Mohan. *Buddhist Monastic Life According to the Texts of the Theravāda Tradition*. Cambridge: Cambridge University Press, 1990.

YDhŚ. *Yājñavalkyadharmaśāstra*. In *Yājñavalkya's Gesetzbuch: Sanskrit und Deutsch*, edited by Adolf Friedrich Stenzler. Osnabrück: Biblio Verlag, 1970 [1849].

CHAPTER 10

Plagues and Brahmins: Did a Combination of Epidemics and Ideology Empty India's Cities?

Johannes Bronkhorst

The *Economist* of July 19th 2014 points out (p. 47) that in India "[t]oday, out of every 100 children, 1.7 more Muslim than Hindu ones survive to five years, a big gap."* The difference, it adds, has much to do with toilet habits: "Hindu tradition, seen for example in the 'Laws of Manu', a Hindu text some 2,000 years old, encourages defecation in the open, far from home, to avoid ritual impurity."[1] As a result, "India fares worse on sanitation than a host of poorer places including Afghanistan, Burundi and Congo." Resistance in India against toilet use continues today, in spite of pressure from above, evoking the comment that "[i]t is easier to build a toilet than to get people to use it" (*Economist* of August 19th 2017, p. 40). The same newspaper refers to the effects of widespread diarrhea in its issue of March 24th 2018 (p. 11): "One reason Indian children are shorter than sub-Saharan African children from families of similar means is that they fall sick more often."[2]

Brahmanical ideology, this example suggests, had (and has) effects "on the ground." These effects went beyond mere toilet habits. This paper will first concentrate on the effects Brahmanism had on issues as important as urbanization and economics.

---

* Thanks are due to Gregory Schopen, Harry Falk, Andrey Klebanov, Claire Maes, and Veena Mushrif-Tripathy for their advice, and to Ernst Prets for drawing my attention to the thesis of Angermeier.

1 The Laws of Manu (*Mānavadharmaśāstra*) does not appear to contain a rule of this nature, but other texts on Dharma do; see Pandurang Vaman Kane, *History of Dharmaśāstra*, second edition (Poona: Bhandarkar Oriental Research Institute, 1968–1977), vol. II.1, 649 ("One should answer calls of nature away from a human habitation towards the south or south-west"). Note, however, Klaus Karttunen, "Ancient Traditions and Modern Challenges: South Asian Studies Today," *Politeja* 1, no. 40 (2016), 14: "In the first half of the 20th century it was still common to explain social relations in modern South Asia with theoretical quotations from the works such as the *Mānavadharmaśāstra*. Now, of course, such an attempt would no longer pass as scholarship."

2 For details about toilet use in modern India and its consequences, see Diane Coffey and Dean Spears, *Where India Goes: Abandoned Toilets, Stunted Development and the costs of Caste* (Noidi: HarperCollins India, 2017).

 | DOI:10.1163/9789004438224_011

Brahmanism did not like urban life.[3] The *Baudhāyanadharmasūtra* states:[4] "'A man who keeps himself well under control will attain final bliss even if he lives in a city with his body covered with the city dust and his eyes and face coated with it'—now that is something impossible." The *Āpastambadharmasūtra* prescribes, similarly:[5] "He should also avoid visiting cities." The *Gautamadharmasūtra* points out that, "according to some, Vedic recitation is always suspended in a town."[6] A pericope that occurs a few times in the *Saṃnyāsopaniṣads* states:[7] "He shall avoid [...] capital cities as he would the Kumbhīpāka hell."

These few quotations come from early texts[8] that actually mention cities. Many more adopt a different strategy altogether: they do not mention them; they do as if there are no towns and cities.[9] Where possible, they depict themselves in a world that is no longer there.[10] In this respect (and to avoid misunderstanding, let me add: only in this respect), their behavior is not dissimilar to that of another group that was obsessed with the past: the National-Socialists of the

3 See, in general, Johannes Bronkhorst, *Greater Magadha: Studies in the Culture of Early India* (Leiden and Boston: Brill, 2007), § III.5.

4 BaudhDhS 2.6.33; ed. trans. Patrick Olivelle, *Dharmasūtras: The Law Codes of Āpastamba, Gautama, Baudhāyana, and Vasiṣṭha* (Delhi: Motilal Banarsidass, 2000), 264–265.

5 ĀpDhS 1.32.21; ed. trans. Olivelle, *Dharmasūtras*, 72–73.

6 GautDhS 16.45. Similarly VasDhS 13.8–11.

7 *Nāradaparivrājakopaniṣad*, ch. 7, ed. Dikshitar, 116, ed. Schrader, 199–200; *Bṛhat-saṃnyāsa Upaniṣad*, ed. Schrader (Madras: The Adyar Library, 1912), 268: *tyajet* [...] *rājadhānīṃ kumbhīpākam iva*; trans. Patrick Olivelle, *Saṃnyāsa Upaniṣads: Hindu Scriptures on Asceticism and Renunciation* (New York and Oxford: Oxford University Press, 1992), 214, 253–254. These Upaniṣads know different terms for towns of various sizes, such as *pattana*, *pura* and *nagara*; see *Nāradaparivrājakopaniṣad*, ed. Dikshitar, 81, ed. Schrader, 159: "A mendicant may spend one night in a village, two in a burg, three in a town, and five in a city," trans. Olivelle, *Saṃnyāsa Upaniṣads*, 187.

8 The above-mentioned texts are actually a lot younger than has often been supposed; see Patrick Olivelle, "Patañjali and the Beginnings of Dharmaśāstra: An Alternate Social History of Early Dharmasūtra Production," in *Aux abords de la clairière: Études indiennes et comparées en l'honneur de Charles Malamoud*, eds. Silvia D'Intino and Caterina Guenzi (Turnhout: Brepols, 2012); Johannes Bronkhorst, *How the Brahmins Won: From Alexander to the Guptas* (Leiden and Boston: Brill, 2016), §1.2. The same can be said of much of middle- and late-Vedic literature; see Bronkhorst, *Greater Magadha*, 175–262 ("Part III: Chronology").

9 Bronkhorst, *Greater Magadha*, 251 f.

10 For this reason, Brahmanical ideology may be better characterized as 'reactionary' (as used, e.g., in Mark Lilla, *The Shipwrecked Mind: On Political Reaction* (New York Review Books, 2016)) than as 'conservative'.

Third Reich. In the House of German Art in Munich, opened by Hitler in 1937, there were hundreds of paintings; not one depicted urban and industrial life.[11] This in spite of the fact that already by 1910, "Germany had almost as many large cities as the rest of Europe put together."[12]

Brahmanical literature of the period concerned, then, presents an image of an ideal world that no longer existed (even if we assume that it ever existed). The presentation of this imaginary world was not only meant to preserve the memory of an idealized past; its other, equally important, purpose was to shape the future in accordance with Brahmanical wishes. It is in this period of transition that a new notion pops up in Brahmanical literature, that of the Brahmanical hermitage (*āśrama*). Brahmins are depicted as living in these simple yet idyllic places, dedicating themselves to their Vedic ritual duties, reciting mantras in the process. These hermitages appear in the literature right at the time when also gifts of land to Brahmins—the so-called *agrahāra*s—begin to appear in literature and in the epigraphic record. It only makes sense to connect the two institutions: The literary *āśrama*s functioned as encouragement for rulers and others close to the centers of political power to provide Brahmins with *agrahāra*s, an encouragement that became extraordinarily successful in subsequent centuries. As was to be expected, Brahmanical literature never suggests that *āśrama*s were an innovation. Quite the contrary, literature presents us with the idea that *āśrama*s had always been there. This illustrates the fact that Brahmanism, even where it innovates, never admits that it does so. Brahmanism projects the image of preserving the past, even in cases where historical scholarship can show that it doesn't.

Brahmanism made a concerted effort to project a certain image of Brahmins and the world they live in. This world was partly based on an idealized memory of the past; partly it represented the present interests of those Brahmins. The ideal Brahmin had no truck with the corruptions of modern life, such as city dwelling, the use of debased dialects (i.e., of languages other than Sanskrit), and much else. Instead he lived (or presented himself as living, or as wishing to live) in a pure and idyllic *āśrama*, he used Sanskrit, i.e. the original and pure language, and he did not use modern inventions such as writing.

---

11 Peter Watson, *A Terrible Beauty: The People and Ideas That Shaped the Modern Mind: A History* (London: Phoenix, 2004), 311–312.

12 Peter Watson, *The German Genius: Europe's Third Renaissance, the Second Scientific Revolution, and the Twentieth Century* (London: Simon & Schuster, 2010), 437.

The Brahmanical dislike for urban life had profound consequences, at least according to certain scholars. According to Giovanni Verardi:[13]

> From the 3rd century onwards, the crisis of the trade economy became increasingly profound. The decline of the Indian cities began, which caused the de-urbanization of the country already in Gupta times (…). This was the moment in which the Brahmanic social and economic model, based on land, regained [I would prefer *gained*, JB] the upper hand. I would like to underline, in this regard, that I do not believe that the new social order that India was preparing depended on the general, changed state of the economic fundamentals (the rise of Islam demonstrates that the conditions of economic stalemate could be successfully overcome by relying on trade, to the point of transforming a very large number of regions very different from each other into a great urban and mercantile civilization). On the contrary, I think that Brahmanic ideology, which had always been hostile to anything that questioned the social equilibrium attained in the rural areas, exerted a fundamental function in determining the decline of the urban and mercantile economy of the subcontinent, the struggle against which coincided basically with the struggle against the Buddhists and Jains. The 'Brahmanic model' did not prevail because of objective and uncontrollable factors; on the contrary, it was actively pursued and constructed.

I think that Verardi is right, that Brahmanical ideology was one of the factors behind the urban decline from Gupta times onward.

Having said this, it is only fair to point out that not all Brahmins shunned cities. The most ardent adherents of the Vedic sacrificial tradition no doubt lived in the countryside, far from the cities. However, not all Brahmins did so. From around 500 BCE onward, kings began to rule their kingdoms from courts and capitals, and these courts and capitals attracted Brahmins, i.e., certain Brahmins, as well as others.

Consider the following passage from the *Mahābhārata* (12.77.2–4):

> Those men who manifest perfectly the marks of learning, who look to the Vedic texts on every matter, are the equivalent of Brahmā, [oh] king, and they are celebrated as 'Brahmins'. Those who are perfectly accom-

13 Giovanni Verardi, "Religions, Rituals, and the Heaviness of Indian History," *Annali* (*Istituto Universitario Orientale, Napoli*) 56 (1996): 239.

> plished as ritual priests or teachers and carry out their proper works are the equivalent of the Gods among Brahmins. Those who serve as priests, court priests (*purohita*), advisors (*mantrin*), ambassadors (*dūta*), or finance managers, [oh] king, are the equivalent among Brahmins of Kṣatriyas.
>
> trans. FITZGERALD

This passage shows that there were Brahmins who sold their services as priests, court priests, advisors, ambassadors, or finance managers. It also shows that these Brahmins were ranked lower than those who concentrated on Vedic learning. Indeed, a few verses later the text specifies that, in case of need, a king can take taxes from Brahmins, with the exceptions of those who are the equivalent of Brahmā or the Gods (v. 9), i.e., with the exception of those who concentrate on Vedic learning. The Brahmins of the cities, as is clear from this passage and many others, aspired to various positions, such as that of *purohita* or councilor to the king. These were the Brahmins who wrote, and read, the *Arthaśāstra*, the *Kāmasūtra*, the courtly literature which has been preserved, and no doubt much besides. Information about these urban Brahmins, and about the privileges they felt entitled too, can be obtained from the *Arthaśāstra*.[14] Kangle sums it up in the following words:[15]

> [S]pecial privileges are intended for [the Brahmin], particularly for a Śrotriya, that is, a Brahmin learned in the Vedas. It is recommended, for example, that land free from taxes and fines should be granted to a Śrotriya, just as such lands are to be granted to the priests and preceptors of the ruler (2.1.7). It is also laid down that the property of a Śrotriya, even when he dies without an heir, cannot escheat to the state like the property of other citizens (3.5.28). Brahmins in general are, it seems, to be exempted from payment at ferries and pickets (3.20.14). In many cases, punishment for offences is made dependent on the varṇa of the offender. In cases of abuse, defamation, assault etc., an ascending scale of fines is prescribed in accordance with the offender's varṇa (Chapters 3.18 and 3.19). […] Discrimination on the basis of varṇa is referred to in connection with the oath to be administered to witnesses (3.11.34–37), in the matter

14 For the date and composition of the *Arthaśāstra*, see Patrick Olivelle, *King, Governance, and Law in Ancient India: Kauṭilya's Arthaśāstra* (Oxford: Oxford University Press, 2013), 25ff.

15 R.P. Kangle, *The Kauṭilīya Arthaśāstra. Part III: A Study*, first edition (Bombay: Bombay University, 1965; reprint, Delhi: Motilal Banarsidass, 1986), 144f.

> of inheritance by sons born of wives belonging to different varṇas (3.6.17–20) and so on. Again, the varṇas are to occupy different residential areas in the city, the Brahmins in the north, the Kṣatriyas in the east and so on (2.4.9–15). It is also laid down that in social matters seniority shall be fixed from the Brahmin downwards. And the Brahmin is declared to be free to refuse contributions to common festivals and yet entitled to take full part in them (3.10.43–44). There can be no doubt about the high status enjoyed by the Brahmin as such, or about the privileges and concessions reserved for him.

It is more than likely that the *Arthaśāstra* paints far too attractive a picture of the privileges of the Brahmins, but this is undoubtedly due to the fact that Brahmins were involved in trying to influence public life at and around the royal court; they had to convince the king that it was his task to install and maintain "the law laid down in the Vedic lore which is beneficial, as it prescribes the respective duties of the four *varṇas* and the four *āśramas*."[16] They may or may not have obtained all the privileges they wanted, but what is important for us is that they were there, at the courts and in the cities. These were urban Brahmins, and we may be well advised not to confuse them with those other Brahmins who stayed as far as possible from urban centers, in the countryside where they stuck to their Vedic traditions.

Consider next the *Kāmasūtra* of Vātsyāyana. This is clearly a Brahmanical text, which traces its ancestry to the Brahmanical god Prajāpati and the Upaniṣadic seer Auddālaki Śvetaketu (1.1.5–9).[17] It grants certain privileges to Brahmins who know the Veda (*śrotriya*), such as its rule that the wife of such a Brahmin cannot be taken as lover by someone else.[18] Successful courtesans are presented as offering thousands of cows to Brahmins.[19] One of its chapters is called *Catuḥṣaṣṭi* "sixty-four;" the *Kāmasūtra* points out that some see a link with the *Ṛgveda* here: the *Ṛgveda*, too, is called *Catuḥṣaṣṭi*.[20] It is yet

16 *Arthaśāstra* 1.3.4; trans. R.P. Kangle, *The Kauṭilīya Arthaśāstra*, Part II, second edition (Bombay University, 1972; reprint: Motilal Banarsidass, Delhi, 1988), 7, modified.

17 According to the *Mahābhārata* (Mhbh 1.113), Śvetaketu, the son of Uddālaka, laid down the rule that "a woman's faithlessness to her husband shall be a sin equal to aborticide" (trans. van Buitenen), thus changing the earlier habit of faithlessness. According to Mhbh 12.35.22cd, Uddālaka had Śvetaketu fathered by one of his pupils.

18 Vātsyāyana, *Kāmasūtra* 1.5.29–31.

19 Vātsyāyana, *Kāmasūtra* 6.5.28. See further H.C. Chakladar, *Social Life in Ancient India: A study in Vatsyayana's Kamasutra*, second revised edition (Calcutta: Susil Gupta, 1954), 75f.

20 Vātsyāyana, *Kāmasūtra* 2.2.3.

also a text which deals with urban dwellers: the man-about-town (*nāgaraka*; trans. Doniger and Kakar) plays a central role (and provokes the envy of village dwellers[21]).

It seems clear that there were Brahmins both inside and outside the cities. Ideally Brahmins lived outside the cities and disapproved of them. This disapproval, it is here proposed, is one of the factors, perhaps the most important factor, behind the decline in city life that we observe from the time of the Guptas onward.

Not all scholars agree with this. Consider the following passage from Cédric Ferrier's book *L'Inde des Gupta*:[22]

> Grâce à l'amélioration de l'agriculture, capable de dégager des surplus pour nourrir une population non rurale, et grâce à l'expansion du commerce, les villes de l'empire gupta se sont développées. La géographie urbaine de cette époque est en grande partie l'héritage des périodes précédentes. À partir des IIIe et IVe siècles, on constate cependant une certaine désurbanisation en Inde du Nord, résultant d'une contraction du commerce à longue distance mais aussi de la défaillance des pouvoirs politiques. La recherche historique récente a également mis l'accent sur le rôle du milieu naturel. Au cours de la période, la croissance démographique aurait été en partie absorbée par les villes mais cela aurait entraîné, dans certains cas, un déséquilibre entre la population et les ressources. La pression sur les champs, la déforestation relative et le surpâturage paraissent avoir déstabilisé les milieux fragiles, causant inondations et sécheresses. La population urbaine aurait été contrainte de se redéployer dans les campagnes, la mise en valeur agricole étant favorisée par le biais des cessions de terres.
>
> La géographie urbaine sous les Gupta présente en fait des situations très contrastées. Dans la haute vallée du Gange, le déclin urbain paraît fort, sans doute dû aux troubles politiques dans le Nord-Ouest. Des villes comme Śrāvastī et Kanauj sont manifestement en pleine déshérence. Mathurā, peut-être intégrée dans l'empire sous Samudragupta, voit son influence fortement diminuée. Dans la vallée moyenne, le déclin reste

---

21 See Vātsyāyana, *Kāmasūtra* 1.4.36: "A man who lives in a village stirs up his clever and curious relatives, describing to them the lifestyle of the set of men-about-town and inspiring their longing for that life. He emulates it himself." (trans. Wendy Doniger and Sudhir Kakar, *Vatsyayana Mallanaga Kamasutra* (Oxford: Oxford University Press, 2002), 21).

22 Cédric Ferrier, *L'Inde des Gupta* (*IVe–VIe siècle*) (Paris: Les Belles Lettres, 2015), 153–154.

> marqué, comme le montre le cas de Vaiśālī. La situation est similare à Gayā et à Rājagṛha. Cette région subirait, mais dans une moindre mesure, les conséquences des changements économiques évoquées précédemment. Parallèlement, certaines villes croissent comme celles de la basse vallée du Gange.

In other words, economic difficulties led to the decline in city life. The question how the Gupta Empire, despite major economic difficulties, could establish itself and expand, is answered by Fred Virkus in the following words:[23]

> Dies würde bedeuten, dass ein geographischer Bereich, der von einer ökonomischen und, wie Thakur meint, auch ökologischen Krise heimgesucht wurde, politisch für etwa zwei Jahrhunderte eine dominierende Rolle spielen konnte. Dabei wäre allerdings auch in Betracht zu ziehen, dass möglicherweise gerade durch verschärften Ressourcenmangel, Bondeknappheit und ein allgemeines Absinken des materiellen Enwicklungs- und Lebensniveaus in Nordindien ein starker Expansionsdrang hervorgerufen wurde.

As indicated above, the claim to be considered in this article is that economic factors were not the only ones behind the urban decline in the first half of the first millennium CE. Ideological factors, most notably the predominance of Brahmanical ideology, played a role, too.

How could one possibly prove such a claim?

I think that an article by Harry Falk that came out in 2006 and is called "The Tidal Waves of Indian History: Between the Empires and Beyond" comes as close to a proof as one might reasonably hope. In this article Falk draws attention to an alternating succession of extroverted and introverted phases in Indian history from the Mauryas to the Guptas. In brief:[24]

> the Mauryas made use of the Hellenistic brain drain; the succeeding Śuṅga Dynasty and its successors shunned intense foreign contacts; instead, traditional grammar and ritual were revived.
>
> The succeeding foreign rulers of Śaka, Pahlava, and Kushana stock neglected the indigenous way of life and built centers of learning, e.g.,

---

23 Fred Virkus, *Politische Strukturen im Guptareich* (Wiesbaden: Harrassowitz, 2004), 27.

24 Harry Falk, "The Tidal Waves of Indian History: Between the Empires and Beyond," In *Between the Empires: Society in India 300 BCE to 400 CE*, ed. Patrick Olivelle (Oxford: Oxford University Press, 2006), 146.

> at Ujjain, and made a living through intensive trade relations with East and West. The succeeding Gupta Dynasty reverted to indigenous values, to sciences of the word; inspiration from foreign sources began to come to a close.

The same alternation finds expression in the currency. To quote Falk:[25]

> The oscillating model is obvious: the foreign or foreign-inspired dynasties aim at a high-value currency of precious metals, whereas the first indigenous dynasties, the Śuṅgas and their contemporaries, tend to neglect the issue, and the second indigenous wave, the Guptas and their affiliates, soon lose their way economically.
>
> The attitude toward large-scale enterprise apparent in this scheme reminds us of the traditional archenemies, i.e., the Paṇis in the Ṛgveda or the Asuras in the Upaniṣads: the Paṇis are the rich merchants, as are the Asuras. The people responsible for the Vedas and Upaniṣads maintained ideals of a very different nature: having a thousand cows was treasured, but making riches through trade and barter was disdained. Acquiring riches as *dakṣiṇā* at a sacrifice was accepted, but organizing production units involving possibly impure cooperators was not the core interest of those defining what Vedic *dharma* is and what is not.

Other scholars, too, have noticed the economic consequences of the "closing down of the open society," i.e., the predominance of Brahmanism. Giovanni Verardi puts it as follows:[26]

> There is little doubt that the closing down of the open society of the Buddhists and the resulting weakening of the religion of Dharma coincides with the fall in international trading activities, and in particular with the much decreased demand for Indian goods from Rome. Kuṣāṇa currency, circulating over a vast territory, had been linked to the Roman currency system. The collapse of the Han dynasty in China (AD 221) contributed to changing the picture in Central Asia. By that time, we observe a change in the Indian landscape, namely, a rapid process of de-urbanization. It is every archaeologist's experience that even in the case of continuous human occupation, post-Kuṣāṇa levels display much poorer

25 Falk, "Tidal Waves," 154.

26 Giovanni Verardi, *Hardships and Downfall of Buddhism in India* (Singapore: Institute of Southeast Asian Studies / India: Manohar, 2011), 106.

> building techniques and reuse of earlier building material. A great number of small and large towns were abandoned in the third century, and in certain areas, as is shown by territorial surveys, the collapse of a whole network of roads and small settlements, which had been kept functioning by Buddhist monasteries, is observable.

In other words, Brahmanism, when and where it was accepted, could and did exert an influence on economic life. It was in this way one of the factors, perhaps the most important one, responsible for the urban decline from Gupta times onward.

Along with this, another contributing factor must be considered: in pre-modern days, cities were places where diseases flourished, they were "population sinks."[27] To cite McNeill:[28]

> Until very recently cities were unable to maintain their numbers without a substantial inflow of migrants from surrounding countrysides. Urban health hazards were simply too great, for, in addition to infectious person-to-person diseases transmitted as childhood diseases usually are—by breathing in droplets of infectious matter sneezed or coughed into the atmosphere—ancient cities suffered from an intensified circulation of diseases transmitted through contaminated water supplies, plus a full array of insect-borne infections. Any breakdown of transportation bringing food from afar threatened famine, and local crop failures were often difficult to compensate for. In view of all this it is not surprising that cities could not maintain themselves demographically, but had to depend on migrants from the countryside to replenish the losses arising from famine, epidemic, and endemic diseases.

What this implies is that without sufficient encouragement for country dwellers to move to the cities, or with active discouragement to do so, cities would dwindle all by themselves. Brahmanical ideology may—indirectly, through royal policies—have reduced the attraction of moving into the cities of northern India, for example by changing the conditions in which typically urban activities, such as commerce, could be carried out.

There is a further way in which diseases may have played a role in India's urban decline, presumably once again in tandem with the growing Brahmani-

---

27 David P. Clark, *Germs, Genes, and Civilization: How Epidemics Shaped Who We Are Today* (Upper Saddle River, New Jersey: FT Press, 2010), 213.

28 William H. McNeill, *Plagues and Peoples* (New York: Anchor Books, 1977), 80.

cal cultural influence: through epidemics.[29] Epidemics are to a large extent an urban phenomenon: "You cannot have an epidemic without a large, vulnerable population to infect: that's what makes it an urban disease."[30]

We know that epidemics have had profound effects elsewhere in the world, to the point of bringing down empires and destroying economies. It has even been argued that infections may have contributed to the collapse of the Indus Valley culture: cholera or a closely related waterborne infection according to some,[31] plague outbreaks according to others.[32] And perhaps the Greeks in Alexander's army turned back from northwestern India because they may have been suffering from malaria and dysentery.[33] Richard Gombrich wonders whether infectious diseases may have been a contributing cause to the early Buddhist conviction that life is suffering.[34]

We have seen above that India before the Guptas had gone through a period of relative openness toward the outside world, with many and frequent commercial contacts. With this in mind, we may look at the following passage from McNeill's classic *Plagues and Peoples*, 124:

> We may infer that by about the beginning of the Christian era, at least four divergent civilized disease pools had come into existence [in Eurasia], each sustaining infections that could be lethal if let loose among

---

29 The observations at the beginning of this article may have some relevance here. The custom of pilgrimage may have made India particularly vulnerable, as the colonial authorities around 1900 CE realized by banning Muslim pilgrimage to the Holy Lands (Sandhya L. Polu, *Infectious Disease in India, 1892–1940: Policy-Making and the Perception of Risk* (London: Palgrave Macmillan, 2012), 31ff.); on pilgrimage in Hinduism, see Knut Jacobsen, *Pilgrimage in the Hindu Tradition: Salvific space* (New York: Routledge, 2012).

30 Brenna Hassett, *Built on Bones: 15,000 Years of Urban Life and Death* (London: Bloomsbury Sigma, 2017), 211. Cf. Jared Diamond, *Guns, Germs, and Steel* (London: Jonathan Cape, 1997), 205f.

31 Clark, *Germs, Genes, and Civilization*, 71ff.

32 Tony Joseph, *Early Indians: The Story of Our Ancestors and Where We Come From* (New Delhi: Juggernaut, 2018), 159–160, with a reference to Simon Rasmussen et al., "Early Divergent Strains of Yersinia Pestis in Eurasia 5,000 Years Ago," *Cell* 163, no. 3 (2015). Rita P. Wright, *The Ancient Indus: Urbanism, Economy, and Society* (Cambridge: Cambridge University Press, 2010) does not include epidemics among the list of causes that may have contributed to the decline of the Indus civilization. J.R. McNeill and William H. McNeill, *The Human Web: A Bird's-Eye View of World History* (New York and London: W.W. Norton, 2003), 62, suggest that the onset of malaria may have played a role.

33 R.S. Bray, *Armies of Pestilence: The Effects of Pandemics on History* (Cambridge: The Lutterworth Press, 1996), 10.

34 Richard F. Gombrich, *Theravāda Buddhism: A Social History from Ancient Benares to Modern Colombo* (London and New York: Routledge and Kegan Paul, 1988), 58–59.

> populations lacking any prior exposure or accumulated immunity. All that was needed to provoke spill-over from one pool to another was some accident of communication permitting a chain of infection to extend to new ground where populations were also sufficiently dense to sustain the infection either permanently, or at least for a season or two. The plague of Athens [in 430–429 BCE] seemed to have been such an episode; others undoubtedly occurred in India, China, and elsewhere without leaving any trace we can discover today.
>
> When, however, travel across the breadth of the Old World from China and India to the Mediterranean became regularly organized on a routine basis, so that thousands of individuals began to make a living by travelling to and fro, both on ship-board and by caravan, then conditions for the diffusion of infections among the separate civilizations of the Old World altered profoundly. The possibility of homogenization of those infections, whose most critical limit was defined by the number of new human hosts available day in and day out, opened up. It is my contention that something approximating this condition did in fact occur, beginning in the first century A.D.

McNeill continues (ibid., 128–129):

> During the two centuries that followed the beginning of the Christian era ... it seems certain that trade between the eastern Mediterranean, India and China operated on a regular basis and attained a scale that dwarfed all earlier exchanges across such distances. Caravans passed overland across the oases and deserts of central Asia by regular stages, while ships travelled freely across the Indian Ocean and its adjacent waters.
>
> Regular movement to and fro across such distances implied exchange of infections as well as goods. Chances of an unfamiliar infection spreading among susceptible populations certainly multiplied, and there is reason to suppose that before the end of the second century A.D. epidemic disasters in fact struck severe blows to Mediterranean populations, and probably afflicted the population of China as well.

What about India? Did it too suffer epidemic disasters during that period? India certainly had its share of them in recent centuries, about which information is available.[35] Lack of information about the earlier period makes McNeill waver (ibid., 129):

35 See, e.g., Polu, *Infectious Disease in India*.

> Either the populations of Middle Eastern and Indian cities had little to fear from diseases previously established among the Chinese and Mediterranean populations ...; or surviving records are so imperfect that disease disasters in Middle Eastern and Indian landscapes cannot now be detected.
>
> Indirect evidence suggests that exposure to new infection had little effect in either India or the Middle East. ... In India, the political consolidation and cultural efflorescence of the Gupta age (A.D. 320–535) ... suggests (though scarcely proves) that no particularly severe demographic disasters afflicted that country as a result of the merging of previously separate disease pools in the first Christian centuries.[36]

David Christian states, similarly:[37]

> New diseases had their greatest impact at the extremes of the Eurasian world system, in the Mediterranean and China, where earlier contacts had been most restricted. They had less impact in Mesopotamia and India, regions that lay closer to the hub of the Eurasian network of exchanges and were therefore more disease-hardened.

These passages should make us reflect. It seems impossible not to think of the de-urbanization considered above as at least in part the result of demographic disasters such as the arrival of major epidemics. Since no direct evidence is

36 About the Plague of Justinian and its aftermath, Bray (*Armies of Pestilence*, 27; cited in Ronald Findlay and Mats Lundahl, "Demographic Shocks and the Factor Proportions Model: From the Plague of Justinian to the Black Death," in *Eli Heckscher, International Trade and Economic History*, eds. Ronald Findlay et al (Cambridge, Massachusetts & London, England: MIT Press, 2006), 159) states: "Its distribution to the east of Syria is something of a puzzle. It would seem that the Arabs did not bring the plague back to the Hejaz ... Equally it would seem that the Arabs did not bring the plague to Afghanistan, the Indus, Ferghana and Transcaucasia and thus into the Indian subcontinent and China, as all observers have the plague ceasing at about the present western border of Iran and the Caucasus." About the Black Death, George D. Sussman, "Was the Black Death in India and China?" *Bulletin of the History of Medicine* 85, no. 3 (2011): 322 observes: "Today, over thirty years after McNeill's *Plagues and Peoples*, we still cannot state with any degree of assurance whether the Black Death, which marked such a caesura in European history, even visited China or the Indian subcontinent in the fourteenth century, much less what impact it might have had in either of these civilizations." In spite of this, Sussman then argues that "the Black Death almost certainly did not appear in India in the fourteenth century."

37 David Christian, *Maps of Time: An Introduction to Big History* (Berkeley: University of California Press, 2004), 315.

available from India,[38] it will be useful to consider the Roman world, about which there is more evidence. As McNeill (*Plagues and Peoples*, 130 ff.) points out, there were at least eleven cases of pestilential disaster in republican times, the earliest dated 387 BCE. Another epidemic struck the city of Rome in 65 CE, followed by a far worse one from 165 to 180 CE, which killed in affected places probably as much as a quarter to a third of the entire population. An epidemic of similar magnitude hit the Roman world in 251–266 CE: five thousand a day are said to have died during its height in the city of Rome alone. The next conspicuously significant pestilence, called Justinian's plague, arrived in 542 CE and raged intermittently until 750.[39] Cities, with their high densities of population, were of course most prominently hit by epidemics.

The results of all this in the Western world are described as follows (ibid., 135):

> Political, economic, and cultural consequences of the intensification of micro- and macroparasitism[40] in the Mediterranean lands are too familiar to need much emphasis here. Repeated waves of barbarian invasion accompanied by the decay of cities, migration of artisans to the countryside, loss of skills (including literacy) and the breakup of imperial administration are the familiar hallmarks of the so-called Dark Ages in the West.

Indeed, after Justinian's plague "it is a fact ... that all around the Mediterranean, the cities, as they had existed in antiquity, contracted and then practically

---

38 Veena Mushrif-Tripathy reminds me that little skeletal evidence is available because cremation was the main way to dispose of bodies. See V. Mushrif-Tripathy et al., "Where Are They Now? The Human Skeletal Remains from India," in *A Companion to South Asia in the Past*, eds. Gwen Robbins Schug and Subhash R. Walimbe (Chichester, West Sussex: John Wiley & Sons, 2016). Further Oskar von Hinüber, "Cremated Like a King: The Funeral of the Buddha within the Ancient Indian Cultural Context," *Journal of the International College for Postgraduate Buddhist Studies* 13 (2009): "The hope to find consolation in the archaeological evidence is almost immediately destroyed: Bones or ashes thrown into a river don't leave very significant traces for later archaeologists to discover. The same is true for the bones deposited at the foot of a tree near the site of a cremation." Charles Allen (*Coromandel: A Personal History of South India* (London: Little, Brown, 2017), 38) adds that also "the Indian porcupine's fondness for chewing old bones may be partly responsible" for the scarcity of human remains.

39 See William Rosen, *Justinian's Flea: The First Great Plague and the End of the Roman Empire* (Penguin Books, 2007).

40 Cf. McNeill, *Plagues and Peoples*, 24: "one can properly think of most human lives as caught in a precarious equilibrium between the microparasitism of disease organisms and the macroparasitism of large-bodied predators, chief among which have been other human beings."

disappeared."[41] Even without documents to support it, it seems reasonable to assume that epidemics hit India, too, during the early centuries of the first millennium.

Little is known about epidemics that hit the Indian subcontinent before relatively recent times.[42] Indeed, "absence of records permits no account whatever of Indian encounters with infectious disease before 1200."[43] For example, judging by the entry *disease* in Bollée's *Cultural Encyclopaedia of the Kathāsaritsāgara*,[44] the *Kathāsaritsāgara* does not mention epidemics.[45] But this does not mean that India had none. Consider, for example, the following verse from the *Atharvaveda*:

> To the Gandhāris, the Mūjavants, the Aṅgas, the Magadhas, like one sending a person a treasure, do we commit the fever (*takmán*).

This is *Atharvaveda* 5.22.14 in the Śaunakīya version, in Whitney's translation.[46] The association of the "fever" to regions and their inhabitants suggests that some kind of epidemic is meant.[47] The *Yugapurāṇa* (verse 92) mentions an epidemic (*janamāra*; trans. Mitchiner) as one of the afflictions that hit the country at the end of the Kali-Yuga, which for this text means during the second half of the first century BCE.[48] The *Rājataraṅgiṇī* (4.524–525) speaks of "a plague

---

41 Findlay and Lundahl, "Demographic Shocks," as cited in Rosen, *Justinian's Flea*, 261.

42 On China, see A. Morabia, "Epidemic and Population Patterns in the Chinese Empire (243 B.C.E. to 1911 C.E.): Quantitative Analysis of a Unique but Neglected Epidemic Catalogue," *Epidemiology & Infection* 137 (2009). It appears that the Romans played a role in spreading tuberculosis into Asia, including South-Asia; see Wilson, "Romans Helped TB Conquer the World," *New Scientist* 239, no. 3185 (7 July 2018), 10.

43 McNeill, *Plagues and Peoples*, 157; cited Ralph W. Nicholas, "The Goddess Śītalā and Epidemic Smallpox in Bengal," *Journal of Asian Studies* 41, no. 1 (1981): 23.

44 Willem Bollée, *A Cultural Encyclopaedia of the Kathāsaritsāgara in Keywords: Complementary to Norman Penzer's General Index on Charles Tawney's Translation*, Studia Indologica Universitatis Halensis 8 (Halle an der Saale: Universitätsverlag Halle-Wittenberg, 2015), 122.

45 To my knowledge there are no other relevant entries in this book.

46 In the Paippalāda versions, as Bahulkar ("The Tradition of the *Atharvaveda* in Varanasi," in *Vedic Investigations*, eds. Asko Parpola and Petteri Koskikallio (Delhi: Motilal Banarsidass, 2016), 3) points out, the word *aṅgebhyaḥ* "to the Aṅgas" has been replaced by *kāśibhyaḥ* "to the Kāśis."

47 Possibly malaria; Kenneth G. Zysk, "Fever in Vedic India," *Journal of the American Oriental Society* 103, no. 3 (1983). Polu (*Infectious Disease in India*, p. 82ff.) calls malaria "India's true plague."

48 John E. Mitchiner, *The Yuga Purāṇa: Critically Edited, with an English Translation and a Detailed Introduction*, second revised edition (Kolkata: Asiatic Society, 2002), 93, 108, xxii.

caused by the *lūtā*-disease [which is] contagious and fatal" (*lūtāmayakṛtā vyāpad … āmayaḥ sparśasaṃcārī tatra vyāpādakaś ca saḥ*).[49] The Buddhist *Suttapiṭaka* appears to refer to an epidemic that hit the village called Nādikā.[50] The *Arthaśāstra* (8.4.1) includes disease (*vyādhi*) and epidemic (*maraka*) among the divine afflictions (*daivapīḍana*). It then presents the following discussion (trans. Olivelle):[51]

> "Between disease and famine, disease hurts undertakings by impeding the activities of workers who die or are sick or afflicted, whereas famine does not hurt undertakings and yields taxes in money and farm animals," say the teachers.
>
> "No," says Kauṭilya. "A disease afflicts a single region and remedies (*pratīkāra*) can be found for it, whereas a famine afflicts all the regions and deprives living beings of their livelihood."
>
> This also explains an epidemic.

Information about the remedies (*pratīkāra*) can be found elsewhere in the text (trans. Olivelle):[52]

> They should counteract the danger of disease through occult remedial measures, physicians through medicines, and thaumaturgic ascetics through pacificatory rites and penances.
>
> That also explains an epidemic. He should have these carried out: bathing at sacred fords, worship of Mahākaccha, milking cows in the cemetery, burning a headless trunk, and a gods' night.
>
> In the case of disease or epidemic affecting farm animals, he should have lustration rites of the stalls and equipment and the worship of their respective gods carried out.

---

49 Vishva Bandhu, *Rājataraṅgiṇī of Kalhaṇa*, Part 1: *Taraṅgas* I–VII (Hoshiarpur: Vishveshvaranand Vedic Research Institute, 1963), 169 (with thanks to James McHugh).

50 DN, II, 91–92; cp. SN, V, 356; Hinüber, "Cremated Like a King," 42, 64.

51 *Arthaśāstra* 8.4.5–8: *vyādhidurbhikṣayor vyādhiḥ pretavyādhitopasṛṣṭaparicārakavyāyāmoparodhena karmāṇy upahanti, durbhikṣaṃ punar akarmopaghāti hiraṇyapaśukaradāyi ca ity ācāryāḥ / neti kauṭilyaḥ / ekedeśapīḍano vyādhiḥ śakyapratīkāraś ca, sarvadeśapīḍanaṃ durbhikṣaṃ prāṇinām ajīvanāyeti/ tena marako vyākhyātaḥ /*

52 *Arthaśāstra* 4.3.13–16: *vyādhibhayam aupaniṣadikaiḥ pratīkāraiḥ pratikuryuḥ, auṣadhaiś cikitsakāḥ śāntiprāyaścittair vā siddhatāpasāḥ / tena marako vyākhyātaḥ / tīrthābhiṣecanaṃ mahākacchavardhanaṃ gavāṃ śmaśānāvadohanaṃ kabandhadahanaṃ devarātriṃ ca kārayet / paśuvyādhimarake sthānārthanīrājanaṃ svadaivatapūjanaṃ ca kārayet /*

Epidemics, we learn from the first of these two passages, afflict a single region at a time. No further details are given.

Medical literature sometimes mentions the "destruction of land and people" (*janapadoddhvaṃsa*), an expression that covers both epidemics and other forms of mass destruction. Particularly relevant in this context is a chapter of the *Carakasaṃhitā* (*Vimānasthāna* 3).[53] Most important for our present purposes is that this text (and certain others) shows awareness of the existence of epidemics, confirming our claim that ancient India did have its share of those. The texts do not provide information about specific epidemics, and indeed, as Ernst Prets reminds me, they may not have recognized epidemics as such, being largely unaware of infection and infectious diseases.[54]

This same chapter of the *Carakasaṃhitā* contains a passage that suggests a link between epidemics and those who engage in commerce, among others. This would imply that those who made their livelihood of commerce were, to at least some extent, held responsible for epidemics. This in its turn would fit in with our thesis that Brahmanical disapproval of commerce and other urban activities was one of the factors that contributed to the emptying of cities. The passage reads, in Wujastyk's translation (my emphasis, JB):[55]

---

53 See Vitus Angermeier, *Über die Auslöschung von Land und Leuten: Die Ursachen von Massensterben und Schutzmaßnahmen gegen diese gemäß Carakasaṃhitā Vi. 3* (Wien 2010); Dominik Wujastyk, *The Roots of Āyurveda: Selections from Sanskrit Medical Writings* (London and New Delhi: Penguin Books 1998), 79–92; Jan G. Meulenbeld, *A History of Indian Medical Literature. Volume IA: Text* (Groningen: Egbert Forsten, 1999), 31–32.

54 Note, however, that Wujastyk (*Roots of Āyurveda*, 83, n. 38) thinks there is "an explicit, though unelaborated, reference to disease contagion through touch" in the sentence *tata uddhvaṃsante janapadāḥ spr̥śyābhyavahāryadoṣāt* (Vi. 3.20) "Then they bring epidemic destruction on the localities, because of the corruption in what one touches, and in what is edible" (trans. Wujastyk); "Daraufhin werden Land und Leute aufgrund der schädlichen Beschaffenheit von Spürbarem und Essbarem ausgelöscht" (trans. Angermeier). On contagion in classical Indian medicine, see further Rahul Peter Das, "Notions of 'Contagion' in Classical Indian Medical Texts," and Kenneth G. Zysk, "Does Ancient Indian Medicine Have a Theory of Contagion?" in *Contagion: Perspectives From Pre-Modern Societies*, eds. Lawrence I. Conrad and Dominik Wujastyk (Aldershot: Ashgate 2000).

55 Dominik Wujastyk, *Roots of Āyurveda*, 83. *Vimānasthāna* 3.20: *yadā vai deśanagaranigamajanapadapradhānā dharmam utkramyādharmeṇa prajāṃ vartayanti, tadāśritopāśritāḥ paurajanapadā vyavahāropajīvinaś ca tam adharmam abhivardhayanti, tataḥ so 'dharmaḥ prasabhaṃ dharmam antardhatte, tatas te 'ntarhitadharmāṇo devatābhir api tyajyante; teṣāṃ tathā 'ntarhitadharmaṇām adharmaṇām adharmapradhānānām apakrāntadevatānām r̥tavo vyāpadyante ... tata uddhvaṃsante janapadāḥ spr̥śyābhyavahāryadoṣāt.*

> [W]hen the leaders in a district, city, guild, or community transgress virtue, they cause their people to live unrighteously. Their subjects and dependents from town and country, and **those who make their living from commerce**, start to make that unrighteousness grow. The next thing is that unrighteousness suddenly overwhelms virtue. Then, those whose virtue is overwhelmed are abandoned even by the gods. Next, the seasons bring calamity on those whose virtue has thus been overwhelmed, on those who have unrighteous leaders, on those who have been abandoned by the gods. ... Then they bring epidemic destruction on the localities, because of the corruption in what one touches, and in what is edible.

Epidemics are mentioned from time to time, but rarely, in Indian narrative literature. The *Avadānaśataka*, a Buddhist text, does so in stories no. 14 (which uses the expressions *īti* and *mahājanamaraka*), 16 (*mahatī ītiḥ*), 31 (*pāṇḍuroga*),[56] as does another Buddhist text, the *Divyāvadāna*.[57] The *Kuvalayamālā*, a Jaina text from the eighth century CE, mentions epidemics in one place (274.9).[58] John Cort (2004: 402–403) mentions some Jaina texts that were presumably composed to remove plague and cholera from local congregations. It is possible, but not certain, that a copper-plate inscription from the first century CE refers to small-pox.[59]The following threat to kings, which occurs in the Laws of Manu (*Mānavadharmaśāstra* 8.22), is also worth noting:

> The entire realm, stricken with famine and pestilence (*vyādhi*), quickly perishes, when it is teeming with Śūdras, overrun by infidels, and devoid of twice-born people.
>
> trans. OLIVELLE

---

56 Avś(V), 36, 42, 78. Cf. Naomi Appleton, "The 'Jātakāvadānas' of the Avadānaśataka: An Exploration of Indian Buddhist Narrative Genres," *Journal of the International Association of Buddhist Studies* 38 (2015): 13–14.

57 Divy, 578, l. 23; Divy(V), 487. See also Gregory Schopen, *Buddhist Nuns, Monks, and Other Worldly Matters: Recent Papers on Monastic Buddhism in India* (Honolulu: University of Hawai'i Press, 2014), 289–292.

58 To be more precise, Christine Chojnacki (*Kuvalayamālā: Roman jaina de 779 composé par Uddyotanasūri*, 2 vols. (I: Étude, II: Traduction et annotations), Indica et Tibetica 50, vols. 1–2 (Marburg: Indica et Tibetica Verlag, 2008), II, 749) uses the term *épidémie* in her translation. Having no access to the original, I do not know what Prakrit word this translates.

59 Harry Falk, "The First-Century Copper-Plates of Helagupta from Gandhāra Hailing Maitreya," *Annual Report of the International Research Institute for Advanced Buddhology at Soka University* (ARIRIAB) 17 (2014): 7–8. Vijay Kumar Thakur (*Urbanisation in Ancient India* (New Delhi: Abhinav, 1981), 297–298) refers to a number of textual passages that deal with what are presumably pestilence and plagues, but none of them are very explicit.

If, then, major epidemics hit the Indian cities—and we have seen that this is a plausible assumption—the effects on at least a number of them may well have been strong and long lasting. In combination with a prevailing Brahmanical ideology, such epidemics might account for the emptying out of certain cities that would subsequently, under pressure from the Brahmanically influenced courts, not be able to re-establish themselves.

We have so far considered two factors that may have contributed to, or even caused, the emptying of certain cities: Brahmanical ideology and epidemics. Are these two factors independent of each other? Or are they somehow linked? A remark in McNeill's already often-cited book makes one wonder. He states (*Plagues and Peoples*, 91): "the taboos on personal contact across caste lines, and the elaborate rules for bodily purification in case of inadvertent infringement of such taboos, suggest the importance fear of disease probably had in defining a safe distance between the various social groups that became the castes of historic Indian society." Sonia Shah, referring to this passage, points out that "[o]f all the various factors that could potentially predict the level of ethnic diversity in a given region, pathogen diversity is one of the strongest."[60] Other authors, too, confirm that fear of disease can affect the ways people interact: "Worrying about parasitic infection correlates with anti-immigrant attitudes, and such biases are heightened at times when people feel more vulnerable to infection;"[61] "pathogen threats strengthen in-group affiliation and solidarity (e.g. ethnocentrism, closeness to family), which creates a supportive network should someone in the group become sick."[62] In view of such observations, it appears that Brahmanism may have found a fertile ground for its hierarchical ideology in a world where infectious diseases took their toll. Brahmanical ideology was perhaps not a response to the threat of infectious diseases (even though this may conceivably have played a role), it certainly could flourish in their presence.

The growing influence of Brahmanism left other traces, too. A recent study of the genomes of a large number of individuals in India comes to the conclusion "that the practice of endogamy was established almost simultaneously,

---

60 Sonia Shah, *Pandemic: Tracking Contagions, from Cholera to Ebola and Beyond* (New York: Sarah Crichton Books, 2016), 194 (with a reference to Elizabeth Cashdan, "Ethnic Diversity and Its Environmental Determinants: Effects of Climate, Pathogens, and Habitat Diversity," *American Anthropologist* 103, no. 4 (2001)).

61 Harriet A. Washington, *Infectious Madness: The Surprising Science of How We "Catch" Mental Illness* (New York, Boston, and London: Back Bay Books, 2015), 175.

62 Ilan Shrira, Arnaud Wisman, and Gregory D. Webster, "Guns, Germs, and Stealing: Exploring the Link between Infectious Disease and Crime," *Evolutionary Psychology* 11, no. 1 (2013), 271; quoted by Washington, *Infectious Madness*.

possibly by decree of the rulers, in upper-caste populations of all geographical regions, about 70 generations before present, probably during the reign (319–550 CE) of the … Gupta rulers."[63] Also an earlier study found evidence for a shift to endogamy, at different moments for speakers of Indo-European languages (on average 72 generations) and those of Dravidian languages (108 generations).[64] The authors of this study translate these numbers of generations into 1900 and 3000 years before present respectively. It is virtually impossible to determine the exact average length of a generation, yet it seems safe to conclude that a shift to endogamy took place during the first half of the first millennium CE, at least in northern India,[65] and it is permitted to assume that the growing influence of Brahmanism played a role in this shift. The question of a possible link between endogamy and rural living remains open for the time being.

What can we conclude from the above? The answer to this question can be brief: in order to correctly assess the cultural, economic, ideological and perhaps political developments in India, it is vital to realize that infectious diseases may have played an important or even determining influence.

Abbreviations

ĀpDhS	*Āpastambadharmasūtra*
Avś(V)	*Avadānaśataka*, ed. P.L. Vaidya, Darbhanga 1958 (BST 19)
BaudhDhS	*Baudhāyanadharmasūtra*
BORI	Bhandarkar Oriental Research Institute, Pune
BST	Buddhist Sanskrit Texts, Darbhanga
Divy	*Divyāvadāna*, ed. E.B. Cowell, R.A. Neil, Cambridge, 1886
Divy(V)	*Divyāvadāna*, ed. P.L. Vaidya, Darbhanga, 1959 (BST 20)

---

63 Analabha Basu, Neeta Sarkar-Roy, and Partha Majumder, "Genomic Reconstruction of the History of Extant Populations of India Reveals Five Distinct Ancestral Components and a Complex Structure." *Proceedings of the National Academy of Sciences* (*PNAS*) 113, no. 6 (2016), 1595.

64 Priya Moorjani et al., "Genetic Evidence for Recent Population Mixture in India," *The American Journal of Human Genetics* 93, no. 3 (2013), esp. 426.

65 Interestingly, the study of Moorjani et al. leaves us with the suspicion that endogamy had been adopted in the Dravidian south a full thousand years before the north, thus suggesting that something like a "caste system" existed there already before and independently of Brahmanical influence. Is it possible that the greater prevalence of infectious diseases in the (warmer) south played a role in this?

DN	*Dīghanikāya*, ed. T.W. Rhys Davids, J.E. Carpenter, 3 vols., 1890–1911 (PTS)
GautDhS	*Gautamadharmasūtra*
Kane, HistDh	Pandurang Vaman Kane, *History of Dharmaśāstra*, 5 vols., second edition, Poona: Bhandarkar Oriental Research Institute, 1968–1977
Mhbh	*Mahābhārata*, crit. ed. V.S. Sukthankar a.o., Poona: BORI, 1933–1941
PTS	Pali Text Society, London
SN	*Saṃyutta-Nikāya*, ed. L. Feer, 5 vols., London: PTS, 1884–1898; vol. 6 (indexes by C.A.F. Rhys Davids), London: PTS, 1904
VasDhS	*Vasiṣṭhadharmasūtra*

References

Allen, Charles. *Coromandel: A Personal History of South India*. London: Little, Brown, 2017.

Angermeier, Vitus. *Über die Auslöschung von Land und Leuten: Die Ursachen von Massensterben und Schutzmaßnahmen gegen diese gemäß Carakasaṃhitā Vi. 3*. Wien, 2010. http://www.academia.edu/209609.

Appleton, Naomi. "The 'Jātakāvadānas' of the Avadānaśataka: An Exploration of Indian Buddhist Narrative Genres." *Journal of the International Association of Buddhist Studies* 38 (2015) 9–31.

Arthaśāstra. Kangle, R.P. *The Kauṭilīya Arthaśāstra*. 3 parts (I: Edition, II: Translation, III: Study). Reprint. Delhi: Motilal Banarsidass, 1986. (See also Olivelle 2013.)

Bahulkar, S.S. "The Tradition of the *Atharvaveda* in Varanasi." In *Vedic Investigations: Papers of the 12th World Sanskrit Conference*. Vol. 1, edited by Asko Parpola and Petteri Koskikallio, 1–15. Delhi: Motilal Banarsidass, 2016.

Bandhu, Vishva. *Rājataraṅgiṇī of Kalhaṇa*. Part 1: *Taraṅgas* I–VII. Hoshiarpur: Vishveshvaranand Vedic Research Institute, 1963.

Basu, Analabha, Neeta Sarkar-Roy, and Partha P. Majumder. "Genomic Reconstruction of the History of Extant Populations of India Reveals Five Distinct Ancestral Components and a Complex Structure." *Proceedings of the National Academy of Sciences* (*PNAS*) 113, no. 6 (2016): 1594–1599.

Bollée, Willem. *A Cultural Encyclopaedia of the Kathāsaritsāgara in Keywords. Complementary to Norman Penzer's General Index on Charles Tawney's Translation*. Studia Indologica Universitatis Halensis 8. Halle an der Saale: Universitätsverlag Halle-Wittenberg, 2015.

Bray, R.S. *Armies of Pestilence: The Effects of Pandemics on History*. Cambridge: The Lutterworth Press, 1996.

Bronkhorst, Johannes. *Greater Magadha: Studies in the Culture of Early India*. Handbook of Oriental Studies 2, vol. 19. Leiden and Boston: Brill, 2007.

Bronkhorst, Johannes. *How the Brahmins Won: From Alexander to the Guptas*. Handbook of Oriental Studies 2, vol. 30. Leiden and Boston: Brill, 2016.

Cashdan, Elizabeth. "Ethnic Diversity and Its Environmental Determinants: Effects of Climate, Pathogens, and Habitat Diversity." *American Anthropologist* 103, no. 4 (2001): 968–991.

Chakladar, H.C. *Social Life in Ancient India: A Study in Vatsyayana's Kamasutra*. Second revised edition. Calcutta: Susil Gupta, 1954 [1929].

Chojnacki, Christine. *Kuvalayamālā: Roman jaina de 779 composé par Uddyotanasūri*. 2 vols. (I: Étude, II: Traduction et annotations). Indica et Tibetica 50, vols. 1–2. Marburg: Indica et Tibetica Verlag, 2008. (Unfortunately I have not had access to the Prakrit original.)

Christian, David. *Maps of Time: An Introduction to Big History*. Berkeley: University of California Press, 2004.

Clark, David P. *Germs, Genes, and Civilization: How Epidemics Shaped Who We Are Today*. Upper Saddle River, New Jersey: FT Press, 2010.

Coffey, Diane and Dean Spears. *Where India Goes: Abandoned Toilets, Stunted Development and the Costs of Caste*. Noida: HarperCollins India, 2017.

Conrad, Lawrence I., and Dominik Wujastyk, eds. *Contagion: Perspectives from Pre-Modern Societies*. Aldershot: Ashgate, 2000.

Cort, John E. "How Jains Know What They Know: A Lay Jain Curriculum." In *Jambhū-Joti (Munivara Jambūvijaya Festschrift)*, edited by M.A. Dhaky and J.B. Shah, 399–413. Ahmedabad: Shresthi Katurbhai Lalbhai Smarak Nidhi, 2004.

Das, Rahul Peter. "Notions of 'Contagion' in Classical Indian Medical Texts." In *Contagion: Perspectives from Pre-Modern Societies*, edited by Lawrence I. Conrad and Dominik Wujastyk, 55–78. Aldershot: Ashgate, 2000.

Diamond, Jared. *Guns, Germs, and Steel*. London: Jonathan Cape, 1997.

Dikshitar, T.R. Chintamani, ed. *The Saṃnyāsa Upaniṣad-s with the Commentary of Śrī Upaniṣad-Brahmayogin*. Reprint. Adyar, Madras: Adyar Library and Research Centre, 1966 [1929].

Doniger, Wendy and Sudhir Kakar, trans. *Vatsyayana Mallanaga Kamasutra: A New, Complete English Translation of the Sanskrit Text with Excerpts from the Sanskrit Jayamangala Commentary of Yashodhara Indrapada, the Hindi Jaya Commentary of Devadatta Shastri, and Explanatory Notes by the Translators*. Oxford: Oxford University Press. 2003.

Falk, Harry. "The Tidal Waves of Indian History: Between the Empires and Beyond." In *Between the Empires: Society in India 300 BCE to 400 CE*, edited by Patrick Olivelle, 145–166. Oxford: Oxford University Press, 2006.

Falk, Harry. "The First-Century Copper-Plates of Helagupta from Gandhāra Hailing Maitreya." *Annual Report of the International Research Institute for Advanced Buddhology at Soka University* (ARIRIAB) 17 (2014): 3–26.

Ferrier, Cédric. *L'Inde des Gupta (IVe–VIe siècle)*. Paris: Les Belles Lettres, 2015.

Findlay, Ronald and Mats Lundahl. "Demographic Shocks and the Factor Proportions Model: From the Plague of Justinian to the Black Death." In *Eli Heckscher: International Trade and Economic History*, edited by Ronald Findlay, Rolf G.H. Henriksson, Håkan Lindgren, and Mats Lundahl, 157–198. Cambridge, Massachusetts, and London, England: MIT Press, 2006.

Fitzgerald, James L., trans. *The Mahābhārata. 11: The Book of Women. 12: The Book of Peace, Part One*. Chicago and London: The University of Chicago Press, 2004.

Gombrich, Richard F. *Theravāda Buddhism: A Social History from Ancient Benares to Modern Colombo*. London and New York: Routledge and Kegan Paul, 1988.

Hassett, Brenna. *Built on Bones: 15,000 Years of Urban Life and Death*. London: Bloomsbury Sigma, 2017.

Hinüber, Oskar von. "Cremated Like a King: The Funeral of the Buddha within the Ancient Indian Cultural Context." *Journal of the International College for Postgraduate Buddhist Studies* 13 (2009): 33–66.

Jacobsen, Knut A. *Pilgrimage in the Hindu Tradition: Salvific space*. New York: Routledge, 2012.

Joseph, Tony. *Early Indians: The Story of Our Ancestors and Where We Come From*. New Delhi: Juggernaut, 2018.

Kangle, R.P. *The Kauṭilīya Arthaśāstra. Part I: A Critical Edition with a Glossary*. Second edition. Bombay: Bombay University, 1969. (Reprint: Delhi: Motilal Banarsidass, 1986.)

Kangle, R.P. *The Kauṭilīya Arthaśāstra. Part II: An English Translation with Critical and Explanatory Notes*. Second edition. Bombay: Bombay University, 1972. (Reprint: Delhi: Motilal Banarsidass, 1988.)

Kangle, R.P. *The Kauṭilīya Arthaśāstra. Part III: A Study*. First edition. Bombay: Bombay University, 1965. (Reprint: Delhi: Motilal Banarsidass, 1986.)

Karttunen, Klaus. "Ancient Traditions and Modern Challenges: South Asian Studies Today." *Politeja* 1, no. 40 (2016): 13–20.

Lilla, Mark. *The Shipwrecked Mind: On Political Reaction*. New York: New York Review of Books, 2016.

McNeill, J.R. and William H. McNeill. *The Human Web: A Bird's-Eye View of World History*. New York and London: W.W. Norton, 2003.

McNeill, William H. *Plagues and Peoples*. New York: Anchor Books, 1977.

Meulenbeld, G. Jan. *A History of Indian Medical Literature. Volume IA: Text*. Groningen: Egbert Forsten, 1999.

Mitchiner, John E. *The Yuga Purāṇa: Critically Edited, with an English Translation and a Detailed Introduction*. Second revised edition. Kolkata: Asiatic Society, 2002.

Moorjani, Priya, Kumarasamy Thangaraj, Nick Patterson, Mark Lipson, Po-Ru Loh, Peryasamy Govindaraj, Bonnie Berger, David Reich, and Lalji Singh. "Genetic Evidence for Recent Population Mixture in India." *The American Journal of Human Genetics* 93, no. 3 (2013): 422–438.

Morabia, A. "Epidemic and Population Patterns in the Chinese Empire (243 B.C.E. to 1911 C.E.): Quantitative Analysis of a Unique but Neglected Epidemic Catalogue." *Epidemiology & Infection* 137 (2009): 1361–1368.

Mushrif-Tripathy, V., K.S. Chakraborty, and S. Lahiri. "Where Are They Now? The Human Skeletal Remains from India." In *A Companion to South Asia in the Past*, edited by Gwen Robbins Schug and Subhash R. Walimbe, 496–533. Chichester, West Sussex: John Wiley & Sons, 2016.

Nicholas, Ralph W. "The Goddess Śītalā and Epidemic Smallpox in Bengal." *Journal of Asian Studies* 41, no. 1 (1981): 21–44.

Olivelle, Patrick. *Saṃnyāsa Upaniṣads: Hindu Scriptures on Asceticism and Renunciation, Translated with Introduction and Notes*. New York and Oxford: Oxford University Press, 1992.

Olivelle, Patrick. *Dharmasūtras: The Law Codes of Āpastamba, Gautama, Baudhāyana, and Vasiṣṭha, Annotated Text and Translation*. Delhi: Motilal Banarsidass, 2000.

Olivelle, Patrick. *Manu's Code of Law: A Critical Edition and Translation of the Mānava-Dharmaśāstra*. Oxford: Oxford University Press, 2005.

Olivelle, Patrick. "Patañjali and the Beginnings of Dharmaśāstra: An Alternate Social History of Early Dharmasūtra Production." In *Aux abords de la clairière: Études indiennes et comparées en l'honneur de Charles Malamoud*, edited by Silvia D'Intino and Caterina Guenzi, 117–133. Bibliothèque de l'École des Hautes Études, Sciences religieuses 154. Turnhout: Brepols, 2012.

Olivelle, Patrick. *King, Governance, and Law in Ancient India: Kauṭilya's Arthaśāstra, a New Annotated Translation*. Oxford: Oxford University Press, 2013.

Polu, Sandhya L. *Infectious Disease in India, 1892–1940: Policy-Making and the Perception of Risk*. London: Palgrave Macmillan, 2012.

Rasmussen, Simon et al. "Early Divergent Strains of Yersinia Pestis in Eurasia 5,000 Years Ago." *Cell* 163, no. 3 (2015): 571–582.

Rosen, William. *Justinian's Flea: The First Great Plague and the End of the Roman Empire*. New York: Penguin Books, 2007.

Schopen, Gregory. *Buddhist Nuns, Monks, and Other Worldly Matters: Recent Papers on Monastic Buddhism in India*. Honolulu: University of Hawai'i Press, 2014.

Schrader, F. Otto, ed. *The Minor Upaniṣads Critically Edited*. Vol. 1: Saṃnyāsa-Upaniṣads. Madras: The Adyar Library, 1912.

Shah, Sonia. *Pandemic: Tracking Contagions, from Cholera to Ebola and Beyond*. New York: Sarah Crichton Books, 2016.

Shrira, Ilan, Arnaud Wisman, and Gregory Webster. "Guns, Germs, and Stealing: Exploring the Link between Infectious Disease and Crime." *Evolutionary Psychology* 11, no. 1 (2013): 270–287.

Sussman, George D. "Was the Black Death in India and China?" *Bulletin of the History of Medicine* 85, no. 3 (2011): 319–355.

Thakur, Vijay Kumar. *Urbanisation in Ancient India*. New Delhi: Abhinav, 1981.

van Buitenen, J.A.B., trans. *The Mahābhārata*. 3 vols. (books 1–5). Chicago: The University of Chicago Press, 1973, 1975, 1978.

Vātsyāyana. *Kāmasūtra*, with Yaśodhara's Jayamaṅgalā Sanskrit commentary and a Hindi commentary, edited by Devduṭṭa Śāstrī. Third edition. Kashi Sanskrit Series 29. Varanasi: Chaukhambha Sanskrit Sansthan, 1982.

Verardi, Giovanni. "Religions, Rituals, and the Heaviness of Indian History." *Annali* (Istituto Universitario Orientale, Napoli) 56 (1996): 215–253.

Verardi, Giovanni. *Hardships and Downfall of Buddhism in India*. Singapore: Institute of Southeast Asian Studies / India: Manohar, 2011.

Virkus, Fred. *Politische Strukturen im Guptareich*. Wiesbaden: Harrassowitz, 2004.

Washington, Harriet A. *Infectious Madness: The Surprising Science of How We "Catch" Mental Illness*. New York, Boston, and London: Back Bay Books, 2015.

Watson, Peter. *A Terrible Beauty: The People and Ideas That Shaped the Modern Mind: A History*. London: Phoenix, 2004.

Watson, Peter. *The German Genius: Europe's Third Renaissance, the Second Scientific Revolution, and the Twentieth Century*. London: Simon & Schuster, 2010.

Whitney, William Dwight, trans. *Atharva-Veda-Saṃhitā*. Reprint. Delhi: Motilal Banarsidass, 1984 [1905].

Wilson, Clare. "Romans Helped TB Conquer the World." *New Scientist* 239, no. 3185 (July 7, 2018): 10.

Wright, Rita P. *The Ancient Indus: Urbanism, Economy, and Society*. Cambridge: Cambridge University Press, 2010.

Wujastyk, Dominik. *The Roots of Āyurveda. Selections from Sanskrit medical writings*. London and New Delhi: Penguin Books, 1998.

Zysk, Kenneth G. "Fever in Vedic India." *Journal of the American Oriental Society* 103, no. 3 (1983): 617–621.

Zysk, Kenneth G. "Does Ancient Indian Medicine Have a Theory of Contagion?" In *Contagion: Perspectives from Pre-Modern Societies*, edited by Lawrence I. Conrad and Dominik Wujastyk, 79–95. Aldershot: Ashgate, 2000.

CHAPTER 11

Allegory and History, Life and Embodiment

Anthony Cerulli

1 Introduction

Since the early 11th century, the Bṛhadīśvara Temple[1] in Thanjavur has stood as an extraordinary example of Dravidian architecture from the late Chola Dynasty (c. 300–1279 CE) and a major lodestone for south Indian devotees of the Hindu god Śiva. Around the turn of the 17th–18th century, a play was staged at the temple in conjunction with one of its festivals. That play was the *Jīvānandanam* or, in English, *Life Delighting*. Its author, Ānandarāyamakhin (henceforth Ānandarāya), belonged to a literati cohort supported by the Maratha king of Thanjavur, Śāhajī, who famously gifted the entire village of Śāhajīrājapuram to 46 scholars and artists in his polity.[2]

Life Delighting is a comical and multilayered drama (*nāṭaka*), which makes it both an enjoyable play to read and a challenging text to translate. Its dialogue is not unique to Sanskrit *nāṭaka*-literature, mixing Sanskrit and Prakrit amongst the men and women and upper and lower classes, respectively. Many of the characters' witty and subversive rants openly challenge and even flout political and religious norms, masterfully drawing in and holding the attention of its audience throughout its progressively expanding seven acts. Confronting civic and popular issues of his day, just like skilled authors throughout history, Ānandarāya's ability to entertain springs from his knack to speak to multiple social concerns simultaneously without imposing an agenda in ham-fisted or moral-

1 Established in 1010 CE, the Bṛhadīśvara Temple has also been known as Rājarājeśvaram Mandir and, in Tamil, Peruvuḍaiyār Kōvil.

2 Today this village in Tamil Nadu is typically known as Thiruviyalur (or Tiruviśainallur). In the text of the *Jīvānandanam*, the play is attributed to Ānandarāyamakhin. Yet ever since T.S. Kuppuswamy Sastri's allusion to the text ("Ramabhadra-Dikshita and the Southern Poets of His Time," *The Indian Antiquary* 33 (May 1904): 181–184) as having been written by someone named Vedakavi, some scholars have followed his lead and claimed that Ānandarāya was not the author, but the patron of the playwright Vedakavi. Neither the text itself nor the editor and commentator on the *Jīvānandanam*, Duraiswami Aiyangar (*Jīvānandanam of Ānandarāya Makhin: A Drama Embodying Teachings of Āyurveda* (Madras: Adyar Library, 1947), 6–12) share Sastri's logic, however. Given the lack of prosopographical evidence for Vedakavi as the play's author, I refer to Ānandarāya as the author of the *Jīvānandanam* throughout this study.

 | DOI:10.1163/9789004438224_012

izing ways. He achieved this by making *Life Delighting* an allegory—a mode of writing with skillful indirectness and double meaning—that poses some challenges for anyone attempting to understand its production, reception, and place in the history of Sanskrit literature. The text's use of language and its layers of meaning open up several questions about the relationship between allegory and historiography in India, and the possibilities of reading the allegorized life of this play as a kind of embodied history of the relatively new Maratha community in Thanjavur, which at the time of Ānandarāya's production had recently replaced the Nayaks as the ruling power in the region. Allegory in general has functioned as a literary vehicle that encapsulates multiple modes for both communicating and preserving historical data.[3] In addition to subtle and symbolic references, *Life Delighting* also contains some prose reporting about the author and his family, aspects of courtly life in Thanjavur, direct references to classical and medieval Āyurvedic texts, allusions to alchemical treatments, and displays of religious principles like *dharma*, *karma*, and *bhakti*.

In this chapter, I reflect on Ānandarāya's play and argue that the overt medicalizing lens of *Life Delighting* presents an epidemiology of an early Maratha-ruled Thanjavur, and as the play unfolds it dramatizes assorted therapeutics this young state might use to bolster its wellbeing, including ethics, political diplomacy, religious devotion, and alchemical autopoiesis. These therapies are not always easy to reconcile with one another, however, and it is not clear whether any one of the play's conceptual registers provides a consistent trajectory to achieve the goal of "making life delightful" (*jīva* + *ānandana*). This fact points to the limitations of allegory as a literary device to provide a unified narrative of history. The play's shifting frames of knowledge (e.g., religious, political, alchemical, etc.) make sense within the parameters of allegory, and they may indeed reflect the texture of lived reality. But allegorical layering, indirectness, and double meaning are difficult to square with the interpretive project and methodological compulsion of historiography to get the facts straight.

In what at first blush might appear to be an incongruous setting for a dramatic production about physical and religious wellbeing, the play centers on an episode of war, a conflict the playwright dramatizes allegorically: its pro-

3 Well known allegories in English language literature, for example, include John Bunyan's allegory, *The Pilgrim's Progress* (1678), about what constitutes the life, thoughts, and deeds of the good Christian's journey through life; Nathaniel Hawthorne's allegory of academic corruption and scientific competition concerning developments in inoculation and alternative (non-Galenic) medicines in *Rappaccini's Daughter* (1844); and George Orwell's allegory of the corruption that power wreaks among political leaders in the events leading up to the Stalin-era Soviet Union in *Animal Farm* (1945).

tagonist is King Life (Jīvarāja) and his adversary is King Disease (Rājayakṣman).[4] Supporting characters are, on one side, concurrently members of King Life's imperial court and personified elements of the king's body, intellect, or emotions. On the other side, supporting characters are members of King Disease's royal entourage and personified illnesses and grotesque bodily disorders. The setting of the play is King Life's *puram*, a Sanskrit term that carries the meanings of fortress (and city) as well as body. In King Life's retinue, the principal actors are his wife, Queen Intellect (Buddhi); his two advisors, Social-knowledge (Vijñānaśarman) and Ascetic-knowledge (Jñānaśarman); and his guru, Devotion-to-Śiva (Śivabhakti). On the other side, King Disease is married to Queen Cholera (Viṣūcī), and he oversees an army of disease-soldiers, including Crown Prince Pallid (Pāṇḍu), Goiter (Galagaṇḍa), Sprue (Grahaṇī), Diarrhea (Atīsāra), Vomiting (Chardi), and a host of other noxious villains.

Life Delighting is an obvious work of fiction. Unlike many works of Sanskrit literature, the text does not tell the story or imagine the court of a mythic (Euhemerised) or heroic king of the past. There was never a "real" King Life or a "real" King Disease. Nor is the play an account of a set of events understood to have transpired chronologically at some point in time. So, then, what kind of (re)source is this text for the historian? To read an allegory like *Life Delighting* with an eye on modes of representation in historical writing (historiography) in India occasions the reader to confront the nature and methods of the historiographic project, and in particular, the ways that historians and historiography might make use of narrative literature. The allegorical nature of this play illuminates what we do when we write history; it invites reflection on the filters we use to classify textual materials (such as "historical," "factual," "realistic," and "mythic"); and it offers novel interpretive frameworks within which an author might imagine his or her present.

The structure of this chapter proceeds in the following way: In section 2, I consider historiography as allegoresis and address some of the major questions I have encountered while exploring allegory and the potential uses of narrative literature in Indian historiography. In section 3, I present a brief history of allegory in Sanskrit literature and sketch out an abridged account of *Life Delighting's* seven acts. Section 4 examines the aspects of the play that nod to

4 In Sanskrit medical literature, the term *rājayakṣman* is a general kind of wasting disease, though it has often been translated as consumption and tuberculosis by modern lexicographers. For a detailed discussion of *rājayakṣman*, see chapter 5 in Anthony Cerulli, *Somatic Lessons: Narrating Patienthood and Illness in Indian Medical Literature* (Albany: State University of New York Press, 2012), 105–122.

Ānandarāya's vision of the Thanjavur Maratha State that supported his artistry and literary output. I suggest in this concluding section some potential allegorical interpretations of Ānandarāya's play and, given the nature of the genre, point to portions of the text that appear to undermine interpretation.

2 Embodying History

What does it mean to propose that modern historiography might be seen as allegoresis, or the interpretation of historical events and texts as allegories? At bottom, I take allegory in this study to be a way of saying one thing while signifying something else. In general, the polysemy of language and the tendency of texts to encode multiple layers of meaning inspire the notion that historiography—the writing of history—may be understood as allegoresis. In some cases, historians seek to discover the "literal" level of historical events, what we might call the "plot" of history: what exactly happened when, where, with whom, and how? Even this task involves acts of allegoresis. But most works of historiography also seek to draw out and interpret other layers, whether political, economic, religious, psychoanalytic, social, ideological, etc. that are encoded in historical events and texts. In any given work of historiography, it is common to see just one of these layers brought to the foreground and emphasized. There are sound, historical reasons that a hermeneutic strategy like allegoresis developed as part of the historiographic project: take, for example, the medieval and early modern task of reading ancient sources allegorically so as to make them fit the so-called true history of the Christian story, or the Pauline reading of the spirit versus the letter of the law. Sometimes the historian's work is to discern the spirit when documents provide the letter, and other times it is to discern the letter when documents offer the spirit.

In earlier stages of history writing in the western world, the overarching frameworks according to which previous events of history were interpreted were Christian: all history, whether pagan or Christian, was ultimately regarded as the unfolding of Divine Providence. In the post-Christian context, heavily marked by cultural relativism, we are interested in emic frameworks of understanding. These frameworks, too, are frequently revealed through acts of allegoresis: by applying alternate emic concepts as keys for interpretation, one can discover emic layers of meaning encoded in historical materials. These kinds of projects most often treat emic frameworks as "ideology," which then becomes one allegorical layer of history or historical text that the historian has uncovered. This emic-ideological layer is routinely subordinated to a more transcendent—or anagogical or mystical—layer, which is the epistemological framework of the historian her- or himself. The historian's epistemology then deciphers the supposed real forces that account for why events unfolded as they did, the secular version of Divine Providence.

"Allegorical interpretation," literary critic Northrup Frye famously asserted, "begins with the fact that allegory is a structural element in narrative: it has to be there, and is not added by critical interpretation alone."[5] Frye's claim suggests that allegorical elements within texts themselves impel and justify the response of allegoresis in the people who encounter them. In such an understanding, any text becomes an appropriate source for the allegorical interpretation of the historian: all texts encode not just the literal level of historical events (Dante's 1st level of interpretation—quotidian meaning), but also the various additional levels of meaning interpreted by the historian (Dante's 2nd, 3rd, and 4th: moral—educational lessons, allegorical—conceptual/abstract meaning, and anagogical—mystical meaning).

Historians may thus situate themselves variously *vis-à-vis* the ways that historical knowledge is encoded in the documents they use for evidence. The historian can, for example, assume that "history" has "snuck into" the text without the author being aware of it. The idea here is an assumption that all texts are grounded in some historical reality that will find a certain mode of expression in the text. The present-day historian ostensibly knows something about the "real" history surrounding a text, and therefore looks for some reflection of this reality in it, with the assumption that the historical *mentalité* of the author accounts for the specific encoding of historical reality in other terms. On this perspective, the historian's allegoresis is inherently anachronistic. It applies back to historical actors the "real" meanings of their actions, meanings of which they were unlikely to have been aware.

Alternatively, the historian can assume that past authors, whose works we study today, *were* conveying history, that they not only understood constructions of the past and how people situated contemporary reality in relation to it, but also that they deployed and may even have modified conventional narrative modes for representing historical events to give new interpretive frameworks for their present moments. This type of undertaking requires that the historian interrogate emic understandings, institutional constructions, and organizational models of history and their relationship to narrative forms, as well as the various devices in literary texts that encouraged their audiences to interpret those texts historically.

---

5 H. Northrop Frye, *Anatomy of Criticism: Four Essays* (Princeton: Princeton University Press, 1957), 14.

Historiographical allegoresis is dependent upon otherness (from the Greek *allos*, "other"), and while it insists that "they"—"those people"—had an alternate historicity, it still, by placing the authors of texts in particular historical circumstances, insists upon putting them within "our" linear model. This is what Walter Benjamin called the historicist "stream of becoming."[6] That there is resistance to our historiographic efforts in-built in texts (and this resistance is something that has been noted time and again with respect to the tradition of Sanskrit literature) might signal that the effort to understand historiography as allegoresis is an important enterprise. What if they/those people/earlier authors call into question the entire historiographic enterprise, and assert, for example, that epochs "we" see as distant from one another can be close, or that people inhabiting the same epoch might dwell in different times and spaces? What if the mode of past writing we study today presents an alternate model of temporality that allows for "untimely" proximity and affinity of past, present, and future as well as orderly distance and difference? And, crucially, what if we do not need to replace the latter pair of relations with the former but, rather, allow for both? Are we then not presented with nonlinear, and even reversible, connections between past, present, and future? Does such an arrangement not call into question our assumptions about what is historical and what is literary? After all, isn't historiography a narrative that reshapes and reinvents the past for the present, a narrative of things past that it is guided by present interests?

Just as Nietzsche did in "On the Use and Abuse of History for Life,"[7] Benjamin railed against the antiquarian spirit that insists on collecting historical facts for the express purpose of organizing them in orderly temporal sequences "as things really were." Both thinkers invite us to question the pervasive logic of tidy temporal difference and distance that informs mainstream historicism. Benjamin qualified historicism's investment in ordering temporal distance by insisting as well on the strategic proximity of past and present, on the way that the layering of multiple temporalities in narrative texts has the potential to destabilize the teleological narrative of progress that makes up "official history." In doing so, he hoped to make room for alternate imaginative and material possibilities in the world of the reader/audience. While we might view all narrative texts as "anachronistic assemblages," along with Jonathan Gil Harris, Benjamin

6 Walter Benjamin, *The Origin of the German Tragic Drama* (New York: Verso, 2009 [1963]), 45.

7 Friedrich Nietzsche originally published this essay ("Vom Nutzen und Nachtheil der Historie für das Leben") in 1874 as the second essay in his *Untimely Meditations* (*Unzeitgemässe Betrachtungen*).

furthermore suggested that allegory, due to its morphology, calls into question the ability of any transcendent framework to account satisfactorily for the disparate layers it amasses.[8]

In my ongoing study and translation of *Life Delighting*, I am interested to probe the text's relationship to and articulation of the historical context of early 18th century Tamil Nadu. In the limited scope of the present chapter, while I follow this general line of inquiry, in a narrower way I ask the following two questions: What are the relationships set up in Ānandarāya's play between its own narrative and its paratextual reality? And, how does *Life Delighting* facilitate the ability of its audience to understand and reproduce those relationships? To explore these questions, the genre or literary mode of allegory itself demands some more consideration, especially within the expansive ambit of Sanskrit literature. I therefore now turn to allegory in India, describe the allegorical storyline of *Life Delighting*, and then speak to what Ānandarāya's decision to write this allegory might suggest about the realities of the Maratha Thanjavur state of his day.

3 Allegorizing Life

I said earlier that allegoresis may be understood as an interpretive response to allegorical structures encoded in texts, and perhaps even in language itself. In an overview of the history of allegory in western literature, Jeremy Tambling notes that "allegorical *writing* is prompted by the ability to perform or undertake allegorical *interpretation*."[9] In other words, texts that are self-consciously allegorical post-date the critical mode of allegoresis. Although it is beyond the scope of this chapter to explore in great detail, it's worth noting that there is good evidence to suggest that the same is true in South Asia: "true" allegories, that is, texts that deliberately structure themselves to have extended multiple and simultaneous registers of meaning, arrive relatively late on the literary scene, perhaps prompted by the rich tradition of reading other modes of simultaneous composition (such as *śleṣa*) and more generally of reading allegorical elements of texts, such as simile, metaphor, and personification.

In Sanskrit literary criticism, only recently do we find a term describing a structural feature of literature that is equivalent to the western category of allegory (from the Greek *allēgoria*, meaning "speaking about something else."). In

8 Jonathan Gil Harris, "Untimely Mediations," *Early Modern Culture*, special issue: "Timely Mediations" 6 (2007), no pagination, online only: https://emc.eserver.org/1-6/issue6.html.

9 Jeremy Tambling, *Allegory: The New Critical Idiom* (New York: Routledge, 2010), 167.

1962, Saroja Agravāla ostensibly coined the Sanskrit neologism *rūpaka-nāṭaka*, which translates to something like symbolic or metaphorical drama, to categorize Kṛṣṇamiśra's 11th century allegorical opus, *Prabodhacandrodaya*.[10] More recently, in the introduction to his translation of the *Prabodhacandrodaya*, Matthew Kapstein cites another recent coinage, *pratīka-nāṭaka*, that Gāyatrī Devī Bakhśī used in 1993 to classify Vedāntadeśika's 14th-century allegorical classic, *Saṃkalpasūryodaya*.[11] Bakhśī's term conveys the meaning that the drama (*nāṭaka*) may be understood by its appearance or on its surface (*pratīka*, in nominal form), hence: a *pratīka-nāṭaka* is a "surface drama." But that meaning itself may in fact be contrary to or inverted from (*pratīka*, in adjectival form) the appearance of the surface storyline, hence: a *pratīka-nāṭaka* is also an "inverted drama."

The *Prabodhacandrodaya* is generally recognized as the earliest known *continuous* allegorical drama in Sanskrit. But Kṛṣṇamiśra was not the first Indian writer to use allegorical techniques in Sanskrit writing. Allegorization occurs quite early in fact. The *Śvetāśvataropaniṣad* (4.6–7) and *Muṇḍakopaniṣad* (3.1), for example, allegorize the story of the two birds in Book 1 of the *Ṛgveda* (1.164.20) to characterize the differing but entangled natures of matter and spirit (*prakṛti* and *puruṣa*). We find a description of allegorical writing in the *Kuvalayānandakārikā*, where Appaya Dīkṣita cast *samāsokti*, highly condensed speech or a "punning verse where the actions of human beings are inferred from a description of nature," as Steven Hopkins recently translated it, to indicate the act of describing something that's present (the birds in the Upaniṣadic allegorization of the Ṛgvedic story) while conveying a message about a concept or something that isn't present (matter and spirit).[12] For Sheldon Pollock, *samāsokti* is a "trope of abbreviation" that involves the implicit characterization of something to be signified while overtly describing only the literal source.[13] We might additionally compare these renderings of *samāsokti* with Monier-

10 Saroja Agravāla, *Prabodhacandrodaya aur uksī Hindī paramparā* (Prayāj: Hindī Sāhitya Sammelan, 1962).

11 Kapstein (trans., *The Rise of Wisdom Moon by Krishna-mishra* (New York: New York University Press and JJC Foundation, 2009), xxxii–xxxiii, lviii, fn. 7) cites Gāyatrī Devī Bakhśī, *Saṃskṛt ke pratīk nāṭak ke rūp meṃ śrī Vedāntadeśika kṛt Saṃkalpasūryodaya ek adhyayan* (Jaipur: Saṃghī Prakāśan, 1993).

12 Steven Hopkins, *The Flight of Love: A Messenger Poem of Medieval South India by Venkatanatha* (New York: Oxford University Press, 2016), 33.

13 Sheldon Pollock, *The Language of the Gods in the World of Men: Sanskrit, Culture, and Power in Premodern India* (Delhi: Permanent Black, 2007 [2006]), 361.

Williams's definition of *upalakṣaṇa*: "the act of implying something that has not been expressed, implying any analogous object where only one is specified; using a term metaphorically or elliptically or in a generic sense; synecdoche (of a part for the whole, of an individual for the species, or of a quality for that in which it resides)."[14] Both of these cases—the trope of abbreviation or punning technique of *samāsokti* or the metaphorical pointing of *upalakṣaṇa*—are usually brief or isolated tactics in Sanskrit composition, however, not narrative-length structural features with which to describe an entire work, which is a defining feature of allegory.

Early suggestions of more sustained allegorical composition might be traced to the 1st–2nd centuries CE in the fragments of Aśvaghoṣa's *Buddhacarita*, where the three central characters Fame (Kīrti), Firmness (Dhṛti), and Wisdom (Buddhi) resemble the kinds of dual-natured characters in Kṛṣṇamiśra's *Prabodhacandrodaya* and Vedāntadeśika's *Saṃkalpasūryodaya*. Yet it's not until much later, in the 9th century CE with Jayanta Bhaṭṭa's didactic and philosophical play, *Āgamaḍambara*, that substantial allegorical ornamentation clearly begins to define entire Sanskrit compositions. After Kṛṣṇamiśra's *Prabodhacandrodaya* in the 11th century, other Sanskrit allegories appeared, such as Yaśapāla's *Mohaparājaya*, Karṇapūra's *Caitanyacandrodaya*, and two works attributed to Ānandarāya, *Jīvānandanam* (*Life Delighting*) and *Vidyāpariṇayam* (*Marriage of Knowledge*).

To test historical truth-claims and allusions to historical events and people, reliable archives and intertextual references greatly facilitate investigations and interpretations of symbolism in complex literary modes like allegory. More often than not, however, the primary investigators of allegories are literary critics, not historians or philologists. What's more, in the history of Sanskrit criticism, allegory has customarily been viewed as an inelegant mode of writing that, Matthew Kapstein argued, "suffers from the constraint of its major premise, for it must tell a story that is in fact a second story, a double task restricting the author's free creation and often lending to allegorical works a rigid, contrived quality."[15] Perceived inelegance born of literary contrivances (e.g., symbolic allusion, punning verse to tell multiple stories at once, and personifying body parts and diseases) might, nevertheless, actually be useful for us

14 Monier Monier-Williams, *Sanskrit-English Dictionary: Etymologically and Philologically Arranged with Special Reference to Cognate Indo-European Languages* (Oxford: Clarendon Press, 1899), s.v., *upalakṣaṇa*.

15 Kapstein, *Wisdom Moon*, xxxiv.

today to recognize and understand the presence of the factual and historical in the so-called fictional and figurative. Instead of dismissing the symbolism in Ānandarāya's story as mere fiction, delivered in an unsophisticated way, and then be done with it, we might ask the text: In the multiple stories permeating *Life Delighting*, are there evident signs of social consciousness or markers signaling how Ānandarāya and his audience might have understood (or misunderstood) the cultural links between the knowledge systems that support the institutions of medicine, religion, and the state in a complicated and sometimes volatile world? There are contrivances in Ānandarāya's drama, yes, and we may see them as stylistic constraints on the playwright's creativity. Even still, that his text dramatizes the human body; tells multiple stories at the same time about medicine, religion, and politics; and mixes art and real life together need not rule out the possibility for this (or any) text to be taken seriously as, in Sheldon Pollock's words, a "real source of social and political action" during the time of its production and perhaps even for other subsequent audiences. In doing this, we must be careful not to exoticize Indian conceptions of myth and history as somehow exceptional in human history. The distinction "between narratives of things that really occurred and narratives of things that have been invented," Pollock reminds us, "is as intelligible in India as elsewhere."[16] To acknowledge that the texts we study in India were meaningful to their composers and consumers and to identify and situate that meaning as part of the social and cultural history (which we then interpret), is an effort to ensure that today we clearly understand the past and, crucially, that we acknowledge that history remains useful to us now.

Even if we see inelegance in allegory as a literary mode because of its in-built setups and authorial constraints, I see *Life Delighting* as an inventive work by a writer who succeeds to entertain and elucidate abstract subjects for implementation by the citizens of Maratha-era Thanjavur. Moreover, as Sanskrit *nāṭaka*s characteristically do, this work also presents some straightforward reporting about the time and place in which it was produced, and as such it is useful as an historical document. Of note, we learn about the artistic community in Thanjavur of which Ānandarāya was a member and about Ānandarāya's family and literary lineage. At the outset of the play, the stage manager explains that *Life Delighting* was composed for live performance at the Bṛhadīśvara Temple Festival in Thanjavur. His remarks convey information about the location for the dramatic performance as well as data about the author of the play and his audience:

16 Sheldon Pollock, "Pretextures of Time," *History and Theory* 46 (October 2007): 380.

Stage Manager:	Listen! Here in the city of Thanjavur, country and city folk from many places have come together hoping to see the chariot festival of Maheśvara. These assembly members have become touchstones to identify the gold that's called poetry that moves the heart. They are receptacles of priceless judgment and playgrounds for the six *darśanas*. They are the rewards for my austerities, and they have made me eager. My heart wants to honor them here with this play.
Asst. Stage Manager:	(shaking his head) On what will the play be based?
Stage Manager:	I am producing a new play called, "Life Delighting."
Asst. Stage Manager:	Who wrote the script?
Stage Manager:	Ānandarāyamakhin. He is the wish-granting tree of the learned poets.[17]

In these opening exchanges between the stage manager and his assistant, the playwright is introduced and the context for the performance of his play is described. Ānandarāya, the stage manager then explains, was something of a celebrity in Thanjavur because of his family and his literary productivity. Under the patronage of the Thanjavur kings, especially Śāhajī, there was a rather large output of literature, on a number of topics and in a variety of languages, chiefly Sanskrit and Telugu, but also Marathi and Tamil.[18] Ānandarāya's family contributed a great deal to that corpus. His patriline looks like this:

---

17 All references to the Sanskrit text of the *Jīvānandanam* are based on Aiyangar's 1947 edition. *Jīvānandanam* 1.5ff.:

Sūtradhāraḥ: śṛṇu tāvat. atra tañjāpure paurajānapadā deśāntarād āgatāś ca
bṛhadīśvararathotsavadidṛkṣayā saṅghībhūtāḥ.
sarasakavitānāmno hemnaḥ kaṣopalatāṃ gatāḥ
viharaṇabhuvaḥ ṣaḍdarśanyā vivekadhanākarāḥ /
vidadhati tapolabhyāḥ sabhyā ime mama kautukaṃ
tad iha hṛdayaṃ nāṭyenaitānupāsitumīhate // 5
Pāripāśvakaḥ: (saśiraḥkampam) kaṃ punaḥ prabandhamavalambya?
Sūtradhāraḥ: nanvasti mama vaśe jīvānandanaṃ nāma navīnaṃ nāṭakam.
Pāripārśvakaḥ: kastasya prabandhasya kaviḥ?
Sūtradhāraḥ: vidvatkavikalpataruḥ ānandarāyamakhī.

18 On this see Dominik Wujastyk, "La bibliotheque de Thanjavur," in *Lieux de savoir—tome 1: espaces et communautés*, ed. Christian Jacob (Paris: Albin Michel, 2007).

Gaṅgādharādhvarin (aka, Kākojīpaṇḍita) had three sons: Nṛsiṃharāya, Tryambakarāya, and Bhagavantarāya. Gaṅgādhara was a minister to King Venkojī (reign: 1674–1684), younger half-brother of Śivajī Bhosle and the first Maratha ruler of Thanjavur.

Nṛsiṃharāya was Ānandarāya's father. He served as a minister in the royal courts of Venkojī and Śāhajī (reign: 1683–1711—the king when *Life Delighting* was produced). Nṛsiṃharāya composed the *Tripuravijaya-champū.*

The middle son, Tryambakarāya, was a minister in the royal courts of Venkojī, Śāhajī, and Sarabhojī (aka, Serfojī I, reign: 1711–1728). He had a son, Gaṅgādharādhvarin II, and a grandson, Nārāyaṇa (who composed the *Vikramasenachampū-kāvya*). Tryambakarāya composed two works, *Strīdharma* and *Dharmakuṭa.*

The youngest son, Bhagavantarāya, was a stepbrother of Nṛsiṃha and Tryambaka, and he wrote the *Mukundavilāsa-kāvya, Rāghavābhyudaya-nāṭaka,* and *Uttarachampū.*

Ānandarāya, son of Nṛsiṃharāya and grandson of Gaṅgādhara, was a factotum in the royal courts of Śāhajī, Sarabhojī, and Tukkojī. He likely died in the final year of Tukkojī's reign (reign: 1728–1736).[19]

We learn about this impressive pedigree in piecemeal throughout *Life Delighting*, and I have augmented this information with extant prosopographical scholarship.[20] About Ānandarāya, the assistant stage manager adds the following:

---

19 The *Jīvānandanam* was written during the reign of Śāhajī, while his other allegory, *Vidyāpariṇayam*, was composed during the reign of Sarabhojī.

20 See, for example: M.S. Mate, *Maratha Architecture (1650 A.D. to 1850 A.D.)* (Poona: University of Poona Press, 1959); V. Raghavan, ed., *Śāhendravilāsa (A Poem on the Life of King Śāhaji of Tanjore, 1684–1710) of Śrīdhara Veṅkaṭeśa (Ayyāvāl)*, Tanjore Saraswati Mahal Series 54 (Tiruchi: The Kalyan Press, 1952); S. Ramanathan and Rajagopal Sarma, eds., *Sāhavilāsagītam of Ḍhuṇḍi Vyāsa*, Tanjore Sarasvati Mahal Series 198 (Thanjavur, 1985); M.G. Rande, *Rise of the Maratha Power & Other Essays* (Bombay: University of Bombay Press, 1961); T. Venkasami Row, *A Manual of the District of Tanjore in the Madras Presidency* (Madras: W.H. Moore, 1883); T.S. Kuppuswami Sastri, "Ramabhadra-Dikshita and the Southern Poets of His Time," *Indian Antiquary* 33 (Bombay: Popular Prakashan, May, 1904); S. Subrahmanya Sastri, ed. and trans., *Varivasyā Rahasya and Its Commentary Prakāśa by Śrī Bhāskararāya Makhin* (Madras: Adyar Library and Research Institute, 1934); K.R. Subramanian, *The Maratha Rajas of Tanjore* (Madras: K.R. Subramanian, 1928).

Asst. Stage Manager: Unlike others, excellence came unexpectedly for Ānandarāyamakhin, as an explosion of eloquence like it did for the yogi Vālmīki.

The stage manager steps in here, pretending to be King Śāhajī himself:

Stage Manager: Since childhood, Ānandarāyamakhin was nourished and cared for with my love. Consequently, he *should* compose poetry that moves the heart! He was undoubtedly blessed by the goddess of speech, Sarasvatī, who took the form of King Śāhajī and was born from the ocean of the imperial lineage of Venkojī.[21]

A 20th century commentator on *Life Delighting*, Duraiswami Aiyangar, pointed out that the phrase "with my love," *premṇā mayā*, refers to the support and encouragement that Śāhajī gave to the writers and artists in Thanjavur during his reign. Śāhajī was renowned for establishing an artistic village near Thanjavur, called Tiruviśainallur or Śāhjīrājapuram, designed for the production of various forms of art and literature, which he gifted as a land grant to 46 pandits, poets, and artists. Apart from Ānandarāya, other noteworthy people in this artistic community included Rāmabhadra Dīkṣita, Bhaskāra Dīkṣita, Raghunātha, and Śrīdhara Veṅkateśa.[22]

---

21 *Jīvānandanam* 1.9 ff.:
ānandarāyamakhino vālmīker iva yoginaḥ | itarāpekṣaṇāt sāraḥ svataḥ sārasvatodayaḥ // 9
Sūtradhāraḥ: {vihasya} māriṣa tattvaṃ na jānāsi yata evaṃ bravīṣi. śṛṇu tāvat—
ā bālyād api poṣito 'jani mayā premṇā tathā lālitas tenāsau sarasām upaitu kavitām ānandarāyādhvarī |
ityekakṣitipālavaṃśajaladherdevyā girāṃ jātayā śrīśāhāvanināyakākṛtibhṛtā nūnaṃ prasādaḥ kṛtaḥ // 10
The reference to the "imperial lineage of Eka"—*ityekakṣitipālavaṃśa*—is problematic. Duraiswami Aiyangar's gloss on the passage simply says: "the imperial lineage named Eka" (*ekākhyaḥ kṣitipālavaṃśaḥ—rājavaṃśaḥ*). I suggest the text refers to the Maratha king Ekojī (also known as Venkojī outside of Thanjavur), who established Maratha rule in Thanjavur in 1676. Ekojī's reign (1676–1683) was relatively uneventful, and by most accounts Ekojī's position in the Maratha dynasty was perpetually overshadowed by his half-brother Śivajī (Row, *A Manual*, 767–769; Subramanian, *Maratha Rajas*, 21–23).

22 Kuppuswami Sastri, "Ramabhadra-Dikshita," 127–129, 176–196; Subramanian, *Maratha Rajas*, 31–33.

Life Delighting thus presents more than one representational mode for us today to read south Indian history. As such it demands that we avoid rigid classifications of fact and fiction (or myth) as the constitutive markers of history and non-history. Historicity here lies in the consciousness of the Indian communities that produced this work, consumed it, celebrated it, rejected it, and so on. There is also historical value for us today in reading literature that sought to produce social action and contributed to the construction of society. To read *Life Delighting* following the model of Rao, Shulman, and Subrahmanyam, open to the availability of historically useful data in various modes of representation, we are then confronted, much like the *karaṇam* literary culture they discuss in *Textures of Time*, with "a certain kind of 'modernity,'" of which a "new sense of the past belongs to a wider series of diagnostic features which include a striking individualism, a new sense of the human body and its possibilities, a common political culture that cuts through the boundaries of collective ('religious') identities and establishes an aesthetic basis for kingship, and so on."[23] Ānandarāya's multilayered storytelling offers a poetics of practice in which political and religious identity and security are achieved through embodied action, writ large like a king struggling to remain independent and to protect his territory and people. In this way, *Life Delighting* offers us today usable historical commentary and sensibility, located in a particular place and time.

Allegorical texts are marked by a persistent tension: one the one hand, they seem to unite in one simultaneous narrative things that are different from one another and to encourage acts of allegoresis that stabilize the allegorical text within transcendent moral or teleological narratives. But on the other hand, as Benjamin argued, the very structure of allegories disrupts this interpretive imperative. Consequently, there is no natural conjunction between the things that allegory conjoins, and no fragment of the whole can deliver a complete truth: the King is life, the city is the body, and both, with all their unique traits, must be taken together. By persistently rupturing the link between the thing and what it represents, allegory disrupts the chronological narrative of the chain of events by presenting an event as containing simultaneous realities that are real in multiple ways. By highlighting the instability of the sign/signifier/signified relationship, by insisting that everything is always at least one other thing, by superimposing and rendering simultaneous disparate processes, allegory obliges its audience to always apprehend the

---

23 The *karaṇam* were composed by scholar-bureaucrats in Andhra Pradesh whom Rao, Shulman, and Subrahmanyam identify as ushering in a new and specific historical awareness in south India between the 16th–18th centuries (*Textures of Time: Writing History in South India 1600–1800* (New York: Other Press, 2001), 264).

text in the sense or capacity of something else. It encodes a strategy of reading, hearing, or seeing *against* the text as a way of reading, hearing, or seeing *with* the text. To read, hear, and/or see an allegory is a lesson in dismantling it, suggesting that, in the words of Steven Mailloux, allegory is a "self-deconstructing textuality."[24]

What do the textuality of *Life Delighting* and its allegorical representations look like? A brief look at the drama's seven acts will help to illustrate the complex morphology of Ānandarāya's text.

After a formulaic benediction by the stage manager (*sūtradhāra*) praising Ānandarāya and the playwright's ancestry, the story opens in Act 1 in the royal court of King Life, where Social-knowledge speaks with one of his lieutenants, Reconnaissance (Dhāranā). She has just returned from a mission at King Disease's compound, and she is recounting what she observed and heard there. She explains that Disease is conspiring to overthrow Life and destroy his fortress. Her assessment of Disease's power and army paints an alarming picture for the welfare of Life and the people living under his protection.

The play alternates back and forth between the fortress-body (*puram*) of Life and the fortress-body of Disease. Crown Prince Pallid assembles his soldiers in Act 2 to discuss their impending attack on Life. Among his best soldiers are Diarrhea (Atisāra), famous for his overwhelming ability to flow through the battlements of any fort, and Abdominal Tumor (Gulma), whose malignancy is swift and sure once he infiltrates a stronghold. While the assault on Life's fortress-body is fully underway in Acts 3 and 4, one of Disease's spies, Swollen Neck Gland (Karṇamūla), reports that Life absconded to an unknown location. The audience knows that he and Queen Intellect fled to the *sanctum sanctorum* of his fortress-body, the Lotus City (Puṇḍarīkapuraṃ), to avoid the attacks of Disease's army. They went there to meet Life's guru, Devotion-to-Śiva, who taught the royal couple that through rigorous devotion (*bhakti*) to Śiva and Pārvatī they would receive an alchemical elixir of mercury (*rasa*) and sulfur (*gandhaka*), transubstantiated from Śiva's semen and Pārvatī's menstrual blood, that brings immunity to disease and ensures physical longevity. In Act 5, Prince Pallid scrambles to prevent Life from getting the Śaiva elixir. He sends Envy (Matsara), Love and Hate (Kāma-Krodhau), Distraction (Vyākṣepa), and Insanity (Unmāda) to interrupt Life's devotional practices in the Lotus City.

24 Steven Mailloux, "Hermeneutics, Deconstruction, Allegory," in *The Cambridge Companion to Allegory*, eds. Rita Copeland and Peter T. Struck (Cambridge: Cambridge University Press, 2010), 255.

When these soldiers fail, Pallid and Disease decide that they should interfere with Life's diet. They order Insalubrity (Apathyā) to upset Life's eating habits and nourishment, which he attempts with the assistance of Excessive Hunger (Atibubhukṣā) and Morbid Appetite (Bhasmakā).

A lengthy dialogue follows in Act 6 between two of Life's courtiers, Action (Karma) and Time (Kāla), about conversations recently overheard between Life and his two advisors, Social-knowledge and Ascetic-knowledge. We learn that Life's two advisors implore the king to pursue very different livelihoods. Social-knowledge is incensed because Life had become so smitten with Devotion-to-Śiva that he increased his meetings with her in the Lotus City. This is problematic because *bhakti*, he thinks, distracts the king from his primary duty of protecting his fortress-body from Disease's army. Conversely, Ascetic-knowledge advises the king to spend more time in the Lotus City with Devotion-to-Śiva, for he believes the physical needs that so obsess Social-knowledge are inferior to the cultivation of the head and heart. When Life's fortress-body is in near ruins from the battle with Disease, he rethinks his behavior, to the delight of Social-knowledge, and he orders up a round of the mercury-sulfur tonic bequeathed by Śiva as a reward for his devotion. Social-knowledge dispenses the elixir to Life's army, making them immune to Disease's attacks.

In the play's final act, Act 7, Social-knowledge advises Life to revisit Devotion-to-Śiva in the Lotus City to continue his religious exercises, acknowledging that since the fortress-body is secure, spiritual pursuits are appropriate. Life goes to her, and during his devotional practice, Śiva and Pārvatī appear before him. Pleased with the king's devotion, the gods gift Life knowledge of Yoga, which in turn leads to knowledge of the self (*ātman*) and absolute reality (*brahman*). This foundational knowledge, the king learns, paves the way for eventual release (*mokṣa*) from the cycle of rebirth and redeath.

4 Historiography as Allegoresis

Is it possible to read Ānandarāya's work, as Rao, Shulman, and Subrahmanyam read *karaṇam* and other South Indian literatures produced between the 17th and 19th centuries, as belonging to a culture of writing that was "intended for communication rather than mere recording," and examine what that communication might entail?[25] *Life Delighting* demonstrates the most crucial element

25 Rao, Shulman, and Subrahmanyam, *Textures of Time*, 136. Also see page 20, where the authors lay out the four defining characteristics of literary "*karaṇam* culture."

of the new historical consciousness that Rao, Shulman, and Subrahmanyam identify: texture. This texture envelops diverse genres and is not strictly under the province of one representational mode of writing: "A matter-of-fact tone predominates even when high drama transpires—even when gods or mysterious subterranean women or a motivating 'fate' enter the story."[26] And, they say, this texture "always provides a strong assertion about the nature of articulated truth ... [which] has its own characteristic, consistency, integrity, and range."[27] The articulation of historical texture in south Indian literature, which immediately opens the historical archive to a "considerable extension of graphic literacy," masterfully illuminates the historiographical task and the value of narrative literature to it.[28] Using this model, Ānandarāya's narrative, though allegorically complex and therefore complicating the texture of the collective consciousness of King Śāhajī's court, can be useful to understanding the history of the Maratha outpost of Thanjavur amid Mughal military pressure and on the eve of territorial battles between the French and British in the region (1744–1761, culminating with the Battle of Vandavasi [Wandiwash] near Pondicherry).

A major challenge to approaching a text like *Life Delighting* to make sense of south Indian history, as Sheldon Pollock has suggested, is to be able "to confront directly what was in the head of premodern makers and readers of texts (which we must then of course go on to interpret)." How do we discern "a form of consciousness that is historical" that informs our understanding of the past?[29] In many ways this is the very task of allegory. Allegory is commentary on its past as well as its present, and this becomes a marker of history for us when we receive it. It is not a genre proper, as much as it is an act of interpretation: *there is allegoresis in allegory itself!* If, following Hans-Georg Gadamer, all understanding is based in history, then we must treat allegory as a kind of historical record that asserts it historicity in dramatic narrative layers.[30] Allegoresis—the act of reading something as allegorical—illustrates a rationality of representation that moves beyond the dichotomous binaries of myth/history and fiction/fact. There is symbolism in allegory, indeed, and this too reveals an important element of epistemology in the cultural moment in which the allegorical text was produced or performed. "It is enough to say that we under-

26 Rao, Shulman, and Subrahmanyam, *Textures of Time*, 99.
27 Rao, Shulman, and Subrahmanyam, *Textures of Time*, 254.
28 Rao, Shulman, and Subrahmanyam, *Textures of Time*, 20.
29 Sheldon Pollock, "Pretextures of Time," 379–380.
30 Hans-Georg Gadamer, *Truth and Method* (New York: Seabury Press, 1975), 235–341.

stand in a different way," said Gadamer, "if we understand at all."[31] In this famous line, he is often misread as slipping into relativism, or even nihilism. I think we should see this remark as an expression of Gadamer's perceptive conviction that all texts from the past come down to us today with something to teach us, despite the fact that we exist in radically different times and under different circumstances. The point is that we seek to understand what mattered in the production of the text, the text's so-called subject and context, rather than attempt only to discern what might have been the author's intent (which is rarely the impact that a text actually makes in history).

As mentioned above, Ānandarāya's seven-act drama is fictive, meant originally for live entertainment. His characters are personifications of human qualities and conditions. Yet, when set against the context of Maratha Thanjavur at the time of its production and performance, some of the text's allegorical representations and allusions suggest that *Life Delighting* provides a valid and reliable historical voice from an isolated Maratha stronghold in southeast India. Had something changed in the region to merit this narrative? Was it no longer business as usual for the Hindu king in the early decades of Maratha Thanjavur—which lasted from 1674 to 1799, at which time the British took over.[32]

This text is the work of a Brahmin playwright in the court of a young king, Śāhajī, who ascended to the throne at age 12 in 1683, after his father, Venkojī, the first Maratha king of Thanjavur (and younger half-brother of the renowned king Śivajī), abdicated the throne to his son.[33] Śāhajī was a polymath and is remembered as an irrepressible patron of the arts. George Mitchell called him "the most able and distinguished of the Maratha line at Thanjavur,"[34] whose "successes in war against real and imagined foes," Sanjay Subrahmanyam quipped, "are only matched by his success in the boudoir, and his literary and musical gifts are the icing on the cake."[35] At the turn of the 17th–18th cen-

31 Gadamer, *Truth and Method*, 264.

32 The Maratha royals putatively held on to the Palace Complex (Aranmanai) until 1855.

33 There is some uncertainty about whether Śāhajī took the throne in 1683 when his father died, or if, as recounted in the *Dharmakuṭa*, a commentary on the *Rāmāyaṇa*, Venkojī abdicated to his son in 1684. Dominik Wujastyk also suggests that Venkojī might have died as late as 1687 ("The Questions of King Tukkoji: Medicine at an Eighteenth Century South Indian Court," *Indian Journal of History of Science* 41, no. 4 (2006): 362). Śāhajī died on 28 September 1711.

34 George Mitchell, *New Cambridge History of India: Architecture and Art of South India, Vijayanagara, and the Successors* (Cambridge: Cambridge University Press, 1995), 21.

35 Sanjay Subrahmanyam, *Penumbral Visions: Making Politics in Early Modern South India* (Ann Arbor: University of Michigan Press, 2001), 149.

turies, across the Deccan the Marathas were regularly engaged in wars with the Mughals, and in Thanjavur, too, Mughals exerted a heavy hand. One of Aurangzeb's generals, Zul'fiqar Khan, cast an especially powerful influence on Śāhajī, demanding from 1691 onward that Thanjavur pay him a tribute to remain independent. The degree of actual independence the Maratha kings enjoyed in Thanjavur is debatable, since, as Subrahmanyam explains, "the Mughals considered the Maratha Raja a *peshkashi zamīndār*, required to pay a half of his estimated annual *jamaʿ* to the Mughal emperor."[36] Śāhajī paid his tribute, and history shows that his rule in Thanjavur was relatively peaceful.

If there was peace when *Life Delighting* was composed and performed, how can we make sense of the social, ethical, and religious imperatives interlacing the play? Was there an historical need for such didacticism? Generally speaking, in both wartime and peacetime, religious and political institutions throughout human history consistently exhort certain activities, prohibit others, insist on certain social relations, condemn others, and so on. In this case, some historians have argued that the Marathas of Thanjavur experienced severe isolation, since they were situated apart from the major portion of the Maratha Empire on the Deccan, which might have resulted in general feelings of instability and insecurity in the royal Thanjavur courts. George Mitchell remarked that the Thanjavur kings ruled with almost no reference to the greater Maratha Empire because of their relative isolation.[37] Julia Leslie saw the Thanjavur dynasty as culturally isolated in the land of Tamil speakers and religiously and politically isolated amid the Mughal powers in the area.[38] Adding important economic and demographic data to the historical picture, Sanjay Subrahmanyam problematized previous readings of Thanjavur's remoteness from the rest of the Maratha confederacy. His research suggests that while certain periods of Maratha rule in Thanjavur were marred by political turmoil and instability, the Thanjavur kings were fairly unfettered financially. They had steady sources of revenue, and they often boasted about it and supported grand public works and religious events. Yet this prosperity did not last, and by the 1760s–1770s the state of Thanjavur was heavily indebted to a number of English Company employees and private Europeans.[39]

36 Subrahmanyam, *Penumbral Visions*, 149.

37 Mitchell, *New Cambridge History of India*, 21.

38 Julia I. Leslie, *The Perfect Wife: The Orthodox Hindu Woman According to the Strīdharmapaddhati of Tryambakayajvan* (New Delhi, Oxford University Press, 1989), 14–15.

39 Subrahmanyam, *Penumbral Visions*, 174–176.

The matter of Thanjavur's geographical location may point to other aspects embedded in *Life Delighting's* allegorical representations. The number of Marathas in Thanjavur at the time the text was composed was fairly small. The largest waves of Maratha migrants to Thanjavur came later, at the behest of Sarabhojī (reign: 1711–1730), Śāhajī's successor.[40] Underneath the often humorous, grotesque, and poetic dramatizations of Ānandarāya's play, there is strong moralizing and palpable anxiety concerning the health of Śāhajī's court and territory. The play reads like a social epidemiology—from the Greek, meaning a study (*logos*) of what is upon (*epi*) the people or district (*demos*)—of the Thanjavur Maratha community. In the Sanskrit medical classics, epidemics (*janamarakam, uddhvaṃsa*) are said to befall communities with people who do not (or cannot) uphold dharmic lifestyles, and in *Life Delighting* Ānandarāya articulates various strategies for acting in the world when faced with external threats. As a sequestered outpost of the Maratha Kingdom, Ānandarāya's play might thus be read as an allegorization of Thanjavur that offers lessons to the small community about Hindu *dharma*. This reading maps King Life's fortress-body (*puram*) onto Thanjavur and the Śāhajī court, whereas King Disease and his army are allegorizations of their political and military opponents, such as Zul'fiqar Khan and the Mughals, or the then-increasingly troubling French and British presences in the region. Such a reading stresses the "epidemiological warfare" of the play, and suggests uncertainty, even anxiety, about security in Thanjavur. In Ānandarāya's hands human embodiment and somatic vulnerability lend themselves well to military imagery and language, and the politics of otherness.

In the end, King Life overcomes the vulnerability of his body/state. An obvious interpretation of the play is that there are remedies to ensure wellbeing. The play personifies and dramatizes various therapeutics to that end, including social ethics, statecraft, religious devotion, and alchemical transformation.

But there is a productive tension within the text between its vision of a recommended mode of existence in the world, one that balances material and spiritual exercise, and the undermining of this view by its very narrative mode, allegory. The play simultaneously encourages and disrupts the links between political, medical, religious, and ethical facts, and it does so with its primary register of the body. Ānandarāya appears to firmly ground his work on the constancy of being: all the protagonists in the play are embod-

40 Sarabhojī explicitly invited Brahmins to settle in Thanjavur, often with the enticement of property. These waves of migration continued through the 18th century until the official fall of Maratha Thanjavur to the British in 1799, at which point migrations basically stopped.

ied (*dehavat*) in the hero, King Life, whose every act to further his own, and thus also his subjects', welfare is an expression of embodiment (*dehavattvam*). For humankind, embodiment is the ground and mode of being in the world, and it cuts across cultural domains and chronological narratives.[41] Over the course of the play, King Life confronts facts of embodied existence and integrates his individual body within the body politic by embodying his particular responsibilities. Insofar as King Life is *every body*, the audience of the play has a model of the constructive integration of the self within a larger social framework.

Though the hero of *Life Delighting* is King Life, and though his body serves to unite the body politic and the narrative as a whole, it's important to note that Ānandarāya was not mobilizing traditional discourse of kingship in any straightforward way. The play gives the reader/viewer an uneasy sense that one cannot think about kingship as usual, and it is precisely through the facts of embodiment that this sense emerges. The king's body is threatened by disease, decay, and death. The play is unremittingly material, physical, bodily. It can't seem to go long without piling up more evidence of the realities of the flesh, and it is difficult for the reader/viewer to subordinate the realities of the flesh to a triumphant model of kingship.

This is the paradox of allegory: in a fundamental way, meaning is simultaneously concretized (the king and his court are the human body and the social body) and destabilized (the king and his court are not just the king and his court, and not just these two bodies, either). To be sure, the promise of stability is there: if general wellbeing in the world can be achieved only through constant efforts of social integration in the face of threats to it, the play gestures toward *mokṣa* and the transfiguration of the messiness of King Life into religious transcendence. But it seems that this eventual transcendence is set against the accumulative (or layering) style of the play, its heaping up of more and more exempla that are part of an account of life. There is an ongoing and abiding tension between the seeming infinity of the gruesome and sickening and the possibility of that which reverses death in an instant and eventually eliminates life itself.

---

41 Thomas Csordas, ed., *Embodiment and Experience: The Existential Ground of Culture and Self* (Cambridge: Cambridge University Press, 1994). In an earlier essay, "Embodiment as a Paradigm for Anthropology" (*Ethos* 18, no. 1, 1990: 5–47), Csordas offers an extensive theorization of embodiment, where he weds the ideas of perceptual consciousness (from Merleau-Ponty's preobjective) and collective practice (from Bourdieu's habitus) to argue that perception is a bodily process consisting of "somatic modes of attention" (a related idea he developed at length in "Somatic Modes of Attention," *Cultural Anthropology* 8, no. 2 (1993): 135–156) that may be correlated with cultural institutions and practices.

The amassing of more and more tableaux, all of which stand as somehow equivalent, calls into question the ability of any one of the play's dominant aesthetic and topical registers to adequately account for the realities of the state in particular, and of life more generally. Neither *dharma*, nor *bhakti*, nor Āyurveda, nor alchemy, nor Yoga alone is sufficient for Life. Although by the end King Life vanquishes King Disease, the model of human history it imparts is one of fluctuating realities and epistemological frameworks, each corresponding to the next. This layering makes sense within the bounds of allegory, and it likely reflects lived reality. At the very least it challenges the historiographical impulse to unify the facts of existence in to one interpretive framework.

Finally, it seems fitting to end by returning to a quote from *Life Delighting* cited above in section 3. In Act 1 (v. 9) the Assistant Stage Manager suggests that poetry came unexpectedly to Ānandarāya, just as it did to Vālmīki. Which is to say, this play is the irruption of something novel, just as the first *śloka* of *kāvya* was suddenly born of Vālmīki's experience of grief (*śoka*). The Stage Manager proceeds to laugh his assistant aside, presumably given the distinguished nature of Ānandarāya's family and of the court that supported him. But it is a striking comment all the same, which signals a recognition that Ānandarāya was presenting an unconventional narrative, both in terms of content and form, one that in more ways than I have been able to discuss in this chapter challenges traditional interpretive strategies, both emic and etic. Even if the comment is simply a flattering comparison to a great poet of the past, it suggests that we should look afresh at the larger tradition of Sanskrit narrative literature for ways in which authors stabilize and destabilize their models of past, present, and future to create untimely histories, histories that are both proximate and distant to our own, thereby making room for alternate imaginative and material possibilities in the worlds of their audiences, including us today.

References

Agravāla, Saroja. *Prabodhacandrodaya aur uksī Hindī paraṃparā*. Prayāj: Hindī Sāhitya Sammelan, 1962.

Aiyangar, Duraiswami. *Jīvānandanam of Ānandarāya Makhin: A Drama Embodying Teachings of Āyurveda*. Madras: Adyar Library, 1947.

Benjamin, Walter. *The Origin of the German Tragic Drama*. Translated by John Osborne. New York: Verso, 2009 [1963].

Cerulli, Anthony. *Somatic Lessons: Narrating Patienthood and Illness in Indian Medical Literature*. Albany: State University of New York, 2012.

Csordas, Thomas. "Embodiment as a Paradigm for Anthropology." *Ethos* 18, no. 1 (1990): 5–47.

Csordas, Thomas. "Somatic Modes of Attention." *Cultural Anthropology* 8, no. 2 (1993): 135–156

Csordas, Thomas, ed. *Embodiment and Experience: The Existential Ground of Culture and Self*. Cambridge: Cambridge University Press, 1994.

Frye, H. Northrop. *Anatomy of Criticism: Four Essays*. Princeton: Princeton University Press, 1957.

Gadamer, Hans-Georg. *Truth and Method*, edited by Garrett Barden and John Cumming. New York: Seabury Press, 1975.

Harris, Jonathan Gil. "Untimely Mediations," *Early Modern Culture*, special issue: "Timely Mediations" 6 (2007): https://emc.eserver.org/1-6/issue6.html.

Hopkins, Steven. *The Flight of Love: A Messenger Poem of Medieval South India by Venkatanatha*. New York: Oxford University Press, 2016.

Kapstein, Matthew, trans. *The Rise of Wisdom Moon by Krishna-mishra*. New York: New York University Press and JJC Foundation, 2009.

Leslie, Julia I. *The Perfect Wife: The Orthodox Hindu Woman According to the Strīdharmapaddhati of Tryambakayajvan*. New Delhi: Oxford University Press, 1989.

Mailloux. Steven. "Hermeneutics, Deconstruction, Allegory." In *The Cambridge Companion to Allegory*, edited by Rita Copeland and Peter T. Struck, 254–265. Cambridge: Cambridge University Press, 2010.

Mate, M.S. *Maratha Architecture (1650 A.D. to 1850 A.D.)*. Poona: University of Poona Press, 1959.

Mitchell, George. *New Cambridge History of India: Architecture and Art of South India, Vijayanagara, and the Successors*. Cambridge: Cambridge University Press, 1995.

Monier-Williams, Monier. *Sanskrit-English Dictionary: Etymologically and Philologically Arranged with Special Reference to Cognate Indo-European Languages*. Oxford: Clarendon Press, 1899.

Nietzsche, Friedrich. "Vom Nutzen und Nachtheil der Historie für das Leben." In *Unzeitgemässe Betrachtungen*. Vol. 2. Leipzig: E.W. Fritzsch, 1873–1876.

Pollock, Sheldon. *The Language of the Gods in the World of Men: Sanskrit, Culture, and Power in Premodern India*. Delhi: Permanent Black, 2007 [2006].

Pollock, Sheldon. "Pretextures of Time," *History and Theory* 46 (October 2007): 366–383.

Raghavan, V., ed. *Śāhendravilāsa (A Poem on the Life of King Śāhaji of Tanjore, 1684–1710) of Śrīdhara Veṅkaṭeśa (Ayyāvāl)*. Tanjore Saraswati Mahal Series 54. Tiruchi: The Kalyan Press, 1952.

Ramanathan, S. and Rajagopal Sarma, eds. *Sāhavilāsagītam of Ḍhuṇḍi Vyāsa*. Tanjore Sarasvati Mahal Series 198. Thanjavur, 1985.

Rande, M.G. *Rise of the Maratha Power & Other Essays*. Bombay: University of Bombay Press, 1961.

Rao, Velcheru Narayana, David Shulman, and Subrahmanyam Sanjay. *Textures of Time: Writing History in South India 1600–1800*. New York: Other Press, 2001.

Row, T. Venkasami. *A Manual of the District of Tanjore in the Madras Presidency*. Madras: W.H. Moore, 1883.

Sastri, S. Subrahmanya, ed. and trans. *Varivasyā Rahasya and Its Commentary Prakāśa by Śrī Bhāskararāya Makhin*. Madras: Adyar Library and Research Institute, 1934.

Sastri, T.S. Kuppuswamy. "Ramabhadra-Dikshita and the Southern Poets of His Time." *The Indian Antiquary* 33 (May 1904): 181–184.

Sanjay Subrahmanyam. *Penumbral Visions: Making Politics in Early Modern South India*. Ann Arbor: University of Michigan Press, 2001.

Subramanian, K.R. *The Maratha Rajas of Tanjore*. Madras: K.R. Subramanian, 1928.

Tambling, Jeremy. *Allegory: The New Critical Idiom*. New York: Routledge, 2010.

Wujastyk, Dominik. "The Questions of King Tukkoji: Medicine at an Eighteenth Century South Indian Court." *Indian Journal of History of Science* 41, no. 4 (2006): 357–369.

Wujastyk, Dominik. "La bibliotheque de Thanjavur." In *Lieux de savoir—tome 1: espaces et communautés*, edited by Christian Jacob, 616–636. Paris: Albin Michel, 2007.

PART 3

Astral Sciences

CHAPTER 12

The Ritualization of Omens in Late Vedic Ritual: A Consideration of Early Evidence

Marko Geslani

1 Introduction*

According to David Pingree, omenology may be one of the earliest branches of Jyotiḥśāstra implicated in the supposed transmission of astronomy and astrology from west to east. Key to his opinion is what he takes to be one of the earliest Vedic divination texts, the *Adbhutabrāhmaṇa*, which survives in three versions in the Vedic corpus. Regarding this text, Pingree wrote,

> One cannot but be struck by the similarity of these omens with their śānti rituals to the Babylonian omens with their namburbi rituals, though no exact parallels exist since the Indian omens are without apodoses. It is possible, however, though not demonstrable, that the original of these three Sanskrit texts was at least in part dependent on a Mesopotamian prototype that reached India slightly before or after the Achaemenid occupation of Gandhāra in the sixth Century B.C.[1]

Whether or not such a supposition should be proved, the problem of omens remains a strikingly novel development within the context of ancient Indian religion. Although divination rituals appear sporadically in early Vedic texts, a new and systematic approach to omens takes hold only in the late-Vedic period. I have argued that divination formed an important impetus for quite

* An earlier version of this paper was presented at the Institute for the Study of the Ancient World in New York in February 2017, at a workshop organized by Bill Mak. It was there that I first had the honor of meeting Professor Zysk.

1 David Pingree, *Jyotiḥśāstra: Astral and Mathematical Literature* (Wiesbaden: Harrassowitz, 1981), 67. Based on this hypothesis, Pingree elsewhere favored a sixth century date for the prototype of this source, assigning the omens to the Neo-Assyrian period: "As a conservative guess one might suggest the sixth century B.C., though a slightly later date is not impossible. However, it does seem to represent a different tradition of omens than that found in the Sanskrit and Pali texts of the Achaemenid period and their descendants," "Mesopotamian Astronomy and Astral Omens in Other Civilizations," 616.

 | DOI:10.1163/9789004438224_013

significant shifts in ritual practice in the post-Vedic period, with lasting effects on later Hindu ritual practice.[2] If we accept the view of an eastward drift of astral knowledge, the development of new ritual forms can be seen as a case of cultural reception. If we favor a more organic, local development, we still have to account for the relatively late development of these rituals in the Vedic context. In either case, however, Vedic divination rituals appear to have developed over several generations at least, given their stratified occurrence in the ritual texts. The current paper focuses on this process of the "ritualization" of omen material that would eventually find its place in Jyotiḥśāstra. If rites of divination and appeasement would eventually become mainstream in popular Hinduism, how did this process occur? In order to unpack this practical reception of omenology then, I turn to our earliest sources, keeping in mind Varāhamihira, in the sixth century CE, as an endpoint.

The major difficulty in reconstructing this process lies in establishing a clear chronology between the texts of different schools of ritual practice. The most reliable method for this investigation is to look at sequential stages of development within a single Vedic school and compare this to other examples. To that end, here I will summarize my work on the late ritual texts of the Atharvaveda, and then compare the possible historical development of the *Adbhutabrāhmaṇa* mentioned above, following N. Tsuji's 1968 article on that text.[3] Varāhamihira's testimony will supply useful clues along the way. My basic hypothesis could be summarized as follows: The early literature in the Vedic-Hindu tradition suggests an increasingly systematic incorporation of omens within the framework of Vedic ritual. As classificatory schemes for omens grew in complexity, new ritual forms were devised to answer them. This process of ritualization was not linear: the texts supply multiple forms of anti-omen rituals and reference debates about their relative efficacy according to schemes of omen classification. On the whole, however, they allow us to imagine a priestly economy that included ritualists and diviners interacting with increasing proximity over time. Not only was divination eventually incorporated into Vedic ritual practice; in the process those same ritual practices were significantly transformed by the interaction.

2 Marko Geslani, *Rites of the God-King: Śānti and Ritual Change in Early Hinduism* (New York: Oxford University Press, 2018).

3 N. Tsuji, "On the Formation of the Adbhuta-Brāhmaṇa," *Annals of the Bhandarkar Oriental Research Institute* 48–49 (1968): 173–178.

2 Rituals for the Tripartite Categorization of Omens in Classical Jyotiḥśāstra

We can determine a provisional telos for the early history of divination rituals by starting from the end of the story, with two general statements about the relationship between rituals and omens. It is well known that canonical divination texts like the *Utpādādhyāya* of Varāhamihira's *Bṛhatsaṃhitā* (chapter 45) classify omens according to three categories: earth, atmosphere, and heavens.[4]

> *divyaṃ graharkṣavaikṛtam ulkānirghātapavanapariveṣāḥ* /
> *gandharvapurapurandaracāpādi yad āntarikṣaṃ tat* // 4
> *bhaumaṃ carasthirabhavaṃ tat śāntibhir āhataṃ śamam upaiti* /
> *nābhasam upaiti mṛdutāṃ śāmyati no divyam ity eke* // 5
> *divyam api śamam upaiti prabhūtakanakānnagomahīdānaiḥ* /
> *rudrāyatane bhūmau godohāt koṭihomāc ca* // 6

> The heavenly omens are irregularities in the planets and asterisms; those of the atmosphere include meteors, whirlwinds, halos, aerial cities and rainbows. Earthly omens occur in moving and stationary objects. These are appeased when struck with appeasement rituals. The atmospheric ones can be weakened. But not the heavenly ones, according to some. But even heavenly omens become appeased by copious gifts of gold, food, and land, or from milking a cow at a shrine to Śiva, or from the Koṭihoma ceremony.

Here Varāhamihira cites a debate concerning the relative intensity of omens. The higher the source of the omen within the tripartite taxonomy, the more irreversible is the resulting misfortune. While the effects of terrestrial omens can be appeased by *śānti* rituals—and those of atmospheric omens only weakened—according to some, at least, heavenly omens cannot be reversed in any way. Varāhamihira disapproves of this opinion, listing an ecclectic set of rituals efficient for heavenly omens, including gifting (*dāna*), and two other rituals—cow-milking (at a Śaiva temple) and the so-called "Homa of 10,000,000 offerings."

Varāhamihira's opinion is at least partly confirmed by a passage from one of the "appendices" (*pariśiṣṭa*) of the *Atharvaveda*, the very latest of the ritual

4 The early prototype for this omen chapter is found in the *Gārgīyajyotiṣa*. An edition is currently being prepared by Kouji Kumagai.

manuals of this fourth and somewhat peripheral Vedic school, parts of which, as Michio Yano and others have argued, predate Varāhamihira.[5] The passage in question comes from *pariśiṣṭa* number 70 (Bhārgavyāni). This is neither a conventional omen catalogue nor a ritual manual, but rather a theoretical treatise that coordinates various ceremonies from the Atharvan repertoire in relation to omens. At the outset a group of Ṛṣis question the Atharvan sage Bhṛgu:

> *lokatrayaniviṣṭānām utpātānām anekadhā* /
> *bhinnānāṃ śamanaṃ noktaṃ vada tv asaṃśayaṃ mune* // 2
>
> AVPŚ 70.1

> Various are the omens found in the triple world, but the appeasement of those assorted (omens) has not been explained. Tell it to us clearly, O Muni!

Again the essential problem stems from the taxonomic diversity of the omens (*utpāta*) which are said to be manifold (*anekadhā*) and assorted. The term *lokatraya*, "triple world" (in the compound *lokatrayaniviṣṭa*) is also a likely reference to the familiar tripartite categorization of omens. The text thus poses the following problem: How might the appeasement (*śamana*) of these omens be configured in a suitably diverse fashion? Bhṛgu responds by introducing the "three-fold" organization of appeasement rituals—a scheme no doubt intended to mirror the omens of the "triple world." After affirming the essential virtue of Atharvan performers of these rites, the text offers the corresponding ritual scheme.

This presents in some detail ritual solutions to the problem of omens. For the potentially incurable heavenly omens, the text concurs with Varāhamihira in prescribing a ritual of *koṭihoma*, or 10,000,000 oblations. The specification of ritual gifts may also correspond to Varāhamihira's statement that heavenly omens can also be appeased through gifting.[6]

For our purposes, these texts establish an end point in the relationship between omens and ritual. By the 6th century, omens were arranged in a three-

5 Michio Yano, "The Nakṣatra system of the Atharvaveda-Pariśiṣṭa," *Journal of Indological Studies*, 28–29 (2016–2017): 65–74. Shingo Einoo operates on the hypothesis that the Gṛhyapariśiṣṭa genre on the whole predates Varāhamihira, an assumption borne out in my work on *śānti* rituals. See Shingo Einoo, "The Formation of Hindu Ritual," in *From Material to Deity*, eds. Shingo Einoo and Jun Takashima (Delhi: Manohar, 2005); Geslani, *Rites of the God-King*.

6 Note that the term *gosahasra*, mentioned here as a payment for the *koṭihoma*, is also the name of one of the gifting rituals described in AVPŚ 16. The same was later canonized as a *mahādāna* in the *Matsyapurāṇa*.

TABLE 12.1 Scheme of appeasement rituals according to Bhṛgu (AVPŚ 70)

Omen	Appeasement ritual	Gift for the guru
Earth	*mahāśānti*, 100,000 *sāvitrī homas*	milch cow, village, a pair of garments
Atmosphere	*mahāśānti*, 1,000,000 *sāvitrī homas*, *ghṛtakambala*	twelve milch cows, 100 gold coins
Heaven	*mahāśānti*, 10,000,000 *homas*	1000 cows

part scheme, in order of increasing severity. The class of rituals meant to counteract these omens was called *śānti*, or appeasement, and this category seems to have embraced a wide number of ritual forms, of increasing cost and complexity. In what follows I attempt to trace the development of this complexity by summarizing the omen rituals in late-Vedic texts, roughly in chronological order. I will begin with the most likely chronology represented in the Atharvan texts, since these present the clearest development.

3 The Case of the *Atharvaveda*

Perhaps the earliest Vedic text dedicated to omens is the so-called "Book of Omens," the thirteenth section of the *Kauśikasūtra*, the earliest ritual manual of the Atharvaveda, which belongs to the domestic (*gṛhya*) level of the Vedic canon.[7] While Bloomfield suggested that the bulk of the *Kauśikasūtra* belongs to the later Vedic *sūtra* period, he also noted that the omen section (*adhyāya* 13) forms a late addition, and was likely an originally independent text.[8] Thus a late-first millennium BCE guess would not be unreasonable.

For our purposes, three observations can be made with regard to this text. In the first place, while it includes a large number of omens (forty-two precisely), it does not classify them in a clear taxonomic scheme; the omens are listed in miscellaneous order. In the second place, the ritual category is called *prāyaścitti*, or atonement, and not *śānti*, or appeasement, as is the regular cat-

7 Along with the *Adbhutabrāhmaṇa*, the text was edited by Albrecht Weber, *Zwei Vedische Texte über Omina und Portenta* (Berlin: Druckerei der Königl. Akademie der Wissenschaften, 1859).

8 In addition to linguistic differences in this *adhyāya*, the chapter also opens with a table of contents. Maurice Bloomfield, "The Kāuşika-Sūtra of the Atharva-Veda," *Journal of the American Oriental Society* 14 (1890): i-424.

egory in later times. Third, the ritual procedure appears to follow the simple Vedic fire sacrifice of the domestic type. To accompany the main offering, a mantra appropriate to the omen in question is recited. For example, in case the dawn does not appear, a mantra is recited dedicated to the dawn, etc. These observations suggest a state of relative simplicity. At this stage, the occurrence of omens does not appear not to have occasioned any major innovations in Vedic ritual thought. The simple sacrifice and the vast repertoire of Vedic mantras were easily applied to each omen individually. There is no sense of a hierarchy of omens, nor any escalation of ritual forms. Despite the fact that omens occur outside of the bounds of the sacrifice, they are placed somewhat awkwardly into the category of atonements for ritual mistakes.

The next stage in the Atharvan omen literature is the *Śāntikalpa* (ritual manual for *śānti*).[9] The text is also of dubious date. One passage in the text appears to list the planets in the non-weekday order, which could suggest that parts of the text predate the fourth century CE. The ritual prescription on the whole closely follows the conventions of the earlier *Kauśikasūtra*, so it is at least highly continuous with the ancient ritual tradition. Both in terms of omen classification and ritual design, however, the *Śāntikalpa* shows a major advance from the earlier omen section of the *Kauśikasūtra*. It describes a ritual known as the *mahāśānti*, or Great Appeasement, which comes in thirty variants, each named after one of the gods of the Vedic pantheon. The paradigmatic variant, the *amṛtā mahāśānti*, is prescribed for "omens of the earth, atmosphere, and heavens." Not only is this the first time in the Atharvan tradition that omen-reversals are called *śānti*, but this shift of ritual category also accompanies a shift in the organization of omens. It would seem that the implicit claim is that this single ritual "appeases" all three categories of omens. As we have seen, such a claim would be contradicted not only by Varāhamihira, but also by the later Atharvan tradition itself.

Although, as in the earlier atonement rites, the form of the *mahāśānti* follows the conventions for the simple Vedic sacrifice, the ritual incorporates a number of striking innovations. At the apex of the fire offering, the priest recites an uncommonly large number of mantras from the Atharvaveda and places the dregs of these offerings into water pots. These pots are used to bathe the sponsor at the end of the ritual. In this way the mantras of the Atharvaveda are materially inscribed on the sponsor's body; later texts will describe these mantras as a kind of spiritual armor. Hence, following from earlier Atharvan

9 Edited by George Melville Bolling in two parts, "The Çāntikalpa of the Atharva-Veda," *Transactions and Proceedings of the American Philological Association* 35 (1904): 77–127; "The Çāntikalpa of the Atharva-Veda," *Journal of the American Oriental Society* 33 (1913): 265–278.

expertise in sorcery, the overall logic of the rite takes on a sense of "bodily protection" from the ill effects of omens. Put simply, by making use of an earlier ritual instrument known as *śānti* water (*śāntyudaka*), Atharvans have essentially transformed the fire sacrifice into a bathing rite.

The final stage of Atharvan development is presented in a much larger and somewhat miscellaneous text, *Atharvavedapariśiṣṭa*s (AVPŚ).[10] Unlike the *Śāntikalpa*, this text incorporates a large number of omen collections, some attributed to astrological authorities. At times the collection resembles canonical divination texts like the *Bṛhatsaṃhitā*. But unlike Varāhamihira, this text seems much more dedicated to the *mahāśānti* as a singular ritual form. In fact, one of the reasons that we know this text to be later than the *Śāntikalpa* is that it prescribes the aforementioned *mahāśānti* ritual on multiple occasions, especially among its various omen catalogues. In total I count at least twenty-nine such references to the *mahāśānti* and its variants. This preference for the *mahāśānti* seems intelligible if we remember that the core of the ritual features the recitation of mantras from the Atharvaveda; only Atharvans would have been qualified to perform it.

While the numerous prescriptions of the *mahāśānti* seems to confirm the claim for its universal applicability for all omens, the situation is somewhat more complex, for, as we have already seen, the text recommends additional rituals as necessary for appeasing increasingly severe omens of the atmosphere and heavens. As in Varāhamihira's earlier comment, which mentions various rituals for omens, the claim of the universal efficacy of the *mahāśānti* was likely disputed. In parts of the appendices, the Atharvan tradition seems to have responded with further ritual complication. One mode of augmenting the *mahāśānti* paradigm appears to have been numerical multiplication. The text supplies three rituals requiring a large number of fire offerings, each increasing by an order of 10: 100,000 (*lakṣahoma*), 1,000,000 (*bṛhallakṣahoma*), and finally, 10,000,000 (*koṭihoma*). The two more excessive forms of this ritual also introduce the novel element of offering kindling sticks (*samidh*s), rather than the traditional vegetable offerings. As seen in the chart above, these high magnitude offerings, paired with the *mahāśānti* and other rituals, were made to correspond to the three levels of omens. The large set of offerings prescribed in such rites most likely required multiple brahmins. Whereas the *mahāśānti* features Atharvan mantras, these large-numbered homa rituals are accompanied by the much more generic *sāvitrī* mantra, which would have been known to all brahmins.

10 George Melville Bolling and Julius von Negelein, eds., *The Pariśiṣṭas of the Atharvaveda* (Leipzig: O. Harrassowitz, 1909–1910).

4 The Case of the *Adbhutabrāhmaṇa*

However cogent, this sequential summary of the Atharvan approach to omens no doubt belies a somewhat more complicated history. Despite their success in later Hindu sources, the Atharvan rituals did not develop in a vacuum; most likely, they were spurred on by competing ritual forms both within and beyond the Vedic tradition. For example, the *mahāśānti* includes a number of preliminary rituals dedicated to the Vināyakas, Planets, Nakṣatras, and Lokapālas. In other Vedic and astrological texts these deities are worshipped independently. Logically speaking, one could easily see how the worship of the planets alone could directly counteract heavenly omens. The *Śāntikalpa* admits as much. But by relegating such ceremonies to mere preliminaries, the Atharvans could place the primary claim of efficacy on their preferred *mahāśānti* paradigm.

The case of the large-numbered fire offerings presents a similar impulse to manage ritual heterogeneity. By the time of the *Atharvavedapariśiṣṭa*s, the full paradigm of the *mahāśānti* had already been established, so much so that it forms a host ritual structure into which the new homa offerings were integrated. For example, the *koṭihoma* explicitly takes the *mahāśānti* as its paradigm:

> *mahāśāntividhānena nirmathyāgniṃ samāhitaḥ |*
> *tāvat kuryād budhaḥ sarvaṃ yāvan no nairṛtaṃ kṛtam ||*
> AVPŚ 31.5.6

> Having carefully kindled the fire by friction according to the procedure laid out in the *mahāśānti*, the wise one should perform everything up to the completion of the *nairṛti* rite.

That the text is referring to the *mahāśānti* ritual as described in the *Śāntikalpa* is proved by the mention of the ritual to worship the goddess of destruction (Nirṛti), which forms the final preliminary in the *mahāśānti* paradigm. In the original scheme, the central ritual prescription of *amṛtā mahāśānti* should follow immediately afterwards. But here the text prescribes 10,000,000 fire offerings. Afterwards, it says, "When the *koṭihoma* is complete, the wise one worships the planets. And he should perform the *amṛtā* [*mahāśānti*] and the *ghṛtakambala*."[11] Leaving aside the mention of planetary worship and the *ghṛtakambala*,

11 *koṭihome samāpte tu yajed grahagaṇān budhaḥ | kārayed amṛtāṃ caiva ghṛtakambalam eva ca ||* (AVPŚ 31.7.5)

the addition of the *amṛtā* at this point essentially continues the original *mahāśānti* schedule as laid out in the *Śāntikalpa*. In other words, the 10,000,000 homa offerings have been encompassed within the original ritual paradigm. In effect, the *koṭihoma*, forms yet another "preliminary" ritual to the *mahāśānti*. This confirms the Atharvan commitment to the *mahāśānti* paradigm, while perhaps suggesting that the large numbered homa is a new, alien element that the tradition is attempting to circumscribe.

Following this supposition allows us to integrate another text into the story, the so-called *Adbhutabrāhmaṇa* mentioned by Pingree, and studied by Weber and Tsuji. In fact this text occurs in three different versions, in the *Adbhutabrāhmaṇa*[12] found in the Sāmavedic *Ṣaḍviṃśabrāhmaṇa*, the later *Āśvalāyanagṛhyapariśiṣṭa* of the *Ṛgveda*,[13] and, indeed, the *Atharvavedapariśiṣṭas*. Triangulating these three versions, Tsuji attempted to delineate the complex history of this small text. Based on mantric evidence, he determined that it must have originated from an Ṛgvedic prototype, to which further texts were added in sequential stages. These stages are represented in the surviving texts as follows:

[1] *Atharvavedapariśiṣṭa* 67.1–7, *Adbhutabrāhmaṇa* 3–10, and *Āśvalāyanagṛhyapariśiṣṭa* 4.12–17, 19, and 21

This original core features a system of omens organized according to seven deities: Indra, Varuṇa, Yama, Agni, Vaiśravaṇa, Viṣṇu, and Vāyu. AVPŚ 67, while a late, versified text, corresponds exactly to this system, in which each deity is given command of a corresponding set of omens. It is to this deity that the sacrifice is dedicated. Take, for example, the section on Indra from the *Adbhutabrāhmaṇa* 3 (cf. AVPŚ 67.1; ĀśvGPŚ 4.12):

> ... *atha yadāsya maṇimaṇikakumbhasthālīdaraṇam āyāso rājakulavivādo vā yānachatraśayyāsanāvasathadhvajapatākāgṛhaikadeśaprabhañjaneṣu gajavājimukhyā vāpramīyāḥ pramīyanta ity evam ādīni tāny etāni sarvāṇīndradevatyāny adbhutāni prāyaścittāni bhavantīndryāyendro marutvata iti sthālīpākaṃ hutvā pañcabhir ājyāhutibhir abhijuhotīndrāya svāhā śacīpate svāhā vajrapāṇaye svāheśvarāya svāhā sarvapāpaśamanāya svāheti vyāhṛtibhir hutvātha sāma gāyet*

12 In Weber, *Zwei Vedische Texte*.

13 In *Āśvalāyanagṛhyasūtram*, ed. Vinayak Ganesha Apte (Poona: Anandashrama press, 1936).

> … Now when his jewels, jewelry, pots, or dishes break; he is exhausted; or there is discord in the royal family; if there is damage on some part of his chariot, umbrella, bed, throne, dwelling room, banner, flag, or house; or the chief horses or elephants die unexpectedly—all of these and others are signs whose atonement belongs to Indra. Offering the *sthālīpāka* dish with the verse "*indrāyendro marutatve*" (SV 1.472 [RV 9.64.22]), he pours an oblation with five portions of ghee, saying "to Indra *svāhā*, to Śacīpati *svāhā*, to Vajrapāṇi *svāhā*, to Īśvara *svāhā*, to the appeaser of all sin, *svāhā*." He should sing the *sāman* after having made an offering while reciting the *vyāhṛti* syllables.

In this case, omens of kingship are related to Indra. Elsewhere, omens pertaining to bodily harm and sleep are related to Yama, and omens pertaining to wind are related to Vāyu. For each case, the ritual solution is a simple fire sacrifice with a mantra dedicated to the deity in question.

[2] *Adbhutabrāhmaṇa* 1–2 and *Āśvalāyanagṛhyapariśiṣṭa* 4.11

The second stage adds an eighth deity, Soma, and effectively spatializes the system, by associating each deity with one of the four cardinal directions, or four vertical "directions" (earth, atmosphere, heavens, and highest heavens).[14] In addition, two texts are added to the head of the original core (Adb 1–2).[15] The first (Adb 1) is what appears to be an originally independent text:

> *athāto 'dbhutānāṃ karmaṇāṃ śāntiṃ vyākhyāsyāmaḥ / pālāśānāṃ samidhām aṣṭasahasraṃ juhuyād aindrayāmyavāruṇadhānadāgneyavāyavyasaumyavaiṣṇavyety aṣṭāv indrāyendro marutvate nāke suparṇam upa yat patantaṃ ghṛtavatī bhuvanānām abhiśriyābhi tyaṃ devaṃ savitāram oṇyar agniṃ dūtaṃ vṛṇīmahe vāta āvatu bheṣajaṃ somaṃ rājānaṃ varuṇam idaṃ viṣnur vicakrama ity etāni sāmaprabhṛty aṣṭaśataṃ japitvā svasti vācayitvā svasti haiṣāṃ bhavati svastidā viśaspatiḥ svastyayanaṃ tārkṣyam ariṣṭanemim tyam ūṣu vājinaṃ devajūtam āyur viśvāyuḥ śataṃ jīva*

14 Hence each section of the *Adbhutabrāhmaṇa* will begin with the instruction, "he turns to the direction x." In the case of AdB 3, the text begins, "he turns to the east" (*sa prācīm diśam anvāvartate*). According to Tsuji's logic, this would not have been the case in the earliest formulation of the system.

15 Tsuji suggested that the addition of Soma—for a total of eight deities—was a result of the influence of this originally independent text.

> *śarado vardhamāna ity etaiḥ sambhārāṇām upasthānam kṛtvā svasti vācayitvā svasti hāsya bhavati svasti hāsya bhavati* |
>
> Adb 1
>
> Now we shall explain the appeasement of ominous objects. He should offer 1008 *samidhs* made of *pālāśa* wood. There are eight verses belonging to Indra, Yama, Varuṇa, Dhānada, Agni, Vāyu, Soma, and Viṣṇu: "*indrāyendro marutvate* (SV 1.472); *nāke suparṇam upa yat patantaṃ* (SV 1.320); *ghṛtavatī bhuvanānām abhiśriyā* (SV 1.378); *'bhi tyaṃ devaṃ savitāram oṇyar* (SV 1.464); *agniṃ dūtaṃ vṛṇīmahe* (SV 1.3); *vāta āvatu bheṣajaṃ* (SV 1.184); *somaṃ rājānaṃ varuṇam* (SV. 1.91); *idaṃ viṣnur vicakrama* (SV 1.222)." Having recited each of these 108 times quietly prior to singing it, and then announcing the blessing, these offerings will be blessed. Having worshipped the sacrificial requisites with these verses, "*svastidā viśaspatiḥ* (ṚV 10.152.2); *svastyayanaṃ tārkṣyam ariṣṭanemim* (ṚV 5.51.16); *tyam ū ṣu vājinaṃ devajūtam* (SV 1.332 [ṚV 10.178]) *āyur viśvāyuḥ* (ṚV 10.17.4) *śataṃ jīva śarado vardhamāna* (ṚV 10.161.4)," and then having announced the blessing, he becomes blessed.

Here the text seems to amalgamate rituals for all of the deities described later in the text in an extended sacrifice involving kindling sticks (*samidhs*). Given the mantric content, Tsuji suggested that the tract derived from an Ṛgvedic source.

At the same time, an additional preface (AdB 2, ĀśvGPŚ 4.11) was added, supplying a frame story concerning a war between the Devas and Asuras:

> *devāś ca vāsurāś caiṣu lokeṣv aspardhanta te devāḥ prajāpatim upādhāvaṃs tebhya etāṃ daivīṃ śāntiṃ prāyacchatte tataḥ śāntīkā asurān abhajayaṃs tato devā abhavan parāsurā bhavanty ātmanā parāsya bhrātṛvyo bhavati ya evaṃ veda*
>
> So did the gods and the Asuras contend with one another among these worlds. Those gods resorted to Prajāpati. He gave them this divine appeasement. Then possessed of that appeasement they conquered the Asuras. Then were the gods superior to the Asuras. He becomes superior to his rival who knows thus.

The text continues with preliminaries for the subsequent rituals. Note that the passage uses the term *śānti* to refer to the ritual.

[3] *Adbhutabrāhmaṇa* 11–12, *Āśvalāyanagṛhyapariśiṣṭa* 4.18 and 20

At the third stage, Rudra and Sūrya were added to the group of deities along the vertical axis.

If we treated this textual development in historical terms, it supplements the ritual history outlined above. In the first place, the organization of omens in this text follows a novel scheme. The first stage shows a minimal categorical framework, organizing omens according to deity, and not, as Pingree seems to infer, according to earth, atmosphere, and heavens. It is only at the second stage that this theological scheme is explicitly spatialized according to horizontal and vertical directions, and even then, the vertical directions are four (not three). This transition may suggest a period when the omen taxonomy had not been fixed but was perhaps moving in the direction of the vertical tripartite scheme of classical Jyotiṣa. Likewise, even though it mentions the verbal noun *śamana* (appeasing), the "core" portion of the text at stage one uses the term *prāyaścitta* (atonement) to describe the ritual. By contrast, the newer, prefatory sections at stage two use *śānti*. This situation seems to align with the transition between atonement and appeasement seen in the ritual texts of the *Atharvaveda*.

The historical view also presents a shift in ritual technique. The core system takes the form of a simple fire sacrifice, as in the early omen atonements in *Kauśikasūtra* 13. The second level, however, introduces the offering of 1008 kindling sticks. While numerically simpler, this technique mirrors the format of the 100,000, 1,000, 000, and 10,000,000 homa rituals of the *Atharvavedapariśiṣṭa*. Shingo Einoo has discussed how this sort of offering—essentially the concurrence of kindling sticks or *samidh*s in the accusative with the verb "to offer" (*hu*)—is quite foreign to earlier Vedic practice, but becomes common in the latest textual stratum, the Gṛhyapariśiṣṭa level. Of these instances, he writes

> It is worth noticing … that a great number of the rites in which the offering of samidhs is performed seem to belong to certain kinds of rites, those which are called prāyaścitta or expiation, *śānti* or propitiatory rite, and kāmya or rite performed for some particular object.[16]

While it cannot be proved with certainty, we might entertain the possibility that this technique of *samidh* offerings in part emerged as an innovation within rituals to counteract omens. If this were so, the evidence would suggest that the

16 Einoo, "The Formation of Hindu Ritual," 41–49.

TABLE 12.2 Summary of anti-omen rituals

Text	Omen classification	Ritual format	Ritual category
(AV) *Kauśikasūtra* 13	Miscellaneous	Simple homa	Atonement (*prāyaścitti*)
(SV) *Adbhutabrāhmaṇa* (level 1)	Seven deities	Simple homa	Atonement (*prāyaścitti*)
(AV) *Śāntikalpa*	Tripartite omens	Bathing ritual + homa	Appeasement (*śānti*)
(SV) *Adbhutabrāhmaṇa* (level 2)	Eight deities, organized horizontally and vertically	Homa with 1008 kindling sticks + mantras repeated 108 times	Appeasement (*śānti*)
Atharvavedapariśiṣṭas	Tripartite omens	Bathing ritual + large numbered homas with kindling sticks	Appeasement (*śānti*)

technique originated outside of the *Atharvaveda*, in what Tsuji suggests was an originally Ṛgvedic source later adopted by the *Sāmaveda*. In the Atharvan tradition, as we have seen, the offering of *samidh*s appears relatively late and only in a subordinate position with respect to the *mahāśānti* paradigm. Ritual texts like the Ṛg- and Sāmavedic *Adbhutabrāhmaṇa* may thus preserve parallel or competing attempts to design new rituals to counteract omens. Texts like these may have spurred on the competitive innovations we have seen in the Atharvan tradition. For both the aspersive *mahāśānti* ritual and the extended homa rites emphasize a large number of mantric recitations, together with an equal number of fire offerings. The numerical compounding of Vedic expertise seems to lend itself to a scenario of ritual competition within the Vedic fold.

A hypothetical scenario might look as in table 12.2 with the lower rows of the table representing later developments. To simplify, I have omitted the Ṛgvedic variant of the *Adbhutabrāhmaṇa*.

5 Conclusion

Much of the above is of course hypothetical, but it does help us to imagine how the simple Vedic sacrifice could have grown in new and unprecedented directions in the era before Varāhamihira. That Varāhamihira is so familiar with these rituals—and that increasingly voluminous lists of omens show up in Vedic ritual manuals—suggests an increasing interpenetration of ritual and divinatory expertise in the first half of the first millennium CE and perhaps earlier. In other work I have explored in greater detail how non-Vedic actors, including astrologers like Varāhamihira himself may also have contributed to this economy of counter-omen rituals. I hope that future work on the corpus of Garga, whose study has recently been invigorated by the work of Professor Zysk, will provide a livelier and dateable picture of how this ritual economy developed.

While the Vedic ritual, to which I have limited myself in this paper, may have formed one site of the reception of omenology, it was also significantly transformed in the process. The type of Vedic rituals that survive into the Hindu period in medieval India are marked precisely by the aforementioned tension between fire sacrifice and bathing. While these ritual forms present a simplification of the complex structures of the ancient Vedic solemn rites, they also emphasize a performative aspect in their large numbers of mantras and offeratory kindling sticks. Even though they belong in technical and textual terms to the private "domestic" arena of Vedic ritual, their ideal context of performance was courtly. Omenology then, provided a political context for the continued labor of ritual actors who were "Vedic" in a new, qualified sense, long after the period of Vedism is supposed to have declined.

References

Apte, Vinayak Ganesha. *Āśvalāyanagṛhyasūtram*. Poona: Anandashrama press, 1936.

Bloomfield, Maurice. "The Kāuşika-Sūtra of the Atharva-Veda." *Journal of the American Oriental Society* 14 (1890): i-424.

Bolling, George Melville. "The Çāntikalpa of the Atharva-Veda." *Transactions and Proceedings of the American Philological Association* 35 (1904): 77–127.

Bolling, George Melville. "The Çāntikalpa of the Atharva-Veda." *Journal of the American Oriental Society* 33 (1913): 265–278.

Bolling, George Melville and Julius von Negelein. *The Pariśiṣṭas of the Atharvaveda*. Leipzig: O. Harrassowitz, 1909–1910.

Einoo, Shingo. "The Formation of Hindu Ritual." In *From Material to Deity*, edited by Shingo Einoo and Jun Takashima, 7–49. Delhi: Manohar, 2005.

Geslani, Marko. *Rites of the God-King: Śānti and Ritual Change in Early Hinduism*. New York: Oxford University Press, 2018.

Pingree, David. *Jyotiḥśāstra: Astral and Mathematical Literature*. Wiesbaden: Harrassowitz, 1981.

Pingree, David. "Mesopotamian Astronomy and Astral Omens in Other Civilizations." In *Mesopotamien und seine Nachbarn*, edited by H.-J. Nissen and J. Renger, 613–631. Berlin: Dietrich Reiner, 1982.

Tsuji, N. "On the Formation of the Adbhuta-Brāhmaṇa." *Annals of the Bhandarkar Oriental Research Institute* 48–49 (1968): 173–178.

Weber, Albrecht. *Zwei Vedische Texte über Omina und Portenta.* Berlin: Druckerei der Königl. Akademie der Wissenschaften, 1859.

Yano, Michio. "The Nakṣatra system of the Atharvaveda-Pariśiṣṭa." *Journal of Indological Studies* 28–29 (2016–2017): 65–74.

CHAPTER 13

The *Rāmāyaṇa* and the *Yugas*

Luis González-Reimann

In the centuries surrounding the beginning of the Common Era, Epic, Purāṇic and Dharmaśāstric Hinduism was taking shape.* It was a time during which the Brāhmaṇical tradition felt outside pressure from successive foreign incursions in the northwest, as well as from emerging native traditions that didn't accept the Vedas as the supreme religious authority. It was most certainly as a reaction to such external pressures that the Brāhmaṇical tradition found it necessary to define itself more clearly in terms of its beliefs and practices as well as in regards to its place in traditional history. Brāhmaṇical *dharma* had to be distinguished from other *dharma*s, especially the Buddhist and Jain ones, and the attempt to do so is embodied in the two Sanskrit epics, in the early Purāṇas, and in the Dharmaśāstra literature, particularly the *Mānavadharmaśāstra*.

As far as past events were concerned, the system of the *yuga*s, along with those of *manvantara*s and *kalpa*s, emerged as the ideal vehicle for organizing traditional history. The *yuga*s were especially well suited for the purpose because they presented an orderly sequence of mythic-historical periods that defined people's adherence to, or rejection of proper *dharma*. In fact, the *yuga* theory was probably developed precisely in order to provide a Brāhmaṇical explanation for the social, political and religious conditions at the time. The post-Vedic Hindu tradition could now assign its stories and histories to different *yuga*s, in accordance with each *yuga*'s connection to the ideals of Brāhmaṇical *dharma*.

Of special importance for the historical organization of events was the present, that is to say the present of the redactors of the Epics and the early Purāṇas: the times they lived in. The *yuga* theory served to explain the crises they were confronting. They had to be living in the Kali Yuga, the worst *yuga* of all, because that is what their historical circumstances seemed to indicate. Through this optic, the fratricidal war of the *Mahābhārata* had to be connected to the dreadful Kali Yuga, a time in which the low status of *dharma* could allow for such an unseemly confrontation between relatives and for the breaches

* An early version of this chapter was presented at the 222nd meeting of the American Oriental Society in Boston, 2012.

 | DOI:10.1163/9789004438224_014

of *dharma* that occurred during the war. Thus, the *Mahābhārata* became the marker for the beginning of the Kali Yuga, and all other mythical and historical events had to be placed either in previous *yuga*s or within Kali. If it was within the Kali Yuga, it had to be after the war of the *Mahābhārata*. The long-lasting Kali Yuga had by then become the current *yuga*, the one we all live in.

The process of arranging past events by situating them in previous *yuga*s is already at work in the *Mahābhārata*, and it develops fully in the Purāṇas. The lists of Viṣṇu's *avatāras* are especially relevant in this respect because they sequentialize the appearance of the various gods, sages and heroes that Vaiṣṇavism appropriated as manifestations of Viṣṇu.[1] Because Kṛṣṇa, one of Vaiṣṇavism's most important *avatāras*, plays a central role in the *Mahābhārata*, he immediately came to be associated with the beginning of the Kali Yuga, to the extent that in the Vaiṣṇava narrative the date of the start of Kali was considered to be the day of Kṛṣṇa's death.[2]

All the lists of Viṣṇu's *avatāras* place Rāma, the hero of the *Rāmāyaṇā*, before Kṛṣṇa. And the story of Rāma is told in the *Mahābhārata* itself, so from a traditional point of view there can be no question that Rāma lived before Kṛṣṇa. In terms of the *yuga*s, this means that Rāma must have lived in a *yuga* that preceded Kali, that is to say in Kṛta, Tretā, or Dvāpara.[3] There is also an element connected to *dharma* that—from a traditional perspective—would seem to confirm Rāma's placement before the Kali Yuga. In the Story of Rāma, *dharma* is ostensibly adhered to and respected more than it is in the narrative of the *Mahābhārata*. This means that Rāma's adventures had to belong in an earlier and better *yuga*, one that is higher up in the scale of adherence to *dharma*. Where, then, should the lives of Rāma and Sītā be placed in the *yuga* scheme?

In Hinduism, it became customary, and a matter of course, to place Rāma and the events of the *Rāmāyaṇa* in the Tretā Yuga, even if this placement has very tenuous support in our earliest extant version of the poem, the Vālmīki *Rāmāyaṇa*. It is clear by looking at Vālmīki's as well as later versions of the *Rāmāyaṇa*, that the placement of the Rāma story within the sequence of *yuga*s was not considered particularly important. This probably is because—unlike the *Mahābhārata*—the *Rāmāyaṇa* has no relevant connection to the histori-

---

1 There are different lists of *avatāras* in the Purāṇas, some including up to 22 descents of the god, but it was a list of ten that became the most common.

2 It must be noted that the link between Kṛṣṇa's death and the beginning of the Kali Yuga is not made in the *Mahābhārata*. It appears later, and it then becomes commonplace. The earliest mention is probably the one in the *Viṣṇupurāṇa* at 4.24.27 and 4.24.32 in the critical edition; 4.24.35 and 4.24.40 in the Wilson & Singh edition.

3 The Kṛta Yuga would later frequently be called the Satya Yuga.

cal present and its perceived negative characteristics, which were considered to have been imposed on society by the degraded Kali Yuga.

1 The Text

In the Critical Edition of the Vālmīki *Rāmāyaṇa* there are only two instances that can be construed as evidence that the action of the *Rāmāyaṇa* took place in a specific *yuga*. Both are in the last book, the Uttarakāṇḍa, which is generally considered to be a late addition to the text. In fact, the different *yuga*s are mentioned by name in only 20 verses in the entire poem, 16 of which—a full 80%—are in the Uttarakāṇḍa. Of the remaining four verses, two are in Book One, the Bālakāṇḍa, while there is one each in books five and six, the Sundarakāṇḍa and the Yuddhakāṇḍa respectively.[4]

Of the four *yuga*s (Kṛta, Tretā, Dvāpara, and Kali), the one mentioned more often—a total of 10 times—is Kṛta, the Golden Age. The most relevant of these mentions is in Book One, when the seer Nārada narrates an abbreviated version of Rāma's story to Vālmīki. Nārada explains that when Rāma ruled Ayodhyā after rescuing Sītā, it was "just like the Kṛta Yuga" (*yathā kṛtayuge tathā*, 1.1.73), a comparison meant to highlight the dharmic nature of Rāma's rule, the *rāmarajya*. Note that this comparison implies that Rāma could not have lived in the Kṛta Yuga, because his reign was merely "like" the Kṛta Yuga. Most other mentions are in the format of "formerly / long ago in the Kṛta Yuga …," when narrating or referencing past mythological events, such as the time when gods and demons were powerful, when mountains had wings, when Manu ruled, when the demon Madhu lived, or when fragrances were pure and pleasant.[5] Of the other *yuga*s, Tretā is mentioned six times, all of them in the Uttarakāṇḍa, while Dvāpara and Kali only receive two and one mentions respectively.

It is important to point out that many of the mentions of distinct *yuga*s are concentrated in one single narration in the Uttarakāṇḍa: the story of Śambūka (7.64–68). If we don't count the Śambūka episode, the Kali Yuga is mentioned only one time in the entire *Rāmāyaṇa* (6.26.13), the Dvāpara does not appear at all, and the Tretā appears only once, while the Golden Age, Kṛta, still receives nine mentions. The contexts in which Kṛta is mentioned indicate that it is used

---

4 In addition, there is a mention of an unspecified "*yuga* of the gods," *devayuga*, in 1.10.11, much like in *Mahābhārata* 3.92.6. See González-Reimann, *The Mahābhārata and the Yugas* (New York: Peter Lang, 2002), 74. In *Mānavadharmaśāstra* 1.71–72, however, a *yuga* of the gods (*devānāṃ/daivikānāṃ yugam*) is the same as a *catur-* or *mahāyuga*.

5 See Appendix 1.

merely as a rhetorical device for evoking a lost, ideal past, and not as a statement regarding where events should be placed in the succession of the ages. If Books One and Seven are accepted as later additions, we are left with only two mentions of Kṛta, zero references to Tretā and Dvāpara, and a single mention of Kali in the entire *Rāmāyaṇa*.[6] This evinces the lack of importance the developed *yuga* system had for Vālmīki, or maybe even his lack of awareness of it.

2 The Story of Vedavatī

Let us take a look at the only place in which the action of the *Rāmāyaṇa* is linked to the Tretā Yuga. It is in the story of Vedavatī, who is said to be the incarnation of Vāc, the personification of speech. She is performing intense austerities while waiting to someday marry Viṣṇu, as hoped-for by her father. However, Rāvaṇa, the king of the *rākṣasas* (demons), appears before her and tries to seduce her. When she refuses, he grabs her by the hair, thus humiliating her.[7] Vedavatī then frees herself by cutting her own hair, but decides to end her life by entering a fire. Before doing so, she vows to be born again in the future in order to cause Rāvaṇa's death. The narration then says:

> That woman called Vedavatī lived long ago, in the Kṛta Yuga. She appeared in the Tretā Yuga for killing that *rākṣasa* (Rāvaṇa).[8]

It is with this brief, casual statement that the adventures of Rāma and Sītā are placed in the Tretā Yuga by the Vālmīki *Rāmāyaṇa*. Nowhere else in the poem, with the exception of Śambūka's story, is the Tretā even mentioned.

It is noteworthy that the emphasis in the Vedavatī story is on Sītā instead of Rāma. It presents Vedavatī/Sītā as the agent of Rāvaṇa's destruction, because

6 Regarding the lateness of Book Seven, the Uttarakāṇḍa, see Robert P. Goldman and Sally J. Goldman, trans., *The Rāmāyaṇa of Vālmīki: An Epic of Ancient India*, vol. 7: Uttarakāṇḍa (Princeton: Princeton University Press, 2017), 78, 106, who, after carefully reviewing the opinion of Brockington on this issue in John L. Brockington, *The Sanskrit Epics* (Leiden: Brill, 1998) conclude that "at least the second half of the Uttarakāṇḍa was composed by a different hand or hands than the earlier portions of the text." It is precisely in the second half that the story of Śambūka appears.

7 Just as Duḥśāsana humiliates Draupadī when he drags her by the hair toward the gambling hall in the *Mahābhārata* (2.60.22–25, Critical Edition).

8 7.17.31.

it is due to the fact that in a previous life she had vowed revenge, that Rāma would later kill Rāvaṇa. The Vedavatī story follows the common strategy of creating former vows and curses for explaining epic events, but it also gives Sītā some form of agency in the killing of Rāvaṇa, which is a crucial feat of Rāma.[9]

The *Brahmavaivartapurāṇa* (8–16th cent.) retells the Vedavatī episode, but it takes the process of organizing past events according to the *yuga*s still further by adding Draupadī, the heroine of the *Mahābhārata*, to the sequence. It states:

> *kṛtayuge vedavatī kṛśadhvajasutā śubhā* /
> *traitāyāṃ rāmapatnī ca sītetī janakātmajā* // 53
> *tacchāyā draupadī devī dvāpare drupadātmajā* /
> *trihāyaṇīti sā proktā vidyamānā yugatraye* // 54
>
> In the Kṛta Yuga, Vedavatī [was] the beautiful daughter of Kṛśadhvaja, and in the Tretā [she was] Rāma's wife Sītā, Janaka's daughter. In the Dvāpara, the shadow [Sītā] was the goddess Draupadī, Drupada's daughter. Because she is present in three *yuga*s she is called Trihāyaṇī (the three-year old one).[10]

According to the Purāṇa, Vedavatī is a part (*aṃśa*) of the goddess Lakṣmī, effectively making her, Sītā and Draupadī successive incarnations of Lakṣmī in three consecutive *yuga*s: Kṛta, Tretā and Dvāpara.[11]

3 The Story of Śambūka

It is in the episode about the *śūdra* Śambūka in the Uttarakāṇḍa of the *Rāmāyaṇa* that we find our second placement of Rāma in a specific *yuga*, but, in this case, the *yuga* is Dvāpara instead of Tretā. This narration is the only one that

9 Because the story states that Vedavatī will marry Viṣṇu in her next birth, it implies that Rāma is Viṣṇu, something that is not evident in the main narrative of the *Rāmāyaṇa*, see Luis González-Reimann, "The Divinity of Rāma in the *Rāmāyaṇa* of Vālmīki," *Journal of Indian Philosophy* 34, no. 3 (2006): 203–220.

10 *Brahmavaivartapurāṇa* 2.14.53–54. Placing Draupadī in the Dvāpara Yuga seems unusual. It most surely implies that it is at the end of Dvāpara, which would then connect her to the start of the Kali Yuga.

11 Vedavatī is called *kamalāṃśā* or *padmāṃśā* in *Brahmavaivartapurāṇa* 2.14.3, 17.

mentions all of the four *yugas* together. The story is a curious account of how Rāma kills a *śūdra* for having committed the transgression of performing asceticism when he was not allowed to do so. According to this unusual narrative, *śūdras* are only allowed to perform *tapas* in the Kali Yuga. Because Śambūka performed *tapas* in Rāma's realm during his reign, and because it was not the Kali Yuga, the *śūdra*'s flagrant breach resulted in the death of the innocent son of a *brahman*. It was then up to Rāma, as king, to right the wrong by stopping the *śūdra*. When Rāma finds the transgressor engaged in ascetic practices and confirms his low social status, he instantly and unflinchingly dispatches Śambūka to the other world. As a consequence, the dead *brahman* child immediately comes to life again.

The story is relevant because of the way in which the *brahman* Nārada explains to Rāma the rules governing the practice of austerities. Nārada points out that long ago (*purā*, 7.65.9), in Kṛta, only *brahmans* were allowed to practice *tapas*, while in Tretā (7.65.17) the privilege was extended to *kṣatriyas*. In Dvāpara, *vaiśyas* were also included, but not the *śūdras* (7.65.21), who will only be able to do so in the degenerate Kali Yuga (7.65.22). The narration refers to the arrival of the Kṛta, Tretā, and Dvāpara Yugas in the past tense, but it speaks of Kali in the future tense. This means that Nārada's explanation cannot have taken place in Kali, it could only have happened in Kṛta, Tretā, or Dvāpara. However, it could not have been in either Kṛta or Tretā, because they both precede Dvāpara, and all three are spoken of in the past tense (7.65.17–18), implying that Kṛta and Tretā have already expired and Dvāpara has begun. The reasonable conclusion is that Dvāpara is the *yuga* in which Nārada is speaking, and this is confirmed by his statement immediately following his description of the four *yugas*. He declares that it is a severe breach of *dharma* for a *śūdra* to practice *tapas* in the Dvāpara Yuga (*dvāpare*), and that there must surely be a *śūdra* doing so in Rāma's realm, because that would explain the unwarranted death of the *brahman*'s child (7.65.23). It follows from Nārada's words that the conversation is taking place in Dvāpara. The fact that he refers to Dvāpara in the past tense does not preclude the action from taking place in the same *yuga*, because he is referring to the situation that arose when Dvāpara arrived and will presumably last throughout the *yuga*. Further confirmation that the action is not taking place in Tretā comes at the end of Śambūka's story, when the sage Agastya relates to Rāma how he had acquired an ornament he has just presented to him. Agastya narrates what happened long ago, in the bygone Tretā Yuga (*purā tretāyuge gate*, 7.67.18). If the Tretā Yuga has already passed, their conversation cannot be taking place in it.

It is true that the text does not directly state that Rāma or Sītā lived in Dvāpara, but the wording of the narration makes this the logical interpretation.

Commentators of Vālmīki's text were clearly troubled by this possibility. To them, the action needed to have taken place in Tretā, not only because the story of Vedavatī indicated as much, but also because at the time when they themselves were writing their commentaries (centuries after the composition of the Vālmīki *Rāmāyaṇa*) it was widely accepted that Rāma belonged in the Tretā Yuga. In order to get around the contradiction posed by Rāma's placement in Dvāpara in this story, the commentators did what commentators often do, they read the relevant verses in imaginative ways in order to make them conform to their received beliefs.

In the verse that states that *śūdra*s are not allowed to perform asceticism in Dvāpara and that a *śūdra* must be doing so in Rāma's kingdom (7.65.23), some commentators take the straightforward *dvāpare*, "in Dvāpara," to rather mean "even in Dvāpara (*dvāpare 'pi*)," and then interpret the text as saying that if circumstances are so in Dvāpara, they must be even more so in Tretā, because in their understanding that was the *yuga* in which Rāma lived.

Although clever, this is ultimately only a way of reading their preconceived conviction into the text. Likewise, when Agastya talks of the bygone Tretā Yuga, commentators don't read it as a reference to the Tretā Yuga that has just passed, but, instead, to the Tretā Yuga of some different, earlier *mahāyuga*, therefore skirting the contradiction.[12] But there is nothing in the narrative to warrant such a reading.[13]

4 *Yugānta*

A discussion of the *yuga*s in the *Rāmāyaṇa* must address, even if briefly, the use of the compound *yugānta*, "the end of the *yuga*." In the *Mahābhārata*, *yugānta* is used as a poetic device to evoke the image of a mythological end of the world, a time when natural disasters, especially intense fires, take place. If one wants to see a connection between the use of *yugānta* and the developed *yuga* theory, then it must allude to the end of the fourfold or great *yuga*, the *caturyuga* or *mahāyuga*, and not to the end of individual *yuga*s.

12 For other examples of this strategy see below, note 23, and the texts it refers to.

13 See the notes to the above-mentioned verses in Goldman and Goldman, *The Rāmāyaṇa of Vālmīki*, vol. 7: Uttarakāṇḍa for an extensive discussion of the strategies used by the commentators in this case. Curiously, Goldman and Goldman opted for following the commentators in their translation of these *yuga* passages. Goldman and Goldman's detailed and extremely useful analysis of the entire Śambūka episode, including its use through the centuries and up to the present, can be found on pp. 104–113.

In the *Mahābhārata*, *yugānta* is also used for making comparisons when terrifying and awesome warriors, weapons, or events are described, it is not used for placing events in a particular moment of the *yuga* cycle.[14] The situation in the Vālmīki *Rāmāyaṇa* is the same in this respect. The image of *yugānta*—or its synonym, *yugakṣaya*, the destruction of the *yuga*—is employed, for instance, for portraying the powerful and dreadful combat skills of Rāma, as when he confronts the *rākṣasa* Khara, or the monkey king Vālin. But it is also used to depict Vālin himself after he was struck down by Rāma, or to describe Lakṣmaṇa, Hanumān, Akṣa (Rāvaṇa's son), Kumbhakarṇa, Mahāparśva, or Nikumbha.[15] Such comparisons are employed to highlight the devastating power of weapons, whether they belong to Rāma, to his brother Śatrughna, or to Rāvaṇa or some other *rākṣasa*. Both epics use the expression formulaically in this sense, and its use cannot be construed as a reference to any specific *yuga*.

5 Tretā or Dvāpara?

As we have seen, the evidence for placing Rāma in the Tretā Yuga is practically nonexistent in Vālmīki's text, where the only instance is surely a late addition. Later versions of the *Rāmāyaṇa* make the placement explicit, but they do so only occasionally, confirming that the matter does not merit much attention. The *Adhyātmarāmāyaṇa* (13–15th cent.), for instance, places the action in Tretā on only three occasions, all of them in books three and four, where no such mentions occur in Vālmīki's text. They are presented as predictions made long before Rāma was born.

In the first of these instances, the *rākṣasa* Kabandha is told that Rāma will free him from a curse when the Tretā Yuga arrives. In the second, we hear that Rāma will alleviate the burden of the earth in Tretā. And in the third, we learn that Rāma will kill Rāvaṇa in the Tretā Yuga. This last prediction was made by a sage to the vulture king Saṃpāti when, in the distant past, his wings were burned while flying too close to the sun. The sage explains to Saṃpāti that he will recover his wings in the future, at the time when Rāma searches for Sītā after Rāvaṇa kidnaps her. When the time comes, Saṃpāti will inform the monkey search party led by Hanumān that Sītā is being held by Rāvaṇa in the island of Laṅka. Upon doing this, Saṃpāti's wings will grow back. In Vālmīki's telling,

---

14 For the use of *yugānta* in the *Mahābhārata* see González-Reimann, *The Mahābhārata and the Yugas*, ch. 2, "Yugānta: the End of the Yuga," 51–85.

15 For textual references see Appendix 2.

the sage simply states that Saṃpāti needs to wait for the proper time and place for this to happen, without indicating when that will be.[16] But the *Adhyātma* version specifies that it will be in the Tretā Yuga. These predictions in the *Adhyātmarāmāyaṇa* concerning Rāma's appearance in Tretā are accompanied by a half verse, repeated in all three instances, which declares that:[17]

> *tretāyuge dāśarathir bhūtvā nārāyaṇaḥ 'vyayaḥ* |
>
> In the Tretā Yuga, the immortal Nārāyaṇa will be born as the son of Daśaratha.

This is evidently no more than a formula inserted at these places in the narrative to reinforce the notion that Rāma lived in Tretā.

The Hindi *Rāmcaritmānas* of Tulsidas (16–17th cent.) follows the *Adhyātma* in stating that the time when Saṃpāti recovers his wings is the Tretā Yuga.[18] Tulsi also adds a mention of Tretā when describing Rāma's righteous rule in the last book of his rendering of Rāma's story. It picks up Vālmīki's comparison of Rāma's rule to the Kṛta Yuga, but adds that his righteous rule took place in Tretā.[19] The authors of versions of the Rāma story that were composed after Vālmikī's text apparently felt the need to add clarification regarding the *yuga* of Rāma's actions.

If the Vālmīki *Rāmayaṇa* is largely silent regarding the *yuga* in which Rāma and Sītā lived, and if the two mentions found in the text place the events in two different *yugas*, one in Tretā and the other in Dvāpara, the question is: how was Rāma's story definitely assigned to Tretā? In order to answer this question we must look outside of the *Rāmāyaṇa*. It was probably the mention of Rāma's story in the *Mahābhārata*—coupled with the placing of Rāma before Kṛṣṇa in the lists of Viṣṇu's *avatāras*—that established that Rāma had lived before

16 4.61.12.

17 *Adhyātmarāmāyaṇa* 3.9.19, 4.6.55, and 4.8.48. In 3.9.19 the text reads *svayam* (himself) instead of *avyayaḥ* (immortal).

18 *Rāmcaritmānas* 4, *caupāī* 28, W. Douglas P. Hill, *The Holy Lake of the Acts of Rāma: A Translation of Tulasī Dās's Rāmacaritamānasa* (Bombay: Oxford University Press, 1971), 337. For a recent translation see Philip Lutgendorf, *Tulsidas: The Epic of Ram*, vol. 5 (Cambridge, Massachusetts: Harvard University Press 2020), 191.

19 7, *caupāī* 21, Hill, *The Holy Lake of the Acts of Rāma*, 444. In Vālmīki, the comparison to Kṛta is made in Book One, in the synopsis of the story, but Tulsi places it, along with the added mention of Tretā, in Book Seven, when detailing the positive characteristics of the rule of Rāma. The *Bhāgavatapurāṇa*, in its very brief rendering of the Rāma story, had made the same comparison of both *yugas* before Tulsidas, *Bhāgavatapurāṇa* 9.10.52.

Kṛṣṇa. When Kṛṣṇa—along with the story of the *Mahābhārata*—was accepted as marking the beginning of Kali, Rāma's time had to be assigned to an earlier *yuga*. That *yuga* could not be Kṛta, since Rāma's adventures in the forest and his subsequent rescue of Sītā include situations not compatible with a time when *dharma* was followed by everyone. It was therefore natural to place it in either Tretā or Dvāpara, which are precisely the two options offered in Vālmīki's Uttarakāṇḍa. But which one?

There is a lack of clarity, even a confusion in the early sources concerning this. The confusion probably stems from the similarity between the name of Rāma Dāśarathi, the hero of the *Rāmāyaṇa*, and Rāma Jāmadagnya, the famed warrior and descendant of Bhṛgu, who appears in both Sanskrit epics. Rāma Jāmadagyna was later known as Paraśurāma, conceivably precisely in order to avoid confusion between the two Rāmas, who appear consecutively in the *avatāra* lists. In the Nārāyaṇīya of the *Mahābhārata*, in one of the earliest lists of *avatāras*, Rāma Jāmadagnya is placed firmly in the Tretā Yuga, while our Rāma Dāśarathi is located in the *sandhi* between Treta and Dvāpara.[20] But in Book One of the *Mahābhārata* it is Rāma Jāmadagnya, not Dāśarathi, who is placed in the *sandhi* between Tretā and Dvāpara.[21] As Rāma Jāmadagnya always precedes Rāma Dāśarathi in the *avatāra* lists, this would imply that the Rāma of the *Rāmāyaṇa* lived in Dvāpara. Some early Purāṇas solved the conflict by using a strategy often used by commentators when confronting contradictions. They assigned the two Rāmas to Tretā Yugas from different *mahāyugas*: Jāmadagnya to the 19th, and Dāśarathi to the 24th.[22] Neither of these two is our current *mahāyuga*, as we live in the 28th, according to standard Purāṇic reckoning.

When deciding between placing Rāma in Tretā or Dvāpara, the contrast between the *Mahābhārata* and the *Rāmāyaṇa* probably played an important role. If Rāma had lived in Dvāpara, this would put him uncomfortably close to the events of the *Mahābhārata*, with its frequent breaches of *dharma*, and the fratricidal war that transpired as Dvāpara was closing. Tretā, on the other hand, was noticeably higher up in the scale of *dharma*, and this made it a more appropriate *yuga* for Rāma. The connection made between the Rāma *avatāra* and a Tretā Yuga in some early Purāṇas, together with the mention of Tretā in the story of Vedavatī in the received version of Vālmīki's text—even though it is most certainly a late interpolation—surely played a fundamental role in finally

20 *Mahābhārata* 12.326.77–78.

21 *Mahābhārata* 1.2.3–8.

22 *Vāyupurāṇa* 2.36.90–91, *Matsyapurāṇa* 47.243–245, *Brahmāṇḍapurāṇa* 2.73.90–92. This strategy was also used with the *kalpas*, by applying the notion of *kalpabheda*, "kalpa difference," thanks to which contradicting events were placed in different *kalpas*.

establishing the poem's traditional placement in Tretā, a placement that was then repeated and thus reinforced in subsequent literature.

Later texts, especially the many retellings of the Rāma story, did not bother to distinguish between Tretā Yugas belonging in different *mahāyugas*, and simply considered that the Rāma of the *Rāmāyaṇa* lived "in the Tretā Yuga." Similarly, later lists of *avatāras* are not always concerned with specifying their placement in the *yugas*, although a connection between the Tretā Yuga and both Jāmadagnya and Dāśarathi appears sometimes.[23] The correspondence between *yugas* and *avatāras* is never fully resolved in the literature, as some *avatāras* are assigned not only to different *yugas* or *mahāyugas*, but also to different *kalpas* and *manvantaras*.

6 Conclusion

Both epics probably existed in some form before the theory of the *yugas* took shape, but they were soon given a position in the *yuga* system. The *Mahābhārata* was placed at the beginning of the Kali Yuga, and this became crucial in Hindu traditional history.[24] The *Rāmāyaṇa*, on the other hand, was clearly not so important in terms of the *yugas*. The only requirement, it would seem, was that it came before the Kali Yuga. The assumption that the events of the *Mahābhārata* marked the beginning of our present *yuga* turned that text into a fundamental historical marker, and traditional dating is often reckoned from the purported date of the beginning of the Kali Yuga. In consonance with this, the *Mahābhārata* is considered to be a text of history, *itihāsa*. By contrast, the *Rāmāyaṇa* has no direct connection to current history as it belongs in an earlier *yuga*. It remained as a narration of events in the distant past, where the portrayal of ideal dharmic characters became more important than their chronology. The *Rāmāyaṇa* was therefore often not considered a text of history, *itihāsa*, so much as a work of poetry, *kāvya*. The purported author of the *Mahābhārata*, Vyāsa, was to be considered by tradition as the composer of all the Purāṇas, while the author of the *Rāmāyaṇa* would be lauded as the first poet, the *ādikavi*. The precise *yuga* of the setting of the story of Rāma was evidently not as important as the *yuga* of the events of the *Mahābhārata*.

---

23 The *Bhāgavatapurāṇa*, for instance, places Dāśarathi squarely in the Tretā (9.10.52), and implies that Jāmadagnya also lived then by stating that he is a 13th descendant of Purūravas and Ūrvaśī, who lived at the beginning of Tretā (9.15.1–13).

24 For the *yugas* in the *Mahābhārata* see González-Reimann, *The Mahābhārata and the Yugas*.

Appendix 1: Mentions of the Four *Yugas* in the Vālmīki *Rāmāyaṇa* (CE)

TABLE 13.1 Table of mentions of the four *yugas* in the Vālmīki *Rāmāyaṇa* (CE)

Quotation	Locus	Context
yathā kṛtayuge tathā just like in the Kṛta Yuga	1.1.73	Rāma's rule is like the Kṛta Yuga (Nārada to Vālmīki in the synopsis of the story)
pūrvaṃ kṛtayuge formerly, in the Kṛta Yuga	1.44.14	In the Kṛta, *daityas* and *ādityas* (two types of demons) were powerful (Viśvāmitra to Rāma)
pūrvaṃ kṛtayuge formerly, in the Kṛta Yuga	5.1.108	Mountains had wings in the Kṛta Yuga (Maināka to Hanumān)
purā kṛtayuge long ago, in the Kṛta Yuga	7.2.4	Pulastya lived in the Kṛta Yuga
pūrvaṃ kṛtayuge formerly, in the Kṛta Yuga	7.53.3	The *daiteya* (demon) Madhu lived in the Kṛta Yuga
purā kṛtayuge long ago, in the Kṛta Yuga	7.65.9	Story of Śambūka
purā kṛtayuge long ago, in the Kṛta Yuga	7.70.5	In the Kṛta Yuga, Manu was the ruler & Ikṣvāku his son (Agastya to Rāma)
pūrvaṃ kṛtayuge yathā like formerly, in the Kṛta Yuga	7.88.8	A fragrance (when Rāma again defends Sītā)
eṣā vedavatī nāma pūrvam āsīt kṛte yuge / tretāyugam anuprāpya vadhārthaṃ tasya rākṣasaḥ // "The woman called Vedavatī lived long ago, in the Kṛta Yuga. She appeared in the Tretā Yuga for killing that *rākṣasa* [Rāvaṇa]."	7.17.31	Story of Vedavatī
kṛta vs. *tiṣya* (Kali)	6.26.13	On *dharma* (the *rākṣasa* Mālyavān speaking to Rāvaṇa)
devayuge In the *yuga* of the gods	1.10.11	Sumantra to Daśaratha (Sanatkumāra's prophecy)
tretāyugaṃ nāma called the Tretā Yuga	7.65.11	Tretā Yuga in the Story of Śambūka
tasmiṃs tretāyuge yuge In that yuga, the Tretā Yuga	7.65.12	
tretāyuge In the Tretā Yuga	7.65.17	

TABLE 13.1 Table of mentions of the four *yugas* in the Vālmīki *Rāmāyaṇa* (CE) (*cont.*)

Quotation	Locus	Context
purā tretāyuge gate long ago, in the bygone Tretā Yuga	7.67.18	
purā tretāyuge long ago in the Tretā Yuga	7.68.1	
dvāparasaṃkhyā with the name Dvāpara	7.65.19	Dvāpara Yuga in the Story of Śambūka
dvāparasaṃkhye known as Dvāpara	7.65.20	
dvāparasaṃkhyāte known as Dvāpara	7.65.21	
Dvāpare In Dvāpara	7.65.23	
kalau yuge in the Kali Yuga	7.65.22	Kali Yuga in the Story of Śambūka

Appendix 2: Mentions of *Yugānta* and *Yugakṣaya*[25] in the Vālmīki *Rāmāyaṇa* (CE)

TABLE 13.2 Table of mentions of *yugānta* and *yugakṣaya* in the Vālmīki *Rāmayaṇa* (CE)

Quotation	Locus	Context
yugāntāgnir iva like the fire at *yugānta*	3.23.26	Rāma fighting Khara
yugānta iva bhāskaraḥ like the sun at *yugānta*	4.11.2	Sugrīva describing Rama
yugāntāgnir iva like the fire at *yugānta*	4.15.15	Rāma described to Vālin
ādityam iva ... yugānte bhuvi pātitam like the Sun fallen to earth at *yugānta*	4.17.10	Vālin struck by Rāma
kālamṛtyuyugāntābhaṃ like Time, death and *yugānta*	4.30.20	Lakṣmaṇa

25 *Yugakṣaya*, literally "the destruction of the *yuga*," is the same as *yugānta*, "the end of the *yuga*."

TABLE 13.2 Table of mentions of *yugānta* and *yugakṣaya* in the Vālmīki *Rāmayaṇa* (*cont.*)

Quotation	Locus	Context
yugānte ... hato ... ivāgninā as if destroyed by fire at *yugānta*	4.60.11	The world (Saṃpāti describing how he lost his wings)
yugāntāgnir iva like the fire at *yugānta*	5.34.12	On Rāma's abilities
yugāntasūryapratimaṃ like the sun at *yugānta*	5.35.67	Rāma
yugāntakālāgnim iva like the fire of Time at *yugānta*	5.45.8	Akṣa, Rāvaṇa's son
yugāntāgnir iva like the fire at *yugānta*	5.56.135	Hanumān describing the burning of Laṅkā
yugāntāgnir iva like the fire at *yugānta*	6.14.14	Rāma
yugāntam iva like yugānta	6.31.10	Rāma to Lakṣmana, on visible omens before attacking Laṅkā
iva ... yugānte like at *yugānta*	6.53.50	Kumbhakarṇa (Rāvaṇa's brother) compared to Yama
yugāntameghastanitasvanopamam like the thundering sound of a cloud at *yugānta*	6.55.13	Hanumān's cry when wounded
yugāntāgnir iva like the fire at *yugānta*	6.55.71	Kumbhakarṇa
yugāntāgnir iva like the fire *at yugānta*	6.58.47	Mahāpārśva, Rāvaṇa's brother
yugāntāgnir iva like the fire at *yugānta*	6.64.10	Nikumbha, Kumbhakarṇa's son
yugāntāgnisamaprabham radiant like the fire at *yugānta*	6.66.29	Makarākṣa's (a *rākṣasa*) lance
kāleneva yugakṣaye Like Time at *yugānta* (*yugakṣaya*)	6.82.38	Said by frightened *rākṣasīs* who feel unprotected
yugāntādityasaṃnibhaiḥ blazing like the sun at *yugānta*	6.83.10	Rāvaṇa describing his arrows
yugāntasamaye vāyuḥ like the wind at *yugānta*	6.84.10	Hanumān killing *rākṣasas*
vātā iva yugakṣaye Like the wind at *yugānta* (*yugakṣaya*)	6.88.4	Describing loosed weapons
yugānta iva pāvakaḥ like the fire at *yugānta*	6.88.37	Rāma

TABLE 13.2 Table of mentions of *yugānta* and *yugakṣaya* in the Vālmīki *Rāmayaṇa* (*cont.*)

Quotation	Locus	Context
yugāntāgnicayopamam like a ritual fire at *yugānta*	6.91.11	Rāvaṇa's spear
yugāntāgniṃ ... iva like the fire at *yugānta*	6.91.21	Rāma's arrows vs. Rāvaṇa's spear
yugāntolkena saprabhā like a meteor at *yugānta*	6.91.25	Rāma's spear
yugāntādityavarcasaḥ radiant like the sun at *yugānta*	6.92.8	Rāma's arrows
yugāntavaiśvānaratulya equal to the fire at *yugānta*	7.6.55	*Rākṣasa* weapons thrown at Viṣṇu
bhīmanihrādo yugānte jalado yathā like the terrible sound of a cloud at *yugānta*	7.7.10	Viṣṇu's conch when fighting *rākṣasas*
yugāntāgnim iva like the fire at *yugānta*	7.15.9	Daśagrīva (Rāvaṇa)
yugānta iva pāvakaḥ like the fire at *yugānta*	7.32.38	Arjuna Kārtavīrya
Dīptam iva kālāgniṃ yugānte like the flaming fire of Time at *yugānta*	7.61.20	Śatrughna's arrow
lokakṣayo ... vā yugasaṃkṣayaḥ the end of the world or *yugānta*	7.61.22	On the power of Śatrughna's arrow
yugāntāgnim iva like the fire at *yugānta*	7.61.31	Śatrughna's arrow/shaft
vartamāne yugakṣaye during/towards *yugānta*	7.65.20	Story of Śambūka

References

Sanskrit Texts

Bhagavatapuranam. 2006. Electronic version Gretil: http://gretil.sub.uni-goettingen.de/gretil/1_sanskr/3_purana/bhagp/bhp1-12u.htm. Viewed July 27, 2012

Bhatt, G.H. and U.P. Shah, eds. *The Vālmīki Rāmāyaṇa: Critical Edition*. 7 vols. Baroda: Oriental Institute, 1960–1975. Electronic version Gretil, Tokunaga and Smith, 2017: http://gretil.sub.uni-goettingen.de/gretil/1_sanskr/2_epic/ramayana/ram_1-7u.htm. Viewed December 18, 2018.

Brahmaṇḍapurāṇa. Electronic version at Gretil, based on the Venkatesvara Steam Press edition, Bombay: http://gretil.sub.uni-goettingen.de/gretil/1_sanskr/3_purana/brnd p2_u.htm. Viewed December 18, 2018.

Matsyapurāṇa. Adhyāyas 1–176. Electronic version Gretil, based on the Caukhamba Vidyabhavan edition, Calcutta, 1954: http://gretil.sub.uni-goettingen.de/gretil/1_san skr/3_purana/mtp176pu.htm. Viewed December 18, 2018.

Pathak, M.M., ed. *The Critical Edition of the Viṣṇupurāṇam*. 2 vols. Vadodara: Oriental Institute, 1997, 1999. Electronic version Gretil, input by Peter Schreiner: http://gretil .sub.uni-goettingen.de/gretil/1_sanskr/3_purana/vipce_pu.htm. Viewed December 5, 2018.

Sukthankar, V.S. and S.K. Belvalkar, eds. *The Mahābhārata: For the First Time Critically Edited*. 19 vols. Poona: Bhandarkar Oriental Research Institute, 1933–1959.

The Vāyumāhāpurāṇam. Delhi: Nag Publishers, 1983.

Texts with Translation

Lutgendorf, Philip, trans. *Tulsidas: The Epic of Ram*. Vol. 5. Murty Classical Library of India. Cambridge, Massachusetts: Harvard University Press, 2020.

Nagar, Shanti Lal, trans. *Brahmavaivarta Purāṇa: Text with English Translation*. 2 vols. Delhi: Parimal Publications, 2001.

Olivelle, Patrick, ed. *Manu's Code of Law: A Critical Edition and Translation of the Mānava-Dharmaśāstra*. South Asia Research Series. New York: Oxford University Press, 2005.

Tapasyananda, Swami. *Adhyātma Rāmāyaṇa*. Chennai: Ramakrishna Matt, 1985.

Wilson, H.H., trans. *The Vishṇu Purāṇa: A System of Hindu Mythology and Tradition, Enlarged & Edited by Nag Sharan Singh: Text & Translation*. 2 vols. Delhi: Nag Publishers, 2003 [1980].

Translations

Goldman, Robert P., trans. *The Rāmāyaṇa of Vālmīki: An Epic of Ancient India*. Vol. 1: Bālakāṇḍa. Princeton: Princeton University Press, 1984.

Goldman, Robert P., Sally J. Sutherland Goldman, and Barend A. van Nooten, eds. and trans. *The Rāmāyaṇa of Vālmīki: An Epic of Ancient India*. Vol. 6: Yuddhakāṇḍa. Princeton: Princeton University Press, 2009.

Goldman, Robert P., and Sally J. Goldman, trans. *The Rāmāyaṇa of Vālmīki: An Epic of Ancient India*. Vol. 7: Uttarakāṇḍa. Princeton: Princeton University Press, 2017.

Hill, W. Douglas P., trans. *The Holy Lake of the Acts of Rāma: A Translation of Tulasī Dās's Rāmacaritamānasa*. Bombay: Oxford University Press, 1971.

Tagare, Ganesh Vasudeo, trans. *The Bhāgavata Purāṇa*. Delhi: Motilal Banarsidass, 1976–1978.

Tagare, Ganesh Vasudeo, trans. *The Vāyu Purāṇa*. 2 vols. Delhi: Motilal Banarsidass, 1987–1988.

Taluqdar of Oudh, A., trans. *The Matsya Puranam*. 2 vols. Allahabad: The Pāṇiṇi Office, 1916–1917.

Secondary Sources

Brockington, John L. *The Sanskrit Epics*. Leiden: Brill, 1998.

González-Reimann, Luis. *The Mahābhārata and the Yugas: India's Great Epic Poem and the Hindu System of World Ages*. New York: Peter Lang, 2002. (Reprint Delhi: Motilal Banarsidass, 2010.)

González-Reimann, Luis. "The Divinity of Rāma in the *Rāmāyaṇa* of Vālmīki." *Journal of Indian Philosophy* 34, no. 3 (2006): 203–220. Reprinted in Will Sweetman, ed. *Hinduism*, vol. 2. New York: Routledge, 17–31.

CHAPTER 14

The Size of the Universe in Indian Astronomy

Michio Yano

1 Āryabhaṭa's Cosmology*

In *Āryabhaṭīya* 3.15–6, Āryabhaṭa says:

> *bhānām adhaḥ śanaiścarasuragurubhaumārkaśukrabudhacandrāḥ* |
> *teṣām adhaś ca bhūmir medhībhūtā khamadhyasthā* || 15
> *saptaite horeśāḥ śanaiścarādyā yathākramaṃ śīghraḥ* |
> *śīghrakramāc caturthā bhavanti sūryodayād dinapāḥ* || 16

> Below the asterisms are Saturn, Jupiter, Mars, the sun, Venus, Mercury, and the moon. Below them is the earth that stays in the middle of the sky like a central pillar (of everything).

> These seven are the Lords of hours, beginning with Saturn, in the order of [increasing] speed. The [every] fourth from this order of speed are the Lords of the days [which begin] from sunrise.

In order to get the order of planetary weekdays, the seven planets must be arranged in the order prescribed here by Āryabhaṭa. This order presupposed that the concentric orbs of the planets were below the sphere of asterisms. In Greek cosmology, the idea of concentric spheres was first conceived by Eudoxus, then supported by Aristotle, and it continued to be the main cosmography until the time of Copernicus. However, it took a long time before the order of the planets was fixed around the beginning of the first millennium. The order was established shortly before Ptolemy (ca. 150 CE), but he says nothing about the weekday, even in his book on astrology, *Tetrabiblos*. Ārybabhaṭa associated this order with the angular velocity of the planets, Saturn being the slowest and the moon the fastest. The linear speeds of all the planets were considered to be the same (see below). When we count every fourth planet thus

* This paper was prepared for the international workshop *Cosmos, East and West: Astral Sciences in South and East Asia and Their Interaction with the Greco-Roman World*, held on February 27, 2017, at New York University.

 | DOI:10.1163/9789004438224_015

arranged, we get the order of weekdays beginning with Saturday. This is because the seven planetary gods, beginning with Saturn, are assigned as the Lord of 24 hours (*horeśa*), and the Lord of the first hour of the day is also the Lord of the day (*dinapa*).

The weekday became one of the most important elements in Indian astrology after the introduction of Hellenistic astronomy and astrology. In Sanskrit astronomical texts, planets are always enumerated and described in the order of the weekdays. Thus, when they say "those beginning with Mars" (*bhaumādi*), they mean the five star-planets in the order of the weekdays.

The size of the sky is well defined by Āryabhaṭa. The basic size is that of the moon's orbit. In *Āryabhaṭīya* 1.6 he says:

> *śaśirāśayaṣ ṭha cakraṃ te aṃśakalāyojanāni yavañaguṇāḥ /*
> *prāṇenaiti kalāṃ bhaṃ khayugāṃśe grahajavo bhavāṃśe 'rkaḥ //* 6
>
> A circle multiplied by 12 is the moon's signs. These multiplied by 30, 60, 10 are [respectively,] degrees, minutes, and *yojanas*. The earth[1] moves one minute in one *prāṇa*. The circumference of the sky [in *yojanas*] divided by the rotations of a planet in a *yuga* is the [orbit of the] planet's motion.[2] The [orbit of the] sun is a sixtieth part of the circumference of the asterism.

This verse is better understood with *Āryabhaṭīya* 3.12:

> *śaṣṭyā sūryābdānāṃ prapūrayanti grahā bhaparināham /*
> *divyena nabhaḥparidhiṃ samaṃ bhramantas svakakṣyāsu //* 12
>
> By sixty solar years planets fulfill the circumference of the asterism. Revolving equally in their own orbits, they [fulfill] the circumference of the sky by one divine [year].

First of all, the circumference of the lunar orbit (C_m) is $12 \times 30 \times 60 \times 10 = 216{,}000$ *yojanas*. The moon's sidereal rotations in a *yuga* (R_m) are 57,753,336. Therefore, the moon travels $C_m \times R_m = 12{,}474{,}720{,}576{,}000$ *yojanas* in a *yuga* consisting of 4,320,000 human years (called "one divine [year]" in 3.12). This is also the cir-

1 Our text reads *bhaṃ* ("asterism") but in Brahmagupta's quotation (BSS 11.17) of this verse he reads *bhūḥ* ("earth"). This is one piece of evidence supporting Āryabhaṭa's theory of the rotation of the earth.

2 I followed Shukla's 1976 translation when translating *grahajava*.

TABLE 14.1 Circumference and distance of the planets in *yojanas* according to Āryabhaṭa

	R_i	C_i	Distance
Moon	57,753,336	216,000	34,377
Mercury	17,937,020	695,473	110,688
Venus	7,022,388	1,776,421	282,726
Sun	4,320,000	2,887,667	459,585
Mars	2,296,824	5,431,292	864,415
Jupiter	364,224	34,250,133	5,451,065
Saturn	146,564	85,114,493	13,546,361
Asterism	72,000	173,260,008	27,575,122
Sky (C_k)		12,474,720,576,000	

cumference of the sky (C_k) (*nabhaḥparidhi*) in *yojanas*. It is presupposed that all the planets travel the same linear distance in a *yuga*. In other words, $C_i \times R_i = C_k$ is constant. Since the number of civil days in a *yuga* is 1,577,917,500 according to Āryabhaṭa's school (Āryapakṣa), the constant daily motion in *yojanas* of all the planets is 12,474,720,576,000/1,577,917,500 = 7,905.

Thus, for example, in the case of the sun, whose rotations in a *yuga* are 4,320,000, its circumference (C_s) is C_k/R_s = 2,887,666 4/5 *yojanas*. This means that the distance of the sun from the earth is $C_s/2\pi \approx$ 459,585 *yojanas* (with Āryabhaṭa's value of π = 62832/20000 = 3.1416).

In table 14.1 I have shown the rotations of planets in a *yuga* (R_i) according to Āryabhaṭa, the resulting circumference (C_i, rounded) of their orbits, and their distance ($C_i/2\pi$) from the earth. These are remarkably different from the "distances" of planets in the Purāṇas which are shown in table 14.2.[3]

According to Āryabhaṭa, the radius of the earth is 1,050/2 = 525 *yojanas* (*Āryabhaṭīya* 1.7). Therefore, using the radius of the earth (r_e) as a unit, we can express the distance of the moon as 34,377/525=65.48 r_e and that of the sun as 459,585/525= 875.4 r_e. The circumference of the sphere of the asterisms (*bhapariṇāha*) is 60 times that of the sun (cf. *Āryabhaṭīya* 1.6 and 3.12 above) (i.e., 173,260,008 *yojanas*). This means that the distance of the fixed stars from the earth is 173,260,008/2π = 27,575,122 *yojanas* = 52,524 r_e.

3 See Willibald Kirfel, *Die Kosmographie der Inder nach Quellen dargestellt* (Hildesheim-Zurich-New York: Georg Olms Verlag, 1990): 128.

TABLE 14.2 Distance of the planets in *yojanas* according to the Purāṇas

Sun	10,000
Moon	20,000
Asterism	30,000
Mercury	50,000
Venus	700,000
Mars	900,000
Jupiter	1,100,000
Saturn	1,300,000
Saptarṣi	1,400,000
Pole Star	1,500,000

TABLE 14.3 Distances from the earth in r_e

	Moon	**Sun**	**Asterism**
Ptolemy	48	1,210	20,000
Āryabhaṭa	65.48	875.4	52,524

Now we can compare the Indian numbers with those given, for example, by Ptolemy (table 14.3). According to his *Planetary Hypothesis*,[4] the mean distance of the moon from the earth in the unit of the earth's radius (r_e) is 48. The mean distance of the sun is 1,210 r_e, and that of the fixed stars of the first magnitude is 20,000 r_e. These are not much different from the Indian values mentioned above. The only difference is that in Greek cosmology nothing seems to be beyond the sphere of the asterism, while in India they wanted to keep the huge size of the sky (*ākāśa* or *kha*) as the dwelling place of Brahman.

Bhāskara I (fl. 629 CE), the earliest commentator of the *Āryabhaṭīya*, thought that the sky (*ākāśa*) is limitless (*aparimita*) but that one can put a boundary on it. Commenting on *Āryabhaṭīya* 1.6, he says:

4 See Bernard R. Goldstein, "The Arabic Version of Ptolemy's Planetary Hypotheses," *Transactions of the American Philosophical Society, New Series* 57, no. 4 (1967): 11, Table I, and Otto Neugebauer, *History of Ancient Mathematical Astronomy* (Springer-Verlag: 1975), 922, Table 26.

> As far the place of the sky that the rays of the sun illuminate entirely, so far is the place of the circumference of the globe of the sky (*khagola*), [namely] the orb of the sky (*khakakṣyā*). Otherwise, since the sky is limitless, telling the size of the sky is not appropriate.[5]

Āryabhaṭa declares at the end of the first chapter of the *Āryabhaṭīya* (1.13):

> *daśagītikasūtram idaṃ bhūgrahacaritaṃ bhapañjare jñātvā* /
> *grahabhagaṇaparibhramaṇaṃ sa yāti bhittvā paraṃ brahma* // 13

> One who knows the aphorism of [the chapter called] *Daśagītikā* (consisting of ten Gīti meter verses) on the motion of the earth and the planets in the cage of the asterisms, after breaking through the passage of the planets and asterisms, arrives at the highest Brahman.

As mentioned above, the diameter of the earth is only 1,050 *yojanas* according to Āryabhaṭa. *Yojana* is a unit of length that is most commonly used in astronomical texts as well as in Purāṇas. Āryabhaṭa defines it saying "a *yojana* is 8,000 *puruṣas*."[6] The *puruṣa* (here *nṛ* means "man") is a unit of length that was used since the time of the Śulbasūtras, manuals for constructing sacrificial altars. Its length is measured from the tip of the extended hands of a standing man, for whom the sacrifice is performed, to the ground. It depends on the person, but let us assume that a *puruṣa* is about 1.8 m. Then, one *yojana* is about 14.4 km. Thus, Āryabhaṭa's diameter of the earth would be about 15,120 (1,050×14.4) km, which is not much different from the modern value (12,756 km). Such a small size for the earth was very much contradictory to the cosmology of the Purāṇas. Thus, when commenting on *Āryabhaṭīya* 1.7, Bhāskara I rejects the Purāṇic point of view.

Since the size of the earth is very close to reality, the mountains on it should also be also close to reality. Thus, Āryabhaṭa says that Mt. Meru is only one *yojana* high[7] on two occasions, namely, *Āryabhaṭīya* 1.7 and 4.11.[8] Stating the

5 *yāvantam ākāśapradeśṃ raver mayūkhāḥ samantād dyotayanti tāvān pradeśḥ khagolasya paridhiḥ, khakakṣyā. anyathā hi aparimitatvād ākāśasya parimānākhyānaṃ na upapadyate.* (translation of Shukla, 26–7).

6 *Āryabhaṭīya* 1.7a: *nṛṣi yojanaṃ*.

7 This is less than two times the height of the highest mountains in the Himalayas.

8 *Āryabhaṭīya* 1.7b: *ka meroḥ*, "(The height of) Meru is one *yojana*." *Āryabhaṭīya* 4.11: *meruḥ yojanamātraḥ*, "Meru is only one *yojana*." Mt. Meru was thought to be spherical in Āryabhaṭa's school.

same import twice is very exceptional in Sanskrit verse texts, and Āryabhaṭa in particular was greatly concerned with the brevity of expression.[9] He probably wanted to stress the inevitable difference from the traditional view.

In Purāṇic cosmography, the height of the cosmic mountain Meru is 84,000 *yojana*s above and 16,000 *yojana*s below the earth.[10] Followers of Āryabhaṭa should have been seriously concerned with the difference between the cosmography of the Purāṇas and that of mathematical astronomy. Bhāskara I was the first defender of Āryabhaṭa's view. In his commentary on *Āryabhaṭīya* 1.7 he says, "I will explain it (later) in the commentary on (the verse) beginning with 'Meru is only one *yojana*'" (*Āryabhaṭīya* 4.11). Unfortunately, all the manuscripts of Bhāskara I's commentary break off toward the end of *Āryabhaṭīya* 4.6. But we have Someśvara's commentary (eleventh–twelfth century CE), which is a kind of summary of Bhākara I's text. Commenting on *Āryabhaṭīya* 4.11, Someśvara says:

> Meru is said to be of the size of 100,000 *yojana*s by those who believe in Purāṇas. This is devoid of reason. [From Laṅkā to the center of Meru there are not even 1,000 *yojana*s].[11]

Further he adds:

> Further, if Meru is of a big size the stars north of Meru can't be seen because they are blocked by the summit of Meru.[12]

Āryabhaṭa does not say how he calculated the diameter of the earth, but Bhāskara I, commenting on *Āryabhaṭīya* 1.7, gives two methods. One is by "the elevation of the pole" (*akṣonnati*), namely, the geographical latitude. He discusses the actual distance in *yojana*s between Laṅkā, Kanyakumārī, and Ujjayinī. They were regarded as on the same meridian, and the differences in

---

9 Michio Yano, "Oral and Written Transmission of Exact Sciences in Sanskrit," *Journal of Indian Philosophy* 34, no. 1–2 (2006): 147.

10 For example, *Agnipurāṇa* 108.3, *Viṣṇupurāṇa* (202.8 in the e-text; translation of Wilson, 116), and *Varāhapurāṇa* 75.54. See Kirfel, *Die Kosmographie der Inder*, 93.

11 *atha paurāṇikair lakṣayojanapramāṇo meruḥ paṭhyate tad yuktirahitam/* [*laṅkāto yāvad merumadhyaṃ tāvad yojanasahasram api na ast*]. The part in the square brackets is supplied in Shukla's text and translation. A similar argument is given in Bhāskara I's commentary on *Āryabhaṭīya* 1.7.

12 *kiṃ ca yadi mahāpramāṇo meruḥ syād meruśikharāntaritatvād bhāvād uttareṇa tārakā na dṛśyeran* (translation of Shukla, 262).

their geographical latitudes were known by the altitude of the north pole. This method should have been the same as that of Eratostenes.[13]

The other method is by "mathematical computation" (*gaṇita*).[14] According to Bhāskara I, the maximum longitudinal parallax (π_{max}) of the moon is 52′30″. The mean distance of the moon (ρ_m) from the earth is 34,377 *yojanas* (table 14.1). The radius of the earth (r_e) is obtained by: $r_e = \sin \pi_{\max} \times \rho_{\mathrm{m}} = \sin 52'30'' \times 34{,}377 = 525$. Therefore, the diameter of the earth is 1,050 *yojanas*.

In this way, Indian astronomers had to reduce the size of the earth as well as that of Mt. Meru to an extreme degree.

On the other hand, they wanted to keep, or even increase, the huge size of the sky or Brahman's world. Of course, there is no astronomical problem whichever size the sky or heaven may be. Brahman's world is sometimes referred to as the Brahmāṇḍa, or the Egg of Brahman, in the traditional cosmography.[15] Astronomers also use this word.

2 Brahmagupta

In his *Brāhmasphuṭasiddhānta*, composed in 628 CE, Brahmagupta (b. 598 CE) gives the number of rotations of planets in a *kalpa* consisting of 4,320,000,000 years,[16] which is 1,000 times that of a *yuga*. The rotations of the moon in a *kalpa* is R_m=57,753,300,000.[17]

Moreover, he says (*Brāhmasphuṭasiddhānta* 21.11):

> *ambarayojanaparidhiḥ śaśibhagaṇāḥ śūnyakhakhajināgniguṇāḥ* |
> *yasya bhagaṇair vibhaktās tatkakṣā'rko bhaṣaṣṭyaṃśaḥ* || 11
>
> The rotations of the moon (in a *kalpa*) multiplied by 324,000 is the circumference of the sky in *yojanas*. When it is divided by the rotations of a planet, [the result is] the orbit (*kakṣā*) of the planet. The [orbit of the] sun is one sixtieth of [that of] the asterism.

13 Neugebauer, *History of Ancient Mathematical Astronomy*, 652 ff.

14 Bhāskara I and *Āryabhaṭīya* 1.7 (translation of Shukla, 29): *bhūvyāsaḥ gaṇitenāpi ānetuṃ śakyate*, "The diameter of the earth can be obtained by mathematical computation, too."

15 For example, *Sūryasiddhānta*, 12.29.

16 *Brāhmasphuṭasiddhānta* 1.15–24.

17 *Brāhmasphuṭasiddhānta* 1.16.

He thought that one minute of the moon's orbit is 15 *yojanas*. Thus, the moon's orbit C_m is $360 \times 60 \times 15 = 324{,}000$ *yojanas*. Accordingly, the circumference of the sky (*ambara*) became $C_k = R_m \times C_m = 18{,}712{,}069{,}200{,}000{,}000$ *yojanas*, which is about 1,500 times that of Āryabhaṭa. This number is found in the *Paitāmahasiddhānta* of the *Viṣṇudharmottarapurāṇa* 3.5.[18] The same number is called the "*yojanas* of the circumference of the orbit of the sky"[19] in Pṛthūdakasvāmin's commentary on the *Brāhmasphuṭasiddhānta*.

With the modern value of π,[20] the radius of the cosmic sphere, or the distance from the earth to the end of the sky, according to Brahmagupta, is $C_k/2\pi = 2{,}978{,}111{,}344{,}537{,}815$ *yojanas* = 42,884,803,361,344,536 km.[21]

The *Sūryasiddhānta* (SS 1.30) gives a slightly different number of rotations of the moon in a *kalpa*: $R_m = 57{,}753{,}336{,}000$. The moon's orbital circumference is 324,000 *yojanas*. Thus, $C_k = 18{,}712{,}080{,}864{,}000{,}000$ *yojanas*.[22]

3 Al-Bīrūnī

Al-Bīrūnī (973–1050 CE),[23] referring to Brahmagupta, says that "the number of the *yojana* of the sphere of zodiac"[24] is 18,712,069,200,000,000. On another

18 Pingree, "The *Paitāmahasiddhānta* of the *Viṣṇudharmottarapurāṇa*," *Adyar Library Bulletin* 31–32 (1967–1968) offers an English translation based on the text published by P.V. Dvivedin. He introduced new numbering of chapters 3–9 which correspond to chapters 168–174 of the second *khaṇḍa* of the *Viṣṇudharmottarapurāṇa*. The relevant part of the latter (printed by Nag publishers) is very corrupt. Pingree argues that the *Paitāmahasiddhānta* is dated at about 450 CE and that it is the oldest text of the Brāhmapakṣa; therefore, it must predate Brahmagupta. But his argument is not very convincing. There is some evidence that the *Viṣṇudharmottarapurāṇa*, or at least some parts of it, is posterior to Āryabhaṭa.

19 *khakakṣāparidhiyojana*. Cf. Setsuro Ikeyama, "*Brāhmasphuṭasiddhānta* (ch. 21) of Brahmagupta with Commentary of Pṛthūdaka, Critically Edited with English Translation and Notes," *Indian Journal of History of Science* 38, nos. 1–4 (2003): S184 and S185. Here *kakṣā* (orbit) is the same as *kakṣyā*.

20 Brahmagupta uses $\sqrt{10}$ as the value of π.

21 For the sake of comparison, I expressed the size of the sky according to the modern understanding. Brahmagupta's distance would be about 4,532 light years, one light year being 460,730,472,580.8 km.

22 This number is given in Kirfel, *Die Kosmographie der Inder*, 173.

23 See Michio Yano, "al-Bīrūnī," in *The Encyclopedia of Islam Three*, eds. Gudrun Krämer, Denis Matringe, John Nawas, and Everett Rowson (Leiden and Boston: Brill, 2013), 50–56.

24 See al-Bīrūnī's *Kitāb fī taḥqīq mā li-l-Hind*: Eduard Sachau's English translation, I.224; Arabic text in Sachau's edition, 110; Hyderabad edition, 183.

occasion, he says "Brahmagupta calls this number the *yojana*s of the ecliptic."[25] In the next paragraph he refers to the view of a "Pulisa" whose reckoning is based on *yuga*s instead of *kalpa*s. Thus, the product of the lunar orbit and the moon's rotations in a *yuga* is 18,712,080,864,000, exactly one thousandth of the *Sūryasiddhānta*'s number. He calls this "*yojana*s of heaven."[26]

It seems that al-Bīrūnī was impressed by the huge size of the heavens in Indian astronomy, especially since he was extremely familiar with the moderate size of the heavens in Greek astronomy.

In this context, it would be interesting to hear al-Bīrūnī's words:

> Brahmagupta says in the first chapter[27] of the *Brahma-siddhānta*, where he enumerates the heavens, placing the moon in the nearest heaven, the other planets in the following ones, and Saturn in the seventh: "The fixed stars are in the eighth heaven, and this has been created round in order to last for ever, that in it the pious may be rewarded, the wicked be punished, since there is nothing behind it." He indicates in this chapter that the heavens are identical with the spheres, and he gives them in an order which differs from that of the traditional literature of their creed, as we shall show hereafter in the proper place. He indicates, too, that the *round* can only be slowly influenced from without. He evinces his knowledge of the Aristotelic notions regarding the round form and the rotating motion, and that there is no body in existence behind the spheres.[28]

I cannot find any of Brahmagupta's words that correspond to al-Bīrūnī's quotation. What he meant by "heavens" is probably *kakṣās*. *Kakṣā* also means the "orbit" of the planets and can be used in plural form. We have another word, *kha*, for heaven that is always used in the singular form and designates empty space. The "fixed stars" are *bha*s or *nakṣatra*s. Anyway, it is interesting to hear from him that "the heavens" are identical with "the spheres." More interesting is that al-Bīrūnī thought there was an Aristotelian origin to the Indian concept of a circular sky.

---

25 See al-Bīrūnī's *Kitāb fī taḥqīq mā li-l-Hind*: Eduard Sachau's English translation, II.71; Arabic text in Sachau's edition, 236; Hyderabad edition, 402.

26 See al-Bīrūnī's *Kitāb fī taḥqīq mā li-l-Hind*: Eduard Sachau's English translation, II.72; Arabic text in Sachau's edition, 237; Hyderabad edition, 404.

27 What al-Bīrūnī means by the "first chapter" is chapter 21 (*Golādhyā*) of the modern editions. See Ikeyama, "*Brāhmasphuṭasiddhānta* (ch. 21) of Brahmagupta."

28 See al-Bīrūnī's *Kitāb fī taḥqīq mā li-l-Hind*: Eduard Sachau's English translation, I.223–224; Arabic text in Sachau's edition, 110, line 4ff.

The "traditional literature" here means the Purāṇas. In fact, al-Bīrūnī refers to the *Viṣṇupurāna* for the distances of the heavenly bodies, which is exactly the same as table 14.2 above.[29]

4 Survival of Purāṇic and Buddhist Cosmology

In an earlier article[30] I argued that there were five stages of the development of the concept of *graha* as heavenly bodies. The Purāṇic texts belong to the fourth stage, namely, a group of seven or nine *graha*s was established, although the order of enumeration was not yet fixed. This stage is evidenced in table 14.2 above. The relative position of the inferior and superior planets was known, but the sun is closer to the earth than the moon, and the asterisms (*nakṣatra*s) are next to the moon because they are the abode of the moon.

The Purāṇas are of an encyclopedic nature and some of them incorporated new elements of astronomy and astrology that had Hellenistic origins. Still, the older elements were not discarded. The *Viṣṇudharmottarapurāṇa* mentioned above is a typical example. Its date was a little later than Āryabhaṭa and almost contemporary with the establishment of the Buddhist cosmology[31] by Vasubandhu (fl. fourth–fifth century CE) in his *Abhidharmakoṣabhāṣya*. Especially popular among Buddhists was the *Lokaprajñaptyabhidharmaśāstra*, which survived in a Chinese translation (立世阿毘曇論, *Lishiapitanlun*) by Paramārtha (眞諦, Zhendi, 499–569 CE)[32] as well as in a Pali translation, *Lokapaññatti*.[33] The Chinese translation was very popular and extensively used by a Japanese Buddhist Entsu (円通, 1754–1834 CE) when he tried to defend Buddhist cosmology in the *Bukkoku Rekishohen* (仏国暦象編) against the newly introduced Western cosmology in the Edo period.

---

29 See al-Bīrūnī's *Kitāb fī taḥqīq mā li-l-Hind*: Eduard Sachau's English translation, 1.238. Similar numbers are found in the *Vāyupuraṇa* 53.95c–97d, but the text is very corrupt.

30 Michio Yano, "Planet Worship in Ancient India," in *Studies in the History of the Exact Sciences in Honor of David Pingree*, eds. Charles Burnet et al (Leiden and Boston, 2004), 331–348.

31 For Buddhist cosmology, see Akira Sadakata, *Indo Uchuron Taizen* (Tokyo, 2011).

32 Paramārtha was born in Ujjayinī, the center of astronomy, and he knew about Āryabhaṭa's theory of the rotation of the earth on its axis, which he refutes. Cf. Sadakata, *Indo Uchuron Taizen*, 243.

33 Takao Hayashi, "On the Chapter on the Measurements of the World in the *Lishiapitanlun*," *The Science and Engineering Review of Doshisha University* 49, no. 4 (2009): 3–8 and "A Study of the Chapter on the Motions of the Sun and Moon in the *Lishiapitanlun*," *South Asian Classical Studies* 8 (2013): 1–50.

References

Primary Sources

Agnipurāṇa, edited by Āchārya Baladeva Upādhyāya. Kashi Sanskrit Series 174. Varanasi: Chowkhamba Sanskrit Series Office, 1966. E-text prepared by Jun Takashima: http://www.aa.tufs.ac.jp/~tjun/data/gicas/ap1_tr_frame.html.

al-Bīrūnī. *Kitāb fī taḥqīq mā li-l-Hind*, edited by Eduard Sachau. London: Trübner & Co., 1887. (Reprint: Frankfurt, 1993.)

al-Bīrūnī. *Kitāb al-Bīrūnī fī taḥqīq mā li-l-Hind min maqūla maqbūla fī l-ʿaql aw marḏūla*. Hyderabad: Dairatu-l-Maʿārif-il-Oṣmānia, 1958. (Reprint: 1978.)

Āryabhaṭīya of Āryabhaṭa, critically edited by K.S. Shukla. New Delhi: Indian National Science Academy, 1976.

Āryabhaṭīya of Āryabhaṭa, with the Commentary of Bhāskara I and Someśvara, edited by K.S. Shukla. New Delhi: Indian National Science Academy, 1976.

Pitāmahasiddhānta, edited by P.V. Dvivedin, in *Jyautiṣa Siddhānta Saṅgraha: A Collection of Ancient Hindu Astronomical Works*. Part 2. Benares Sanskrit Series 154. Benares: Braj Bhushan Das, 1917.

Brāhmasphuṭasiddhānta and Dhyānagrahopadeśādhyāya, by Brahmagupta, edited with his own commentary by Sudhākara Dvivedin. Benares: Medical Hall Press, 1902.

Varāha Purāṇa, translated and annotated by S. Venkitasubramonia Iyer. Delhi: Motilal Banarsidass, 1985.

Viṣṇupurānam, critically edited by M.M. Pathak. Vadodara: Oriental Institute, 1997 and 1999.

Secondary Sources

Burgess, E. "Translation of the *Sūryasiddhānta*." *Journal of the American Oriental Society* 6 (1858): 141–498.

Goldstein, Bernard R. "The Arabic Version of Ptolemy's Planetary Hypotheses." *Transactions of the American Philosophical Society, New Series* 57, no. 4 (1967): 3–55.

Hayashi, Takao (林隆夫). "立世阿毘曇論数量品について" ("On the Chapter on the Measurements of the World in the *Lishiapitanlun*"). *The Science and Engineering Review of Doshisha University* 49, no. 4 (2009): 3–8.

Hayashi, Takao (林隆夫). "立世阿毘曇論日月行品の研究" ("A Study of the Chapter on the Motions of the Sun and Moon in the *Lishiapitanlun*"). *South Asian Classical Studies* 8 (2013): 1–50.

Ikeyama, Setsuro. *The Brāhmasphuṭasiddhānta Chapter 21 with the Commentary of Pṛthūdakasvāamin*. Doctoral dissertation. Brown University, 2002. Published as "*Brāhmasphuṭasiddhānta* (ch. 21) of Brahmagupta with Commentary of Pṛthūdaka, Critically Edited with English Translation and Notes" in *Indian Journal of History Science* 38, nos. 1–4 (2003): S1-S308.

Kirfel, Willibald. *Die Kosmographie der Inder nach Quellen dargestellt*. Bonn and Leipzig: Kurt Schroeder, 1920. (Reprint: Hildesheim-Zurich-New York: Georg Olms Verlag, 1990.)

Neugebauer, Otto. *A History of Ancient Mathematical Astronomy*. 3 parts. New York: Springer-Verlag, 1975.

Pingree, David. "The *Paitāmahasiddhānta* of the *Viṣṇudharmottarapurāṇa*." *Adyar Library Bulletin* 31–32 (1967/68): 472–510.

Pingree, David. "History of Mathematical Astronomy in India," *Dictionary of Scientific Biography* 15 (1978): 533–633.

Sachau, Eduard. *Alberuni's India*. London: Trübner & Co., 1888. (New Edition: 1910.)

Sadakata, Akira (定方晟). *Indo Uchuron Taizen* (インド宇宙大全). Tokyo: Shunjusha, 2011.

Yano, Michio. "Planet Worship in Ancient India." *Studies in the History of the Exact Sciences in Honor of David Pingree*, edited by Charles Burnet et al, 331–348. Leiden and Boston: Brill, 2004.

Yano, Michio. "Oral and Written Transmission of Exact Sciences in Sanskrit." *Journal of Indian Philosophy* 34, nos. 1–2 (2006): 143–160.

Yano, Michio. "al-Bīrūnī." *The Encyclopedia of Islam Three*, edited by Kate Fleet, Gudrun Krämer, Denis Matringe, John Nawas, and Everett Rowson. Leiden and Boston: Brill, 2013: 50–56.

CHAPTER 15

Trigonometry through Geometry: Bhāskarācārya's Generation of Sines

Krishnamurthi Ramasubramanian and Clemency Montelle

1 Introduction

Bhāskarācārya (b. 1114 CE) is an illustrious figure in the history of astronomy in India, celebrated not only for his technical prowess, but also for his accomplished abilities as a poet. Amongst the articulation of many complex and technical results in mathematics and astronomy, Bhāskara is known for his treatment of trigonometric rules and identities, relations which are indispensable for an astronomer. Besides enumerating sine values for a base circle $R = 3438$ and arc multiples of $3°45'$, Bhāskara also included a chapter called the *Jyotpatti* (lit. 'Production of Sines') in his *Siddhāntaśiromaṇi*, which sets out, in 25 verses, various rules to generate sines, including the famous sum-to-product rules.

In addition, we find an exploration of the trigonometrical rules in at least one other place in the *Siddhāntaśiromaṇi*, in a lesser known passage at the beginning of chapter 5 of the *Golādhyāya*. Bhāskara calls this section too as *jyotpatti* and commences the passage with a beautiful metaphor to impress upon his readers the importance of the trigonometric relations in the study of the motion of celestial objects. Then, in only five verses, he articulates some of the key rules required to construct a sine table with 24 values. Furthermore, Bhāskara's own commentary, the *Vāsanābhāṣya*, on this passage, sheds light on the scope of these rules, and how he conceived them diagrammatically and derived them algebraically. In order to elucidate this aspect of Bhāskara's conception of trigonometry, we present a translation of these verses and their accompanying commentary, along with a thorough description of their meaning and significance.

Highlighting, as it were, the general attitude of Indian astronomers and mathematicians in presenting their results in a very concise manner without feeling the need to provide elaborate scaffolding (as often found in certain other traditions), towards the end of his *Bījagaṇita*, Bhāskara notes:

upadeśalavaṃ śāstraṃ kurute dhīmato yataḥ |
tat tu prāpyaiva vistāraṃ svayam evopagacchati || 6

 | DOI:10.1163/9789004438224_016

jale tailaṃ khale guhyaṃ pātre dānaṃ manāgapi |
prājñe śāstraṃ svayaṃ yāti vistāraṃ vastuśaktitaḥ || 7

An iota of instruction of *śāstra* does [the trick (i.e., is sufficient)] for the intelligent ones, since that itself having lodged [in their mind] burgeons in its own accord. Because of the inherent potential (*vastuśakti*) [in brilliant minds], the *śāstra* instilled in them proliferates, as a drop of oil in the water, as a piece of secret in the crooked, or as an act of charity [done] to a deserving person.

Despite being an advocate of brevity, Bhāskara is still highly attuned to the importance of explanation for more general audiences so that they too can appreciate the line of thought that underlies these various concepts associated with trigonometry. In fact, it is precisely for this reason Bhāskara writes his *Vāsanābhāṣya* wherein he elaborates how to demonstrate the validity of certain mathematical results, as well as conceptualize certain geometrical objects. In this *jyotpatti* passage, he emphasizes diagrams as being key to understanding. Indeed, when he wraps up this chapter (verse 44), Bhāskara himself declares that the entire section is for the 'keen witted.' In other words, according to him, the geometrical picture implied by the procedure outlined in the verses will immediately resound in those that possess such qualities of intelligence:

ye darbhagarbhāgradhiyo 'tra teṣāṃ
syāc chedyakārthaḥ paramāṇurūpaḥ |
ye 'nye jaḍāḥ kuṇṭhadhiyaś ca teṣāṃ
syād indravajrāhatapakṣatulyaḥ || 44

For those who possess keen wits as sharp as the tip of the innermost blade of *darbha* grass, the content explained by geometrical representation (*chedyaka*) is just like an atom (*paramāṇu*). But those students, who are dull-witted and have blunt understanding, find this subject like [an impenetrable mountain] that has been shorn of its wings by the thunderbolt of Indra.

In this spirit, we offer a scholarly analysis of the six verses on the generation of sines to Professor Zysk, an eminent researcher of noted intellect and sharpness of mind, the very candidate Bhāskara seems to have directed his text to. Indeed, we have no doubt he will esteem this small (*paramāṇu*) contribution of ours as a big (*śaila*) one, and, even more, that he will enjoy the diagrammatic explanation of trigonometric relations, finding it as informative as it is interesting!

2 The Significance of Sines in the Study of Spherics

Siddhāntas are typically made up of two parts: *gaṇita* (lit. 'computation'), which deals with astronomical rules and computations, and *gola* (lit. 'spherics'), which concerns explanations of the various assumptions and geometric models on which the astronomical computations rest. While Bhāskara's sine values are presented early on in the *gaṇita* part of the *Siddhāntaśiromaṇi*, as they are essential for computing the true longitudes and velocities of the planets, his account of the rationales and rules behind these trigonometric relations are treated much later in the *gola*. The fifth chapter of the *gola*, consisting of 44 verses altogether, concerns various geometric and diagrammatic characterizations of the models for computing true planetary positions. The first six of these are dedicated to the *jyotpatti*.

Jyotpatti is a compound formed from the words *jyā*, here 'sine,' and *utpatti*, 'generation' or 'production.' In order to impress upon the need for studying such passages on *jyotpatti*, Munīśvara in his *Siddhāntasārvabhauma* (*Spaṣṭādhikāra*, verses 1–2) notes:[1]

> *svaspaṣṭabhogaiḥ phaladā grahendrāḥ*
> *yato 'khile karmaṇi tena teṣām* /
> *sphuṭakriyāmāgamayuktimūlāṃ*
> *apāstaniryuktimatāṃ pravacmi* // 1

> Since, based on their true positions, the planets bestow [humans] with the desired outcomes in all their activities, I [set forth myself to] explain the correction procedure which is based on the knowledge of the scriptures or oral tradition (*āgama*) and logical analysis (*yukti*) and which has discarded schools that are devoid of rationales.

Then outlining the significance of sine function, he notes:

> *dehaṃ yathā snāyunibaddham asti*
> *tathā grahaspaṣṭamukhakriyā hi* /
> *jyābhis tathā jyāḥ prathamaṃ vadāmi*
> *pratyekabhāgaṃ vivare 'nupātāt* // 2

1 See Gopī Nātha Kavirāja, ed., *The Siddhāntasārvabhauma of Munīśvara* (Benares City: Vidya Vilas Press, 1932).

> As the body is tied together [i.e., completely criss-crossed] with veins, so too are the procedures for the computation of the true positions of the planets with the sines. [Hence,] I present the values of sines for every degree [and] for the interstices [the values are to be known] by interpolation.

3 The Relevance of This Set of Jyotpatti Verses

The verses pertaining to this first *jyotpatti* do not involve any numerical values. Only rules are presented in their generality. Bhāskara encourages the mathematician to construct their own sine table with 'as many values as they please'; his rules can then be implemented to help the mathematician compute the actual values depending on the number of data points they wish. It may be noted that the values of the sines obtained depend on the magnitude of the chosen radius. With the rules Bhāskara has demonstrated here, the production of the sine table is restricted to 24 values. Further rules that allow one to compute for a greater number of divisions than this (i.e., more than 24 values) are found in the *Jyotpatti* chapter later on.[2]

It is significant that these verses open the chapter called *chedyakādhikāra*. *Chedyaka* is often associated with diagrams or geometrical explanations. However, the word is originally derived from the root *chid* which means to split, smash, or destroy. In this way, doubts that may arise in one's own mind regarding the validity of procedures outlined earlier will be metaphorically 'destroyed' by recourse to diagrams.

3.1 *The Distinction between the Jyotpatti Described Here and Elsewhere in the Text*

Bhāskara includes a versified trigonometric table including 24 values for arc multiples of 3°45′ and $R = 3438$ early on in the *Siddhāntaśiromaṇi* (*spaṣṭādhikāra*, 2–9). In his *karaṇa* text, the *Karaṇakutūhala*, he includes a versified trigonometric table with 10 values for arc multiples of 10° and $R = 120$ (2, 6–8). Against this background, the opening verse of the later *Jyotpatti* chapter in the *Siddhāntaśiromaṇi*, with its appeal to generality in constructing sine tables, can be understood:

> *iṣṭāṅgulavyāsadalena vṛttaṃ kāryaṃ digaṅkaṃ bhalavāṅkitañ ca* |
> *jyāsaṃkhyayāptā navater lavā ye tadādyajīvādhanur etad eva* || 2

2 See Clemency Montelle, "The Production of Sines: Bhāskarācārya's *Jyotpatti*," in *Bhāskaraprabhā*, ed. K. Ramasubramanian et al. (Hindustan Book Agency, 2019): 291–300.

> A circle with radius [equal to] the desired digits (*aṅgulas*) is to be drawn. It is to be marked with directions and 360 degrees (*lava*). Whatever [results] from ninety degrees divided by the number of sines [one wishes to generate], that arc length indeed corresponds to the first Rsine.

Just as Bhāskara has done in two separate works, he makes it quite clear that the number of data points for which the Rsine values are to be determined is not fixed, but is completely left to the choice of the individual as per their convenience.

Indeed, it is noteworthy that Bhāskara throughout the discussion in this set of *jyotpatti* verses does not give any prescription whatsoever implicitly or explicitly over the choice of the value of the radius. In fact, he explicitly mentions in his *Vāsanābhāṣya* that here in the *jyotpatti* one can choose the magnitude of the radius to be any desired value

> *atra jyotpattau abhīṣṭā trijyā kalpyate*

> Here, in the study of the *jyotpatti*, any desired value is chosen as the radius.

This gives us a clue that Bhāskara has chosen to write this section purely to explain the trigonometric functions from a pedagogical viewpoint through simple geometrical construction, laying emphasis on the principles and not the magnitudes themselves. For instance, if we choose to divide the quadrant into 30 parts, the first Rsine will be 3° ($= 180'$) and if we choose to divide the quadrant into 24 parts then it will be $3^\circ 45'$ ($= 225'$).

He is not as general elsewhere when he discusses trigonometry. In fact, several rules in this later *Jyotpatti* chapter convey conclusively that Bhāskara had as his foundation a radius value of 3438. For instance, in the quote above, paraphrased as: *dhanur eva ādya jīvā*. "the [measure] of the arc is indeed [the measure] of the first Rsine;" the implication is that the first sine is approximately equal to the arc ($3^\circ 45'$) if $R = 3438$. In addition, this chapter includes several rules that use numerical approximations which are based on this assumption as well; for instance, the rules in verses 8 and 16–20.[3]

3.2 *The Significance of the Commentary in This Section*

While these verses are notable for their clear appeal to a geometrical conception of sines, Bhāskara's commentary is even more explicit, coaxing the reader

3 See Montelle, "Production of Sines," 297–299.

into drawing their own diagrams with careful labelling to delimit various segments and use them to reason diagrammatically. In his commentary, Bhāskara derives rules using symbolic styles of reasoning, explicitly using symbols to stand for various unknown quantities. In a passage which toggles between prose statements and symbolic expressions, Bhāskara manipulates various mathematical expressions according to rules to arrive at a desired result. Such forms of explicit algebraic reasoning of this type are infrequent in the mathematical corpus[4] and they give a glimpse into the ways in which these rules might have been demonstrated and justified by these early mathematicians.

Indeed, the placement of this set of verses before a description of the geometrical aspects of the planetary model indicates an obvious close connection between the two, and in particular facilitates the imagination of circles and epicycles on which planetary motions are conceived. The relation between the two is mentioned by later authors,[5] which may further shed light on the inspiration to conceive of the sine in this particular way. This is emphasized in the latter half of the 6th verse and the commentary associated with it. All in all, this short succinct passage gives us a very rich picture of Bhāskara's conception of the scope and nature of trigonometry.

4 The Sanskrit Text and Translation

We present here the text of the first six verses from chapter five[6] of the *Golādhyāya* of Bhāskara's *Siddhāntaśiromaṇi*. We have included the text in Roman transliteration with a literal but fluent English translation underneath. We have taken the text from the edition prepared by Chaturvedi (1981). After presenting the verses, in the following section, we have included a copy of the *Vāsanābhāṣya* of Bhāskara along with an English translation approached in a similar manner.

4.1 *Preamble*

paṭo yathā tantubhir ūrdhvatiryak-
rūpair nibaddho 'tra tathaiva golaḥ |

4 See Clemency Montelle and K. Ramasubramanian, "Determing the Sine of One Degree in the *Sarvasiddhāntarāja* of Nityānanda," *SCIAMVS* 19 (2018): 1–52.

5 See Dhammaloka et al., "Computation of Sines in Nityānanda's *Sarvasiddhāntarāja*," *SCIAMVS* 17 (2016): 1–54.

6 The chapter numbering and divisions vary in certain editions of the text.

doḥkoṭijīvābhir amuṃ pravaktuṃ
jyotpattim eva prathamaṃ pravakṣye // 1

Just as fabric (*paṭa*) is made of warp and weft threads, so too [the study of] spherics is criss-crossed with Rsines and Rcosines. [Therefore,] I will firstly narrate the *jyotpatti* ('generation of sines') to elaborate on this.

4.1.1 Analysis

Bhāskara opens this chapter with a preamble that includes a beautiful and particularly apt simile for introducing the trigonometric relations of the sine and cosine as applied in planetary computation. The simile invoked is that of the warp and weft of fabric, from which two levels of reading can be extracted. First, cloth is made up of threads woven together perpendicularly; the composition and structural integrity of the cloth relies on these threads woven in this manner. So too then is the study of spherics. At its heart lies the trigonometry interspersed with the computation of sines and cosines.

The second reading is of a more visual nature. Just as the warp and weft threads run vertically and horizontally, in the same manner the sine and cosines run vertically and horizontally too in the circular diagram; their names 'upright' (*doḥ*) and 'lateral' (*koṭi*) parallel the terms 'warp' and 'weft.' The analogy thus captures the relations between the line segments the sine and cosines represent for any given arc. The metaphor then is strikingly graphical in its poetic intent. This sets the scene for the geometric reasoning that is put forth in the following verses.

4.2 *Definition of the Basic Trigonometric Relations*

iṣṭā trijyā sā śrutir dorbhujajyā
koṭijyā tadvargaviśleṣamūlam /
doḥkoṭyaṃśānāṃ kramajye pṛthak te
trijyāśuddhe koṭidorutkramajye // 2

[Draw a circle with] any desired radius; that is [conceived of as] the hypotenuse, the upright (*doḥ*) is the Rsine (*bhujajyā*), and the Rcosine (*koṭijyā*) is the square root of the difference of their squares. These two 'direct sines' (*kramajyā*) corresponding to [the arcs measured in] degrees [relating to] the upright (*doḥ*) and the lateral (*koṭi*) separately subtracted from the radius [produce] the two Rversines (*utkramajyā*) of the lateral and the upright [respectively].

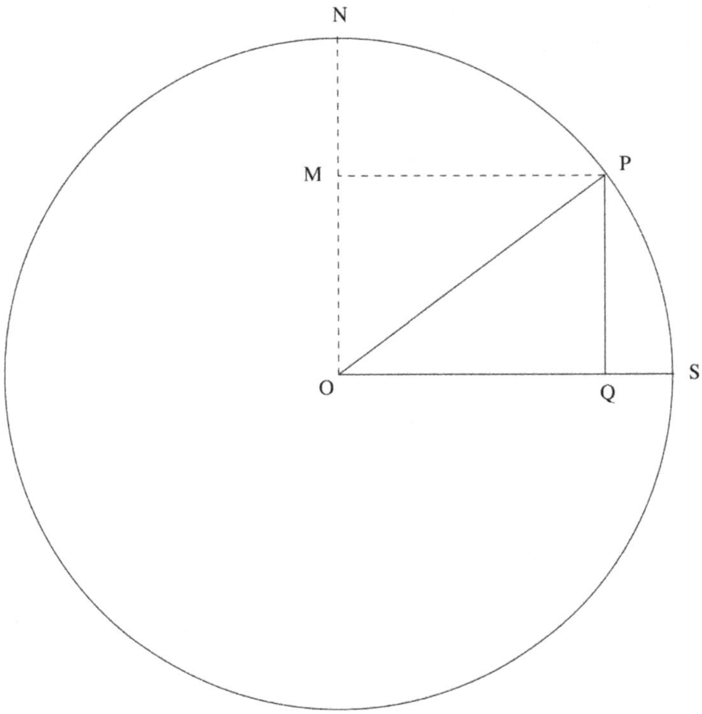

FIGURE 15.1 A right triangle inside a circle to define elementary trigonometric functions

4.2.1 Analysis

The basis of this passage is to prompt the reader to create a graphical representation of a circle with a right-angle triangle OPQ (see Figure 15.1) inscribed inside, whose hypotenuse OP is the radius. Using this, Bhāskara defines various trigonometric fundamentals. The sides of this triangle, their corresponding arcs, and the segments associated with them give visual representation of the different trigonometric relations to be defined. Here, Bhāskara defines four trigonometric quantities, known in modern terminology as the Rsine, the Rcosine, the Rversine, and the Rcoversine. The first two, called 'direct sines' (*kramajyā*), are related to the sides of the triangles as indicated in the table below:

Side of triangle	Sanskrit term	Trigonometric object	Sanskrit term
Upright	*doḥ*	Rsine	*bhujajyā*
Lateral	*koṭi*	Rcosine	*koṭijyā*

Related to these are the two 'versines,' or the *utkramajyā*. They are obtained when the two 'direct sines' are decreased from the radius of the circle. In Figure 15.1, these are the line segments QS and MN. Subtracting the Rcosine *OQ* from the radius gives the Rversine *QS* and likewise subtracting the Rsine *PQ* (= *OM*) gives the Rcoversine *MN*. Bhāskara refers to these as the *utkramajyā* of the *koṭi* and the *doḥ* respectively.

In this verse, he also relates the Rsine to the Rcosine through the familiar Pythagorean relation:

$$OQ = \sqrt{OP^2 - PQ^2}$$

$$R\cos\theta = \sqrt{R^2 - (R\sin\theta)^2}.$$

4.3 Geometrically Defining the Utkramajyā and the Rsines of 30° and 45°

jyācāpamadhye khalu bāṇarūpā
syād utkramajyā tribhamaurvikāyāḥ |
vargārdhamūlaṃ śaravedabhāga-
jīvā tataḥ koṭiguṇo 'pi tāvān || 3

tribhajyakārdhaṃ khaguṇāṃśajīvā
tat koṭijīvā kharasāṃśakānām |

Indeed, the segment [lying] between the chord (*jyā*) and the arc (*cāpa*), in the form of an arrow, is the Rversine. The square-root of half the square of the Rsine of three signs (i.e., the radius) is the Rsine of 45 degrees. It follows (*tataḥ*) that the Rsine of the lateral (i.e., the Rcosine) is also of the same measure.

Half the radius is the Rsine of 30 degrees. That is equal to the Rcosine of 60 degrees.

4.3.1 Analysis

After identifying the two Rversines on the diagram as being the line segments between the chord and the arc, Bhāskara specifies the values of the Rsine of 45° and 30° and relates them to their Rcosines:

$$R \sin 45^\circ = \sqrt{\frac{R^2}{2}} = R \cos 45^\circ, \tag{1}$$

and

$$R \sin 30^\circ = \frac{R}{2} = R \cos 60^\circ. \tag{2}$$

4.4 *Procedure for Determining the Rsines and Rcosines of Half-Angles*

kramotkramajyākṛtiyogamūlāt
dalaṃ tadardhāṃśakaśiñjinī syāt // 4

trijyotkramajyānihater dalasya
mūlaṃ tadardhāṃśakaśiñjinī vā /
tasyāḥ punas taddalabhāgakānāṃ
koṭeś ca koṭyaṃśadalasya caivam // 5

Half the square root of the sum of the squares of the Rsine and the Rversine [of an arc] is the Rsine (*śiñjinī*) of half of the degrees [corresponding to that arc]. Or, the square root of half the product of the radius and the Rversine is the Rsine of half that arc. From the Rcosines, the Rcosines of halves of the arcs can be found in a similar way.

4.4.1 Analysis

Bhāskara then gives two versions of a rule to find the Rsine of half an arc, given the Rsine and Rversine of an arc. These are as follows:

$$R \sin \frac{\theta}{2} = \frac{\sqrt{(R \sin \theta)^2 + (R \operatorname{versin} \theta)^2}}{2}$$

and

$$R \sin \frac{\theta}{2} = \sqrt{\frac{R \cdot (R \operatorname{versin} \theta)}{2}}.$$

While Bhāskara observes the Rcosines of half-arcs can be found in a similar way, he is silent to the way in which this is to be done. Presumably, Bhāskara intends the reader to use the relation between the Rsine and the Rcosine given in verse 2.

4.5 *Sine Values in Their Generality*

evaṃ triṣaṭsūryajinādisaṃkhyā
abhīṣṭajīvāḥ sudhiyā vidheyāḥ /
trijyotthavṛtte bhagaṇāṅkite vā
grāhyā abhīṣṭā vigaṇayya jīvāḥ // 6

In this way, any desired Rsines numbering 3, 6, 12, 24, and so on, are to be determined by the intelligent ones. Or, in a circle drawn with [given] radius, marked with [360] divisions, the desired Rsines may be obtained by measuring (*vigaṇayya*).

4.5.1 Analysis

Bhāskara then concludes the discussion with a reference to computing actual Rsine values. By his reference to the Rsines numbered 3, 6, 12, 24 and so on, it is clear he has in his mind Rsine tables of varied lengths. The number 3 corresponds to three Rsine values, namely $R \sin 30^\circ$, $R \sin 60^\circ$, and $R \sin 90^\circ$ $(= R)$. The number 6 corresponds to Rsine values for every 15°. The number 24 corresponds to listing Rsine values for the quadrant by dividing it into arc multiples of $3^\circ 45'$.

Having set out the various rules to be able to compute a sine table of such scope, Bhāskara concludes the passage with an alternative way to produce values: drawing out a circle with desired radius and simply 'measuring' (*vigaṇayya*) the resulting half-chord. This is a very interesting alternative approach, as it simply appeals to the reader's graphical and geometric skills and also challenges their abilities to make fine measurements; no mathematical computation is involved but for physical measurement. This crude yet very practically oriented approach to building a sine table is quite instructive; it is in direct contrast to teaching sines as mere ratios of two sides of a right triangle or as a mere number.

5 Bhāskara's Commentary on This Section on *Jyotpatti*

Bhāskara commences his commentary to these verses with a detailed description of the geometrical construction that is required to introduce the various trigonometric functions. He then deduces their values for certain arguments.

5.1 Establishing the 24 Chords

"atra trijyotthavṛtte bhagaṇāṅkite vā" ity etadantyavṛttasya uttarārdhamādau vyākhyāyate jyotpattau abhīṣṭā trijyā kalpyate samāyāṃ bhūmau trijyāmitāṅgulena sūtreṇa vṛttaṃ vilikhya digaṅkitaṃ cakrāṃśakaiś ca aṅkitaṃ kṛtvā, tatra ekasminn ekasmin vṛttacaturthāṃśe navatir navatir bhāgā bhavanti / tato yāvanti jyārdhāni kāryāṇi tāvadbhir vibhāgaiḥ ekaikaṃ vṛttacaturthāṃśaṃ vibhajya tatra cihnāni kāryāṇi / tad yathā—yatra caturviṃśatirjīvāḥ sādhyāḥ tatra caturviṃśatir bhavanti / evaṃ dvitīyacaturthāṃśe 'pi / tato dikcihnādubhayataścihnadvayopari gataṃ sūtraṃ jyārūpaṃ bhavati / evaṃ caturviṃśatir jyā bhavanti / tāsām ardhāni jyārdhāni / tatpramāṇāny aṅgulair mitvā grāhyāṇi /

An explanation for the latter half of the last verse, that is "*atra trijyotthavṛtte bhagaṇāṅkite vā*" is being provided first. In the study of *jyotpatti*, any desired radius is selected. Drawing a circle on the flat earth with a thread whose measure is the radius in digits (*aṅgulas*), having marked the cardinal directions as well as the 360 degree marks, [it can be seen that] in each one of the four quadrants, there are exactly 90 degrees.[7] Divide each of the four quadrants into as many parts as the number of sines required to be obtained, and mark them. This is how it is [done]—if 24 Rsines, then there will be 24 [markings]. In the same way [make markings] in the second (i.e., subsequent) quadrant also. Then, the line passing through the point on either side of the direction point will be the chord. Thus there are 24 chords. Their halves are the Rsines (*jyārdha*, lit. half-chords). The lengths of them in digits (*aṅgulas*) have to be measured and noted.

5.1.1 Analysis

In an interesting exegetical approach, Bhāskara tackles the end of the set of verses describing the *jyotpatti* first. The intent of Bhāskara in doing so is to provide an explanation for the trigonometric relations given by him so that they immediately resound or appeal to reasoning from a diagram. In other words, he wants to teach trigonometry through geometry and thereby discover the intrinsic connections between them.

Before getting into the other details we may highlight some of the salient features of the diagrammatic representation that is outlined here:

---

7 The repetition of the word *navati* in the text is technically referred to as *vīpsā*, and is generally meant for emphasis.

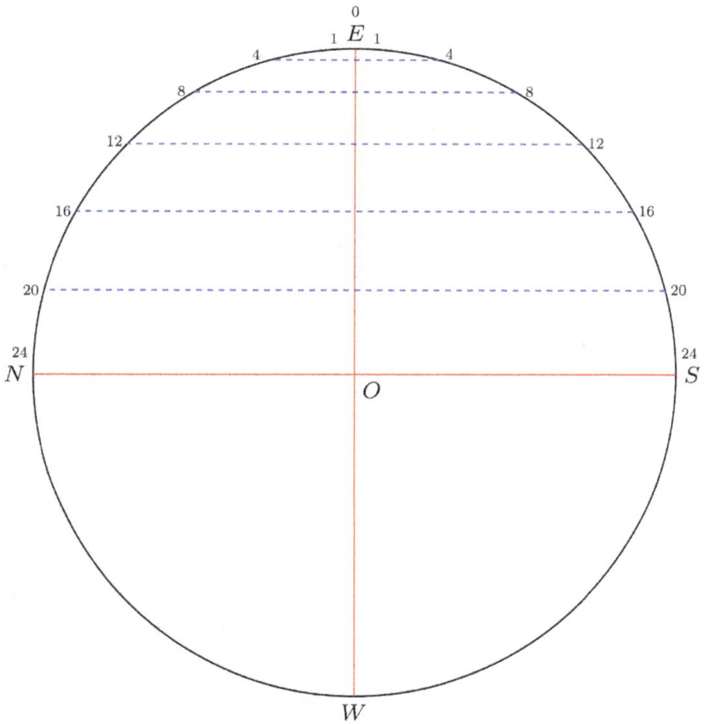

FIGURE 15.2 Demarcating the base circle into 24 segments

(i) the radius is to be of whatever length in *aṅgula*s ('digits') the reader desires,
(ii) having demarcated the four directions, and consequently dividing it into four quadrants, the circumference of the circle is to be marked with 360 degrees,
(iii) the number of desired Rsine values determines the number of segments into which each quadrant is divided.

Though the method outlined here is generic, Bhāskara chooses a particular example of constructing an Rsine table with 24 values. Having marked the circumference of the first and second quadrants into 24 (the precise way to achieve this is not explicitly described by Bhāskara; the diligent steps of construction required to achieve this are not entirely trivial), one is to join the markings, producing 24 chords. Bhāskara appeals to the reader's visual inference which is quite straight-forward that these half-chords will be the Rsines.

He reaffirms, as was stated in the verse, that their magnitudes, namely, their linear measures or numerical values (*pramāṇa*s), can be obtained by physically 'measuring' (*mitvā*) them. Indeed, this remark may sound a bit puzzling as it is a very crude way of establishing Rsine values. What perhaps Bhāskara intends is some sort of empirical demonstration that might be useful in comprehending what can be otherwise computed precisely with the rules that follow.

5.2 *Identifying the Trigonometric Objects on the Diagram*

athādito vyākhyāyate | yā iṣṭā trijyā sa karṇaḥ kalpyaḥ | yā bhujajyā sa bhujaḥ | tayoḥ karṇabhujayor vargāntarapadaṃ koṭiḥ | koṭijyety arthaḥ | tatra ye bhujakoṭijye te bhujakoṭyaṃśānāṃ kramajye jñātavye | bhujajyā trijyāto yāvadviśodhyate tāvat koṭyaṃśānām utkramajyāvaśiṣyate | evaṃ koṭijyonā trijyā bhujāṃśānāmutkramajyā syāt | atha utkramajyāsthānaṃ darśayati | tatra pūrvalikhite vṛtte cihnayor upari gataṃ sūtraṃ kila jyā | tadupari tayoś cihnayormadhye yadvṛttakhaṇḍaṃ taccāpaṃ dhanuḥ | cāpamadhyasya jyāmadhyasya ca yadantaraṃ bāṇākāraṃ sā utkramajyety ucyate | "tribhamaurvikāyā" ity agre sambandhaḥ |

Now, it is explained from the beginning. Whatever is the desired radius, that is to be conceived of as the hypotenuse. Whatever is the upright, that is the Rsine (*bhujajyā*). The lateral (*koṭi*) is the square root of the difference of the squares of the hypotenuse (*karṇa*) and the upright. That is, it is the Rcosine (*koṭijyā*). Here, whatever these Rsines and Rcosines are, they are to be understood as the direct Rsines (*kramajyā*). By the measure by which the Rsine is subtracted from the radius, that measure turns out to be the Rcoversine (*utkramajyā*). Similarly, the measure obtained by subtracting the Rcosine from the radius will be the Rversine. Now, the position of the *utkramajyā* is shown. There in the previously-drawn circle, the thread passing over the two points is indeed the chord. Furthermore (*tadupari*), whatever is the segment of the circle that passes between the same two points, the arc corresponding to that [is called] the 'bow' (*dhanuḥ*). Whatever is the interstice between the midpoint of the arc and the midpoint of the chord, which resembles the arrow, that is called *utkramajyā*. The phrase "*trimaurvikāyāḥ*" is connected with the later part.

5.2.1 Analysis

Having first explained the end of the section on *jyotpatti*, Bhāskara returns to the beginning. He provides a detailed explanation of various technical terms that were introduced in verse 2. He restates the relations between the hypotenuse, upright, and lateral of a triangle with their corresponding trigonometric identifications, and spells out the derivation of the two direct Rsines, or *utkramajyās*, namely:

$$R \operatorname{versin} \theta = R - R \cos \theta$$

and

$$R \operatorname{coversin} \theta = R - R \sin \theta.$$

Having succinctly defined the way to compute the Rversine and the Rcoversine, Bhāskara directs the reader's attention to their diagrammatic instantiation. By a graphic description of the diagram, he elaborates on the very concisely expressed phrase in the verse (first quarter of verse 3), namely, that the line segment delimited by the midpoint of the chord and the midpoint of the arc is the 'arrow,' which in the verse he calls *bāṇa*. For referring to the bow he used the word *dhanu*, along with *cāpa* which are synonyms. This description is to be understood for both the Rcoversine, as well as the Rversine.

5.3 *Determining Specific Rsine Values by Numerical Computation*

evaṃ sādhāraṇyena jyākṣetraṃ darśayitvā atha nirdiṣṭāṃśānāṃ gaṇitena jyānayanam / tribhamaurvikāyā yadvargārdhasya mūlaṃ, sā pañcacatvāriṃśadaṃśānāṃ jyā syāt / tasyā yāvat koṭijyā sādhyate tāvat tāvaty eva bhavati, yataḥ tatra koṭyaṃśā api pañcacatvāriṃśat /

Having thus illustrated [this] bow-chord figure in general, the computation of the Rsines for the specific degree [values of the arguments] have to be obtained by mathematics. Whatever is the square root of half the square of the Rsine of three signs (i.e., the radius), that is the Rsine of 45 degrees. By the time (*yāvat*) the Rcosine is mathematically computed (*sādhyate*), [it becomes apparent that] this is also of the same magnitude [as the Rsine of 45 degrees], since, there, the measure relating to the lateral (*koṭi*) is 45 degrees also.

5.3.1 Analysis

Bhāskara next elaborates on the derivation for the Rsine of 45°. The commentary on this begins with a note that the Rsine values that were identified geometrically in the diagram themselves have to be found 'by mathematics' (*gaṇita*, i.e., numerical computation). The two statements are to be taken as complementary to each other and not to be viewed as contradictory. Indeed, the two approaches are no doubt designed to help the reader give diagrammatic and empirical significance to values which can be derived by using geometrical rules, ultimately based on the so-called Pythagorean theorem.

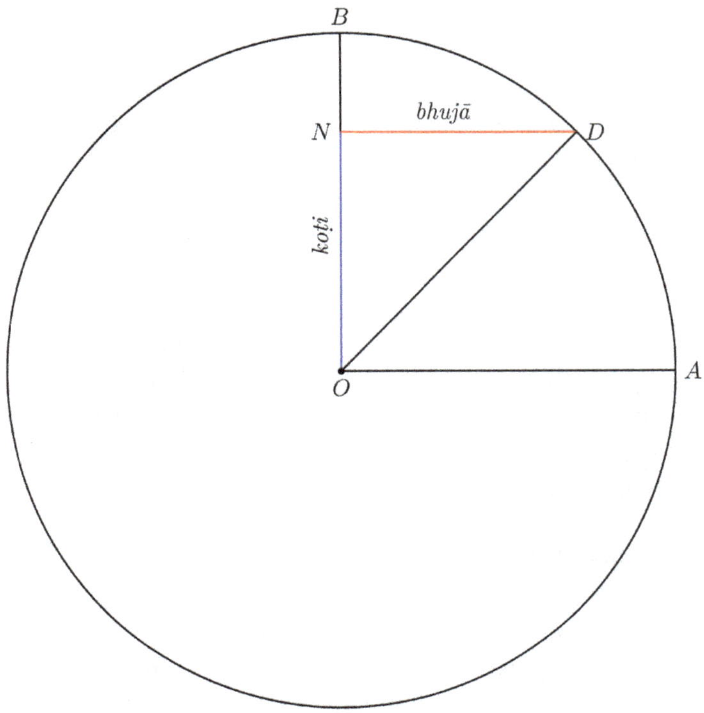

FIGURE 15.3 The diagram to accompany the first explanation for determining the Rsine of 45°

After restating the mathematical expression given in equation 3, to derive the Rsine of 45°, it seems that Bhāskara is appealing to an isosceles triangle that can be constructed (see Figure 15.3), namely the triangle *ODN* whose hypotenuse (*OD*) is the radius, and whose two sides (*ON* and *DN*), the upright and the lateral, are equal. Now angles $NOD = ODN = 45°$. Considering the arc *BD*, the Rsine of it is given by *ND* and the Rcosine will be *ON*. It is this fact that he uses to confirm to the reader that the Rsine of 45° is equal to the Rcosine of the same arc. From the geometrical construction then, it is obvious that:

$$OD = R$$

and

$$DN = ON = \frac{R}{\sqrt{2}} = \sqrt{\frac{R^2}{2}}.$$

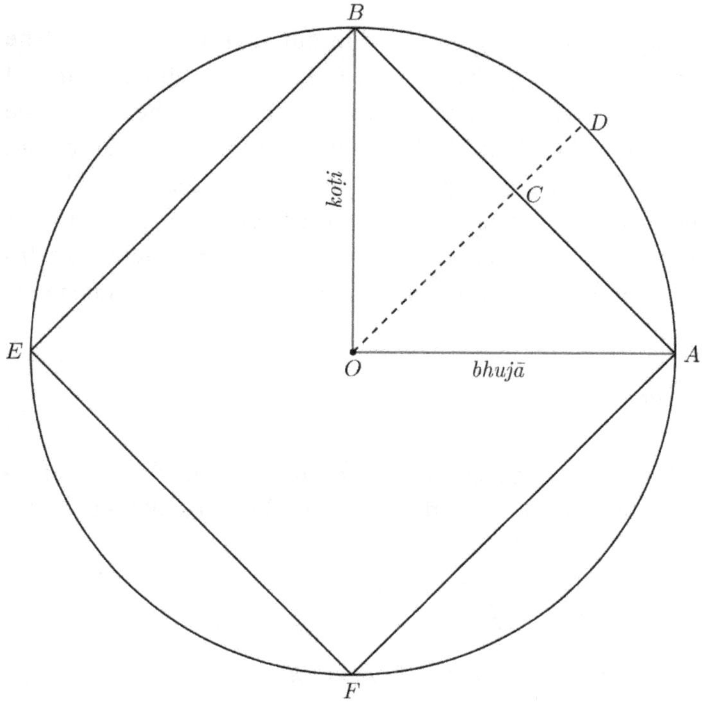

FIGURE 15.4 The diagram to accompany the second explanation for determining the Rsine of 45°

5.4 *Rationale behind the Expression for the Rsine of 45°*

atropapattiḥ—trijyā bhujaḥ trijyā ca koṭiḥ tadvargayogapadaṃ vṛttāntaḥ samacaturasrasya bhujaḥ syāt / saiva navatibhāgānāṃ jyā / tadardhaṃ grāhyam / ato vargayogasya caturthāṃśaḥ kṛtaḥ / tad eva trijyāvargārdham / ataḥ tanmūlaṃ śaravedabhāgajyetyupapannam /

Here is the rationale: The upright (*bhuja*) is the radius and the lateral (*koṭi*) is also the radius. The square root of the sum of the two squares will be the side (*bhuja*) of the square which is inscribed in the circle. That indeed is the chord (*jyā*) of 90 degrees. Half of it is to be taken. Therefore, a fourth part of the sum of the squares (of the *bhuja* and the *koṭi*) is found. That is exactly (*eva*) half the square of the radius. Thus, [it is] demonstrated that the Rsine of 45 degrees is the square root of that.

5.4.1 Analysis

Having set up the appropriate diagram for determining the Rsine of 45°, he now uses a different triangle to derive the rule. Consider now the right-angled triangle AOB in Figure 15.4. Here both the *bhuja* and the *koṭi* are themselves the radii of the circle, that is OA and OB. The hypotenuse AB forms one of the sides of the square inscribed in the circle (*vṛttāntaḥ samacaturasra*).

Reorienting Figure 15.3, one can see AOB is also an isosceles triangle with the upright and lateral now the radii of the circle. Obviously, the hypotenuse of this triangle is the square-root of the sum of the squares of the upright and lateral. That is,

$$AB = \sqrt{R^2 + R^2} = \sqrt{2R^2}.$$

Thus, to produce the Rsine or indeed the Rcosine, one is to divide this expression by two. This results in the Rsine or identically, the Rcosine that are given by:

$$R\sin(45^\circ) = R\cos(45^\circ) = \frac{1}{2}\sqrt{2R^2} = \sqrt{\frac{R^2}{2}}.$$

5.5 *Rationale for the Expression for the Rsine of 30°*

atha triṃśadbhāgānāṃ jyā trijyārdhamitā syāt | tasyāḥ koṭijyā ṣaṣṭibhāgānāṃ jyā syāt | atropapattiḥ—vṛttāntaḥpātisamaṣaḍasrasya bhujo vyāsārdhamitaḥ syād iti prasiddham | gaṇite 'pi kathitam | ataḥ trijyārdhaṃ triṃśadbhāgajyety upapannam |

Now, the Rsine of 30 degrees is measured by half of the radius. The lateral (*koṭi*) of that will be the Rsine of 60 degrees. Here is the rationale: It is well known that the side of a hexagon (*ṣaḍasra*) which is inscribed in a circle is equal to the radius. This is also stated in mathematical texts. Therefore, that half the radius is the Rsine of thirty degrees is [easily] demonstrated.

5.5.1 Analysis

Having provided the rationale in the form of a geometrical demonstration for the Rsine of 45°, Bhāskara next proceeds to present the same for the rule to compute the Rsine of 30°. As he observes, it is a simple mathematical fact that the side of a hexagon inscribed in a circle is equal to the radius of that circle. This side is of course the chord of 60° as can be directly observed off the diagram. Dividing it by two gives the Rsine of 30°. The diagram Bhāskara draws our attention to, which illustrates this fact can be seen in Figure 15.5.

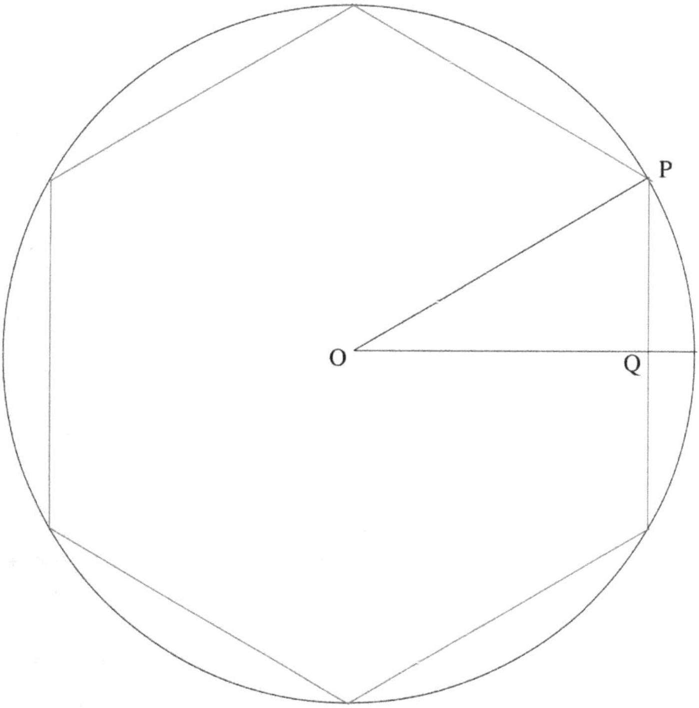

FIGURE 15.5 The diagram to explain the derivation of the Rsine of 30°

5.6 *Determining the Rsine of 15° from the Rsine of 30° and Its Rversine*

ataḥ prāgvadutkramajyā / ṣaṣṭibhāgajyayonā trijyā rāśer utkramajyā / sā koṭirūpiṇī / kramajyā bhujarūpiṇī / tadagrayor nibaddhasūtraṃ tat karṇaḥ / tat triṃśadbhāgānāṃ jyārūpam / ataḥ tadardhaṃ pañcadaśabhāgānāṃ jyārdhamityupapannam / evaṃ sarvatra tadardhāṃśakaśiñjinīnām upapattir jñeyā /

As [mentioned] earlier, the *utkramajyā* [is considered next]. The *utkramajyā* of a sign (*rāśi*, i.e., 30 degrees) is the radius diminished by the Rsine of 60 degrees. This is of the form of a lateral (*koṭi*). The *kramajyā* is of the form of an upright (*bhuja*). The thread which is tied to their [open] ends is the hypotenuse of them. That form is the chord of 30 degrees. Thus, it is demonstrated that half of this [chord] is the Rsine of 15 degrees. In all instances of finding the Rsines corresponding to half-arcs, the demonstration is to be understood in this way.

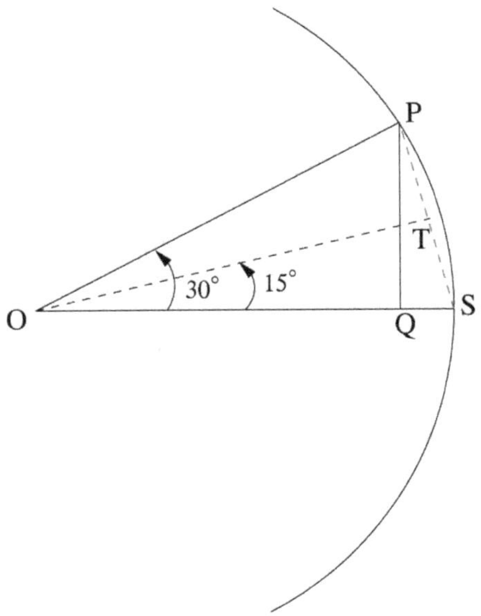

FIGURE 15.6
The diagram to explain the half-arc formula

5.6.1 Analysis

After demonstrating the geometrical way of obtaining the Rsine of 30°, Bhāskara prescribes how to obtain $R\,\text{versin}\,30^\circ$, again directing the reader's attention to the diagrammatic representation of this problem. Furthermore, he uses this to set up a visual precedent for obtaining sines of half-arcs. He first notes, by way of the general formulation given above, that the Rversine is given by:

$$R\,\text{versin}\,30^\circ = R - R\sin 60^\circ$$

where $R\sin 60^\circ = R\cos 30^\circ$. He then points out (see Figure 15.6) that this $R\,\text{versin}\,30^\circ$ is in fact the lateral of a smaller triangle *PQS*, and whose upright *PQ* is $R\sin 30^\circ$. Bhāskara entreats the reader to manually join a thread to each end of these sides, namely at P and S, to form a hypotenuse. Visual inspection reveals that this hypotenuse *PS* is indeed the chord of 30°. Bisecting the chord gives the Rsine of half the arc; in this case $R\sin 15^\circ$ (*TS*).

From such carefully laid out geometrical steps, Bhāskara observes that applying the same procedure over and over again can produce further half-arcs. In this way, he sets the scene geometrically for the analytic rule which he proceeds to work through using symbolic styles of reasoning in the next section of his commentary.

5.7 *Another Method to Demonstrate the Half-Arc Formula*

atha prakārāntareṇa tadardhāṃśakaśiñjinīmāha—"trijyotkramajyānihater" ityādi / asyopapattiḥ—tatra ādyākṣaracihnaiḥ bījaprakāreṇa kathyate / tatra utkramajyonā trijyā kila koṭijyā / tasyā vargo 'yam / uva 1 *utribhā* 2̇ *triva* 1 *anenonā trijyākṛtir dorjyākṛtiḥ syāt / uva* i *utribhā* 2 */ ayaṃ kramajyāvargaḥ utkramajyāvargayuto jātaḥ / utribhā* 2 */ asya caturthabhāgaḥ utribhā* $\frac{1}{2}$ */ asya mūlaṃ grāhyam / ata uktaṃ "trijyotkramajyānihater ityādi" / evaṃ tasyā apyanyā tadardhāṃśakaśiñjinīti / evaṃ koṭijyāyā api yāvadabhimatakhaṇḍāni syuḥ /*

Now, by another method, he states [how to obtain] the sine of the degrees of half-arcs, commencing with "*trijyotkramajyānihateḥ.*" Its rationale: there, by [making] markings using the initial letters, by [resorting to] an algebraic method (*bījaprakāreṇa*) this is being stated. There, the Rcosine is indeed the radius diminished by the *utkramajyā*. This is the square of that. *uva* 1 *utribhā* 2̇ *triva* 1. The square of the radius diminished by this will give the square of the Rsine. *uva* i *utribhā* 2. This is the sum of the square of the *utkramajyā* and the square of the Rsine. *utribhā* 2. A fourth part of this is *utribhā* $\frac{1}{2}$. The square root of this is to be taken. Therefore, it was stated, from the product of the radius and the *utkramajyā* (*trijyotkramajyānihateḥ*) and so on. In the same way from that, *tadardhāṃśakaśiñjinī* (i.e., the other half sines). In a similar way even from the Rcosine also, one can obtain it for as many segments [as desired].

5.7.1 Analysis

Having guided the reader through a visual demonstration of finding the Rsine of the half-arc considering $R \sin 30^\circ$ as an example, Bhāskara now claims to explain it 'by means of another method.' This other method employs a derivation of the rule using symbolic styles of reasoning, or, as Bhāskara puts it below, using 'an algebraic method' (*bījaprakāra*). This method involves dealing with algebraic equations. For doing so, the line segments are denoted by the initial letters (*adyākṣara*) of the term used to describe them. For instance, '*u*' represents *utkramajyā*. The table below presents the signifier and the physical or mathematical quantity represented by them.[8]

8 Such symbolic styles of reasoning have been discussed in other contexts in Montelle and Ramasubramanian, "Sine of One Degree."

Signifier	Signified quantity	Mathematical equivalent
u	*utkramajyā*	$R \operatorname{versin} \theta$
uva	*utkramajyā-varga*	$(R \operatorname{versin} \theta)^2$
tri(*bhā*)	*tribhājyā*	R
tri(*bhā*)*-va*	*tribhājyā-varga*	R^2

The coefficients are represented in their numeral form, with a dot on the top if they are subtractive. Thus, the symbolic reckoning

uva 1 *utribhā* 2̇ *triva* 1

when expressed using current mathematical notation, is equivalent to:

$$1 \cdot (R \operatorname{versin}(\theta))^2 - 2 \cdot (R \operatorname{versin} \theta) \cdot R + 1 \cdot R^2.$$

Likewise, the expression

uva 1̇ *utribhā* 2

is equivalent to

$$(-1) \cdot (R \operatorname{versin} \theta)^2 + 2 \cdot (R \operatorname{versin} \theta) \cdot R.$$

To derive the half-arc formula, Bhāskara wants to find an expression for half of the hypotenuse of the triangle PQS, namely PS/2 (see Figure 15.6). It is evident from the figure that:

$$R \sin \frac{\theta}{2} = \frac{PS}{2} = \frac{\sqrt{(R \sin \theta)^2 + (R \operatorname{versin} \theta)^2}}{2}. \qquad (3)$$

Essentially, Bhāskara has in mind to recast the right-hand side of this equation using various identities to establish its equivalence with

$$\sqrt{\frac{(R \operatorname{versin} \theta) \cdot R}{2}}.$$

He begins with the basic identity

$$R\cos\theta = R - R\,\text{versin}\,\theta. \tag{4}$$

Then, he squares it to produce

$$\begin{aligned}(R\cos\theta)^2 &= (R - R\,\text{versin}\,\theta)^2\\ &= (R\,\text{versin}\,\theta)^2 - 2\cdot R\cdot R\,\text{versin}\,\theta + R^2.\end{aligned} \tag{5}$$

Subsequently, Bhāskara states 'the square of the radius diminished by this will give the square of the Rsine,' namely

$$\begin{aligned}(R\sin\theta)^2 &= R^2 - (R\cos\theta)^2\\ &= R^2 - ((R\,\text{versin}\,\theta)^2 - 2\cdot(R\,\text{versin}\,\theta)\cdot R + R^2)\\ &= -(R\,\text{versin}\,\theta)^2 + 2\cdot(R\,\text{versin}\,\theta)\cdot R.\end{aligned} \tag{6}$$

Rearranging this, we get

$$2\cdot R\cdot R\,\text{versin}\,\theta = (R\sin\theta)^2 + (R\,\text{versin}\,\theta)^2. \tag{7}$$

Taking a fourth part and the square-root of the result gives

$$\begin{aligned}\sqrt{\frac{R\cdot R\,\text{versin}\,\theta}{2}} &= \sqrt{\frac{(R\sin\theta)^2 + (R\,\text{versin}\,\theta)^2}{4}}\\ &= \frac{\sqrt{(R\sin\theta)^2 + (R\,\text{versin}\,\theta)^2}}{2}.\end{aligned} \tag{8}$$

The simple step required for the reader is to connect this back to the half-chord of the hypotenuse of the triangle *PQS*. It is evident from the figures that

$$PS = \sqrt{(R\sin\theta)^2 + (R\,\text{versin}\,\theta)^2}.$$

Hence,

$$\frac{PS}{2} = R\sin\frac{\theta}{2} \tag{9}$$

as required. Bhāskara rounds off this derivation noting that it can be applied repeatedly to obtain as many half-arc segments as required. He also mentions that the corresponding Rcosines can be found. Thus, in an interplay between mathematical operations expressed in prose and symbolic reckoning, Bhāskara algebraically derives the half-arc formula.

5.8 *The Sequence in Which Rsine Values Can Be Determined*

tadyathā—yatra caturviṃśatiḥ khaṇḍāni tatra rāśerjyā aṣṭamaṃ khaṇḍam 8 / *tatkoṭijyā ṣoḍaśam* 16 / *śaravedabhāgajyā dvādaśam* 12 / *asmāt khaṇḍatrayāt kathitaprakāreṇa caturviṃśatiḥ khaṇḍāny utpadyante* / *tatra aṣṭamāt tadardhāṃśakaśiñjinī caturtham* 4 / *tatkoṭijyā viṃśam* 20 / *evaṃ caturthāddvitīyam* 2 / *dvāviṃśaṃ ca* 22 / *dvitīyāt prathamam* 1 *trayoviṃśaṃ ca* 23 / *evaṃ daśamacaturdaśapañcamaikonaviṃśasaptamasaptadaśaikādaśatrayodaśāni ityaṣṭamāt* 10 / 14 / 5 / 19 / 7 / 17 / 11 / 13 / *atha dvādaśāt ṣaṣṭhāṣṭadaśatṛtīyaikaviṃśanavamapañcadaśāni* 6 / 18 / 3 / 21 / 9 / 15 / *trijyā caturviṃśamiti* 24 / *ato 'vaśiṣṭāṃ jyotpattimagre vakṣyāmaḥ* // 2–6

This is how it is. Wherever there are 24 segments, there, the Rsine of 30 degrees will be the segment corresponding to the 8th division. The Rcosine of that will be the 16th. The Rsine of 45 degrees corresponds to the 12th. From these three segments, as per the method described, the 24 segments are generated. There, from the 8th, the Rsine corresponding to half of that angle is the 4th. The Rcosine of that, the 20th. In the same way, from the 4th, the 2nd and the 22nd. From the 2nd, the 1st and the 23rd. In the same way, the 10th, the 14th, the 5th, the 19th, the 7th, the 17th, the 11th, and the 13th. Now, from the 12th, the 6th, the 18th, the 3rd, the 21st, the 9th, and the 15th. The radius is the 24th. Whatever is remaining [to be described with respect to] the production of Rsines, we will state [this] later.

5.8.1 Analysis

In this last section of the commentary, Bhāskara outlines the order in which the various Rsines are derived starting from the radius. In the case that 24 Rsines are required, the various segments are numbered off; the Rsine of $3^\circ 45'$ is the first, the Rsine of $7^\circ 30'$ the second, and so on and so forth. In this way, the numbers which count off the segments act as an index.

From the radius R, the Rsine of 30° is found, which is the 8th segment. Correspondingly, the Rcosine of 30° can be immediately determined, which is equivalent to the Rsine of 60°, or the 16th segment. The Rsine of 45° is the 12th. Now, using repeated application of the half-arc rule, and symmetry for the Rcosine, from the 16th, Bhāskara states the 8th, the 4th, and correspondingly the 20th. Proceeding in this way, all 24 Rsines can be determined.

Bhāskara wraps up the discussion by promising more details later. Here he is certainly referring to the *Jyotpatti* chapter where he gives many more rules that would be required if one wishes to construct a sine table with more than 24 divisions.

References

Chaturvedi, Murali Dhara, ed. *Siddhāntaśiromaṇi of Bhāskarācārya with his autocommentary Vāsanābhāṣya and Vārttika of Nṛsiṃha Daivajña*. Sampurnanand Sanskrit University, 1981.

Dhammaloka, Jambugahapitiye, Clemency Montelle, and K. Ramasubramanian. "Computation of Sines in Nityānanda's *Sarvasiddhāntarāja*." SCIAMVS 17 (2016): 1–54.

Kavirāja, Gopī Nātha, ed. *The Siddhāntasārvabhauma of Muniśvara*. The Princess of Wales Saraswatī Bhavana Texts 41. Benares City: Vidya Vilas Press, 1932.

Montelle, Clemency. "The Production of Sines: Bhāskarācārya's *Jyotpatti*." In *Bhāskaraprabhā*, edited by K. Ramasubramanian, Takao Hayashi, and Clemency Montelle, 291–300. Hindustan Book Agency, 2019.

Montelle, Clemency and K. Ramasubramanian. "Determining the Sine of One Degree in the *Sarvasiddhāntarāja* of Nityānanda." SCIAMVS 19 (2018): 1–52.

CHAPTER 16

Mathematics and Presentation of Eclipse Diagrams in Sanskrit Astronomy

Kim Plofker

1 Introduction: Eclipse Phases in Ancient Astral Science

The identification of certain crucial moments or phases in an eclipse appears as far back as Babylonian eclipse records.[1] Standard phases for a total lunar eclipse are pictured in figure 16.1, showing the moon passing through the earth's shadow from west to east. The phases are qualitatively the same in a solar eclipse, except for their direction: the disk of the sun appears to pass from east to west in the course of its eclipse behind the eastward-moving moon. In either configuration, the phases proceed from the moment of first contact of the disks of the eclipsed and eclipsing bodies, through the beginning of totality when the eclipsed body becomes entirely obscured and the subsequent moment of mid-eclipse or greatest obscuration, to the end of totality and the final separation of the disks.

It is not known when or where astronomers first adopted the practice of drawing ephemeral diagrams, somewhat similar in form to figure 16.1, representing the progress of these phases in a predicted eclipse. (Note, by the way, the persistence of such representations in modern forms such as the image from the online NASA Lunar Eclipse Page reproduced in figure 16.2.) The numerous detailed Babylonian eclipse records surviving on cuneiform tablets are numerical and verbal rather than graphical, but diagrams illustrating the geometric theory of eclipses in general were a staple of Hellenistic astronomy.[2]

1 See, for example, the references in F.R. Stephenson and J.M. Steele, "Astronomical Dating of Babylonian Texts Describing the Total Solar Eclipse of S.E. 175," *Journal for the History of Astronomy* 37, no. 1 (2006): 55–69 to the directions in which a solar eclipse "began" and "threw off" the shadow, and in F.R. Stephenson, "Babylonian Timings of Eclipse Contacts and the Study of the Earth's Past Rotation," *Journal for the History of Astronomy* 9, no. 2 (2006): 146 to Babylonian timings of "first and second contacts," i.e., contact and beginning of totality, as well as the analyses in Clemency Montelle, *Chasing Shadows* (Baltimore, MD: Johns Hopkins, 2011).

2 See, e.g., Nathan Sidoli, "What We Can Learn from a Diagram: The case of Aristarchus's *On The Sizes and Distances of the Sun and Moon*," *Annals of Science* 64, no. 4 (2007): 525–547.

 | DOI:10.1163/9789004438224_017

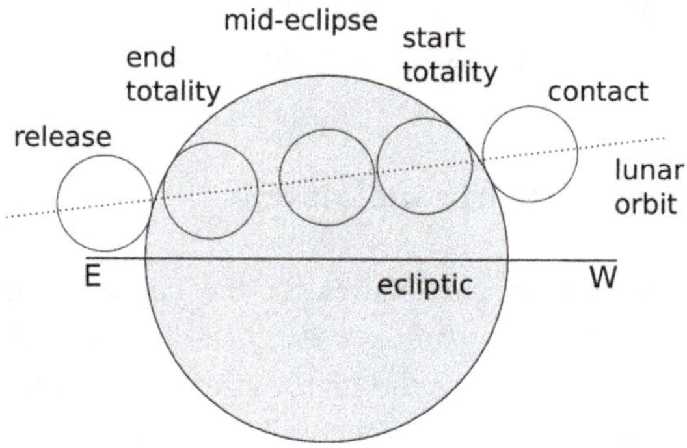

FIGURE 16.1 Lunar eclipse diagram showing successive eclipse stages from right to left as the disk of the moon moves eastward through the earth's shadow in the southern half of the observer's visible celestial hemisphere

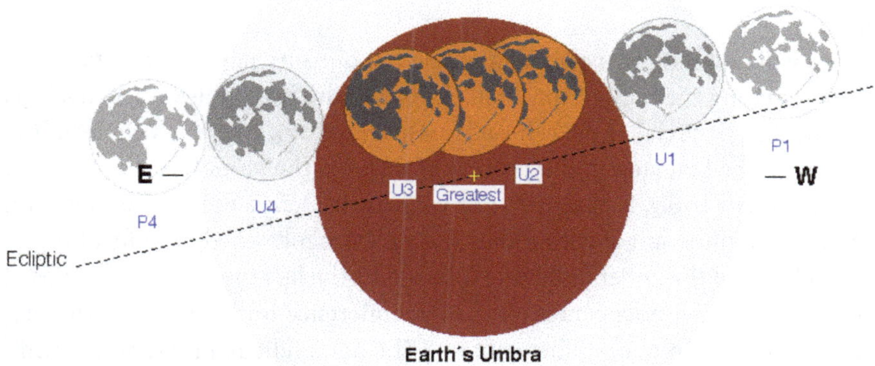

FIGURE 16.2 NASA diagram of the stages of the 2019-01-21 lunar eclipse, reproduced from https://eclipse.gsfc.nasa.gov/LEplot/LEplot2001/LE2019Jan21T.pdf

Later works such as Ibn al-Haytham's *On the Shape of the Eclipse* investigated in depth the geometric characteristics of different eclipse configurations.[3] None of these texts, however, describe or prescribe procedures for drawing stylized illustrations of a particular eclipse.

The creation of such diagrams to represent with reasonable accuracy a given eclipse's physical appearance—the amount of obscuration, direction of contact and release with respect to the cardinal directions, and so on—was an important tool for practitioners of medieval Sanskrit astronomy. Its development must have begun quite early in the first stages of Indian spherical astronomy near the start of the Common Era, since surviving texts from the mid-first

3 See Dominique Raynaud, *A Critical Edition of Ibn al-Haytham's* On the Shape of the Eclipse (Cham: Springer, 2016).

millennium CE already exhibit highly refined and standardized procedures for it. The following section examines some of these in more detail.

2 Early Eclipse Diagram Construction in Sanskrit Texts

The earliest extant Sanskrit exposition of procedures for the *parilekha* or eclipse diagram is found in the sixth-century *Pañcasiddhāntikā* of Varāhamihira.[4] Its rules already exhibit to a large extent the highly streamlined form of this technique in later texts, especially in their generalization of the diagram description to represent either a lunar or a solar eclipse by referring to the bodies generically as the "eclipsed" and the "eclipser" (although this generalization is not always consistently maintained in the *Pañcasiddhāntikā*).

We will maintain this general approach in our summary of this procedure by designating the radii of the circular discs of the two bodies respectively as $r_{\bigcirc}$ (the "bright circle" of the eclipsed luminary) and $r_{\bullet}$ (the "dark circle" of the eclipser). The first step in the process of drawing the *parilekha*, illustrated in part **a.** of figure 16.3, is to lay out two concentric circles, the smaller one with radius $r_{\bigcirc}$ and the larger with radius $r_{\bigcirc} + r_{\bullet}$, both measured in units of digits. (The choice of the larger radius is explained by the fact that when the eclipsed body is shown with its center on the circumference of this larger circle, that represents the moment of first contact.) The perpendicular lines of the cardinal directions are drawn on the figure.

The second step involves the calculation of the so-called "deflection" or *valana*. This quantity accounts for the degree to which the local east-west direction straight across the flat plane of the observer's horizon differs from the direction of the visible half of the celestial equator. Although the celestial equator also extends from the due east point of the observer's horizon to its due west point, its visible semicircle sags southward in the sky (assuming the observer is in the earth's northern hemisphere), as illustrated in figure 16.4.[5]

In figure 16.4, consider the moon above the horizon at *M*. At that point, the deflection of the northeasterly path of the celestial equator away from the due

---

4 T.S. Kuppanna Sastry, *Pañcasiddhāntikā of Varāhamihira* (Madras: PPST Foundation, 1993) and Otto Neugebauer and David Pingree, *The Pañcasiddhāntikā of Varāhamihira*, 2 vols. (København: Munksgaard, 1970–1971).

5 The figure is reproduced from Clemency Montelle and Kim Plofker, "*Karaṇakesarī* of Bhāskara: A 17th-Century Table Text for Computing Eclipses," *History of Science in South Asia* 2 (2014): 25.

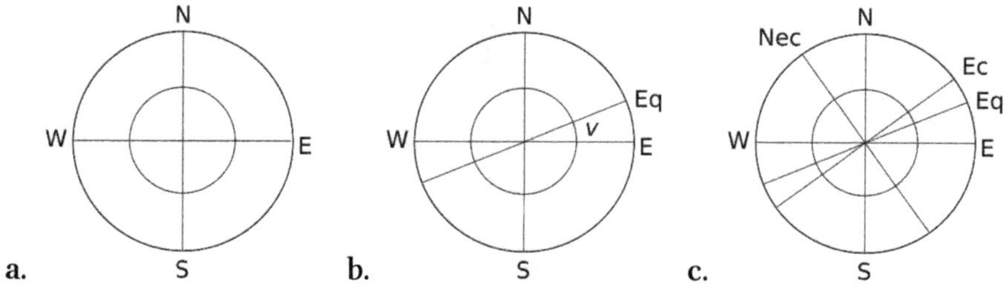

FIGURE 16.3 Initial sequence of steps in the *parilekha* construction procedure of the *Pañcasiddhāntikā*. **a.** The two concentric circles with radii $r_{\bigcirc}$ and $r_{\bigcirc}+r_{\bullet}$, and the observer's local cardinal directions. **b.** The angle ν of deflection between the true east-west line and the northeasterly local direction of the celestial equator separates the due east point E from the equator's path through E_q. **c.** The angle between the local directions of the equator and ecliptic marks off E_q from the ecliptic path through E_q, which is one quadrant away from the corresponding "north" point N_{ec}.

east-west direction will be $\nu = \angle NMP$, the angle at M between the observer's due north point and the celestial pole. When this angle is marked off from E to E_q on the *parilekha* circle north of the east point as in part **b.** of figure 16.3, the resulting diameter represents the local trajectory of the celestial equator at the site of the eclipse.

But the progress of an eclipse does not actually follow the direction of the celestial equator, but more nearly that of the ecliptic (see again figure 16.4), which is inclined to the equator by an angle of obliquity denoted ε. The moon M is considered to have an ecliptic longitude, i.e., a position on the ecliptic where it intersects the great circle containing the arc MP, and an ecliptic declination, which is the arc of that same great circle intercepted between the ecliptic and the equator. The angle between the "eastward" directions of the ecliptic and the equator at that position is equal to the declination of a point on the ecliptic exactly 90° away from it. (To avoid a spherical-trigonometry explanation, we will content ourselves with justifying this statement qualitatively by an appeal to the extreme cases: i.e., where the equator and ecliptic cross the declination is zero and the angle between their directions is ε, while at their furthest separation the declination is ε but the angle between their (parallel) directions is zero.)

Therefore the next step in the *parilekha* construction is to lay off that angle from the "equatorial-eastward" direction E_q to the "ecliptic-eastward" direction E_c, and draw its "north"-"south" perpendicular direction marked N_{ec} in part **c.** of figure 16.3. (Note that if the point M were in the western half of the observer's sky rather than the eastern, the "eastward" direction along the south-sagging equator arc would be south of due east rather than north of it; likewise, if the

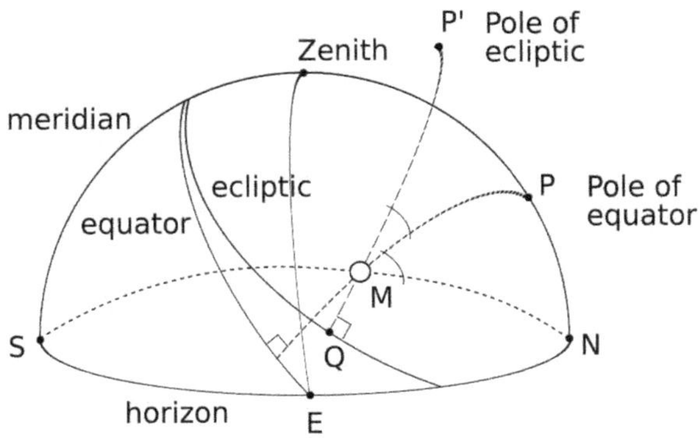

FIGURE 16.4 The observer's celestial hemisphere with its local zenith at *Z* and the poles *P* and *P'* of the celestial equator and ecliptic, respectively. The prime vertical circle passes through the zenith and the local due east point *E* perpendicular to the great circle through the due south point *S* and due north point *N* and the moon at *M*.

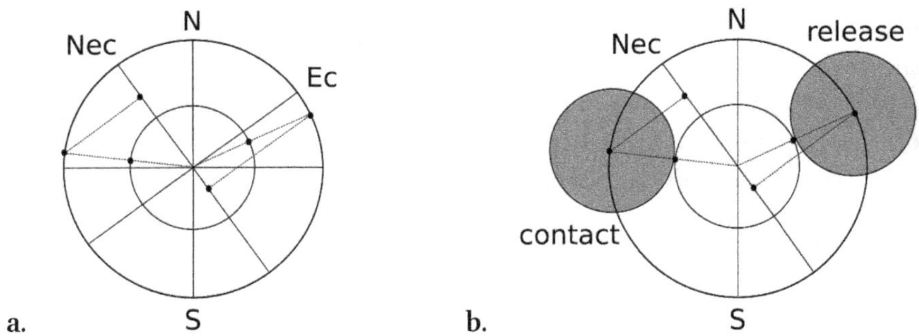

FIGURE 16.5 Subsequent steps in the *parilekha* construction procedure of the *Pañcasiddhāntikā*. **a.** Line segments parallel to the ecliptic path are drawn at distances representing the lunar latitude values at the moments of contact and release. Radial lines connecting the endpoints of those segments to the center of the diagram mark the points on the circumference of the eclipsed body where contact and release will occur. **b.** The dark disk of the eastward-moving new moon in a solar eclipse first touches the western edge of the sun's disk and then separates from its eastern edge when the eclipse is over.

given longitude happened to be on the half of the ecliptic lying south of the celestial equator, E_c would be marked to the south of E_q rather than to its north.)

Finally, the *parilekha* must represent the moon's celestial latitude or distance from the ecliptic, shown as arc MQ in figure 16.4. As illustrated in figure 16.5, northern or southern lunar latitude values for the moments of contact and release are laid off in the appropriate direction along the N_{ec} line perpendicular to the ecliptic. The line segments drawn from their endpoints parallel to the ecliptic path end on the outer circle of the diagram, showing where the center of the eclipsing body must be placed to depict the tangent circles of eclipser and eclipsed at contact and release. The occurrence of first contact on the western rim of the eclipsed body indicates that this is a solar eclipse: in a lunar eclipse, the directions at contact and release would be reversed.

The next known Sanskrit text on the construction of the *parilekha* after the *Pañcasiddhāntikā* is found in the supplementary chapter on eclipses in the 628 CE *Brāhmasphuṭasiddhānta* of the renowned astronomer-mathematician Brahmagupta. His introduction to the subject emphasizes its esoteric nature, and rather alludes to than expounds its techniques:[6]

> *grahaṇagrahasaṃyogagraharkṣatantreṣu sarvavidām |*
> *ācāryaś chedyakavid yatas tataś chedyakaṃ vakṣye ||* 1
>
> *durjanakṛtaghnaśatrupratikañcukakāripatitamūrkhebhyaḥ |*
> *chedyakam adeyam ebhyo dadataḥ sukṛtāyuṣor nāśaḥ ||* 2
>
> *uṣitāya dīrghakālaṃ śiṣyāya guṇādhikāya bhaktāya |*
> *pātre vā suhṛde vā sutāya vā chedyakaṃ deyam ||* 3
>
> *viṣuvadapamaṇḍaladiśo valanajyābhis trigṛhavṛtte*
> *samparke grāsaṃ vā yo vetti chedyakajñaḥ saḥ ||* 4
>
> *samparkamaṇḍale yaḥ pragrahamokṣau pṛthak svavikṣepāt |*
> *madhyān madhyagrāsaṃ parilikhati chedyakajñaḥ saḥ ||* 5
>
> *parilikhatīṣṭagrāsaṃ tātkālikabāhukoṭikarṇair yaḥ |*
> *atha vā nimilanonmīlanadvayaṃ chedyakajñaḥ saḥ ||* 6
>
> *grāhyaṃ parilikhyaikyaṃ parilikhati grahagṛhādikaṃ tatra |*
> *bhūmau yaḥ phalake vā parivarttya chedyakajñaḥ saḥ ||* 7
>
> Since one who knows diagram-construction is the master of those who know everything in treatises on eclipses, planetary conjunctions, and occultations, I will explain diagram-construction.

6 S. Dvivedi, *Brāhmasphuṭasiddhānta* (Benares: The Pandit, 1902), 2.275–276.

> Diagram-construction is not to be given to [someone] evil or ungrateful, or a rival or slanderer, or [someone] degraded or stupid. [Anyone who] gives [it] to them [will have] happiness and life destroyed.

> Diagram-construction is to be given to a devoted student of long standing and exceptional merit, or to a protector or a friend or a son.

> One who knows [how to draw] the directions of the equatorial and ecliptic circles in the Radius-circle, or the obscuration in the contact[-release circle], is expert in diagram-construction.

> One who draws the contact and release in the contact[-release] circle separately according to their own latitude [values], and the mid[-eclipse] obscuration from the mid[-eclipse latitude], is expert in diagram-construction.

> One who draws a desired obscuration [at an arbitrary time] with the [right-triangle] sides and hypotenuses for that [given] instant, or the two [configurations of] beginning and end of totality, is expert in diagram-construction.

> One who draws a combined eclipse [diagram showing] the eclipser and eclipsed and so forth there on the ground or, having reversed it, on a board, is expert in diagram-construction.

These and other verses Brahmagupta devotes to this subject indicate that *parilekha* techniques in his time were highly thought of, and perhaps in some cases closely guarded, as a professional specialty of *jyotiṣī*s. Varāhamihira may be subtly critiquing such an attitude when he says at the beginning of his treatise (*Pañcasiddhāntikā* 1.26),

> Having put aside envy/selfishness, I will state in this [work] that chief secret in which the minds of the authors of treatises are perplexed: the eclipse of the sun.

Doubtless the ominous significance and astrological implications of eclipses made it desirable to be able to predict their appearance in as much detail as possible. But even if *parilekha* methods were prized as a sort of esoteric knowledge to be used advantageously to impress clients and patrons, they did not ultimately remain secret. Treatises and commentaries over the next thousand years described a remarkable variety of procedures for refining or approximating eclipse diagrams, adding features such as a circular-arc path for the continuous progress of the moon through the phases, debating which mathematical formula for *valana*-computation was most correct, and suggesting variations on the basic construction methods. The comparative investigation of these techniques and influences on their development is a work in progress, but it is clear that they retained their status as an intriguing and impressive accomplishment throughout the history of Jyotiṣa.

3 Eclipse Diagrams in Sanskrit Manuscripts

The study of *parilekha* techniques in Sanskrit science would be greatly assisted by comparing their textual descriptions with historical examples of their application to create eclipse diagrams. But the evidence encountered so far suggests that they may have remained in the category of ephemera rather than being reproduced as part of an instructional text. Interestingly, the few figures that could be described as "eclipse diagrams" currently known from Sanskrit astronomical manuscripts reflect a much simpler and apparently more naive approach to the subject than the sophisticated constructions described in the texts.

As the examples in the following figures illustrate, such diagrams typically show a single circle, marked with regularly spaced angle ticks or strokes intended to suggest them, and crossed with perpendicular axes indicating the cardinal directions. The directions are conventionally oriented with East at the top and South on the right. But the accompanying oblique lines do not appear to represent any ecliptic or equatorial deflection from the main directional axes, as depicted in the reconstructed diagram shown in figures 3 and 5. Rather, they merely indicate the intercardinal directions, labeled with their standard abbreviations, which appear to be completely irrelevant to any known *parilekha* techniques.

Both examples in figure 16.6 include some eclipse-related terms labeling the diagram. In part **a.**, clockwise from top: "end of totality," "sky-obscuration," "release," "mid-eclipse," "contact." In part **b.**, clockwise from the top: "end of totality," "latitude," "mid-eclipse," "release," "beginning of totality," "sky-obscuration." While it is plausible to label certain regions on a *parilekha* in accordance with the eclipse phase(s) occurring there (as indeed was done in figure 16.5), there seems to be little coherence in the labels. "Beginning of totality" and "end of totality" are roughly opposite each other in diagram **b.**, as are "contact" and "release" in diagram **a.**, but the rest of the phase labels do not appear consistent with the general eclipse progress direction thus indicated.

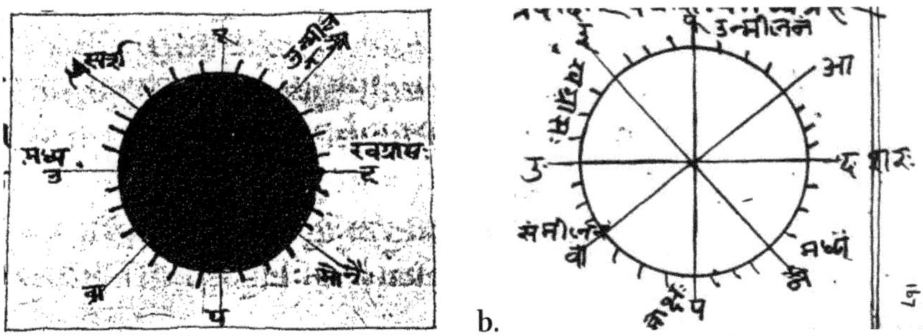

a. b.

FIGURE 16.6 Simple *parilekha* diagrams in manuscripts. **a**. Manuscript of an unidentified *Grahaṇādhikārodāharaṇa* ("worked example on the eclipse chapter"), SRCMI 1.20 i 20 (Sri Ram Charan Museum of Indology, Jaipur).[7] Worked example for Saṃvat 1677. **b**. Manuscript of the *Parvadvayasādhana* of Mallāri, CSS d.800(vii) (Bodleian Library, Oxford), copied Saṃvat 1856
Note: See David Pingree, *Census of the Exact Sciences in Sanskrit*, Series A, vol. 4 (Philadelphia: American Oriental Society, 1970–1994), 9.A4, 365.

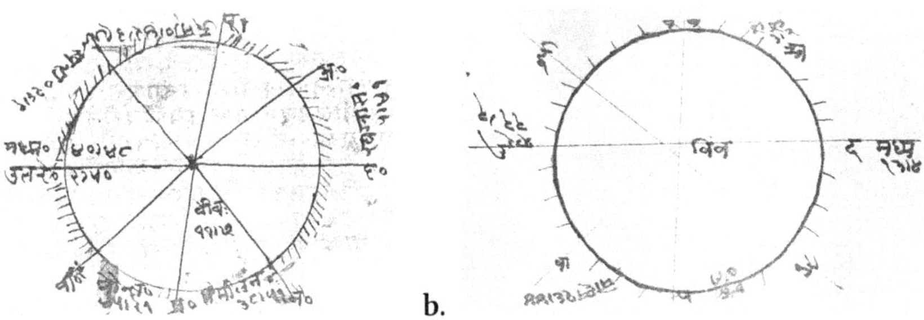

a. b.

FIGURE 16.7 Simple *parilekha* diagrams in manuscripts. Both diagrams from a Jaipur manuscript of a *Sūryacandragrahaṇavivaraṇa* ("commentary on solar and lunar eclipse"), excerpted from the commentary of Viśvanātha on Gaṇeśa's *Grahalāghava* (MS Plofker 52, now in the collection of Jaina Vidya Sansthan, Jaipur). Worked example for Saṃvat 1667.
Note: See Pingree, *Census of Exact Sciences in Sanskrit*, Series A, vol. 5, 669.

The diagram labels in figure 16.7 include numerical values evidently taken from a worked example in the text. Most of them apparently refer to times when various phases occur, as well as sizes of disks and their obscuration. But again, they seem to have little relation to actual *parilekha* construction techniques. The nearly-blank diagram in figure 16.8 may have been left incomplete accidentally, but it suggests the same simplified and uninformative construction as the rest.

---

7 See Ram Charan Sharma, *Vāṅmaya-cayanikā: A Classified Catalogue of Selected Manuscripts from Universal Institute of Orientology*, 6 vols. (Jaipur: Universal Institute of Orientology, 1986–1994).

FIGURE 16.8
Incomplete *parilekha* diagram from another Jaipur manuscript of Viśvanātha's commentary on Gaṇeśa's *Grahalāghava* (MS Plofker 9)

There are other possible sources for physical evidence of *parilekha* diagrams that have yet to be explored, besides the wealth of currently unexamined Sanskrit scientific manuscripts. E.g., it may be that some artworks such as miniature paintings depicting *jyotiṣīs* might include at least a crude rendition of what a working *parilekha* really looked like. Pending further discoveries, we must rest content (as so often in the study of Jyotiṣa) with attempting to reconstruct from text, mostly in the form of more or less rarefied and elliptical Sanskrit verses, the details of actual scientific practice.

References

Chatterjee, Bina. *Śiṣyadhīvṛddhida Tantra of Lalla with the commentary of Mallikārjuna Sūri*. 2 vols. Indian National Science Academy: New Delhi, 1981.

Dvivedi, S. *Brāhmasphuṭasiddhānta*. Benares: The Pandit, 1902.

Geslani, Marko et al. "Garga and Early Astral Science in India." *History of Science in South Asia* 5, no. 1 (2017): 151–191.

Montelle, Clemency. *Chasing Shadows*. Baltimore, MD: Johns Hopkins, 2011.

Montelle, Clemency and Kim Plofker. *Sanskrit Astronomical Tables*. Cham: Springer, 2018.

Montelle, Clemency and Kim Plofker. "*Karaṇakesarī* of Bhāskara: A 17th–Century Table Text for Computing Eclipses." *History of Science in South Asia* 2 (2014): 1–62.

Neugebauer, Otto and David Pingree. *The Pañcasiddhāntikā of Varāhamihira*. 2 vols. København: Munksgaard, 1970–1971.

Pingree, David. *Census of the Exact Sciences in Sanskrit*. Series A, 5 vols. Philadelphia: American Philosophical Society, 1970–1994.

Rao, S. Balacandra and Padmaja Venugopal. *Eclipses in Indian Astronomy*. Bangalore: Bharatiya Vidya Bhavan, 2008.

Raynaud, Dominique. *A Critical Edition of Ibn al-Haytham's* On the Shape of the Eclipse. Cham: Springer, 2016.

Śāstrī, Bāpūdeva, editor. *The Siddhāntaśiromaṇi*. Revised by Gaṇapati Deva Śāstrī. Kashi Sanskrit Series 72. Varanasi: Chaukhambha Sanskrit Sansthan, 1989.

Sastry, T.S. Kuppanna. *Pañcasiddhāntikā of Varāhamihira*. Madras: PPST Foundation, 1993.

Sharma, Ram Charan. *Vāṅmaya-cayanikā: A Classified Catalogue of Selected Manuscripts from Universal Institute of Orientology*. 6 vols. Jaipur: Universal Institute of Orientology: 1986–1994.

Shukla, K.S. *Vaṭeśvarasiddhānta and Gola of Vaṭeśvara*. 2 vols. New Delhi: Indian National Science Academy, 1986.

Sidoli, Nathan. "What We Can Learn from a Diagram: The Case of Aristarchus's *On the Sizes and Distances of the Sun and Moon*." *Annals of Science* 64, no. 4 (2007): 525–547.

Stephenson, F.R. "Babylonian Timings of Eclipse Contacts and the Study of the Earth's Past Rotation." *Journal of Astronomical History and Heritage* 9, no. 2 (2006): 145–150.

Stephenson, F.R. and J.M. Steele, "Astronomical Dating of Babylonian Texts Describing the Total Solar Eclipse of S.E. 175." *Journal for the History of Astronomy* 37, no. 1 (2006): 55–69.

Van Brummelen, Glen. *The Mathematics of the Heavens and the Earth: The Early History of Trigonometry*. Princeton, New Jersey: Princeton University Press, 2009.

CHAPTER 17

Numerical Systems Followed in Grantha Manuscripts

Saraju Rath

1 Introduction*

In the context of the paleographic study of Indian scripts, ancient Indian numerical systems have usually received attention only marginally and mostly only in regard to the earliest development of the numerals, the scripts, and their origins, either as independent creations in ancient India or as derived from scripts and numerical systems from outside India. This remark has become commonplace, since H.D. Sankalia,[1] H. Falk,[2] and R. Salomon[3] pointed in this

---

* This article is an additional outcome of my five-year IIAS project for the study of the ca. 500 South Indian Sanskrit manuscripts of the Johan van Manen collection (2004–2009). A large majority, ca. 81%, of these manuscripts (588 identified texts: S. Rath, "On the Johan van Manen Collection: Its Origin and Background," in *Aspects of Manuscript Culture in South India*, ed. Saraju Rath (Leiden: E.J. Brill, 2012), 55–67a) are in the Grantha script. I am happy to acknowledge here once more the funding received for this project from the Stichting J. Gonda Fonds (KNAW), the support of the IIAS, and the library assistance received from the staff of the then Kern Library (now integrated into the Leiden University Library). For the research on which this article is based, I studied in addition numerous manuscripts in Grantha and other scripts in different libraries in Pune and South India (see S. Rath, *Descriptive Catalogue of Manuscripts*, vol. 5: Vyākaraṇa, Abhidhāna, Ṣaḍdarśanas (Pune: Vaidika Saṁśodhana Maṇḍala, 1998); "A Note on the *Taittirīya* Manuscripts Belonging to the van Manen Collection, Kern Institute, Leiden," *Acta Orientalia Vilnensia* 6, no. 2 (2006), 52–62; "Varieties of Grantha Script: The Date and Place of Origin of the Manuscripts," in *Aspects of Manuscript Culture in South India*, ed. Saraju Rath (Leiden: E.J. Brill, 2012), 187–206; "Observations on Vedic Accents in Grantha Palmleaf Manuscripts," in *The Vedas in Indian Culture and History: Proceedings of the Fourth International Vedic Workshop* (Austin, Texas 2007), ed. Joel P. Brereton (Florence: Società Editrice Fiorentina, 2016), 93–103). I am thankful for the kind cooperation of the librarians and staff of these libraries. My observations regarding the use of numerals, mainly in numbering the folios, are synthetically presented in various tables in this article. I also thank Prof. Jan E.M. Houben for translating and explaining a few relevant passages in German and for providing several useful suggestions for the presentation of my observations in this article.

1 H.D. Sankalia, preface to Shobhana Laxman Gokhale, *Indian Numerals* (Poona: Deccan College, 1966).

2 Harry Falk, *Schrift im alten Indien: Ein Forschungsbericht mit Anmerkungen* (Tübingen: Gunter Narr Verlag, 1993), 168.

3 Richard Salomon, *Indian Epigraphy: A Guide to the Study of Inscriptions in Sanskrit, Prakrit, and other Indo-Aryan Languages* (New York: Oxford University, 1998), 63, note 204.

 | DOI:10.1163/9789004438224_018

respect mainly to A.H. Dani who left out numerals in his discussion of Indian scripts,[4] but added a brief sketch of his view on the character and development of Indian numerals in the preface to the second edition of his text.[5] G. Bühler dealt with numerals only in an appendix in the second edition of his *On the Origin of the Indian Brāhma Alphabet*.[6] His *Indische Palaeographie* (1896), however, does contain a reasonably extensive section on numerals.[7] Also, T.V. Mahalingam's *Early South Indian Palaeography*[8] (1967) deals with numerals adequately. On the other hand, D. Diringer,[9] taking up the much wider topic of the alphabet and alphabetic writing in the world, has an extensive section on Indian writing, but he does not discuss numerals anywhere in it. In a concluding paragraph, Diringer says: "The origin and the development of the numerals ... are other problems connected with the alphabet, which belong rather to the 'history of writing,' and will be dealt with in my book on writing."[10] In his book on writing, however, Diringer writes in the introduction: "Such subjects as numerical and musical notation ... are of course relevant to the history of writing and worthy of treatment in their own right. The exigencies of a book of this size have made it impossible, however, to deal with them in a systematic fashion."[11]

Among older publications, Gokhale, *Indian Numerals*, which is entirely devoted to Indian numerals, is the only real exception confirming the rule. It is still the most detailed discussion of numerals in Prakrit and Sanskrit inscriptions found throughout India from 300 BCE to 700 CE. The first comprehensive study specifically devoted to numerical notations of the world is S. Chrisomalis, *Numerical Notation: A Comparative History* (New York: Cambridge University Press, 2010). Chapter 6 is devoted to "South Asian Systems" which, according to the author, "includes all systems that derive from the Brāhmī numerals used on the Indian subcontinent, including Western numerals." Chrisomalis's typology of numerical systems (section on five pure types and hybrid systems) and worldwide perspective are useful, but the paleographic background in the sec-

4 Ahmad Hasan Dani, *Indian Palaeography* (Oxford: Clarendon Press, 1963).

5 Ahmad Hasan Dani, *Indian Palaeography*, second edition (Delhi: Munshiram Manoharlal, 1986).

6 Georg Bühler, *On the Origin of the Indian Brāhma Alphabet* (Strassburg: K.J. Trübner, 1898).

7 Georg Bühler, *Indische Palaeographie* (Strassburg: K.J. Trübner, 1896), 73–83 and Georg Bühler, *Indian Palaeography*, ed. J.F. Fleet (appendix to *Indian Antiquary* 33 (1904)), 76–87.

8 T.V. Mahalingam, *Early South Indian Palaeography* (Madras: University of Madras, 1967).

9 David Diringer, *The Alphabet: A Key to the History of Mankind*, second revised edition (New York: The Philosophical Library, 1953).

10 Diringer, *The Alphabet*, 570.

11 David Diringer, *Writing* (London: Thames and Hudson, 1962), 25.

tion on South Asia is not always reliable; a few times we read, for instance, that "the Grantha script ... is ancestral to modern Tamil, Malayalam and Sinhalese," which hardly does justice to their complex relationship (cp. now I. Mahadevan, *Early Tamil Epigraphy* (Cambridge, Mass.: Harvard Univ. Press, 2003), 215).[12]

The present article deals with the actual use and development of numerical systems in Grantha, one of the scripts derived from Brāhmī, especially in Grantha manuscripts up to the early twentieth century. Although there is insufficient evidence for a detailed analysis of earlier steps in its evolution, the numerical systems, especially the digital numerals found in Grantha manuscripts, do have an earlier history in those systems found in the earliest Indian scripts, especially early Brāhmī and Kharoṣṭhī inscriptions. I will therefore first briefly deal with the numerical system attested to in these early inscriptions, with their decipherment, and with some of the theories proposed about their origin.

2 Numerals in the Earliest Readable Indian Scripts: Kharoṣṭhī and Brāhmī

One of the earliest attempts to identify and decipher ancient Indian characters for numerals—i.e., digits, or, as we will also call them, digital numerals—is found in the work of James Prinsep, who succeeded in deciphering numerous inscriptions in Brāhmī on stones and coins. In this context, he deciphered several digits, but mistook some characters representing decades (for instance 2x10, 8x10, 9x10) for the corresponding numeral in the decimal place-value system, 2, 8, 9. In an article that appeared in 1838, Prinsep emphasized that his results were preliminary and needed to be corroborated by more observa-

12 Two more recent publications, R. Grünendahl, *South Indian Scripts in Sanskrit Manuscripts and Prints: Grantha Tamil-Malayalam-Telugu-Kannada-Nandinagari* (Wiesbaden: Harrassowitz, 2001) and P. Visalakshi, *The Grantha Script* (Thiruvananthapuram: Dravidian Linguistics Association, 2003), have mainly a pedagogic aim: the first as a "guide" providing a "basis for the self-reliant study of documents written in South Indian scripts" (Grünendahl, *South Indian Scripts*, vii), including Grantha, viz., "manuscripts and inscriptions" and "printed editions"; the second as a work that makes available "to all those who are interested" in learning the Grantha script "suitable learning materials" (Visalakshy, *The Grantha Script*, 9). The former is, as far as Grantha is concerned, elementary and, for its numerals, sketchy. The latter provides the student with a didactically well-organized general introduction to the script, and it has useful samples from printed texts, copper plates, and palm-leaf manuscripts. However, although several Grantha numerals (digits) show a striking similarity with their Tamil counterparts, we cannot say, with Visalakshy (*The Grantha Script*, 164), that they are "the same as [those] of the Tamil script."

tions: "I am far from being confident of having succeeded; but having once as it were broken the ice, we may soon hope for a more perfect solution of the curious problem [of numerals in ancient India], through the multitude of new, or rather old, monuments which seem to emerge from oblivion just at the time they are wanted, under the united efforts of the Society's associates in central India."[13] The problem of numerical notation was dealt with later on in studies by, among others, A.C. Burnell,[14] Bh. Indraji,[15] G. Bühler,[16] W.E. Clark,[17] Mahalingam,[18] A.H. Dani,[19] H. Falk,[20] and R. Salomon.[21]

Numerals—that is, characters functioning specifically as digits—in Brāhmī and other derived scripts usually strongly resemble syllabically organized letters in the same alphabet,[22] and there is evidence that some of them were at one time also actually pronounced as such.[23] With regard to the syllabic nature, or at least the syllabic outlook, of these ancient numerals, S.L. Gokhale concluded in her meticulous study that some numerals keep a similar shape

13 J. Prinsep, "Examination of the Inscriptions from Girnár in Gujerát, and Dhauli in Cuttack, continued," *Journal of the Asiatic Society of Bengal* 7, part 1 (1838): 353.

14 A.C. Burnell, "The Oldest Known South Indian Alphabet," *Indian Antiquary* 1 (1872): 229–230; *Elements of South Indian Palaeography* (London: Trübner & Co., 1874); "The Origin of the Indian Alphabets," *Academy* 21 (1882): 433. Reprint: *Indian Antiquary* 11 (1882): 239–240.

15 Bh. Indraji, "On the Ancient Nāgarī Numerals," *Indian Antiquary* 6 (1877): 42–47.

16 Bühler, *Indische Palaeographie*.

17 W.E. Clark, "Hindu-Arabic Numerals," in *Indian Studies in Honor of Charles Rockwell Lanman* (Cambridge, Mass.: Harvard Univ. Press, 1929), 217–236.

18 Mahalingam, *Early South Indian Palaeography*.

19 Dani, *Indian Palaeography*, second edition [with brief additional remarks and discussion of numerals in Preface].

20 Falk, *Schrift im alten Indien*.

21 Salomon, *Indian Epigraphy*.

22 The term "alphabet" remains justified in light of the observation that "Scripts of the type of Brāhmī, Grantha and Devanāgarī, which give—ideally—a one to one representation of each phoneme, consonant and vowel, are typologically far removed from other (non-alphabetic) syllabaries, where characters do not show any particular resemblance to each other if the syllable referred to shares certain phonemes (vowels or consonants). That is why the term 'alphasyllabary' (Bright 1999), which is perhaps suitable for old-Persian cuneiform and Aramaic, is not a very felicitous one for Brāhmī and derived scripts, as it would suggest the script is basically a syllabary that shares some characteristics with the alphabet .... From a functional point of view, however, Brāhmī etc. are better characterized as 'syllabically organized alphabetic writing systems', or, more compactly, as 'syllabic alphabets' ..." (J. Houben and S. Rath, "Manuscript Culture and its Impact in 'India': Contours and Parameters," in *Aspects of Manuscript Culture in South India*, ed. S. Rath (Leiden: E.J. Brill, 2012), 9, note 23).

23 Bühler, *Indische Palaeographie*, 75.

when the script changes and that certain numerals "do not show any similarity with the given syllable of the period."[24] It has, therefore, been proposed that the numerals found in Brāhmī had some status of their own before they became part of this script.[25]

Brāhmī has distinct digits for all units up to 10 *and* for all decades up to 100, as well as for 100 and 1000—20 different digits (basic characters) in total, similar to one of the systems that remained in use in Grantha manuscripts until quite recently. In terms of Chrisomalis's typology,[26] it is cipher additive for numbers up to 100. For numbers above 100, Brāhmī is multiplicative-additive as it employs here modified or ligatured digits, for instance a ligature of 100 and 4 for 400. Early Kharoṣṭhī follows a cumulative-additive system, with digits for only 1, 10, 20, 100, and 1000; later on, it also got a special character for four, a cross, instead of the earlier four single strokes, making the total number of different digits (characters) six. It is widely agreed that Kharoṣṭhī digits and letters are directly derived from a Semitic script, namely, Aramaic.[27] The origin of Brāhmī digits is still disputed.

In most of the earlier attempts to understand the shape, development, and system of digits and numerals in early Brāhmī, authors tried to see these as letters or syllables of the script of which it is part, of an older form of the script, or of a different script. A.H. Dani, however, perceived an underlying system of semicircles, hooks, and stripes or dashes, in which the symbol for 10, a semicircle open to the right with an extra "handle," is central.[28]

H. Falk[29] gives a detailed discussion of preceding views and concludes, on the basis of the evidence that was available at the time, that the numerals were invented in Aśoka's time, a theory that matches his position regarding the origin of the Brāhmī script as an invention, likewise in Aśoka's time. Falk explicitly rejects the widely accepted earlier view of Bühler,[30] who proposed that the Brāhmī script is an improved version of an originally Semitic script that was already current among Indian merchants for several centuries before Aśoka. Almost 15 years before Falk, S.R. Goyal also studied numerous pieces of evidence and reached a similar conclusion: the Brāhmī script was invented during

24 Gokhale, *Indian Numerals*, 48.

25 Through very different arguments: G.R. Kaye, "The Asoka Numerals," *Indian Antiquary* 40 (1911): 55–58; Mahalingam, *Early South Indian Palaeography*, 68; Salomon, *Indian Epigraphy*, 60.

26 Chrisomalis, *Numerical Notation*, 12.

27 Falk, *Schrift im alten Indien*, 174; Salomon, *Indian Epigraphy*, 64.

28 Dani, *Indian Palaeography*, second edition, preface.

29 Falk, *Schrift im alten Indien*.

30 Bühler, *On the Origin of the Indian Brāhma Alphabet*.

or just before the reign of king Aśoka.[31] In Falk's view, the script was created during Aśoka's reign, on the basis of two already existing scripts in use in Aśoka's empire since his grandfather Candragupta had acquired extensive territory to the west of the Indus, including Taxila (Takṣaśilā in Gandhāra) and Kandahar: the Kharoṣṭhī script, which emerged in Gandhāra not long before Aśoka, and the Greek script. Both these scripts, Greek and Kharoṣṭhī, are independently derived from Semitic scripts in order to suit Indo-European languages, Greek, and Prakrit.[32]

Although there had been wide consensus among archaeologists up to the early 1990s that there is no evidence of Brāhmī writing that can be dated to a period before Aśoka, claims have in the meantime been put forward regarding specimens of writing in the form of potsherds that have brief inscriptions and that are dated to a pre-Aśokan period through radiocarbon dating.[33] These early radiocarbon datings have not been independently confirmed on the basis of archaeology and paleography, as Mahadevan and Falk have shown.[34] In another, comparable context it has been pointed out that "among the variables of carbon dates, variation in script and linguistic variation, the first is the most objective but still much in need of calibration for relatively recent, historical dates" and that "the 'hardest' evidence to judge the date ... would ... be the palaeographic evidence."[35]

---

31 S.R. Goyal, "Brāhmī: An Invention of the Early Mauryan Period," in *The Origin of Brahmi Script*, eds. S.P. Gupta and K.S. Ramachandran (Delhi: D.K. Publishers, 1979), 1–52; now also S.R. Goyal, *Brāhmī Script: An Invention of the Early Maurya Period* (Jodhpur: Kusumanjali Book World, 2006).

32 Falk, *Schrift im alten Indien*, 337–340.

33 S.U. Deraniyagala, "Radiocarbon Dating of Early Brahmi Script in Sri Lanka: 600–500 B.C.," *Ancient Ceylon* 11 (1990): 149–168; S.U. Deraniyagala and M. Abeyratne, "Radiocarbon Chronology of Anuradhapura, Sri Lanka: A Revised Age Estimate," in *South Asian Archaeology 1997, Proceedings of the Fourteenth International Conference of the European Association of South Asian Archaeologists, Rome 7–14 July 1997*, eds. Maurizio Taddei and Giuseppe De Marco (Rome: Istituto Italiano per l'Africa e l'Oriente, 2000), 759–791; R. Coningham, F.R. Allchin, C.M. Battt, and D. Lucy, "Passage to India? Anuradhapura and the Early Use of the Brahmi script," *Cambridge Archaeological Journal* 6, no. 1 (1996): 73–97.

34 Mahadevan, *Early Tamil Epigraphy*, 167–215; H. Falk, "Owners' Graffiti on Pottery from Tissamaharama," *Zeitschrift für Archäologie aussereuropäischer Kulturen* 6 (2014): 45–94.

35 J.E.M. Houben, "Linguistic Paradox and Diglossia: On the Emergence of Sanskrit and Sanskritic Language in Ancient India," *Open Linguistics* 4, no. 1 (2018): 7. Falk, "Owners' Graffiti on Pottery," rightly pointed out that Coningham and his colleagues (Coningham et al., "Passage to India?") had omitted to take into account the distinct conditions of background radioactivity in Sri Lanka, as it is much closer to the equator than India and Pakistan. In addition, simply in light of differences in altitude, it is to be expected that on Sri Lanka, as elsewhere (Houben, "Linguistic Paradox and Diglossia," 7, note 12), there must be considerable variation in natural background radioactivity.

As for the numerals, Falk highlights the value of the system of Brāhmī as, what he calls, a "proto-decimal" system where only the zero is still lacking: This was, compared with contemporaneous systems, an important historical innovation. In Falk's view, the "inventor" of Brāhmī had selected the best from the numeral systems of the scripts of three contemporaneous cultures, Greek, Gandhāra (Kharoṣṭhī), and Chinese, for his script.[36] The structure of a proto-decimal system in Brāhmī was, according to Falk, probably taken from the Chinese, although formal similarities in the numerals are not with Chinese numerals but with those in Greek, and especially those in Kharoṣṭhī. In the section devoted to Brāhmī, Falk proposes, therefore, that the Brāhmī numeral system was created on the basis of these three existing systems.[37] When referring to Falk's position, Salomon leaves out Falk's reference to the other two systems and speaks only of "a possible influence from the early Chinese system of numerical notation," adding the comment that "the similarities are in fact less striking than those with Egyptian."[38] That individual similarities in characters are not there was, however, also what Falk emphasized, who spoke of an affinity with the Chinese system only with respect to its "proto-decimal" nature. Similarly, Chrisomalis attributes to Falk the position that Brāhmī numerals have "a Chinese origin" in view of "structural and paleographic resemblances between the Brāhmī and the earliest Chinese numerical notations,"[39] although this is precisely what Falk did not say.

Brāhmī numerals show a "proto-decimal" system (which continues as one of the systems in Grantha), or, in terms of Chrisomalis's typology, a ciphered-additive system, at least for the numbers from 1 to 100 (with bases 1 and 10), and for higher numbers a multiplicative-additive system (400 indicated through a ligature of the digit for 100 and for 4, etc.). Is there indeed sufficient reason to attribute this to influence from the contemporary Chinese system of numerals? Closer to home, the numerical system followed in the Vedic language was based on 10 and powers of 10. Although no written records are available for the earlier periods, this must have constituted a more direct source of influence for the "proto-decimal" system in early Brāhmī.[40] This does not exclude the hypothesis that the encounter between Indian and Chinese traders brought a confirmation of a "proto-decimal" system.

36 Falk, *Schrift im alten Indien*, 176.
37 Falk, *Schrift im alten Indien*, 168–176.
38 Salomon, *Indian Epigraphy*, 60.
39 Chrisomalis, *Numerical Notation*, 192.
40 Even if early Brāhmī was designed for Prakrit and not for Sanskrit, some form of early Prakrit was at that time, after all, also the generally spoken language of Brahmins or Vaidikas (Houben, "Linguistic Paradox and Diglossia").

In the context of oral transmission and performance in ritual, fingers or sticks were used to count or to calculate (not to write) the number of utterances to be recited or chanted. In *Taittirīyasaṃhitā* 7.2.20.1,[41] we find the earliest textual reference to a sequence of numbers as powers of 10, starting with *éka* "one," and continuing with *dáśa* "ten," *śatá* "hundred," *sahásra* "thousand," etc., up to *parārdhá* (1 with 11 zeros after it, i.e., 100,000,000,000). Following the decimal base system, numbers such as 11, 27, etc., were referred to in the form of cumulative (*dvandva*) compounds as *ékādaśa* and *saptáviṁśati* respectively. Fractional numbers were already in the *Ṛgveda* referred to as *ardhá* "half," *pā́da* "quarter," *śaphá* "one-eighth," and *kalā́* "one-sixteenth," etc. The authors of Vedic texts understood a number such as 19 as the result of the subtraction from the nearest decade: *ekona-viṁśatí* (i.e., one less than twenty, as an alternative for *návadaśa* "nine-teen"). The *Taittirīyasaṁhitā* also expresses other numbers with reference to the next decade: *ékān ná catvāriṁśát* "40 less 1" (i.e., "39"), *ékān nā́śīti* "79," *ékān ná śatám* "99." The *Āpastambaśulbasūtra* refers to 972 as *aṣṭaviṁśatyūnam sahasram* (28 less than 1000), etc. Thus, the decimal numerical system was fully accepted and had been in use since Vedic times. More extensive use of decimal base numerals is found in mathematical calculations for the construction of the *yajñavedi*, the sacrificial altar, for instance for the shapes and sizes of special types of bricks used in it, in the context of astronomical calculations in the ritual calendar in the auxiliary discipline of Jyotiṣa, etc., and in ancient texts like the *Vājasaneyasaṁhitā*, the *Śatapathabrāhmaṇa*, and the *Baudhāyanaśulbasūtra*.

3 Numerical Notation in (Classical) India: A General Overview

Apart from the numerical system in the originally oral tradition of ancient Vedic texts mentioned above, not one but, confusingly, many distinct systems of numerals are found in use in inscriptions attested from the third century BCE onward and then in astrological, astronomical, and mathematical texts in manuscript form. These numerical systems need to be addressed before continuing with a more specific discussion of numerical systems in Grantha. From a systematic point of view, these systems can be categorized as follows:

3.1 Word or, better, "reality" numerals (*bhūtasaṅkhyā*)
3.2 Syllabic numerals (*akṣarapalli*), also known as letter numerals

41 *Taittirīyasaṁhitā*. 7.2.20.1, *śatā́ya svā́hā, sahásrāya svā́hā'yútāya svā́hā níyutāya svā́hā* etc.

3.2.1 numerals similar to (seemingly) arbitrary syllables;
3.2.2 the *kaṭapayādi* system and systems starting at *ka* ("*kādi*");
3.2.3 the *bārākhaḍi* and related systems;

3.3 Digital numerals or digits (*aṅkapalli*).

3.1 *Word or "Reality" Numerals*

Numerals are here expressed not through digits but through words referring to some "object" or "reality" characterized by the relevant number, for instance "eye" standing for "two," "(co-uterine) brother" standing for "three"—since among the five Pāṇḍava's there are three (co-uterine) brothers, the remaining two brothers being twins from another mother—or "seer" standing for "seven" (since Vedic mythology knows a group of seven primeval seers). This system is commonly known as *bhūtasaṅkhyā* in Sanskrit and is often referred to as the system of "word numerals," although the term *bhūtasaṅkhyā* is rather to be understood as "reality numeral."[42] One numeral may be represented by various "realities," for instance, "seven" can be referred to through several words meaning "seer," such as *ṛṣi* and *muni*, but also through other "realities," such as *aśva* "horse," because the sun-god's chariot is pulled by seven horses, a mythical "reality." The numeral system of "reality" numerals has been employed in Chandas (metrics), Jyotiḥśāstra (astronomy), Gaṇitaśāstra (arithmetic, mathematics), and also in the dates of inscriptions and manuscripts. In its developed form, as found for instance in Varāhamihira's *Bṛhatsaṁhitā*, the system of word numerals has "reality" words for the numbers 1–49 and 100 and 1000 and is used with reference to a place-value notation that includes the zero, referred to as *śūnya* "a void," *kha*, or *ambara* "space." *Kha-khāṣṭi-yamāḥ* is then *kha*(0)-*kha*(0)-*aṣṭi*(16)-*yama*(2), which represents, when read from right to left,[43] 21600 (Bühler 1904: 86). The system not only presupposes familiarity

42 The term "word numeral" is, in fact, not very felicitous, as it could refer to a numeral written in words such as *catur-viṁśati* (in digits: 24 or २४ etc.). "Word-numeral" is the term Bühler used in the English version of his work on Indian paleography that appeared posthumously, edited by J.F. Fleet, in 1904 (Bühler, *Indian Palaeography*, 83–86). Where the German version (Bühler, *Indische Palaeographie*) has, on p. 80, the headings "Zahlenbezeichnung durch Wörter und Buchstaben—A. Die Wortbezeichnung," the English version has, on p. 83, "Numeral notation by words and letters—A. The word-numerals." Kim Plofker, *Mathematics in India* (Princeton: Princeton University Press, 2009), 47, speaks of "concrete" numbers, apparently interpreting *bhūta* in *bhūta-saṅkhyā* as "object," but this would cover only *bhūtasaṅkhyā* numbers such as "eye," not the numerals that are referred to through a mythological "reality" (seven seers) or through an epic "reality" (three co-uterine brothers among the Pāṇḍavas or the hundred sons of Dhṛtarāṣṭra).

43 This "reading from right to left" is, of course, for those accustomed to modern numerals, where the position for higher powers of 10 come first. The traditional rule of thumb, to

with a decimal, ciphered-positional notation of numbers (as in *khendu*, "space and moon," i.e., 0 and 1; positionally: 0x1 and 1x10 or 10), but also with various Indian knowledge systems: *dhātu* is associated with the number 7 as seven *dhātus* are acknowledged in Āyurveda; *la-kāra* is associated with the number 10 as traditional grammar accepts ten sets of verbal endings for different tenses and moods; *aṣṭi* stands for 16 because in metrics *aṣṭi* is the name of a meter having 16 syllables in each *pāda*, etc. A. Weber found early traces of the "reality numeral" system in the *Kātyāyana-* and *Lātyāyanaśrautasūtras*.[44] The system is also employed in the mathematical text in the Bakhshali manuscript (seventh-twelfth century CE[45]). Here (Table 17.1), by way of illustration, I give some frequent "reality" numerals for the numbers 1–10, 100, and 1000;[46] different "realities" are separated by a comma and conceptual synonyms are separated by a slash.

3.2 *Syllabic Numerals* (Akṣarapalli) *also Known as Letter Numerals*

3.2.1 Numerals Similar to (Seemingly) Arbitrary Syllables

In the case of Brāhmī we saw that characters or digits used as numerals were also similar to syllables. Although it is not possible to identify any underlying system, Bühler (*Indische Palaeography*, 75) pointed to evidence that some of these were at some time also actually pronounced as such. We will see that a set of (seemingly) arbitrary syllables are also used in Grantha to express numerical values, and they are used next to another, later, set of digits that seem to be more differentiated from characters with a phonetic value.

3.2.2 The *Kaṭapayādi* System and Systems Starting at *Ka* ("*Kādi*")

One of the earliest Indian mathematicians, Āryabhaṭa I, born in 476 CE, designed his own system of numerical notation through letters, which was based

understand these numerals from left to right (*aṅkānāṁ vāmato gatiḥ*), proceeds, on the contrary, from the position of lower powers of 10 to higher ones, as is also done in numerals as they are commonly expressed in the language: *ekāśīti* (81), *pañcaviṁśatiśatam* (125), etc.

44 Bühler, *Indische Palaeographie*, 80.

45 T. Hayashi, *The Bakhshālī Manuscript: An Ancient Indian Mathematical Treatise* (Groningen: Egbert Forsten, 1995); Kim Plofker, Agathe Keller, Takao Hayashi, Clemency Montelle, and Dominik Wujastyk, "The Bakhshālī Manuscript: A Response to the Bodleian Library's Radiocarbon Dating," *History of Science in South Asia* 5, no. 1 (2017): 134–150; Houben, "Linguistic Paradox and Diglossia," § 3.

46 More extensive lists in Bühler, *Indische Palaeographie*, 80–81, *Indian Palaeography*, 84–86 (for numbers 1–49), and D.C. Sircar, *Indian Epigraphy* (Delhi: Motilal Banarsidass, 1965), 230–233 (for numbers 1–49, 100, and 1000).

TABLE 17.1 Word numerals or "reality" numerals

Numerals	Word numerals, "reality" numerals
0	*śūnya, pūrṇa, kha/ambara/viyat/ākāśa …*
1	*rūpa, ādi, tanu, indu/śaśin/śaśāṅka/candra …, bhū/bhūmi/kṣiti …, nāyaka, pitāmaha*
2	*yama, aśvin, pakṣa, kara, nayana/netra/akṣi …, bāhu, karṇa, kuṭumba, ravicandrau*
3	*agni/vahni/vaiśvānara/hotṛ …, rāma, guṇa, loka/jagat …, trikāla, trigata, sahodara, trinetra,* also *netra*; *veda* (but normally = 4)
4	*aya/āya, veda/śruti, abdhi, diś, yuga, bandhu, koṣṭa, varṇa*
5	*indriya, artha/viṣaya, bhūta, iṣu, pāṇḍava, prāṇa, ratna*
6	*rasa, ṛtu, aṅga, māsārdha, darśana, rāga, ari/ripu, kāya*
7	*ṛṣi/muni, svara, aśva, aga, dhātu, chandas, dhī, kalatra* (seventh house in astrology)
8	*anuṣṭubh, vasu, ahi, gaja, maṅgala/bhūti, siddhi*
9	*aṅka, nanda, chidra, graha, nidhi, pavana*
10	*diśaḥ, rāvaṇaśiras, avatāra, karman, khendu* (*kha+indu*: 0–1 gives 10), *lakāra*
100	*dhārtarāṣṭra, puruṣāyus, śakrayajña*
1000	*śeṣaśīrṣa, ravibāṇa, jāhnavīvaktra*

on powers of ten, but without zero. This system was convincingly explained and apparently well understood by later commentators on the work of Āryabhaṭa I. No later generation mathematician seems to have adopted the system of Āryabhaṭa I for his own work, and hence it cannot be regarded as a general system of letter numerals either for India or for the Grantha script. Did the definition of this system by Āryabhaṭa I imply a demarcation from the *kaṭapayādi* system, which got much wider adherence? There is no proof, either for this interpretation, in spite of the suggestion by a later commentator,[47] or, more generally, for the existence of the *kaṭapayādi*-system before Āryabhaṭa I. In the present context, it is not necessary to give all the details of Āryabhaṭa

47 Sūryadeva Yajvan, twelfth–thirteenth century, in his commentary on the *Āryabhaṭīya*, quoted according to S.R. Sarma, "The *kaṭapayādi* System of Numerical Notation and Its Spread outside Kerala," *Revue d'histoire des mathématiques* 18 (2012): 41, note 13: *nanu vargākṣarāṇāṁ svarūpavat kakārāditvasyāpi lokasiddhatvāt "kāt" ity etad anarthakam. na. vargākṣarāṇāṁ saṁkhyāpratipādane kaṭapayāditvaṁ nañayoś ca śūnyatvam api prasid-*

I's system,[48] except one aspect of it: It started counting at *ka* (value = 1), continuing with *kha* (value = 2), etc., in the sequence of the traditional alphabet (*akṣarasamāmnāya* or *varṇamālā*), which was already presupposed in Pāṇini's grammar when he used the term *ku* in the sense of *ka-varga*, etc. (although Pāṇini gives, in addition to this, an inventory of phonemes of his own in the *pratyāhārasūtras*). We can, hence, speak of "*kādi*" systems—in use, as we will see, also in Grantha manuscripts—in contradistinction to the *kaṭapayādi* system, although only the latter term is attested.

The method known as *kaṭapayādi*, which represents numbers, including huge ones, by using the *akṣaras* of the alphabet, became a popular and convenient syllabic notation system to refer to numbers. The earliest datable attestation appears in the seventh century.[49] Advantages of the *kaṭapayādi* method are its precision and remarkable brevity, combined with a sufficient flexibility to combine the expression of high numbers with the construction of meaningful words. Here, the vowelless consonants signify a specific numerical value, from 1 to 9 and 0, only when followed by a vowel, but the vowel has no numerical value. In comparison with the system designed by Āryabhaṭa I, this system is simpler or at least easier to understand for those familiar with the modern decadic place-value notation. It expresses numerical value through a small set of *akṣaras* of two series (twice two *vargas* of consonants): both the *ka* and *ca* series and the *ṭa* and *ta* series represent the nine decadic numerals 1–9, while *ña* and *na* represent 0. A line in Sanskrit explains the system as follows: *kādi nava, ṭādi nava, pādi pañca, yādy aṣṭau*—this does not mention the distribution of zero value. A verse in a relatively late text, Śaṅkaravarman's *Sadratnamālā*,[50] explains the principles of the *kaṭapayādi* system as follows:

> *nañāv acaś ca śūnyāni saṁkhyāḥ kaṭapayādayaḥ* /
> *miśre tūpāntyahal saṁkhyā na ca cintyo hal asvaraḥ* // 3.3[51]

---

dham. tannirāsārthaṁ "kād"-grahaṇam. "kāt" prabhṛty eva vargākṣarāṇāṁ saṁkhyā, na ṭakārāt pakārāc ca prabhṛti. "kāt" prabhṛti sarvāṇi saṁkhyāṁ pratipādayanti, na tu ñakāra-nakārayoś ca śūnyatvam ity arthaḥ.

48 The system of Āryabhaṭa I is well explained by Plofker (*Mathematics in India*, 73–75).

49 Sarma, "The *kaṭapayādi* System," 41 refers to the *Graha-cāra-nibandhana* composed by Haridatta in Kerala in 683.

50 K.V. Sarma, "*Sadratnamālā* of Śaṅkaravarman (critically edited)," *Indian Journal of History of Science* 36, nos. 3–4, supplement (2001): 22. The author lived from 1774 to 1839 and composed the *Sadratnamālā* in 1819 CE (K.V. Sarma in his Introduction, "*Sadratnamālā*," 3–4).

51 The verse expresses that two nasals, *na* and *ña*, and the vowels (*ac*-s) standing alone rep-

TABLE 17.2 The *kaṭapayādi* system of numerical notation

1	2	3	4	5	6	7	8	9	0
ka	kha	ga	gha	ṅa	ca	cha	ja	jha	ña
ṭa	ṭha	ḍa	ḍha	ṇa	ta	tha	da	dha	na
pa	pha	ba	bha	ma					
ya	ra	la	va	śa	ṣa	sa	ha		(vowels)

A tabular representation of this can be seen in table 17.2.[52]

The vowels designate zero only when standing alone (or they are not used to designate a value). The reading of the numerals is from left (lower power units first) to right (higher power units), so that *bhavati* represents 4-4-6 (*catuś-catvāriṁśat ṣaṭśatāḥ*), or, in our notation, which puts higher power units first, 644. This is the *kaṭapayādi* system that apparently originated in Kerala in the first millennium CE and has been widely used in South India, especially in Kerala and Tamil Nadu.[53] A few minor variations in the system are observed, but only one justifies us to distinguish, apart from the widely used type (a), a second type (b): the system defined by Āryabhaṭa II (middle of twelfth century?), who inverted the order of reading the numerals and who attributes a numerical vowel to consonants even if they are not the last in a ligature. His system, like the one designed by Āryabhaṭa I, never caught on.[54]

resent zeros. The consonants, beginning with *ka* and *ṭa* (namely, *ka, kha, ga, gha, ṅa, ca, cha, ja, jha* and *ṭa, ṭha, ḍa, ḍha, ṇa, ta, tha, da, dha*) denote numbers (two times 1–9, since it was already said that *na* and *ña* were zero). The series beginning with *pa* and *ya* (*pa, pha, ba, bha, ma and ya, ra, la, va, śa, ṣa, sa, ha*) represent numbers (given the sequence in the *varṇamālā* this must be 1–5 and 1–8 respectively). In a ligature, the last consonant attached to a vowel specifies a number. If the consonant is not attached to a vowel it gets no numerical value.

52 Sarma, "The *kaṭapayādi* System," 39, note 4, gives an overview of major publications on the *kaṭapayādi* system since the first half of the nineteenth century, to which needs to be added only S.R. Sarma's own "*Kaṭapayādi* Notation on a Sanskrit Astrolabe," *Indian Journal of History of Science* 34 (1999): 273–287.

53 Sarma, "The *kaṭapayādi* System."

54 Plofker, *Mathematics in India*, 76–77; Sarma, "The *kaṭapayādi* System," 49–50 and 57.

3.2.3 The *Bārākhaḍi* and Related Systems

This system is also known as the *siddhākṣarasamāmnāya*, *siddhamātṛkā*, or *dvādaśākṣari* numerical system,[55] and it is mainly used for the pagination of manuscripts. In this case, a combination of the consonants with twelve vowels—the well-known vowels of the *varṇamālā*, minus the liquid vowels *ṛ* and *ḷ* and their long forms, plus *kaṁ* and *kaḥ*—is used for the pagination of manuscripts. For example, *ka, kā* up to *kaḥ*, is used to count from 1–12, followed by *kha, khā* up to *khaḥ*, to count from 13–24, etc. This system was in use not only in India but also in Sri Lanka, Thailand, and Myanmar, sometimes with the liquid vowels and their long forms, an *aṁ* and *aḥ*, added, so that their number becomes sixteen instead of twelve: one could speak of "*ṣolākhaḍi*" but the term is not attested to.

3.3 *Digital Numerals or Digits* (Aṅkapalli)

This popular system was known, in South India mostly, as *aṅkapalli*, as it consists of *aṅka*s or digital figures and their combinations.[56] Its frequent use is seen in astronomical and mathematical calculations and in later manuscripts for pagination. The origin of this system is still controversial. Its use became predominant during the last few centuries as a convenient system for the pagination of manuscripts, books, and other documents. In our current system of letters and digits, the two sets are normally kept strictly separate, except in playful or cryptic writing. Just as our numeral 2 distantly resembles the upper-case letter Z and our numeral 5 resembles the upper-case letter S, digital numerals in Indian scripts may resemble syllables, as we have seen in the case of Brāhmī. The characters in a developed set of digits are kept distinct from similar characters with syllabic value.

4 Grantha Numerals

Grantha script is one of the important ancient South Indian scripts used for writing royal documents and literary texts, mainly in Sanskrit, and mainly in inscriptions and in palm-leaf manuscripts from the seventh to the nineteenth century CE. The gradual development of Grantha script shows that its calligraphic style was influenced by the various regional scripts of South India. Something similar happened also in the case of numerals, which show many

---

55 Bühler, *Indian Palaeography*, 2, 87.

56 Varāhamihira (sixth century AD) employs the word *aṅka* for the decimal figures in his text *Pañcasiddhāntikā* 18.33.

variations, even more than in the case of letters, especially in the manuscripts. The numbers on folios are usually given on recto "a" and in the left margin. This pagination style is characteristic of South Indian manuscripts.

As we have observed above (3.2), "reality" words to represent numbers were more in use in texts dealing with mathematics, astronomy, and metrics, and in earlier inscriptions and manuscripts to indicate dates. Otherwise, the "reality" word notation system is generally not in use in Grantha manuscripts.

Several systems of syllabic or letter numerals, or *akṣarapalli*, have been frequently used in manuscripts written in Grantha script. A system of numerals similar to (seemingly) arbitrary syllables was in use mostly in the period between the fifteenth and seventeenth centuries. According to Mahalingam,[57] the letters used to denote the numerals from 1 to 10 are as follows: *na, nna, nya, sra, jhra, ha, gra, pra, dra,* and *ma*. To express 11 an author writes *ma* (10) and *na* (1) = *mana*. In this connection, Sircar, *Indian Epigraphy*, 234, refers to a "Nannādi system popular in Kerala" that is summarized in the following two lines, which show occasional differences from the syllables identified by Mahalingam or me (see table 17.3 below):

> *na-nna-nya-ṣka-jhra-hā(ha)-gra-pra-dre-ma* /
> *tha-li(la)-pta-ba-tra(tru)-tri-ca-ṇa-ña* //

The ten syllables of the first line, with variants as indicated by Sircar, stand for the numbers 1 to 10, whereas the syllables of the second line and their variants stand, respectively, for 20, 30, 40, 50, 60, 70, 80, 90, and 100.

In the following tables, I note the syllabic or letter numerals belonging to this system as I have found them in numerous Grantha manuscripts, and I compare them with the digital numerals that become frequent in later Grantha manuscripts.[58]

The first table (17.3) gives single digit or single character numerals.

Column 1 contains the numerals as given by Mahalingam, *Early South Indian Palaeography*, 68. The examples in columns 3 and 4 are taken from the manuscripts studied by me.[59] Column 3 is Grantha A: letter numerals (seemingly

57 Mahalingam, *Early South Indian Palaeography*, 68.

58 For my research, I studied 588 Grantha manuscripts in the then Library of the Kern Institute, Leiden University, and numerous manuscripts in Grantha and other scripts in different libraries in Pune and South India (see footnote *). I am thankful for the kind cooperation of the librarians and staff of these libraries. My observations regarding the use of numerals, mainly in numbering the folios, are synthetically presented here.

59 Saraju Rath, "A Note on the *Taittirīya* Manuscripts," "On the Johan van Manen Collection," "Varieties of Grantha Script," and "Observations on Vedic Accents."

TABLE 17.3 Illustration for single digits (letter and digital numerals). Manually prepared by the author on the basis of extensive manuscript study (see footnote *)

Column	1	2	3	4	5
Roman	Southern style	My observations	Grantha A letter numerals	Grantha B digital numerals	Tamil numerals
1	na	-			௧
2	nna	nna/ña			௨
3	nya	-			௩
4	ṣra	ṣkra	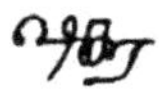		௪
5	jhra	nḍra	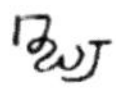	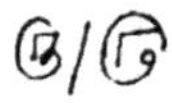	௫
6	ha	ha			௬
7	gra	ga			௭
8	pra	pra			௮
9	dra	dre		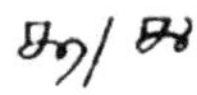	௯
0/10	ma	ma			௦/௰
100	-	-	-		௱
1000	-	-	-	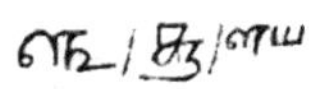	௲

arbitrary syllables), and Column 4 is Grantha B: the digital system. Characters that, according to my study, cannot be identified in the way Mahalingam suggested are 1 and 3, which seem to have become independent of their phonetic value—they seem to have been evolving in the direction of digits, but they are distinct from those of the later set. The Grantha A type of letter numerals are seen in manuscripts belonging to the sixteenth-eighteenth centuries CE, whereas the Grantha B type of numerals show the popular use of only digital numbers in manuscripts belonging to the nineteenth-twentieth centuries CE. These Grantha digital numerals are similar to, but not, as is sometimes suggested,[60] identical with Tamil numerals. The digit for 1 in the variant given here is similar to one of the variants of Grantha *ka*, and, unlike Nāgarī 1, it is in any case not derived from the corresponding Brāhmī numeral[61] but possibly from the "*kādi*" style of counting. The sign for 10 appears together with 2–9 to indicate decades and is as such parallel to the 0 in the place of decades in the decadic place-value notation. Otherwise, it only represents 10 and never zero.

Between the eighteenth and nineteenth centuries, we find variations and combined systems of numerical notation in which different systems are used in parallel or consecutively. They can be placed into three categories:

i. *bārākhaḍi* (in fact in Grantha mansucripts usually "*ṣolākhaḍi*") plus "*kādi*" system;
ii. *bārākhaḍi* (in fact "*ṣolākhaḍi*") plus digital system;
iii. "*kādi*" (*ka-kṣa*, then arbitrary letter numerals, then *kā, ki*, etc.) plus digital system.

i. Among the abovementioned three categories, I found in my paleographical study that the first one is somewhat older than the second and third categories. If different systems are used consecutively, more confusion is likely to occur to the manuscript reader when counting the pages if a few pages are missing. One possible combination in numbering pages is as follows:

a. *a-aḥ* (12 plus 4 is 16 vowels)
b. *ka-kṣa* (34 consonants, "*kādi*")
c. *ka, kā … kaḥ* (consonant k, with 16 *mātrās*/*khaḍi*s= 16)
d. *kha, khā … khaḥ* (16)
e. *ga, gā … gaḥ* (16), etc.

60 Visalakshy, *The Grantha Script*, 164. "The numeral symbols of Grantha are the same as that of the Tamil Script." This may apply to printed texts, but it is not valid for manuscripts.

61 G.H. Ojha, *Bhāratīya prācīna lipimālā*, second revised edition (Ajmer: Scottish Mission Industries Company, 1918), plate LXXXIV.

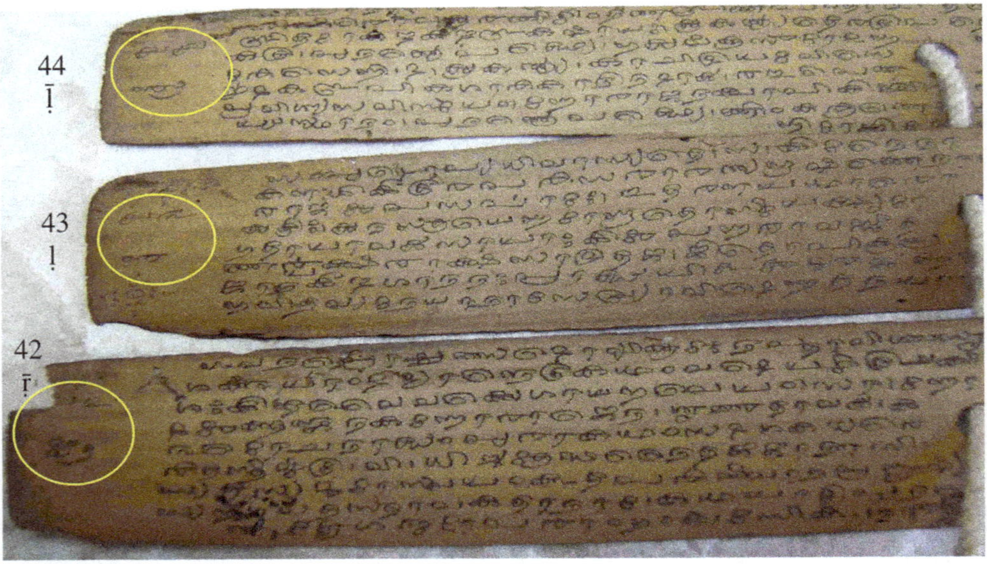

FIGURE 17.1 A palm-leaf ms. in Grantha script
JOHAN VAN MANEN COLLECTION

ii. As for the second combined category: in fact, *bārākhaḍi* consists of 12 vowel *mātrās*, excluding *dīrgha ṝ*, *ḹ*, *aṁ*, and *aḥ*. But in the pagination system in Grantha manuscripts, we usually find all 16 *mātrās* serially.

An illustration (figure 17.1) gives a clear picture of this system.

iii. Also in the third combined category we see both numerals on the same page. This type of dual numbering on the same page is indicative of a transitional phase, when an older, archaic variety of Grantha script disappears and another Grantha variety appears (see Rath, "Varieties of Grantha Script"). This is a confusing state of transformation, a change from the older variety to a later, "middle" style of Grantha. However, numerals for counting verses in the manuscript text follow only the digital system, never the *bārākhaḍi* or the "*kādi*" system.

See the illustrations below (figures 17.2 and 17.3), showing this variety.

Digits in decimal notation is a popular system in relatively recent manuscripts. They are found irregularly in the nineteenth century CE and constantly in the twentieth century CE. However, we see some variations and an inconsistency in the presentation style of these numbers.

Table 17.4 shows the normal style of decimal notation. Variations which we find may include, for example, two digits as three digits: Numeral 24 could be written in two ways, ௨௪ (two digits in normal style) and ௨௰௪ (three digits which is a variant, not a regular form). Similarly, numeral 39, ௩௯ and ௩௰௯; numeral 53, ௫௩ and ௫௰௩.

FIGURE 17.2 Eight folios of a palm-leaf ms. in the Grantha script
JOHAN VAN MANEN COLLECTION

Three-digit numbers are sometimes presented in four digits, showing again the inconsistency of a transitional period. For example, the number 167 is found as ளசஎ (three digits) and ளசயஎ (four digits); similarly the number 195 is presented as ளகூடு and ளகூயடு.

As we show below in table 17.5, three-digit numbers are occasionally presented in four digits, side by side. But table 17.6 shows that the numbers after 200, if presented in four-digit numbers, get split, and 200 is written above and the remainder of the number below. This is the usual style to represent a four-digit number after 200, which continues after 300, 400, 500, and onwards. In the number 1000, a single complex character ளட/௲ is used, alternatively also a multiplied number (100 × 10) ளய in two digits.

83
ha

84
ḷa

85
kṣa

86
ṣke

87
ṣpa

88
iti

89
kaṅ

90
kaḥ

FIGURE 17.3
These clips are taken from the above palm-leaf ms. in Grantha script in order to have a comparison of both varieties in the same margin

TABLE 17.4 Numerals consisting of single and two digits. Manually prepared by author on the basis of extensive manuscript study (see footnote *)

Roman	Grantha digits (11-20)	Roman	Grantha digits
11	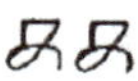	30	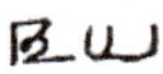
12		40	
13	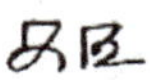	50	
14	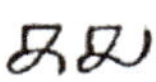	60	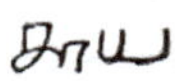

TABLE 17.4 Numerals consisting of single and two digits (*cont.*)

Roman	Grantha digits (11-20)	Roman	Grantha digits
15	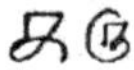	70	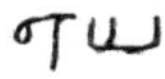
16	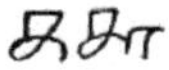	80	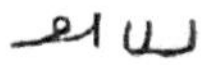
17	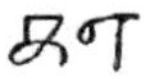	90	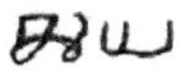
18	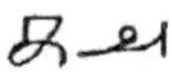	100	
19	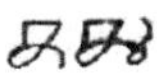	200	
20	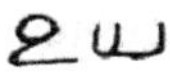	1000	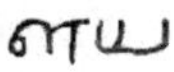
		1000	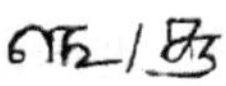

TABLE 17.5 Numerals, 111 onwards: 3 digits, sometimes 4 digits with variations. Manually prepared by author on the basis of extensive manuscript study (see footnote *)

Roman	Devanāgarī	Grantha
111	१११	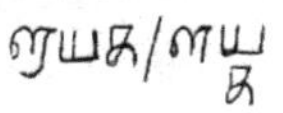
120	१२०	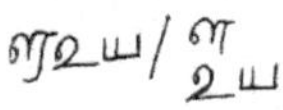
210	२१०	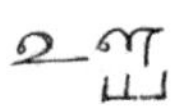
167	१६७	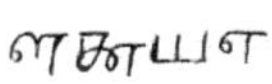
195	१९५	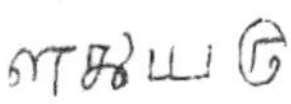

TABLE 17.6 Numerals, 211 onwards: 4 digits. Manually prepared by author on the basis of extensive manuscript study (see footnote *)

Roman	Devanāgarī	Grantha
211	२११	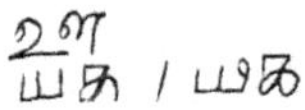
220	२२०	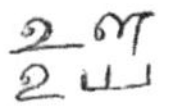

5 Conclusion

In its long history, the Grantha script has used a large number of very different systems to indicate numbers. A set of digits generally accepted by scribes in Grantha, similar to but not identical with Tamil numerals, gets established relatively late. A peculiar set of numerals similar to (seemingly) arbitrary syllables was in use mainly from the fifteenth to seventeenth century, and in the nineteenth century it was still known to older generations in Kerala. Typologically it is similar to the emerging set of digitals in early Brāhmī, which were, in their own way, either derived from or assimilated to seemingly arbitrary syllables. Various kinds of combined systems of counting are found, which can be understood if one is familiar with the sequence of the traditional alphabet (*akṣarasamāmnāya* or *varṇamālā*) and with the Grantha script. The use of various systems, simultaneously or consecutively, must have been quite confusing to scribes and readers. Grantha seems to have been particularly late in establishing for itself a set of unambiguous digital numerals, no doubt because it was never a regional script with a larger public that needed a consistent and practical numerical system.

References

Bright, William. "A Matter of Typology: Alphasyllabaries and Abugidas." *Written Language and Literacy* 2 (1999): 45–56.

Bühler, Georg. *Indische Palaeographie*. Grundriss der Indo-Arischen Philologie und Altertumskunde 1, vol. 11. Strassburg: K.J. Trübner, 1896.

Bühler, Georg. *On the Origin of the Indian Brāhma Alphabet*. Second revised edition of Indian Studies 3. Strassburg: K.J. Trübner, 1898.

Bühler, Georg. *Indian Palaeography*, edited by J.F. Fleet. Appendix to *Indian Antiquary* 33 (1904).

Burnell, A.C. "The Oldest Known South Indian Alphabet." *Indian Antiquary* 1 (1872): 229–230.

Burnell, A.C. *Elements of South Indian Palaeography*. London: Trübner & Co., 1874.

Burnell, A.C. "The Origin of the Indian Alphabets." *Academy* 21 (1882): 433. Reprint: *Indian Antiquary* 11 (1882): 239–240.

Chrisomalis, Stephen. *Numerical Notation: A Comparative History*. Kindle edition. New York: Cambridge University Press, 2010.

Clark, Walter Eugene. "Hindu-Arabic Numerals." In *Indian Studies in Honor of Charles Rockwell Lanman*, 217–236. Cambridge, Mass.: Harvard University Press, 1929.

Coningham, R., F.R. Allchin, C.M. Batt, and D. Lucy. "Passage to India? Anuradhapura and the Early Use of the Brahmi Script." *Cambridge Archaeological Journal* 6, no. 1 (1996): 73–97.

Dani, Ahmad Hasan. *Indian Palaeography*. Oxford: Clarendon Press, 1963.

Dani, Ahmad Hasan. *Indian Palaeography*. Second edition [with brief additional remarks and discussion of numerals in Preface]. Delhi: Munshiram Manoharlal, 1986.

Deraniyagala, S.U. "Radiocarbon dating of Early Brahmi Script in Sri Lanka: 600–500 B.C." *Ancient Ceylon* 11 (1990): 149–168.

Deraniyagala, S.U. and M. Abeyratne. "Radiocarbon Chronology of Anuradhapura, Sri Lanka: A Revised Age Estimate." In *South Asian Archaeology 1997, Proceedings of the Fourteenth International Conference of the European Association of South Asian Archaeologists, Rome 7–14 July 1997*, edited by Maurizio Taddei and Giuseppe De Marco, 759–791. Rome: Istituto Italiano per l'Africa e l'Oriente, 2000.

Diringer, David. *The Alphabet: A Key to the History of Mankind*. Second revised edition. New York: The Philosophical Library, 1953.

Diringer, David. *Writing*. London: Thames and Hudson, 1962.

Falk, Harry. *Schrift im alten Indien: Ein Forschungsbericht mit Anmerkungen*. Tübingen: Gunter Narr Verlag, 1993.

Falk, Harry. "Owners' Graffiti on Pottery from Tissamaharama." *Zeitschrift für Archäologie aussereuropäischer Kulturen* 6 (2014): 45–94.

Gokhale, Shobhana Laxman. *Indian Numerals*. Preface by H.D. Sankalia. Poona: Deccan College Postgraduate Research Institute, 1966.

Goyal, S.R. "Brāhmī: An Invention of the Early Mauryan Period." In *The Origin of Brahmi Script*, edited by S.P. Gupta and K.S. Ramachandran, 1–52. Delhi: D.K. Publishers, 1979.

Goyal, S.R. *Brāhmī Script: An Invention of the Early Maurya Period*. Jodhpur: Kusumanjali Book World, 2006.

Grünendahl, Reinhold. *South Indian Scripts in Sanskrit Manuscripts and Prints: Grantha Tamil-Malayalam-Telugu-Kannada-Nandinagari*. Wiesbaden: Harrassowitz, 2001.

Hayashi, Takao. *The Bakhshālī Manuscript: An Ancient Indian Mathematical Treatise.* Groningen: Egbert Forsten, 1995.

Houben, J.E.M. "Linguistic Paradox and Diglossia: The Emergence of Sanskrit and Sanskritic Language in Ancient India." *Open Linguistics* 4, no. 1 (2018): 1–18. DOI: https://doi.org/10.1515/opli-2018-0001.

Houben, J.E.M. and Saraju Rath. "Manuscript Culture and Its Impact in 'India': Contours and Parameters." In *Aspects of Manuscript Culture in South India*, edited by S. Rath, 1–53. Leiden: E.J. Brill, 2012.

Indraji, Bhagvānlāl. "On the Ancient Nāgarī Numerals." *Indian Antiquary* 6 (1877): 42–47.

Kaye, G.R. "The Asoka Numerals." *Indian Antiquary* 40 (1911): 55–58.

Mahadevan, Iravatham. *Early Tamil Epigraphy: From the Earliest Times to the Sixth Century A.D.* Harvard Oriental Series, 62. Cambridge, Mass.: Harvard University Press, 2003.

Mahalingam, T.V. *Early South Indian Palaeography*. University of Madras, Archaeological Series, no. 1. Madras: University of Madras, 1967.

Ojha, Gaurishankar Hirachand. *Bhāratīya prācīna lipimālā* (in Hindi). Second revised edition. Ajmer: Scottish Mission Industries Company, 1918.

Ojha, Gaurishankar Hirachand. *Bhāratīya prācīna lipimālā* (in Hindi). Reprint (= preceding, with addition of English title: *The Palaeography of India*): Delhi: Munshiram Manoharlal, 1993.

Plofker, Kim. *Mathematics in India*. Princeton: Princeton Univ. Press, 2009.

Plofker, Kim, Agathe Keller, Takao Hayashi, Clemency Montelle, and Dominik Wujastyk. "The Bakhshālī Manuscript: A Response to the Bodleian Library's Radiocarbon Dating." *History of Science in South Asia* 5, no. 1 (2017): 134–150.

Prinsep, James. "Examination of the Inscriptions from Girnár in Gujerát, and Dhauli in Cuttack, continued." *Journal of the Asiatic Society of Bengal* 7, part 1 (1838): 334–356.

Rath, Saraju. *Descriptive Catalogue of Manuscripts*. Vol. 5: Vyākaraṇa, Abhidhāna and Ṣaḍdarśanas (describing 651 Sanskrit manuscripts, pp. 1–182). Pune: Vaidika Saṁśodhana Maṇḍala, 1998.

Rath, Saraju. "A Note on the *Taittirīya* Manuscripts Belonging to the van Manen Collection, Kern Institute, Leiden." *Acta Orientalia Vilnensia* 6, no. 2 (2006): 52–62.

Rath, Saraju. "On the Johan van Manen Collection: Its Origin and Background." In *Aspects of Manuscript Culture in South India*, edited by Saraju Rath, 55–67. Leiden: E.J. Brill, 2012.

Rath, Saraju. "Varieties of Grantha Script: The Date and Place of Origin of the Manuscripts." In *Aspects of Manuscript Culture in South India*, edited by Saraju Rath, 187–206. Leiden: E.J. Brill, 2012.

Rath, Saraju. "Observations on Vedic Accents in Grantha Palmleaf Manuscripts." In *The*

Vedas in Indian Culture and History: Proceedings of the Fourth International Vedic Workshop (Austin, Texas 2007), edited by Joel P. Brereton, 93–103. Florence: Società Editrice Fiorentina, 2016.

Salomon, Richard. "On the Origin of the Early Indian Scripts." Review article. *Journal of the American Oriental Society* 115, no. 2 (1995): 271–279.

Salomon, Richard. *Indian Epigraphy: A Guide to the Study of Inscriptions in Sanskrit, Prakrit, and the Other Indo-Aryan Languages*. New York: Oxford University, 1998.

Sarma, K.V. "*Sadratnamālā* of Śaṅkaravarman (critically edited)." *Indian Journal of History of Science* 36, nos. 3–4, supplement (2001): 1–58.

Sarma, Sreeramula Rajeswara. "*Kaṭapayādi* Notation on a Sanskrit Astrolabe." *Indian Journal of History of Science* 34 (1999): 273–287.

Sarma, Sreeramula Rajeswara. "The *kaṭapayādi* System of Numerical Notation and Its Spread outside Kerala." *Revue d'histoire des mathématiques* 18 (2012): 37–66.

Shukla, K.S. and K.V. Sarma. *Āryabhaṭīya of Āryabhaṭa: Critically Edited with Introduction, English Translation, Notes, Comments and Indexes*. New Delhi: Indian National Science Academy, 1976.

Sircar, D.C. *Indian Epigraphy*. Delhi: Motilal Banarsidass, 1965.

Visalakshy, P. *The Grantha Script*. Thiruvananthapuram: Dravidian Linguistics Association, 2003.

PART 4

Interactions across Culture and Time

CHAPTER 18

Inventing the *Pothi*: The Adoption and Spread of a New Manuscript Format in Indian Buddhism

Stefan Baums

1 Introduction*

The characteristic manuscript format of classical India and Indian Buddhism consists of a stack of prepared palm leaves, measuring about three to five cm in height by 40 to 50 cm in width, and inscribed on both sides, in three to ten lines parallel to the long edge of the writing surface. The stack of leaves is held together by one or two strings passed through punched holes, protected by top and bottom wooden covers, and often additionally wrapped in cloth. This manuscript format—usually though, as we shall see, somewhat imprecisely called a *pothi*—spread from India throughout Southeast Asia, along the Silk Roads and into Tibet. In northwestern India and Central Asia, it was eventually replaced by imitations in birch bark and paper, preserving the general format and method of binding.

The *pothi* was not, however, to our current knowledge, the original manuscript format in which Buddhist texts were committed to writing. This honor goes to birch-bark scrolls that appear to have been used in northwestern India (the area of Gandhāra) since at least the third century BCE, though the earliest surviving examples date to the first century BCE. The earliest preserved palm-leaf manuscripts in *pothi* format have likewise been found in the northwest of the subcontinent and on the Silk Roads, far away from the south Indian cradle of this format, in manuscript deposits dating back to approximately the third century CE. It is these discoveries that provide insight into the general adoption of the *pothi* format and the emergence of a new pan-Buddhist textuality, and that grant at least an indirect glimpse into the mainland-Indian manuscript

* It is a great pleasure to dedicate this article to Kenneth Zysk, who as a colleague at the Asien-Institut, University of Copenhagen, from 2000 to 2002 supported and inspired me with his deep and abiding interest in South Asian manuscripts. I presented an earlier version of my argument at a conference on Buddhist Manuscript Cultures at Princeton University, 15–17 January 2016, and give my warm thanks to Stephen Teiser for organizing the conference and inviting me to participate.

DOI:10.1163/9789004438224_019

tradition that preceded it. This paper provides a contrastive characterization of the *pothi* format and the scroll format that it replaced, and discusses the historical development of the new *pothi*-based manuscript culture in northwest India and Central Asia.

2 Gandhāran Scrolls

The technique of writing reached India from the west, in the form of the Aramaic language and script used by the Achaemenid administration of the province of Gandhāra from the sixth to the fourth centuries BCE.[1] The main writing materials throughout the Achaemenid empire were papyrus for more ephemeral and leather for more durable documents.[2] Remains of both have been found immediately to the north of Gandhāra, most notably leather documents from the archives of the satrapy of Bactria from the late fourth century BCE, just before its conquest by Alexander the Great.[3] Alexander's generals Nearchus and Megasthenes left us accounts of public administration in India, remarking in particular that writing was still not used in the capital of the Mauryan empire, Pāṭaliputra in eastern India, whereas documents on cloth (in addition to the attested leather and papyrus) were in use in the northwest.[4]

Barely fifty years later, in the middle of the third century BCE, the Mauryan emperor Aśoka inscribed his famous series of edicts on rocks and pillars througout his empire, significantly using two different scripts. In his mainland Indian inscriptions, Brāhmī—the ancestor of all later indigenous scripts of India—made its first appearance,[5] but in the northwest (at Shahbazgarhi

---

1 O. v. Hinüber, *Der Beginn der Schrift und frühe Schriftlichkeit in Indien* (Mainz: Akademie der Wissenschaften und der Literatur, 1990), 55–58.

2 Bezalel Porten, "Aramaic Letters: A Study in Papyrological Reconstruction," *Journal of the American Research Center in Egypt* 17 (1980): 39–75.

3 Joseph Naveh and Shaul Shaked, *Aramaic Documents from Ancient Bactria (Fourth Century BCE)* (London: The Khalili Family Trust, 2012).

4 David Diringer, *The Hand-Produced Book* (London: Hutchinson's Scientific and Technical Publications, 1953), 45–46. It is true that Megasthenes referred to the absence of written texts in legal proceedings in particular, and that the use of writing in other contexts cannot be ruled out on his testimony alone: Ludo Rocher, *Studies in Hindu Law and Dharmaśāstra* (London: Anthem Press, 2012), 215–218. In the absence of any positive evidence, however, the scenario sketched here still appears most likely.

5 It has been suggested, but remains uncertain, that Brāhmī may have had pre-Aśokan roots as a traders' script: R.A.E. Coningham, F.R. Allchin, C.M. Batt, and D. Lucy, "Passage to India? Anuradhapura and the Early Use of the Brahmi Script," *Cambridge Archaeological Journal* 6 (1996): 73–97.

and Mansehra not far from Peshawar), Aśoka used the script that we now call Kharoṣṭhī. Many of the letters of Kharoṣṭhī as well as its writing direction and general ductus are clearly derived from the Aramaic tradition that preceded it in this region, but an original vowel notation by diacritical marks on base consonants—as in all the later Indian scripts—made Kharoṣṭhī more suitable for the local language Gāndhārī. The most likely explanation for Aśoka's two-script solution is that in the northwest, a Kharoṣṭhī scribal tradition had already established itself in succession of the foreign Aramaic tradition in the administrative sphere, making it the only practical choice for the Mauryan emperor's edicts in this region. In the rest of his empire, however, Aśoka appears to have felt free to introduce his own new imperial script Brāhmī, improving on the system of the earlier Kharoṣṭhī (particularly in the distinction of short and long vowels) and possibly inspired by the general monumental ductus of the Greek inscriptions that were used by Alexander's successors in Bactria.[6]

Hypothetical third- and second-century-BCE Kharoṣṭhī documents on perishable material presupposed by Aśoka's northwestern inscriptions are lost, but the preserved Kharoṣṭhī birch-bark scrolls from the first century BCE onwards allow some deductions about their physical makeup. I have argued elsewhere that the earliest Kharoṣṭhī documents probably imitated Aramaic short-format scrolls of approximately 20 cm in width by 40 cm in height, substituting locally available birch bark as writing material for the Aramaic leather or papyrus.[7] At some later point—probably after the installation of Aśoka's northwestern edicts, the introduction of a Buddhist relic cult and the establishment of the first Buddhist monasteries in Gandhāra—Buddhist texts were for the first time committed to writing in Gandhāra.

Taking a closer look at the mature Kharoṣṭhī tradition, the following picture emerges.[8] The main manuscript format were vertical scrolls made from the bark of the Himalayan birch (*Betula utilis*). The bark of this tree has many uses unrelated to writing, such as the packaging of goods and the water-proofing of roofs and pipes,[9] and we can thus assume that a production chain and even the occasional use for note-taking were well-established when birch bark began to take the place of imported papyrus and expensive leather. Gandhāran scrolls

6 Harry Falk. *Schrift im alten Indien: Ein Forschungsbericht mit Anmerkungen* (Tübingen: Gunter Narr Verlag, 1993), 111–112.

7 Stefan Baums, "Gandhāran Scrolls: Rediscovering an Ancient Manuscript Type," in *Manuscript Cultures: Mapping the Field*, eds. Jörg B. Quenzer, Dmitry Bondarev, and Jan-Ulrich Sobisch (Berlin: De Gruyter, 2014), 183–225.

8 For a detailed discussion see Baums, "Gandhāran Scrolls."

9 G. Bühler, *Detailed Report of a Tour in Search of Sanskrit MSS. Made in Kaśmîr, Rajputana, and Central India* (Bombay: Society's Library, 1877), 29.

come in two basic formats: a short format of approximately 20 cm in width by 40 cm in height, just as the Aramaic administrative documents that preceded it, and a long format of up to five meters in length.

Short-format scrolls consisted of a single sheet of bark separated from the tree and cut to size such that the shorter edge of the sheet was parallel to the lenticels (dark pores) of the bark. The preferred writing surface (thus typically the *recto* of manuscripts) was the whiter and smoother inner side of the bark. Lines ran parallel to the shorter edge of the sheet and contained—depending of course on the scribal hand—around 40 *akṣara*s per line. As in the Aśokan inscriptions,[10] while there was no separation of words, phrases were set off by punctuation space and sometimes punctuation dots. The end of larger textual units was marked by more elaborate punctuation marks such as circles, stylized lotuses and swastikas. The very end of a text was often noted by a graphical vignette such as a lotus flower or a stūpa. Short-format scrolls were typically used for canonical texts such as small groups of prose Sūtras or short verse collections such as the *Rhinoceros Sūtra*, and sometimes for short original compositions such as Stotras. Scrolls were folded up into narrow strips from the bottom to the top of the *recto* such that the *recto* faced inwards, and the resulting folded package was sometimes folded one more time horizontally (presumably for final deposit rather than later use).

Long-format scrolls are constructed by gluing several sheets of bark together vertically, and usually reinforcing the overlaps by cross-stitching and increasing the overall cohesion of the scroll by sewing threads down the left and right margins. Long-format scrolls are inscribed and folded up the same way as short-format ones, but due to the thickness of the folded-up package never received a final horizontal fold. The vertical orientation of the long-format Gandhāran scrolls stands in contrast to both contemporary Greek and Chinese scroll formats, and it would thus appear that it was an independent innovation within the writing culture of Gandhāran Buddhism when the need arose to commit to writing longer texts.[11] The longest single extant scroll contains around 600 lines of text, and our sole preserved example of a multi-scroll text must have contained at least 450 lines when complete. Long-form scrolls were used for some

10 Klaus Ludwig Janert, *Abstände und Schlußvokalverzeichnungen in Aśoka-Inschriften* (Wiesbaden: Franz Steiner Verlag, 1972).

11 Counter to Lore Sander, "Early Prakrit and Sanskrit Manuscripts from Xinjiang (Second to Fifth / Sixth Centuries C.E.): Paleography, Literary Evidence, and Their Relation to Buddhist Schools," in *Collection of Essays 1993: Buddhism across Boundaries: Chinese Buddhism and the Western Regions*, eds. Jan Nattier and John R. McRae (Sanchung: Fo Guang Shan Foundation for Buddhist & Culture Education, 1999), 72.

canonical verse texts of substantial size (such as the *Anavataptagāthā* and the *Dharmapada*), but especially served the needs of two new genres of literature: Mahāyānasūtras and commentarial texts. Both of these appear to have arisen in tandem with the development of Buddhist writing culture. In the case of the former this is indicated by the repeated reference to manuscripts in the texts themselves, in the case of the latter by a particular style indicating composition by a process of note-taking and subsequent revision.

The early Gandhāran scrolls preserve the full range of genres of Buddhist literature, from canonical poetry, Sūtra and Vinaya texts to scholastic treatises, Mahāyāna texts and Stotras to narrative literature, and even include a number of non-Buddhist compositions such as a Rājanīti text in Sanskrit language and Kharoṣṭhī script. From the range of genres, Sūtras and some of the scholastic treatises stand out in regard to formal characteristics pointing to particular modes of composition and use contexts.

In the case of Sūtras we only have very small-scale collections of three or four Sūtras on a single scroll and, in the case of the Senior collection, around fifteen of such scrolls deposited and apparently commissioned together.[12] What is more, many of the small Sūtra manuscripts in the Senior collection break off mid-sentence as the writing surface came to an end, giving the impression that completing the written text was not deemed important. Nonetheless, the notion of a complete collection of *Ekottarikāgamasūtras* corresponding in size to the collections surviving in Pali and Chinese existed also in Gandhāra, as is evident from a reference in one of the scholastic texts (verse commentary II) to the 'section with sixteen parts in the Ekottarikā.'[13] The most plausible explanation would appear to be that canonical Sūtras continued to be primarily transmitted in oral form by specialized reciters (*bhāṇakas*), and that the extant manuscripts only represent written snapshots of a particular partial recitation taken for later reference (when the *bhāṇaka* might be unavailable) or possibly sometimes—as in the case of the Senior collection—simply to create a physical representation of the text for immediate deposit as a dharma relic.

Among the scholastic texts, three verse commentaries (CKM 5, 9, 11, 15 and 20) and a commentary on the *Saṃgītisūtra* (CKM 17) present quite the opposite

12 Mark Allon, "The Senior Kharoṣṭhī Manuscripts," in *From Birch-Bark to Digital Data: Recent Advances in Buddhist Manuscript Research: Papers Presented at the Conference 'Indic Buddhist Manuscripts: The State of the Field,' Stanford, June 15–19 2009*, eds. Paul Harrison and Jens-Uwe Hartmann (Wien: Verlag der Österreichischen Akademie der Wissenschaften, 2014), 19–33.

13 Stefan Baums, "A Gāndhārī Commentary on Early Buddhist Verses: British Library Kharoṣṭhī Fragments 7, 9, 13 and 18" (PhD diss., University of Washington, 2009), 513.

picture. All are carefully written down on long-format (and in one case multi-volume) scrolls, in a very condensed style that presupposes knowledge of the root texts and intimate familiarity with the exegetical methods employed. Furthermore, at least one of the verse commentaries (just like one of the narrative texts) received two continuations (one by a different hand) after the end of the text had already been indicated by a lotus vignette. I argued elsewhere (in a paper presented at the 2014 conference of the International Association of Buddhist Studies in Vienna) that here we have to do with a process of 'episodic composition,' where written notes are taken in a context of recurring oral instruction and subsequently turned into a clean copy for future reference by either the student or the teacher (or indeed by a student for use in his future teaching career).

The word used by the Gandhāran birch-bark scrolls to refer to themselves is *pustaga*, a loan word from Iranian that originally denoted documents on leather. It had thus become disassociated from a particular materiality and assumed a general meaning of 'book' (in the physical sense: in the case of multi-scroll documents, each of the scrolls individually is called a *pustaga*).[14] The last known Gandhāran birch-bark scrolls date to the second or at most third century CE. There is a gap in the archeological record until the production of the earliest Gilgit manuscripts,[15] further to the northeast, in the fifth or sixth century CE, and it remains unclear whether the Gandhāran scroll format continued to be used for another century or two and what caused its ultimate demise.[16] For the time between the third and the fifth century, the available evidence directs our attention further west to Bamiyan, where palm-leaf manuscripts in so-called *pothi* format reigned supreme.

14 The word *pothi* is itself derived from Gāndhārī *pustaga*, Sanskrit *pustaka*.

15 Oskar von Hinüber, "The Gilgit Manuscripts: An Ancient Buddhist Library in Modern Research," in *From Birch-Bark to Digital Data: Recent Advances in Buddhist Manuscript Research: Papers Presented at the Conference 'Indic Buddhist Manuscripts: The State of the Field,' Stanford, June 15–19 2009*, eds. Paul Harrison and Jens-Uwe Hartmann, (Wien: Verlag der Österreichischen Akademie der Wissenschaften, 2014), 79–136.

16 We can rule out the fading of Hellenistic influence—Lore Sander, "The Earliest Manuscripts from Central Asia and the Sarvāstivāda Mission," in *Corolla Iranica: Papers in Honour of Prof. Dr. David Neil MacKenzie on the Occasion of His 65th Birthday on April 8th, 1991*, eds. Ronald E. Emmerick and Dieter Weber (Frankfurt am Main: Peter Lang, 1991), 172—as a cause since the Gandhāran scrolls were independent of the Greek manuscript tradition from the very beginning.

3 Introducing the *Pothi*

At an indeterminate point of time, but between the introduction of Brāhmī under Aśoka in the third century BCE and, on the evidence of the Sri Lankan chronicles, the first century BCE, also the Buddhists of South India began to record their texts in writing, inscribing them on palm leaves.[17] From this time on, Indian Buddhism had two distinct written traditions separated by writing material, script and languages. In the third century CE in Bamiyan in the northwest and in Central Asia, we witness the meeting of these traditions and the emergence of a new Buddhist manuscript culture. Palm-leaf manuscripts are made from the individual fingers of the fanned leaves of the talipot palm (*Corypha umbraculifera*) native to southern India (up to the latitude of Mumbai on the west coast and of Bengal on the east coast). The alternative use of the palmyra palm (*Borassus flabellifer*) is recent (since the 16th century), as demonstrated long ago by Rudolf Hoernle,[18] and thus outside the scope of the present paper. The leaves of the talipot palm were boiled, dried, polished with pumice and—as noted above—cut to a regular size of three to five cm in height by 40 to 50 cm in width.

While the earliest preserved palm-leaf manuscripts are inscribed using ink and a wooden stylus, the common treatment in modern southern India is to incise the letters into the leaves with a sharp metal stylus and then rub a solution of ink in oil into the incisions to make the letters visible (a procedure that could be repeated as needed). Dominik Wujastyk suggests that the direct application of ink to the writing surface "carried over" from birch bark to palm leaf,[19] presumably intending the northwest in particular and leaving open whether the incising was the original method for palm leaves. Since no early palm-leaf manuscripts are preserved from southern India, it cannot be decided which of the two procedures is more ancient (or whether they are of the same age). It is both the ancient and the modern practice for the palm leaves to be

17 The ultimate origin of the Indian palm-leaf manuscript format remains unclear. Roman account books on wooden codices or parchment (Diringer, *Hand-Produced Book*, 192) may have become known in southern Indian through trade, but were bound on one side and could only have served as a loose inspiration for a type of book consisting of sheets. The practice of writing on leaves (including palm leaves) is reported from many parts of the ancient world, including Egypt, Rome and Arabia (Diringer, *Hand-Produced Book*, 37, 40, 42–45).

18 A.F. Rudolf Hoernle, "An Epigraphical Note on Palm-Leaf, Paper and Birch-Bark," *Journal of the Asiatic Society of Bengal* 69 (1900): 93–134.

19 Dominik Wujastyk, "Indian Manuscripts," in *Manuscript Cultures: Mapping the Field*, eds. Jörg B. Quenzer, Dmitry Bondarev, and Jan-Ulrich Sobisch (Berlin: De Gruyter, 2014), 166.

held together by one or two threads passed through the leaves. In the case of a single thread, it typically passes through the leaves at a distance of between one fifth and one third of the width of the folio from the beginning of lines, but there are also occasional examples of a thread through the center of the folio. No wooden covers or cloth wrappers are preserved among the fragmentary remains of the earliest palm-leaf manuscripts.

3.1 *Central Asia*

While both the Gandhāran writing tradition based on birch-bark scrolls and the Indian one based on palm leaves were exported along the Silk Roads to Central Asia, only the palm-leaf format gained wide currency there and was later imitated in paper.[20] Our sole example of a birch-bark scroll found in Central Asia is a second- or third-century Gāndhārī *Dharmapada* manuscript (CKM 77), discovered in 1892 outside the town of Khotan and now preserved in two parts in Paris and St Petersburg (another third of the manuscript, apparently intact on discovery, is now lost). The Khotan *Dharmapada* scroll differs from the typical Gandhāran scroll format described above by its great length of approximately five meters, and it has been suggested that maybe it was produced locally rather than having been imported. The language of the manuscript does share some peculiarities with the Gāndhārī of the Niya documents,[21] but this cannot be taken as evidence that it was produced in Central Asia, since both the *Dharmapada* and the documents may simply represent the dialect of the particular group of Gāndhārī speakers that settled in the Krorayina kingdom and neighboring Khotan.[22] At least the writing material birch bark was not locally available in Khotan and must have been imported,[23] but as shown by Dieter Schlingloff,[24] birch trees grow not only in Kashmir, but also in the Tian Shan mountain range which could have served as a source for the *Dharmapada*'s writing support as well as later Central Asian birch-bark *pothi*s. The question could potentially be answered if samples of Central Asian birch-bark manuscripts received the kind of material analysis that has recently been applied to Indian palm-leaf manuscripts found in Tibet and Nepal.[25]

20 In addition to these two literary manuscript types, a Central Asian tradition of administrative documents on wood, especially prominent in the early Krorayina and Kucha kingdoms, appears to be based on Indian models described in the *Arthaśāstra*.

21 For instance, the assimilation of original [nt] to [nd] and of original [nd] to [nn].

22 T. Burrow, "The Dialectical Position of the Niya Prakrit," *Bulletin of the School of Oriental Studies, University of London* 8 (1936): 419–435.

23 Sander, "Early Prakrit and Sanskrit Manuscripts," 69.

24 Cf. Lore Sander, "The Earliest Manuscripts," 137.

25 Martin Delhey, Emanuel Kindzorra, Oliver Hahn, and Ira Rabin, "Material Analysis of

Until very recently, the Central Asian manuscript finds of the early twentieth century provided our oldest examples of palm-leaf manuscripts from anywhere in the Indic world (including the celebrated fragments of Aśvaghoṣa's dramas).[26] What is more, for *pothi*-format manuscripts found in Central Asia, palm leaf is the exclusive writing material up to the fifth century CE (with the single exception of the *Subashi Udānavarga* on wood).[27] The oldest Central Asian *pothi* on birch bark, found in Kucha and containing the metrical treatise *Chandoviciti*, was probably imported from Kashmir,[28] though again a material analysis would be needed to rule out an origin of the writing material in the Tian Shan. Also in the fifth century, Chinese paper begins to be used for *pothi* manuscripts.[29] It is not entirely clear why, for more than two hundred years, almost no other writing material than the clearly imported palm leaf was used for Buddhist texts. Lore Sander suggested[30] that particular sanctity attached to the prototypically Indian palm leaf (similar to the use of birch-bark scrolls for amulets in India after the general demise of the format in Gandhāra), making it the preferred medium for Buddhist 'missionaries' that brought with them not only finished manuscripts but also empty palm-leaf stock, just as within India southern palm leaves were brought to the north for manuscript production.[31]

The Indian palm-leaf fragments discovered on the northern and southern Silk Roads belong to at least thirty distinct manuscripts. Between four and seven of these, found in the area of Kucha and dating to approximately the third or fourth century CE, were written in Kharoṣṭhī script and Sanskritized Gāndhārī language and appear to have contained Buddhist narrative (possibly

---

Sanskrit Palm-Leaf Manuscripts Preserved in Nepal," *Journal of the International Association of Buddhist Studies* 36/37 (2013–2014): 119–152. The authors suggest (144) that in their case—manuscripts from the monastery of Vikramaśīla and Nepal—empty palm-leaf folios were transported together with finished manuscripts from one monastery to another.

26 Heinrich Lüders, *Bruchstücke buddhistischer Dramen* (Berlin: Georg Reimers, 1911).

27 Hideaki Nakatani, *Udānavarga de Subaši: edition critique du manuscrit sanskrit sur bois provenant de Subaši : Bibliothèque nationale de Paris, fonds Pelliot* (Paris: Edition-Diffusion de Boccard, 1987).—Nakatani Hideaki [中谷英明], スバシ写本の研究: 亀茲国致隷藍の『ウダーナ・ヴァルガ』[Subashi shahon no kenkyū: Kucha Kocchi Reiran no "Udāna·varuga"] (京都 [Kyōto]: 人文書院 [Jinbun shoin], 1988).

28 Sander, "The Earliest Manuscripts," 137–138.

29 Sander, "Early Prakrit and Sanskrit Manuscripts," 83.

30 Sander, "The Earliest Manuscripts," 138.—Sander, "Early Prakrit and Sanskrit Manuscripts," 78.

31 Lore Sander, *Paläographisches zu den Sanskrithandschriften der Berliner Turfansammlung* (Wiesbaden: Franz Steiner Verlag, 1968), 25.

allegorical) texts.[32] The remainder of the Central Asian palm-leaf manuscripts were written in Brāhmī script and Sanskrit language and received a detailed analysis by Lore Sander.[33] One of her main findings was the absence of Sūtra and Vinaya texts among the early Sanskrit manuscripts from Central Asia (third to fifth century CE),[34] and their almost complete absence among the Central Asian finds of palm-leaf manuscripts.[35] Instead, what we see among the early Central Asian palm-leaf manuscripts is a preponderance of scholastic treatises, court poetry and the *Udānavarga* verse collection.[36] Sander first of all takes this imbalance as confirmation that the *bhāṇaka* system remained intact until the fifth century (as also reported by the Chinese pilgrim Fǎxiǎn), providing an unbroken oral transmission for Sūtra and Vinaya and making manuscripts optional.[37] She further suggests that Kāvya and (I would submit to a lesser degree) Abhidharma were particularly useful for winning the interest and support of local Central Asian elites, and that the *Dharmapada* or *Udānavarga* literature served a similar function with respect to the local laity, comparing how Buddhism first entered China.[38] Her argument remains convincing, especially when considered on a background of general scarcity of the preferred writing material palm leaf in Central Asia (as evidenced by numerous palimpsests).[39]

3.2 *Bamiyan*

The spectacular discovery of a Buddhist manuscript trove at Bamiyan in 1993–1995, apparently a deposit of discarded manuscripts from a Buddhist monastic

32 Richard Salomon, "Kharoṣṭhī Manuscript Fragments in the Pelliot Collection, Bibliothèque nationale de France," *Bulletin d'études indiennes* 16 (1998): 123–160.

33 Sander, "The Earliest Manuscripts."—Sander, "Early Prakrit and Sanskrit Manuscripts."

34 Sander herself dated the earliest manuscripts found in Central Asia to the second century CE on paleographic grounds. On the strength of more recent paleographic arguments—Eli Franco, *The Spitzer Manuscript: The Oldest Philosophical Manuscript in Sanskrit* (Wien: Verlag der Österreichischen Akademie der Wissenschaften, 2004), 29–33—and radiocarbon dating—Eli Franco, "Three Notes on the Spitzer Manuscript," *Wiener Zeitschrift für die Kunde Südasiens* 49 (2005): 109–110—this date probably needs to be adjusted to the third century.

35 The one exception is a *Dharmaguptaka Prātimokṣasūtra* on palm leaf dating from the Gupta period: Sander, "The Earliest Manuscripts," 140. Sūtra manuscripts written on paper appear in large numbers from the Gupta period onward.

36 Sander, "Early Prakrit and Sanskrit Manuscripts," 79.

37 Sander, "The Earliest Manuscripts," 141.

38 Sander, "Early Prakrit and Sanskrit Manuscripts," 80.

39 I remain somewhat less convinced by her attempt to seek a historical explanation for the high manuscript profile of Abhidharma and Kāvya texts in the Council of Kashmir (apocryphally connected with the poet Aśvaghoṣa) and an ensuing Sarvāstivāda mission that chose to emphasize its foundational school texts: Sander, "The Earliest Manuscripts," 137.

library, now takes us one step further back to the historical origin of the *pothi* format. This is obviously true geographically (with one of the main trade routes from India to the north and west running through the Bamiyan valley), but also chronologically. The Bamiyan fragments (numbering around 5,000 substantial and 7,000 smaller ones from an approximate number of 1,000 original manuscripts) date from the second or third to the seventh century CE,[40] making the oldest ones about a hundred years older than those found in Central Asia. But even more important is the fact that in the oldest stratum of the Bamiyan find, Gāndhārī fragments in Kharoṣṭhī script exist side by side with Sanskrit fragments in Kuṣāṇa Brāhmī script. Both are written on palm leaves with precisely corresponding manuscript formats and scribal conventions, presenting the end result of a meeting and amalgamation of the early Gandhāran and mainland Indian textual traditions, and thus a crucial link in the early Buddhist textual transmission from India to Central Asia.

The first thing to note is that the palm-leaf manuscripts from Bamiyan—like the Gandhāran birch-bark scrolls, but unlike the early Central Asian palm-leaf fragments—preserve the full gamut of Buddhist literary genres, from Sūtra and Vinaya texts to scholastic treatises, Mahāyāna Sūtras and collections of stories (see table 18.1). The four volumes of editions of the Bamiyan fragments so far (BMSC I–IV) contain the remains of approximately thirty-two palm-leaf manuscripts, using two languages (Gāndhārī and Sanskrit) and five different scripts.

In the earliest layer (third to fourth century) six Kharoṣṭhī manuscripts (*Mahāparinirvāṇasūtra*, *Ekottarikāgama*, *Bhadrakalpikasūtra*, *Bodhisattvapiṭakasūtra*, *Sarvapuṇyasamuccayasamādhisūtra* and an 'auto-Stotra') outnumber Brāhmī manuscripts in a variety of Kuṣāṇa hands (*Aṣṭasāhasrikā*, *Śāriputrābhidharma* and an unidentified commentary). The manuscripts in both scripts follow very much the same format, however. Folios were from 2.5 to 5 cm high and, as far as we can tell, up to 40 cm wide; they contained between three and five lines of writing.[41] Folios are numbered on the *recto* in the first margin according to the direction of writing (the right margin for Kharoṣṭhī, the left margin for Brāhmī), and even the earliest Bamiyan palm-leaf manuscripts were very voluminous indeed: the highest preserved folio number is 248 on a

40 Lore Sander, "An Analysis of the Scripts," in *Traces of Gandhāran Buddhism: An Exhibition of Ancient Buddhist Manuscripts in the Schøyen Collection*, eds. Jens Braarvig and Fredrik Liland (Oslo: Hermes Publishing, 2010), xxx–xxxii.

41 The two narrowest manuscripts (2.5 cm high with 3 lines of text) are both in Kharoṣṭhī, but it is not clear that this is statistically significant.

TABLE 18.1 Palm-leaf manuscripts from Bamiyan (number of folios, folio format in cm, number of lines per folio, script; Kuṣāṇa Brāhmī = 3rd–4th century, Kharoṣṭhī = 3rd–4th century, Western Gupta Brāhmī = 4th century, Northwestern Gupta Brāhmī = 5th century, Gilgit-Bamiyan type I = 6th–7th century)[42]

Commentary	?	? × 4.8	4–5	Kuṣāṇa	BMSC II
Śāriputrābhidharma	160+	? × 4.3	4	Kuṣāṇa	BMSC II
Aṣṭasāhasrikā	248+	40 × 5	4–5	Kuṣāṇa	BMSC I, II
Mahāparinirvāṇasūtra (I)	?	26 × 2.5	3	Kharoṣṭhī	BMSC I
Ekottarikāgama	?	40 × 2.5	3–4	Kharoṣṭhī	BMSC IV
Bhadrakalpikasūtra	62+	40 × 3.5	5	Kharoṣṭhī	BMSC IV
Bodhisattvapiṭakasūtra (I)	?	? ×4.3	5	Kharoṣṭhī	BMSC IV
Sarvapuṇyasamuccayasamādhi	41+	? × ?	4 or 5	Kharoṣṭhī	BMSC IV
Auto-Stotra	?	?	5	Kharoṣṭhī	BMSC IV
Caṅgīsūtra	?	? × 5	6	W Gupta	BMSC I, II
Prātimokṣavibhaṅga (I)	?	? × 4.6	6	W Gupta	BMSC III, IV
Vinayadhara	80+	17.6 × 3.5	5	NW Gupta	BMSC III
Bodhisattvapiṭakasūtra (II)	?	45 × 4.5	5	NW Gupta	BMSC III
Mahāyānasūtra Collection (I)[43]	550+	38 × 3.5	4	NW Gupta	BMSC I
Huviṣka legend	121+	? × 3.4	3	NW Gupta	BMSC II
Prasādapratibhodbhava	?	? × 3.6	4	NW Gupta	BMSC II
Play	?	? × 3.9	6	NW Gupta	BMSC III
Mahāparinirvāṇasūtra (II)	?	? × 3	3	GB type I	BMSC II
Mahāparinirvāṇasūtra (III)	?	? × 3	4	GB type I	BMSC II
Śalyasūtra	?	?	5–6	GB type I	BMSC IV
Śikhālakasūtra	?	? × 4.9	5–6	GB type I	BMSC III
Ityuktaka	?	?	5	GB type I	BMSC IV
Prātimokṣavibhaṅga (II)	140+	? × 3.5	6	GB type I	BMSC I, II
Karmavācanā	120+	24.5 × 3.5	4	GB type I	BMSC II, III
Mahāsamājasūtra Comm.	?	? × 4.5	7	GB type I	BMSC III
Candrottaradārikāvyākaraṇa	25+	20 × 2.5	3	GB type I	BMSC II
Ratnaketuparivarta	48+	?	5	GB type I	BMSC IV
Saddharmapuṇḍarīkasūtra (I)	?	? × 3.8	6	GB type I	BMSC II
Saddharmapuṇḍarīkasūtra (II)	81+	? × 3	5	GB type I	BMSC II

42 Cf. Lore Sander, "A Brief Paleographical Analysis of the Brāhmī Manuscripts in Volume I," in BMSC I, 285–300.

43 *Śrīmālādevīsiṃhanādanirdeśa, Pravāraṇāsūtra, Sarvadharmāpravṛttinirdeśa, Ajātaśatru-kaukṛtyavinodanāsūtra.*

TABLE 18.1 Palm-leaf manuscripts from Bamiyan (*cont.*)

Mahāyānasūtra Collection (II)[44]	198+	54×4.7	5	GB type I	BMSC II, IV
Aśoka legend	?	? × 3.5	4–5	GB type I	BMSC I
Varṇārhavarṇa	?	? × 2.5	4	GB type I	BMSC II

fragment of the *Aṣṭasāhasrikā*, but judging from the available parallels in other languages, several of the early texts would have taken up well above five hundred folios in total. Manuscripts in both languages have typically one string hole near the beginnings of lines (near the right-hand side for Kharoṣṭhī and the left-hand side for Brāhmī). Last not least, the system of punctuation marks used for both scripts is the same, consisting primarily of small dots to separate phrases and a peculiar crescent-with-prong shape to mark the end of sections (with orientation according to the direction of writing). In addition, verses and other recurring textual units are marked by number signs at their end.

Among the Bamiyan palm-leaf manuscripts are a number of palimpsests, including cases where the original and the new texts were both in the same script (CKM 105)[45] and cases where an original text in Kharoṣṭhī script and Gāndhārī language was replaced by a new text in Brāhmī script and Sanskrit language (fragment MS 2376/99).[46] There are no known palimpsests among the Gandhāran birch-barks scrolls, presumably partly because the fragility of birch bark makes erasure difficult, but also because fresh writing material was readily available in Gandhāra. In Bamiyan, on the other hand, palm leaves were a rare import commodity and had to be made maximum use of.[47]

As in our discussion of the Gandhāran scrolls, a closer look at particular Bamiyan *pothi* manuscripts will serve to bring out further properties of the

44 *Samādhirājasūtra, Ratneketuparivarta.*

45 Richard Salomon, "Thirty-Two Fragments Written by Bamiyan Kharoṣṭhī Scribe 7," in BMSC IV, 367–406.

46 Matsuda Kazunobu [松田和信], "平山コレクションのガンダーラ語貝葉写本断簡について" [Hirayama korekushon no Gandārago baiyo shahon dankan ni tsuite], 印度學佛教學研究 [Indogaku bukkyōgaku kenkyū] 62 (2013): 354–346.

47 Lore Sander, "Dating and Localizing Undated Manuscripts," in *From Birch-Bark to Digital Data: Recent Advances in Buddhist Manuscript Research: Papers Presented at the Conference 'Indic Buddhist Manuscripts: The State of the Field,' Stanford, June 15–19 2009*, eds. Paul Harrison and Jens-Uwe Hartmann (Wien: Verlag der Österreichischen Akademie der Wissenschaften, 2014), 171–186.

writing tradition and use context that they represent (compare table 18.1). To start with, the Bamiyan Gāndhārī *Ekottarikāgama* manuscript[48] stands in stark contrast to the short written 'snapshots' of Sūtras taken from a primarily oral tradition in first- and second-century Gandhāra. The fragments of the Bamiyan *Ekottarikāgama* are from throughout the overall collection (when compared to the Pali *Aṅguttaranikāya*), and even though no folio numbers happen to be preserved, the proven existence of other Bamiyan palm-leaf manuscripts of over five hundred folios makes it very likely that the Ekottarikāgama, too, was intended to be a complete written redaction of that collection of Sūtras. This would make it the earliest known case of a complete Āgama in written form. We have no means of knowing whether such a feat had been attempted before in the northwest or elsewhere (*pace* the accounts of the Pali chronicles of the writing down of the canon under king Vaṭṭagāmaṇī Abhaya—chronicles that happen to be contemporary with the Bamiyan *Ekottarikāgama* manuscript), nor do we know what prompted the production of this particular manuscript—a monastery concerned about a possible interruption of the oral tradition, a local ruler commissioning the manuscript as a technical *tour de force* and source of merit, or yet another reason.

Comparable to the *Ekottarikāgama* is the Bamiyan manuscript of a Gāndhārī version of the *Bhadrakalpikasūtra* (CKM 128).[49] Here too, the fifty-one fragments that we have are from throughout the overall text, on the basis of a comparison with Dharmarakṣa's near-contemporary Chinese translation and with the ninth-century Tibetan translation. One folio number (probably 62) is preserved, and the complete manuscript would have covered more than three hundred folios. The *Bhadrakalpikasūtra*'s comprehensive description of all 1,004 buddhas of our fortunate eon (not to mention its treatment of 350 sets of six perfections) stands in contrast to the much more modest treatment of fifteen buddhas of the past and present in the Library of Congress birchbark scroll (CKM 261) and to comparable accounts in Pali and Sanskrit Buddhist literature (such as the *Mahāvadānasūtra*). Where in the case of the *Ekottarikāgama*, we have a pre-existing text collection assuming physical form, the *Bhadrakalpika* represents the expansion of of an existing scheme to truly cosmical proportions. While the precise motivation, as always, escapes us, here we

48 Chanida Jantrasrisalai, Timothy Lenz, Lin Qian, and Richard Salomon, "Fragments of an Ekottarikāgama Manuscript in Gāndhārī," in BMSC IV, 1–122.

49 Stefan Baums, Andrew Glass, and Kazunobu Matsuda, "Fragments of a Gāndhārī Version of the Bhadrakalpikasūtra," in BMSC IV, 183–266.

seem to be dealing with a genuine intellectual turn represented in (and facilitated by) textual practice.[50]

From the end of the fourth century onwards, only Brāhmī was used at Bamiyan. Two manuscripts (*Caṅgīsūtra* and *Prātimokṣavibhaṅga*) are in an unusual angular Western Gupta style of Brāhmī which, like the earlier Kuṣāṇa Brāhmi fragments, leaves open the question of whether they were produced locally in Bamiyan or imported. This contrasts with a further six manuscripts in the local Northwestern Gupta script that began to take shape in the fifth century, and another fifteen in its local successor script Gilgit Bamiyan type I (sixth to seventh century).[51] Folio sizes stayed, as far as we can tell, stable throughout this later period, but we find a number of manuscripts that fit up to six (and in one case seven) lines into the same space. Another new development during this second phase of the Bamiyan palm-leaf manuscript tradition are new physical collections of related texts—the type of *Sammelhandschriften* that we know from Central Asia. Two cases from Bamiyan involve the bundling of several Mahāyānasūtras—at least four in a fifth-century manuscript, and at least two in a sixth/seventh-century manuscript (see table 18.1). The earlier of these two manuscripts preserves the largest folio number (549) of any of the Bamiyan manuscripts.

4 Conclusions

In the preceding, I have outlined the characteristics of three distinct Buddhist manuscript traditions and textualities:[52] those of Gandhāra, Bamiyan and Central Asia. It remains to recapitulate and to consider their possible historical connections.

The manuscript tradition of early Gandhāra (ca. third century BCE to second century CE) is characterized by birch-bark scrolls, Gāndhārī language

---

50 Peter Skilling, "Notes on the *Bhadrakalpika-sūtra*," 創価大学国際仏教学高等研究所年報 [Sōka daigaku kokusai bukkyōgaku kōtō kenkyūjo nenpō] 13 (2010): 201, in connection with the *Bhadrakalpikasūtra*, colorfully evokes an "'age of exploration' of time and space that led to colonizations of the cognitive map—planting flags of the Dharma […] across numberless universes by creative action of the spirit."

51 From the seventh century onwards and until the closure of the Bamiyan manuscript deposit, the translocal Gilgit-Bamiyan type II (a variety of Siddhamātṛkā) is used, and palm leaf is replaced by imitation *pothis* in birch bark that are outside the scope of this paper.

52 The combined use of texts in any medium—written or oral—by a textual community.

and Kharoṣṭhī script, though at least the last two of these were not used invariably (we have Sanskrit texts in Kharoṣṭhī script, and we have one small Kuṣāṇa Brāhmī scroll containing extracts from a Sanskrit Sūtra). The Gandhāran scroll was derived from the Aramaic scribal tradition introduced by the Achaemenids, it remained a local tradition, and it reduced the full range of Buddhist genres to writing, but not exhaustively. Writing occurred in tandem with a continuing strong oral tradition.

In Bamiyan, one century later and five hundred kilometers to the west, we encounter an entirely different manuscript tradition (ca. third to seventh centuries CE) characterized by palm-leaf *pothi* manuscripts, Sanskrit language and Brāhmī script, but with a significant component of early Kharoṣṭhī manuscripts. Neither palm leaf nor birch bark were locally available as writing material, and the exclusive use of south Indian palm leaf (rather than the more readily importable birch bark) is historically significant. It suggests that the Buddhist textual community that established itself in Bamiyan did not come from neighboring Gandhāra, but rather from some place in mainland India, and that it brought its established manuscript tradition based on palm leaves with it. All textual indications from the manuscripts themselves (including the *Prātimokṣavibhaṅgas*) point to a Mahāsāṃghika community,[53] which agrees well with the Mahāsāṃghika trajectory reflected in the epigraphic record, from first-century Mathura (the Lion Capital) to third- and fourth-century Termez (numerous potsherds).

This means that the Kharoṣṭhī palm-leaf manuscripts from Bamiyan cannot be considered a direct evolution from the earlier birch-bark scrolls, adopted by the same community due to their technical superiority. What we have instead is a scenario where an immigrant textual community established itself as a neighbor of the Gandhāran tradition, beginning to interact with and absorb some (though doubtless not all) of the textual production of Gandhāra. Gāndhārī texts were copied onto the same palm-leaf manuscripts that the Bamiyan immigrant community used for their own Sanskrit texts, and eventually were either translated into Sanskrit (such as the *Bodhisattvapiṭakasūtra*, preserved in Bamiyan in both languages) or fell out of use. We have, unfortunately, no contemporary data of the reverse influence that the Bamiyan textual commu-

53 Sander, "Dating," 174, Vincent Tournier, "Protective Verses for Travellers: A Fragment of the Diśāsauvastikagāthās Related to the Scriptures of the Mahāsāṃghika-Lokottaravādins," in BMSC IV, 407–437 and Vincent Tournier, *La formation du Mahāvastu et la mise en place des conceptions relatives à la carrière du bodhisattva* (Paris: École française d'Extrême-Orient, 2017), 51 and *passim*.

nity must have exerted on Gandhāra, but know from the later production of birch-bark *pothis* in Gilgit that the influence did in fact go both ways.

In Bamiyan as in Gandhāra, we have a full range of Buddhist literary genres, but not so, as we have seen, in Central Asia. Both the Gandhāran scroll tradition and the Indian palm-leaf tradition we see at Bamiyan left their marks in Central Asia, as sporadic exports in the case of the former, and as a sustained cultural transfer and eventual imitation (adopting Chinese paper) in the case of the latter. On the background of the rich Gandhāran and Bamiyan traditions, the absence of early Sūtra and Vinaya manuscripts in Central Asia cannot be explained by lack of precedence for committing these genres to writing. Rather, we should look to the particular conditions of the early Buddhist inroads into Central Asia, which must have been characterized by individual initiative, difficulty of transport and a limited amount of manuscripts and writing material that could be carried. The natural outcome would have been to economize in the transportation and production of written texts, relying on oral transmission wherever most feasible, such as especially in the case of Sūtra and Vinaya.

We are left with a new recognition of the central role of the Bamiyan Buddhist community. After an initial coexistence of local and translocal scripts and languages, this community created a new textuality for Greater Gandhāra and shaped its further development in Central Asia. This textuality was based on the translocal Sanskrit language and Brāhmī script, opening up communication with wider parts of the Indian Buddhist world. But the general adoption of the imported *pothi* format in Greater Gandhāra had a conditioning effect of its own that should not be underestimated. The amount of text that can be committed to a scroll is comparatively limited—the Bajaur *Mahāyānasūtra* contains approximately 24,000 *akṣaras* in 600 lines—and can only be accessed sequentially. A typical *pothi*, on the other hand, would contain around 400,000 *akṣaras* on 500 folios with five lines per page—more than fifteen times the amount of text. Combined with the feature of random access (the ability to jump directly to any desired place in the manuscript), the *pothi* format introduced Greater Gandhāra and Central Asia to entirely new ways of using writing to propagate and interpret the word of the Buddha.[54]

---

54 It is curious that the Gandhāran tradition never developed random-access manuscripts on the basis of its indigenous scroll format. The Chinese manuscript tradition illustrates such a gradual evolution in the invention of the concertina format that in East Asia lives on to this day. From Gandhāra, we have two relic inscriptions on linked copper plates (CKI 442 and 564) that in essence are in concertina format, but the idea was apparently never transferred to manuscripts. In this way, too, the introduction of the *pothi* provided a true disruption from outside.

References

Allon, Mark. "The Senior Kharoṣṭhī Manuscripts." In *From Birch-Bark to Digital Data: Recent Advances in Buddhist Manuscript Research: Papers Presented at the Conference 'Indic Buddhist Manuscripts: The State of the Field,' Stanford, June 15–19 2009*, edited by Paul Harrison, and Jens-Uwe Hartmann, 19–33. Wien: Verlag der Österreichischen Akademie der Wissenschaften, 2014.

Baums, Stefan. "A Gāndhārī Commentary on Early Buddhist Verses: British Library Kharoṣṭhī Fragments 7, 9, 13 and 18." PhD diss., University of Washington, 2009.

Baums, Stefan. "Gandhāran Scrolls: Rediscovering an Ancient Manuscript Type." In *Manuscript Cultures: Mapping the Field*, edited by Jörg B. Quenzer, Dmitry Bondarev, and Jan-Ulrich Sobisch, 183–225. Berlin: De Gruyter, 2014.

Baums, Stefan, Andrew Glass, and Kazunobu Matsuda. "Fragments of a Gāndhārī Version of the Bhadrakalpikasūtra." In BMSC IV, 183–266.

BMSC I = Jens Braarvig, ed. *Buddhist Manuscripts in the Schøyen Collection, Volume I*. Oslo: Hermes Publishing, 2000.

BMSC II = Jens Braarvig, ed. *Buddhist Manuscripts in the Schøyen Collection, Volume II*. Oslo: Hermes Publishing, 2002.

BMSC III = Jens Braarvig, ed. *Buddhist Manuscripts in the Schøyen Collection, Volume III*. Oslo: Hermes Publishing, 2006.

BMSC IV = Jens Braarvig, ed. *Buddhist Manuscripts in the Schøyen Collection, Volume IV*. Oslo: Hermes Publishing, 2016.

Bühler, G. *Detailed Report of a Tour in Search of Sanskrit MSS. Made in Kaśmîr, Rajputana, and Central India*. Bombay: [Royal Asiatic] Society's Library, 1877.

Burrow, T. "The Dialectical Position of the Niya Prakrit." *Bulletin of the School of Oriental Studies, University of London* 8 (1936): 419–435.

CKI = Stefan Baums and Andrew Glass. 2002–. *Catalog of Gāndhārī Texts*. https://gandhari.org/catalog. Part II.

CKM = Stefan Baums and Andrew Glass. 2002–. *Catalog of Gāndhārī Texts*. https://gandhari.org/catalog. Part I.

Coningham, R.A.E., F.R. Allchin, C.M. Batt, and D. Lucy. "Passage to India? Anuradhapura and the Early Use of the Brahmi Script." *Cambridge Archaeological Journal* 6 (1996): 73–97.

Delhey, Martin, Emanuel Kindzorra, Oliver Hahn, and Ira Rabin. "Material Analysis of Sanskrit Palm-Leaf Manuscripts Preserved in Nepal." *Journal of the International Association of Buddhist Studies* 36/37 (2013–2014): 119–152.

Diringer, David. *The Hand-Produced Book*. London: Hutchinson's Scientific and Technical Publications, 1953.

Falk, Harry. *Schrift im alten Indien: Ein Forschungsbericht mit Anmerkungen*. Tübingen: Gunter Narr Verlag, 1993.

Falk, Harry and Ingo Strauch. "The Bajaur and Split Collections of Kharoṣṭhī Manuscripts within the Context of Buddhist Gāndhārī Literature." In *From Birch-Bark to Digital Data: Recent Advances in Buddhist Manuscript Research: Papers Presented at the Conference 'Indic Buddhist Manuscripts: The State of the Field,' Stanford, June 15–19 2009*, edited by Paul Harrison and Jens-Uwe Hartmann, 51–78. Wien: Verlag der Österreichischen Akademie der Wissenschaften, 2014.

Franco, Eli. *The Spitzer Manuscript: The Oldest Philosophical Manuscript in Sanskrit*. Wien: Verlag der Österreichischen Akademie der Wissenschaften, 2004.

Franco, Eli. "Three Notes on the Spitzer Manuscript." *Wiener Zeitschrift für die Kunde Südasiens* 49 (2005): 109–111.

Hinüber, Oskar von. *Der Beginn der Schrift und frühe Schriftlichkeit in Indien*. Mainz: Akademie der Wissenschaften und der Literatur, 1990.

Hinüber, Oskar von. "The Gilgit Manuscripts: An Ancient Buddhist Library in Modern Research." In *From Birch-Bark to Digital Data: Recent Advances in Buddhist Manuscript Research: Papers Presented at the Conference 'Indic Buddhist Manuscripts: The State of the Field,' Stanford, June 15–19 2009*, edited by Paul Harrison and Jens-Uwe Hartmann, 79–136. Wien: Verlag der Österreichischen Akademie der Wissenschaften, 2014.

Hoernle, A.F. Rudolf. "An Epigraphical Note on Palm-Leaf, Paper and Birch-Bark." *Journal of the Asiatic Society of Bengal* 69 (1900): 93–134.

Janert, Klaus Ludwig. 1972. *Abstände und Schlußvokalverzeichnungen in Aśoka-Inschriften*. Wiesbaden: Franz Steiner Verlag.

Jantrasrisalai, Chanida, Timothy Lenz, Lin Qian, and Richard Salomon. "Fragments of an Ekottarikāgama Manuscript in Gāndhārī." In BMSC IV, 1–122.

Lüders, Heinrich. *Bruchstücke buddhistischer Dramen*. Berlin: Georg Reimers, 1911.

Matsuda Kazunobu [松田和信]. "平山コレクションのガンダーラ語貝葉写本断簡について" [Hirayama korekushon no Gandārago baiyo shahon dankan ni tsuite]. 印度學佛教學研究 [Indogaku bukkyōgaku kenkyū] 62 (2013): 354–346.

Nakatani, Hideaki. *Udānavarga de Subaši : edition critique du manuscrit sanskrit sur bois provenant de Subaši: Bibliothèque nationale de Paris, fonds Pelliot*. Paris: Edition-Diffusion de Boccard, 1987.

Nakatani, Hideaki. スバシ写本の研究: 亀玆国致隸藍の『ウダーナ・ヴァルガ』[Subashi shahon no kenkyū: Kucha Kocchi Reiran no "Udāna·varuga"]. 京都 [Kyōto]: 人文書院 [Jinbun shoin], 1988.

Naveh, Joseph and Shaul Shaked. *Aramaic Documents from Ancient Bactria (Fourth Century BCE)*. London: The Khalili Family Trust, 2012.

Rocher, Ludo. *Studies in Hindu Law and Dharmaśāstra*. London: Anthem Press, 2012.

Salomon, Richard. "Kharoṣṭhī Manuscript Fragments in the Pelliot Collection, Bibliothèque nationale de France." *Bulletin d'études indiennes* 16 (1998): 123–160.

Salomon, Richard. "Thirty-Two Fragments Written by Bamiyan Kharoṣṭhī Scribe 7." In BMSC IV, 367–406.

Sander, Lore. *Paläographisches zu den Sanskrithandschriften der Berliner Turfansammlung*. Wiesbaden: Franz Steiner Verlag, 1968.

Sander, Lore. "The Earliest Manuscripts from Central Asia and the Sarvāstivāda Mission." In *Corolla Iranica: Papers in Honour of Prof. Dr. David Neil MacKenzie on the Occasion of His 65th Birthday on April 8th, 1991*, edited by Ronald E. Emmerick and Dieter Weber, 133–150. Frankfurt am Main: Peter Lang, 1991.

Sander, Lore. "Early Prakrit and Sanskrit Manuscripts from Xinjiang (Second to Fifth / Sixth Centuries C.E.): Paleography, Literary Evidence, and Their Relation to Buddhist Schools." In *Collection of Essays 1993: Buddhism across Boundaries: Chinese Buddhism and the Western Regions*, edited by Jan Nattier and John R. McRae, 61–106. Sanchung: Fo Guang Shan Foundation for Buddhist & Culture Education, 1999.

Sander, Lore. "A Brief Paleographical Analysis of the Brāhmī Manuscripts in Volume I." In BMSC I, 285–300.

Sander, Lore. "An Analysis of the Scripts." In *Traces of Gandhāran Buddhism: An Exhibition of Ancient Buddhist Manuscripts in the Schøyen Collection*, edited by Jens Braarvig and Fredrik Liland, xxx–xxxii. Oslo: Hermes Publishing, 2010.

Sander, Lore. "Dating and Localizing Undated Manuscripts." In *From Birch-Bark to Digital Data: Recent Advances in Buddhist Manuscript Research: Papers Presented at the Conference 'Indic Buddhist Manuscripts: The State of the Field,' Stanford, June 15–19 2009*, edited by Paul Harrison and Jens-Uwe Hartmann, 171–186. Wien: Verlag der Österreichischen Akademie der Wissenschaften, 2014.

Skilling, Peter. "Notes on the *Bhadrakalpika-sūtra*." 創価大学国際仏教学高等研究所年報 [Sōka daigaku kokusai bukkyōgaku kōtō kenkyūjo nenpō] 13 (2010): 195–229.

Tournier, Vincent. "Protective Verses for Travellers: A Fragment of the Diśāsauvastikagāthās Related to the Scriptures of the Mahāsāṃghika-Lokottaravādins." In BMSC IV, 407–437.

Tournier, Vincent. *La formation du Mahāvastu et la mise en place des conceptions relatives à la carrière du bodhisattva*. Paris: École française d'Extrême-Orient, 2017.

Wujastyk, Dominik. "Indian Manuscripts." In *Manuscript Cultures: Mapping the Field*, edited by Jörg B. Quenzer, Dmitry Bondarev, and Jan-Ulrich Sobisch, 159–181. Berlin: De Gruyter, 2014.

CHAPTER 19

Indian Physicians in Greco-Roman Literature?

Klaus Karttunen

The question of possible contacts between Greek and Indian medical traditions has long been discussed. There are certain unquestionable similarities in the Hippocratic system of humors and the comparable Āyurvedic system. Many have tried to show a common origin, either in India[1] or in the West.[2] A closer look, however, reveals that there are also important differences. The Indians have bile (*pitta*, corresponding to yellow bile in the Greek system), phlegm (*kapha* or *śleṣman*) and air (*vāta* or *prāṇa*), while the Greek system contains yellow bile (*kholḗ*), black bile (*mélan kholḗ*—completely unknown in India), phlegm (*phlégma*) and blood (*haîma*). Although there is some variation in both systems, the parallelisms are insufficient to establish proof of contact.[3] During the 20th century it has also become clear that real parallels can be due to the influence of ancient Mesopotamia on both systems.[4] There is very little evi-

1 In this case usually accompanied by exaggerated dates for the early Āyurvedic texts. However, the Indian system as we know it cannot be dated with any reliability before the early centuries CE, although some of its roots certainly go back to much earlier times. See K.G. Zysk, *Asceticism and Healing in Ancient India: Medicine in the Buddhist Monastery* (Indian edition. Delhi: Motilal Banarsidass, 1997).

2 Perhaps the most extreme attempt was the one made by Ernst Haas in the 1870s. He preferred very late dates (10th to 15th century) for the then known classics of Āyurveda (*Caraka* and *Suśruta*) and concluded that the whole system arose from Muslim influence. Soon the discovery of the Bower manuscript showed his theory to be untenable. On Haas, see E. Windisch, *Geschichte der Sanskrit-Philologie und indischen Altertumskunde*, part 2 (Berlin and Leipzig: Vereinigung wissenschaftlicher Verleger, 1920), 362 f.

3 This statement is partly based on a comment sent by D. Wujastyk to the INDOLOGY Mailing List on 3.9.1999. He mentions that Suśruta and some of his commentators accept blood as a humor, and that it also took some time before the Greek system was established (by Galen). To this we may add that though Plato once lists *pneûma*, *kholḗ* and *phlégma* (*Timaeus* 84cd), corresponding to the Indian system, they are not necessarily humors, even though presented as causing certain diseases. The early Buddhist *Suttanipāta* 3, 2, 433–434 lists blood alongside air, bile and phlegm, but not in the context of medicine. See also J. Filliozat, *The Classical Doctrine of Indian Medicine*, trans. by D.R. Chamana (Delhi: Munshiram Manoharlal, 1964), 229 ff. on the *Timaeus*.

4 Such parallels can, for example, be shown in early astronomy and astrology (as shown by Pingree, cf. Zysk, *The Indian System of Human Marks*, 2 vols., Sir Henry Wellcome Asian Series 15

 | DOI:10.1163/9789004438224_020

dence of any direct contact between Greece and India in this field, although the possibility of contact existed from the late 6th century BCE.

The Hippocratic corpus knows pepper as medicine,[5] and also calls it (in *Morb. mul.* 1, 81 and *Nat. mul.* 32, 1) "Indian eye medicine," which corresponds to the Āyurvedic use of pepper, although only attested much later.[6] But as we also read that pepper (*péperi*) is the Persian name of the medicine, it seems clear that there is no direct contact with the Indian tradition (for which, in any case, a single parallel would be insufficient). The word "Indian" thus only tell us the origin of the drug. It was one of the main ideas of Filliozat (*The Classical Doctrine*, 238 ff.) that the parallels could hail from contacts, partly indirect, taking place in the Achaemenid Empire.[7]

Be that as it may, the first attested example of direct contact with Indian physicians is found in the histories of Alexander's expedition. Their skill in the treatment of serpent bites is highly praised, but unfortunately we are not provided with any details.[8] Slightly later, Megasthenes reports that in India physicians belong to the class of ascetic Sarmanes (*śramana*). In their cures, diets have an important place alongside drugs.[9] This certainly corresponds to Indian traditions. In the Hellenistic times there was already some intellectual exchange between India and the West, notably in the field of astronomy-astrology,[10] but possibly even in medicine. In the following I shall analyse a few

(Leiden and Boston: Brill, 2015): 25 ff.) and in narrative literature. For medicine, see Filliozat, *The Classical Doctrine*, 258 ff.

5 The concordance of G. Maloney and W. Mohn, *Concordantia in corpus Hippocraticum*, Alpha–Omega, Reihe A, LXXV (Hildesheim, Zürich, and New York: Olms-Weidmann, 1986) knows fifteen passages referring to pepper. Beyond these, India is only mentioned once in the *Morb. mul.* 2, 185 where a mouth-wash for bad-smelling breath is referred to as "the Indian mixture" (*Indikòn phármakon*); the mixture, however, does not contain any Indian ingredients.

6 Filliozat, *The Classical Doctrine*, 253 ff. lists parallels from the *Uttaratantra* of the *Suśrutasaṁhitā* (especially in ch. 17).

7 In the late 5th century BCE, a Greek physician, Ctesias of Cnidus, spent long time in Persian service and among other works composed the first known Greek monograph about India. Admittedly he was more interested in fabulous peoples and marvels of nature—and certainly reveals no knowledge of Indian medicine—but he also notes medical uses of some Indian products.

8 Nearchus F 10a in Arrian, *Indica* 15, 11 f. Aristobulus F 38 in Strabo 15, 1, 45 praises Indian roots and drugs against snake-bites, and Theophrastus, *Historia plantarum* 9, 15, 2 describes two of them. Curtius 9, 1, 2 briefly notes that snake-bites in India are fatal if remedy is not obtained from local people.

9 Megasthenes in the end of F 33 in Strabo 15, 1, 60.

10 This has been much studied by Pingree and others based on Indian evidence. In the West we hear very little of it. In the so-called *Parapegma Milesium*, a calendar fragment pre-

references culled from Greek medical literature.[11] The harvest is sparse, indeed, but a few cases in the Galenic corpus deserve closer consideration, especially because of certain problematic references to them in earlier studies.

In a short article,[12] F.W. Thomas mentions a certain Orbanus (*Orbanós*), who, according to Galen, was an Indian physician who invented an antidote to ease childbirth (or cause abortion; this is not quite clear). Referring to the medical Sanskrit work *Arkaprakāśa,* containing a dialogue between the demon king of Laṅkā and his queen, Thomas suggested that this Orbanus could be Rāvaṇa. In a note on this, J. Jolly[13] pointed out that the Greek form of the name was probably affected by Latin *Urbanus* (and perhaps also by Greek *orphanos*, 'orphan').

The passage in question is found in Galen's *De antidotis* (2, 1, XIV, p. 109 K). According to this work, Orbanus, who is called Indian,[14] had invented the antidote.[15] However, it seems more fitting to explain the name as Greek (or rather Latin), since, as we shall soon see, there are two parallel cases of physicians with Greek names who are somehow connected with India. Finally, the very late date now given to the *Arkaprakāśa* rules out any early connection with Greek medicine.[16]

---

served in a 2nd century BCE inscription in the theater of Miletus, "Indian Callaneus" is mentioned as an authority on the supposed connection between storms and zodiacal signs. But it seems that Callaneus is just Calanus of Alexander's histories, and the source in question is probably a Hellenistic forgery in his name. See H. Diels and A. Rehm, "Parapegmenfragmente aus Milet," *Sitzungsberichte der Königlich Preussischen Akademie der Wissenschaften zu Berlin* (Jan-Jun 1904): 92–111.

11 J. André and J. Filliozat, *L'Inde vue de Rome: Textes latins de l'Antiquité relatifs à l'Inde* (Paris: Les Belles Lettres, 1986) duly note all references to India in Latin medical literature, while E. Salza Prina Ricotti in R.M. Cimino, ed., *Ancient Rome and India: Commercial and cultural contacts between the Roman world and India* (New Delhi: Istituto Italiano per il Medio ed Estremo Oriente & Italian Embassy Cultural Centre, 1994), 88 ff. offers a survey of medicines with an Indian origin. But although coming from India, their use has little to do with Indian medicine.

12 F.W. Thomas, "'Ορβανός = Rāvaṇa?," *Journal of the Royal Asiatic Society* (1904): 749 f.

13 J. Jolly, "Ginger," *Journal of the Royal Asiatic Society* (1905): 168.

14 *Legoménē toû Indoû.*

15 Thomas, "'Ορβανός = Rāvaṇa?," thinks that the mention of ginger among the ingredients supports the alleged Indian origin of Orbanus. In fact, there is quite a number of Indian products used in the same receipt, including spikenard, amomum, black, white and long pepper, and cardamom. Indian aromatics were much appreciated as medicines—Galen, Dioscorides and other medical authors refer to them in numerous cases—but usually their uses have nothing to do with Āyurveda. The only exception I have noted is the abovementioned use of pepper for collyrium, mentioned in both Greek (beside Hippocrates also in Galen, *De compositione medicamentorum secundum locos* 4, XII, p. 780 K) and Indian traditions.

16 G.J. Meulenbeld, *History of Indian Medical Literature*, vol. 2A, Groningen Oriental Stud-

In another passage (*Commentarius II in Hippocratis librum III Epidemiorum*, XVII, 1, p. 603 K) Galen mentions a certain Pamphilo as a medical author and adds that he comes from an Indian town.[17] This has been noted by A. Weber, "On the History of Religion in India: A Brief Review," *Indian Antiquary* 30 (1901): 287, note 57, who adds that this "Indian physician ... lived and studied in Alexandria." Though the same is never stated by Galen, it is, indeed, very likely, as Galen's book is said to be dedicated to Ptolemy Euergetes. Living in Alexandria, however, does not make Pamphilo Indian.[18]

The third case (*De compositione medicamentorum secundum per genera*, XIII, p. 741 f.) presents the "Indian plaster" of the surgeon Tharseas[19] used for oozing ulcerous wounds; the plaster, however, is elsewhere (by Scribonius Largus) called the "black plaster." Both Tharseas and Pamphilo have pure Greek names, also attested elsewhere, which makes their Indian origin highly unlikely.

In *De temperamentis* (3, 1, p. 618 K) we read that Egyptians, Arabs and Indians are black because they live in hot countries—a traditional Greek theory.[20] Beside this, the only references I have found in Galen (admitting that I have not gone through the entire corpus) are about drugs of Indian origin. Indian ingredients, such as pepper, malabathrum and spikenard, were standard items in international trade and accordingly often mentioned in medical receipts. Especially Dioscorides often makes reference to them.[21] However, their uses typically have nothing to do with Indian medicine.

---

ies 15 (Groningen: Egbert Forsten, 2000), 467 ff. puts it in the 17th century or later. Among other things, it mentions syphilis as *phiraṅgaroga* or "Frankish disease."

17 *Tò génos ek póleōs Indikēs.*

18 We now possess good archaeological evidence for the presence of Indians on the Egyptian Red Sea coast, but as for Alexandria, we only have the rhetorical passage of Dio Chrysostomus 32, 40 listing Indians among the members of his audience. This does not suffice as proof, though many have used it as such to argue for the presence of a permanent Indian colony in the city.

19 Tharseas (elsewhere referred to by the inferior variants Thraseas and Tharrias) is briefly discussed by M. Michler, "Tharseas," *Realencyclopädie der Klassischen Altertumswissenschaften*, Supplementband 11 (1968): 1259 f., who dates him to before 100 BCE.

20 With regard to Indians, it was first mentioned by Ctesias; with regard to Aethiopians, by Aristotle. See the discussion and list of passages in P. Schneider, *L'Éthiopie et l'Inde. Interférences et confusions aux extrémités du monde antique* (VIIIe *siècle avant J.-C.*–VIe *siècle après J.-C.*), Collection de l'École française de Rome 335 (Rome, 2004), 88 ff.

21 On Indian drugs, see E. Salza Prina Ricotti in R.M. Cimino, ed., *Ancient Rome and India*, 88 ff. In Dioscorides I have counted 18 passages where a drug (including one animal and one jewel) is specified as Indian, but he does not have any other references to "India." The same applies to all Latin medical and veterinary authors collected in J. André and

As we can see, Indian medicine was probably completely unknown to Galen. Also on the Indian side there is very little to show for a knowledge of the Western tradition.[22] Zysk presents an interesting case of possible interaction.[23] In the *Suśrutasaṁhitā* (3, 5, 49)[24] there is a curious and rather macabre method of dissection where a corpse is left to decay in flowing water for some time, after which the tissues are removed one by one with a bundle of grass and the location of the various organs revealed. The method is never mentioned elsewhere in Āyurveda, but *Suśruta* is the only text specializing in surgery. Its original version can roughly be dated to the beginning of the Common Era, and Zysk suggests that perhaps this un-Indian interest in anatomy was borrowed from Hellenistic medicine. As he points out, dissections were practised in Alexandria from the end of the 4th century BCE by Herophilus and others, and considerably increased the anatomic knowledge. But as the method was certainly different, there is no concrete evidence for such a borrowing, and it must therefore remain a hypothesis.[25]

References

André, J. and J. Filliozat. *L'Inde vue de Rome: Textes latins de l'Antiquité relatifs à l'Inde.* Paris: Les Belles Lettres, 1986.

Cimino, R.M., ed. *Ancient Rome and India: Commercial and cultural contacts between the Roman world and India.* New Delhi: Istituto Italiano per il Medio ed Estremo Oriente & Italian Embassy Cultural Centre, 1994.

J. Filliozat, *L'Inde vue de Rome* (up to 600 CE): Celsius (40 f.), Scribonius Largus (48 f.), Pelagonius and Chiron (200 f.), Theodorus Priscianus (250 f.), Marcellus (270 f.), Caelius Aurelianus and Cassius Felix (272 f.), the *Physica Plinii* (274 f.), and Dioscorides Latinus (314 f.). They mostly mention Indian spikenard, though sometimes also lycium and other drugs.

22 The few references to Yavanas I have found in Sanskrit medical literature, collected in K. Karttunen, *Yonas and Yavanas in Indian Literature*, Studia Orientalia 116 (Helsinki: Finnish Oriental Society, 2015), 109 ff., have nothing to do with Greek medicine.

23 K.G. Zysk, "The Evolution of Anatomical Knowledge in Ancient India, with Special Reference to Cross-Cultural Influences," *Journal of the American Oriental Society* 106 (1986): 687–705

24 Before this passage the text offers a full summary of the anatomical knowledge then possessed. Zysk, "The Evolution," 697 ff. contains a commented translation of the whole chapter 3, 5.

25 Much later, Arabs learned Indian medicine, and so the method travelled with them to the west where we find it in the 12th-century Salernitan anatomical text *Anatomia magistri Nicolai phisici* (Zysk, "The Evolution," 696 f.).

Diels, H. and A. Rehm. "Parapegmenfragmente aus Milet." *Sitzungsberichte der Königlich Preussischen Akademie der Wissenschaften zu Berlin* (Jan-Jun 1904): 92–111.

Filliozat, Jean. *The Classical Doctrine of Indian Medicine*, translated by D.R. Chamana. Delhi: Munshiram Manoharlal, 1964 (from the French original 1949, with some additions by the author).

Jolly, Julius. "Ginger." *Journal of the Royal Asiatic Society* (1905): 167 f.

Karttunen, Klaus. *Yonas and Yavanas in Indian Literature*. Studia Orientalia 116. Helsinki: Finnish Oriental Society, 2015.

Maloney, Gilles and Winnie Mohn. *Concordantia in corpus Hippocraticum*, edited by G.M. and W.F. in collaboration with P. Potter. Vols. 1–5. Alpha–Omega, Reihe A 75. Hildesheim, Zürich, and New York: Olms-Weidmann, 1986.

Meulenbeld, G.J. *A History of Indian Medical Literature*. Vol. 2A. Groningen Oriental Studies 15. Groningen: Egbert Forsten, 2000.

Michler, M. "Tharseas," *Realencyclopädie der Klassischen Altertumswissenschaften*. Supplementband 11 (1968): 1259 f.

Schneider, Pierre. *L'Éthiopie et l'Inde. Interférences et confusions aux extrémités du monde antique* (*VIIIe siècle avant J.-C.–VIe siècle après J.-C.*). Collection de l'École française de Rome 335. Rome: École française de Rome, 2004.

Thomas, F.W. "'Ορβανός = Rāvaṇa?," *Journal of the Royal Asiatic Society* (1904): 749 f.

Weber, Albrecht. "On the History of Religion in India: A Brief Review," *Indian Antiquary* 30 (1901): 268–288.

Windisch, E. *Geschichte der Sanskrit-Philologie und indischen Altertumskunde*. Part 2. Berlin and Leipzig: Vereinigung wissenschaftlicher Verleger, 1920.

Zysk, K.G. "The Evolution of Anatomical Knowledge in Ancient India, with Special Reference to Cross-Cultural Influences." *Journal of the American Oriental Society* 106 (1986): 687–705.

Zysk, K.G.. *Asceticism and Healing in Ancient India: Medicine in the Buddhist Monastery*. Indian edition. Delhi: Motilal Banarsidass, 1997.

Zysk, K.G.. *The Indian System of Human Marks*. 2 vols. Sir Henry Wellcome Asian Series 15. Leiden and Boston: Brill, 2015.

CHAPTER 20

Greco-Indian Astrological Theories on Conception and Childbirth in Mīnarāja's *Vṛddhayavanajātaka*

Bill M. Mak

1 Introduction*

The development of the medical and astronomical sciences in ancient India share many interesting parallels, the implication of which is yet to be fully explored. Āyurveda, the science of longevity, and Jyotiṣa, the astral sciences, were the two large receptacles of knowledge for almost all the important branches of traditional science in India.[1] The theoretical framework of both bodies of knowledge was largely established during the first half of the first millennium of the Common Era through the production of texts that were considered authoritative for the following millennium, thus laying out the definitive course of development for many traditional sciences before their encounter with their modern counterparts. This process of canonization, completed largely by the end of the Gupta period in the sixth century CE, was characterized by a generally empirico-rational epistemology on one the hand and Hinduisation on the other.[2] Among Zysk's many erudite contributions to the study of traditional Indian sciences is his investigation of early Indian medical theories and his insight into the paradigmatic shift from a magico-religious

---

* The research on the *Vṛddhayavanajātaka* began in September 2015, in collaboration with Yano Michio as part of a project to reevaluate the extant materials on Greco-Indian astral science, supported in part by JSPS KAKENHI Grant 15K01118, "Overlapping Cosmologies of Asia," and JSPS International Collaborative Research Accelerator Grant 15KK0050. I thank Yano for his extensive advice on the content of the text, as well as Charles Greifenstein, William Monroe, Timothy Engels, and Kim Plofker for providing me access to all the related materials. In addition, I thank Ronnie Dreyer for her comments on the translation and an earlier draft of this paper.

1 Michio Yano, "Medicine and Divination in India," *East Asian Science, Technology, and Medicine* 24, no. 4 (2005): 44.

2 Some scholars suggest that the "religious over-coding" was a secondary process that early Āyurvedic works, characterized by a "world-affirming, pragmatic realism," were subjected to. See Dominik Wujastyk, "Indian Medicine," in *Companion Encyclopedia of the History of Medicine*, eds. W.F. Bynum and Roy Porter (London: Routledge, 1993), 758.

 | DOI:10.1163/9789004438224_021

approach to healing in the Vedic period to an empirico-rational one that was eventually absorbed into Brāhmaṇic orthodoxy.[3] Furthermore, as a result of his historical and philological investigation, Zysk came to an unconventional conclusion that, contrary to the traditionalist view, the canonized medical knowledge in Āyurveda had a heterodox origin; it was originally shunned by Brāhmaṇic orthodoxy in the early Vedic period. Buddhists, who were open to new and even foreign ideas, in fact played a key role in the advancement of medical knowledge and the development of empirical methodology.[4] The way this body of heterodox knowledge was absorbed and transformed into an orthodox Hindu science may be seen by comparing the Buddhist texts and the earliest extant Āyurvedic treatises such as the *Carakasaṃhitā*, dated likely to c. 100 BCE–200 CE and the *Suśrutasaṃhitā* of a later date.[5]

What about the Indian astral sciences? Did a similar paradigmatic shift take place, and if so, who were the agents? What can the parallels between the two sciences tell us? From a historical point of view, the Sanskrit medical theories from around the beginning of the Common Era appear to systematically eliminate elements of magico-religious beliefs.[6] The Vedic astral science, on the other hand, continued to be practiced by the Brahmins, though at the same time it was also gradually supplanted by a more rational body of knowledge. This new knowledge, including concepts such as the spherical cosmological model, mathematical parameters for the computation of planetary motion based on the eccentric-epicyclic model (with *maṇḍa* and *śīghra* corrections), cyclical planetary weekdays, sexagesimal units, and horoscopy, all hitherto unknown and mostly incompatible with the early Vedic beliefs, was promoted and developed not by the religious mendicants but by another heterodox group, the Yavanas, or the Greco-Indians.[7] Their cosmological, astro-

3 Kenneth G. Zysk, *Asceticism and Healing in Ancient India: Medicine in the Buddhist Monastery* (New York, Oxford: Oxford University Press, 1991), 26–27. For a review of earlier scholars, especially P.V. Sharma, on the Buddhist influence on Āyurvedic works such as the *Carakasaṃhitā*, see G. Jen Meulenbeld, *A History of Indian Medical Literature*, vol. 1A (Groningen: E. Forsten, 1999), 110–112.

4 Zysk, *Asceticism*, 117–119.

5 Meulenbeld, *A History of Indian Medical Literature*, 1A, 114, 342–352.

6 Yano, "Medicine and Divination," 51. According to Zysk, this is the intervening period between Vedic medicine and the absorption of Indian medicine into Brāhmaṇic orthodoxy, that is, c. eighth century BCE to the early centuries CE. See Zysk, *Asceticism and Healing*, 26.

7 Here, I emphasize their Indianness rather than their Greekness due to their command of the Sanskrit language and the lack of evidence regarding their knowledge of Greek. As I discussed elsewhere, it is uncertain whether they are distant descendants of the Indo-Greeks who migrated from the Hellenistic world after or even prior to the military campaign of Alexander the Great or if they were members of a community of unknown ethnicity

nomical, and astrological ideas were eventually embraced by Brāhmaṇic orthodoxy. Though, by modern standards, Jyotiṣa, just like Āyurveda, contains elements of belief and philosophical speculations that are not purely empirical, when compared to their early Vedic antecedents, the knowledge of the new astral science, similar to its medical counterpart, was detached from the Vedic magico-ritualist worldview and was tied instead to a body of empirical knowledge. The foreign elements in the astral works of the Yavanas, as exemplified by works such as the *Vṛddhayavanajātaka* and the *Yavanajātaka*, were so well acculturated and disguised that later Hindu authors saw only the continuity in development, with no awareness of their Hellenistic origin or the paradigmatic shift.[8] Their success lies also in the fact that some of these Yavana authors were both eager to and skilful at Indianizing these works.

Some scholars such as Pingree consider the Indian elements in the Yavana texts as ad hoc adaptations and scribal accretions to what were believed to be essentially Sanskrit translations of some Greek exemplars. However, the way these elements were interpolated into the text suggests otherwise: These works were conceived from the outset as a thorough amalgamation of materials from various sources, including some significant materials of non-Hellenistic origins.[9] Zimmerman, while commenting on Pingree's overly Hellenocentric interpretation of the *Yavanajātaka*, observed the pervasive presence of indigenous materials of Āyurveda in the work, and remarked thus "bien qu[e le *Yavanajātaka*] ait été traduit du grec, ce texte est dans le détail plus hindou qu'on l'aurait pensé."[10] In fact, as we shall also see in the *Vṛddhayavanajātaka*,

---

who had access to both cultures. See Bill M. Mak, "The First Two Chapters of Mīnarāja's *Vṛddhayavanajātaka*," *Zinbun* 48 (2017): 12.

8 Varāhamihira seems barely aware of the foreign origin of many of these concepts when he explained the meaning of the word *horā* in BJ 1.3: *horety ahorātravikalpam eke vāñchanti pūrvāparavarṇalopāt*. According to some (*ity eke*), the word *horā* ("half-sign" / "horoscopy") is an abbreviation of *ahorātra* ("nychthemeron") instead of a derivation of the Greek ὥρα ("hour/ascendant"). For a list of Greek technical vocabulary found in Sanskrit Jyotiṣa works, see Michio Yano 矢野道雄, "Greek Words Borrowed in Sanskrit Astronomical and Astrological Texts" (in Japanese) インドの占星術・天文学書に見られるギリシア語からの借用語について, *The Bulletin of the International Institute for Linguistic Sciences Kyoto Sangyo University* 京都産業大学国際言語科学研究所所報 8 (1987): 74–85. See also, more recently, Klaus Karttunen, *Yonas and Yavanas in Indian Literature* (Helsinki: Finnish Oriental Society, 2015), 368–375.

9 See Mak, "The First Two Chapters," 10–12.

10 Francis Zimmermann, "Les Aspects Médicaux du *Yavanajātaka* (traité sanskrit d'astrologie)," *Sudhoffs Archiv für Geschichte der Medizin (und der Naturwissenschaften) Archiv* 65, no. 3 (1981): 305.

FIGURE 20.1 A folio from a manuscript of the *Vṛddhayavanajātaka*. British Museum, London Or. 5244 (Bendall 484), labeled "L" in Pingree's edition. Ch. 4 begins with *lagnaṃ yadā* on line 10

there are clear traces of Āyurveda in the work, and there is no evidence that the work was a translation from any known or unknown Greek texts in spite of its title, "Ancient Greek Genethlialogy."

The interactions between Jyotiṣa and Āyurveda from the historical and philological points of view present many curious problems. While references to astral materials are scant and limited in the *Caraka-* and the *Suśrutasaṃhitās*, within Jyotiṣa medical concepts remain crucial to astrological topics such as conception, birth, longevity, and death. Whenever medical references are made, the readers are given a glimpse of how these two bodies of knowledge intersected in practice, and in some cases, how the two might have been subsumed under a broader philosophical framework or worldview.[11]

The *Vṛddhayavanajātaka* is a popular Greco-Indian *horā* (i.e., horoscopy) text that is dated, most likely, to sometime after the fourth century CE and is among the oldest surviving Jyotiṣa texts.[12] It is a compilation of earlier Greco-Indian and indigenous Indian materials on divination from a tradition that was later shared by almost all major Jyotiṣa writers, including Varāhamihira in the sixth century CE and Sphujidhvaja (date unknown, but before the seventh cen-

11 An example of how medicine and astrology could have been connected as a broad therapeutic system is found in the *Vīrasiṃhāvaloka* by Vīrasiṃha, composed in 1383 CE. The work treats the etiology and therapy of groups of diseases from three distinct points of view: astrology, religion, and medicine. See *Wujastyk*, "Indian Medicine," 769.

12 Mak, "The First Two Chapters," 2–3.

tury CE).[13] Like most *horā* texts, the *Vṛddhayavanajātaka* contains individual chapters devoted to the topics of conception, birth, longevity, and death. The third and fourth chapters of the *Vṛddhayavanajātaka*, which deal with the topics of conception and birth respectively, are exceptionally elaborate. Here, I would like to provide an analysis of their content, together with an annotated English translation, with the goal of understanding the medical content therein and the role they played within the text.

2 Textual Treatment of Conception and Birth in the *Vṛddhayavanajātaka*

Very little is known about Mīnarāja, the purported author of the *Vṛddhayavanajātaka*, and practically nothing is known about Maya, the ultimate progenitor of the original, expanded version of the text.[14] The materials found in the text are remarkably rich, with a bewilderingly large amount of textual variants found across the extant manuscripts.[15] A total of 4270 verses in 71 *adhyāyas* are reported in Pingree's 1976 edition of the text, making it by far one of the largest extant Jyotiṣa works. Some passages of the work are cited in Sphujidhvaja's *Yavanajātaka*.[16] The work was most likely known to al-Bīrūnī in the early eleventh century,[17] and it was cited as late as the fifteenth century in Śivarāja's *Jyotirnibandha*,[18] thus attesting to the wide circulation of the work and its popularity in India for over a millennium. Here, I shall first briefly

13 On the relationship between the *Vṛddhayavanajātaka* and the *Yavanajātaka*, and its implications for the dating of both, see ibid., 10–12.

14 *yad uktavān pūrvamunis tu śāstraṃ horāmayaṃ lakṣamitaṃ mayāya / tan mīnarājo nipuṇaṃ svabuddhyā vicintya cakre 'ṣṭasahasramātram //* (1.2) On Maya, ibid.

15 Pingree listed a total of no less than 58 manuscripts (including incomplete ones), with the oldest one SOI 9503 dated Śaka 1449 = 1527 CE. CESS A4, 427–429, A5, 310. All specimens are in Devanāgarī script as in the sample shown in fig. 20.1.

16 Ibid. Note that Pingree claimed the opposite and suggested that the *Vṛddhayavanajātaka* cited the *Yavanajātaka*, thus dating the former to the fourth century, which was some decades after the dating he assigned to the *Yavanajātaka*. As I have shown elsewhere, both his dating and analysis of the texts' relation are untenable on philological grounds.

17 "There is a larger book than this (the *Sārāvalī*), a compendium on every subject of astrology, known as the *Yavana* ..." See Edward C. Sachau, *Alberuni's India*, vol. 1 (London: Trübner, 1888), 158. Though some uncertainties remain, on the basis of this description, Pingree identified the work al-Bīrūnī referred to as the *Vṛddhayavanajātaka*. See David Pingree, *The Yavanajātaka of Sphujidhvaja*, Harvard Oriental Series 48, vol. 1 (Cambridge: Harvard University Press, 1978), 24, note 76.

18 See Pingree, *The Yavanajātaka*, vol. 2, 450–451, vol. 1, 32, 37 (for other late references).

TABLE 20.1 Topics on conception in VYJ, BJ, and YJ

	VYJ	YJ	BJ
Introduction	3.1		
Theory of conception		5.1–4	4.1
Condition for favorable conception		5.5–6	4.3
Condition for unfavorable conception	3.2–5	5.7–8	4.3–4
Nature of sexual union	3.6–8		4.2
Day/night conception		5.35	4.5
Death of pregnant mother and/or child			4.6–10
Sex of child	3.20–24	5.20–28	4.11–13
Twins and other multiple births		5.29–33	4.14–15
Gestation of ten months	3.9cd–18	5.9–13	4.16
Child deformity			4.17–20
Time of childbirth			4.21–22
Other theories		5.34, 36–37	

discuss the textual treatment of the topics of conception and birth in the *Vṛddhayavanajātaka*, paying attention to the similarities and differences among the three works: *Vṛddhayavanajātaka*, *Yavanajātaka*, and *Bṛhajjātaka*.

The third chapter of the *Vṛddhayavanajātaka* on conception (*ādhāna*) is concerned with a variety of subjects and activities that take place between conception and childbirth, from sexual union, conception, gestation, and pregnancy up to the moment of childbirth and the postdelivery puerperium. These topics are discussed within the framework of *horā*, that is, how planetary influences affect the outcome of these events in a schematic manner: good outcomes result from planetary configurations (*yoga*) associated with the benefics, that is, the good planets such as Jupiter and Venus, while bad outcomes result from those associated with the malefics, that is, the bad planets such as Mars and Saturn. Often, the purported medical prognoses of the planets parallel the assumed characters of the planets in a fairly arbitrary manner and should not be assumed to be based on actual medical knowledge. The topics of this chapter (40 verses) resemble those of the fifth chapter of the *Yavanajātaka*, Ādhānavidhi (37 verses), and the fourth chapter of the *Bṛhajjātaka*, Niṣeka (22 verses). While similar topics are presented, their treatments vary in terms of both content and structure (table 20.1).

The fourth chapter of the text deals with the birth of the native, as well as a number of other topics following birth, such as character of the new-

TABLE 20.2 Topics on birth in VYJ, BJ, and YJ

Topics	VYJ	YJ	BJ	Other topics
Condition for birth	4.1			
Illegitimate birth	4.2	6.3	5.6	
Inauspicious birth	4.3		5.7–10	Imprisonment of father, birth in unusual places
Absence of father	4.4	6.1	5.1–2	
			5.3	Serpent birth
	(3.38–39)		5.4	Twins
Door of maternity chamber	4.5ab	6.2ab	5.18b	
Lamp, wick, and oil	4.5cd	4.12	5.18a	
Bodily features	4.6	6.5	5.23–24	
Bodily defects	4.7–8		25–26	
Abandonment	4.9		5.14–15	
Inverted birth	4.10a–c			
Coiling of umbilical cord	4.10d		5.5	
Special marks	4.11–12		5.24	
Missing/extra limbs	4.13		4.19	
Birth place	4.14–26	6.4–12	5.11–13, 16–17, 19	
			5.20–22	Direction and midwives

born and place of birth. Parallel materials, as shown in table 20.2, are found in the fifth chapter of the *Bṛhajjātaka* on "Characteristics surrounding the time of birth" (*janmakālalakṣaṇa*) and the sixth chapter of the *Yavanajātaka* on "Rules related to birth" (*sūtikavidhi*, i.e., rules related to women before and after delivery). Some verses, such as those on the length of the wick and the amount oil in the lamp in the maternity chamber (*sūtikāgṛha*),[19] are highly specific but laconic and appear to be excerpted from some more explicit sources.[20] As with most other chapters, the materials here in the third and fourth chapters

19 The Sanskrit term is not given in the text but can be understood from the context. *Sūtikā* refers to a woman who has just given birth to a child and the period referred to here should cover from before the delivery up to the end of puerperium, that is, six weeks according to *Aṣṭāṅgasaṃgraha* 2.3.9–41. Some texts give the translation as "lying-in woman" and the place as "lying-in room" to refer to the special construction (*sūtikāgṛha*/*sūtikāgāra*) where the woman would stay during this period.

20 Possibly of Maya, as mentioned at the beginning of the work, and possibly from a chapter or work titled "Disquisition on Offspring" (*jantuviniścaya*). See VYJ 3.1.

of the *Vṛddhayavanajātaka* do not show any exact correspondence with their counterparts in the other two works.

The text of the two chapters, based on the manuscripts available to us, is characterized by hypermetrical *padas* (twelve instead of eleven syllables in *upajāti*), which are scattered across the two chapters (marked h.m. in my edition of the Sanskrit text). Since the additional syllable is always short in the penultimate position of a *pada*, the result may be described as an uneven mixture of *upajāti* and *vaṃśastha*. Leaving aside the possibility of corrupt transmission, which, oddly, affected these chapters more than others, such mixed meters may be the vestige of an older text or an older stratum of the text that resisted standardization, as noted in the cases of the *triṣṭubh-jagatī* mix found in the *Mahābhārata* and some Pāli and Sanskrit Buddhist texts.[21] Despite the title of the text, no Greek texts can be identified as the sources of these chapters. As Pingree has shown, some of the medical descriptions, such as those on conception and gestation, are in fact closer to the descriptions found in Āyurvedic works such as the *Carakasaṃhitā* and the *Suśrutasaṃhitā* than to those in any known Greek works.[22] Although Indian medical theories, such as the *tridoṣa* humoral theory featured in the *Vṛddhayavanajātaka*, carry some resemblance to their Greek counterparts, the two are in general incommensurable.[23] In all likelihood, just as the astral elements found in the *Yavanajātaka* may be best characterized as an amalgamation of foreign Hellenistic astral science and its Vedic counterpart,[24] the medical elements, too, could be the synthesis of two bodies of medical knowledge from the Hellenistic world and India, the ultimate sources of which are yet to be identified.

21 I thank Seishi Karashima for the suggestion. According to Karashima, such mixed meters might have been considered stylistically more elaborate in earlier times (personal communication, 8 July 2017). For a discussion of examples of mixed meters in Buddhist texts, see Seishi Karashima, "The Triṣṭubh-Jagatī Verses in the *Saddharmapuṇḍarīka*," ARIRIAB (*Annual Report of the International Research Institute for Advanced Buddhology at Soka University*) 19 (2016): 193–210.

22 Pingree cited *Carakasaṃhitā* and *Suśrutasaṃhitā* for close parallels in YJ 5.2–4 and 5.9–12 on conception and gestation respectively. See Pingree, *The Yavanajātaka*, vol. 2, 259–260.

23 Hence, the three Indian *doṣas*, wind, bile, and phlegm (Skt. *vāyu, pitta, śleṣman*), are not compatible with the four Greek humors: blood, phlegm, yellow bile, and black bile.

24 Bill Mak, "The Date and Nature of Sphujidhvaja's *Yavanajātaka* Reconsidered in the Light of Some Newly Discovered Materials," *History of Science in South Asia* 1 (2013): 1–20; "The Last Chapter of Sphujidhvaja's *Yavanajātaka* Critically Edited with Notes," *SCIAMVS* 14 (2013): 59–148; "The 'Oldest Indo-Greek Text in Sanskrit' Revisited: Additional Readings from the Newly Discovered Manuscript of the *Yavanajātaka*," *Journal of Indian and Buddhist Studies* 印度學佛教學研究 62, no. 3 (2014): 1101–1105.

3 Characteristics of the Medical Elements Described in the *Vṛddhayavanajātaka*

3.1 *Positioning of the Medical Knowledge*

The presentation of medical knowledge in the *Vṛddhayavanajātaka* is schematic, that is, subservient to astrological concepts, categories, and theories, which are largely arbitrary as typical in Jyotiṣa texts. Such presentation is different from those of the Āyurvedic texts, which generally work with a different set of schematics, such as the *tridoṣas*, the *mahābhūtas*, or materialist theories, that pertain to human physiology. Thus, in the *Vṛddhayavanajātaka*, we find arbitrary explanations, such as predictions based on the mode of sexual union—that is, specific outcomes of the sexual act taking place during the day and at night when the astrological configurations are associated with the Sun and Saturn respectively (3.5cd–6). Similar symbolism and sympathetic beliefs abound in predictions concerning the character of the native and the place of nativity (4.14–26).[25]

The *Vṛddhayavanajātaka*, moreover, hints at a number of overarching beliefs and assumptions under which the prognoses are framed. Arguably, the magico-religious elements of the earlier Vedic world never completely disappeared, as malefic influences by the *grahas* (as not only "planets," but also demonic forces) are reckoned with in Jyotiṣa as in Āyurveda.[26] Among other beliefs, there are theories on memory and past lives. On the one hand, the planets are thought to be the agents that cause the native to be born within a particular caste or under a given circumstance in a completely deterministic manner (3.2–4ab). On the other hand, the native also carries memory of his past, presumably in his past lives, including things of positive and negative natures associated with his past actions (3.14cd–17ab).[27] Similar descriptions of the fetus's recollection of previous lives (*jātismara*) are also found in the *Suśrutasaṃhitā* and other Indian texts.[28]

---

25 The description in VYJ is particularly elaborate, as noted by Pingree. For a general comparison with other *horā* works, see Pingree, *The Yavanajātaka*, vol. 2, 267.

26 In particular, *grahas* are believed to be the causes of nine types of infant diseases, and both medicinal and propitiatory rituals are prescribed as remedies. See Swami Sada Shiva Tirtha, *The Āyurveda Encyclopedia* (Delhi: Sri Satguru, 1998), 518–522.

27 Such a form of "soft determinism" may be seen from the beginning of the text (1.3): *yā pūrvakarmaprabhavasya dhātrī dhātrā lalāṭe likhitā praśastiḥ / tāṃ śāstram etat prakaṭaṃ vidhatte dīpo yathā vastughane 'ndhakāre //* ("This teaching reveals clearly the destiny, the fulfiller of the power of actions done in the past (*pūvakarma-*), written on [one's] forehead by the Creator, just as a lamp in darkness reveals a multitude of things.")

28 *Suśrutasaṃhitā* 3.2.57–58.

3.2 *Theory of Ten-Month Gestation*

The description of gestation in ten successive months, each assigned to one of the seven planetary lords, plus the Ascendant and the repetition of Sun and Moon, is standard across all extant *horā* texts.[29] While not specified in the text, the month (*māsa*) here should be understood as sidereal (c. 27.3 days), a scheme of Greco-Babylonian origin; the total period of gestation is thus about nine civil or synodic months.[30] A comparison of the three *horā* texts (table 20.3) reveals that the *Vṛddhayavanajātaka* and the *Bṛhajjātaka* are identical in terms of the planetary assignment, beginning with Venus and Mars in the first and second months respectively, and possibly originated from a work attributed to Garga; the source of the planetary scheme in the *Yavanajātaka* is unknown.[31]

While the description of the gestation period of ten sidereal months and the planetary assignment are certainly of foreign origin, the text given here in the *Vṛddhayavanajātaka* includes many unique Indian elements. There is no conclusive evidence so far to suggest that Greek and Indian embryology have influenced each other or arisen independently.[32] The description of fetal development given in these *horā* texts bears greater resemblance to Indian works such as the *Carakasaṃhitā* and the *Suśrutasaṃhitā*, including technical terminology such as *kalala* and *ghana* unique to Āyurvedic texts.[33] Some speculative elements characteristic of Āyurveda, such as the *śakti* theory of the embryo,[34] are echoed in the concept of *śabdabrahman* in the seventh month of gestation (3.14). Similarly, the nutritive fluid *rasa* mentioned in the text (3.16) is one of

---

29 For some preliminary remarks, see Yano, "Medicine and Divination in India," 55–56.

30 Pingree opined that the theory of the gestation period of ten sidereal months goes back from the Greeks to Babylonia. Pingree, *The Yavanajātaka*, vol. 2, 260, citing O. Neugebauer, "Decem tulerunt fastidia menses," *The American Journal of Philology* 84 (1963): 64–65. It may be noted that in both *Caraka* and *Suśruta*, pregnancy is described generally as nine months, hence synodic or civil, not sidereal. For criticism of Neugebauer's suggestion and a survey of ten-month gestation in classical and ancient Near Eastern Sources, see Marten Stol, *Birth in Babylonia and the Bible: Its Mediterranean setting* (Groningen: Styx, 2000), 23–25.

31 Pingree, *The Yavanajātaka*, vol. 2, 260–261.

32 Joseph Needham, *A History of Embryology* (Cambridge: Cambridge University Press, 1959), 25–27. Though Needham believes it is "more likely" that one has influenced the other, and was hesitant to say in which direction the influence took place, he does suggest, for example, that the Āyurvedic description of the embryo as formed of a mixture of semen and blood is "a very Greek idea."

33 *Carakasaṃhitā* 4.4.5–29, *Suśrutasaṃhitā* 3.3.18–30. For a summary, see Meulenbeld, *A History of Indian Medical Literature*, 1A, 41–42 (*Caraka*), 246–247 (*Suśruta*).

34 Charles Eliot, *Hinduism and Buddhism: An Historical Sketch*, vol. 2 (London: Edward Arnold & Co., 1921), 319–320.

the six factors of classical Indian embryogenesis.[35] The *Mahāsaṃnipātasūtra*, a Mahāyāna text dated to the early centuries of the Common Era, of which a Chinese translation from the sixth century survived, includes a similar description of fetal development, expressed in cycles of seven days with a total gestation period of thirty-eight weeks or 266 days.[36]

3.3 *Sexuality*

Description of the sexuality of the native is a prominent feature of *horā* texts such as the *Yavanajātaka* and is somewhat less prominent in the *Vṛddhayavanajātaka*. In various horoscopes in the *Yavanajātaka*, the native is described as skilful, lucky, failing, fond of, indulgent, averse to, preoccupied with, or addicted to his sexual proclivities.[37] Other unusual forms of sexuality, such as promiscuity (*godharma*),[38] homosexuality (*puṅguhyaratipriya*, *klīva*),[39] and adultery, are also noted in the *Yavanajātaka* and are generally associated with vileness (*adharma*, *anārya*). A set of four verses in the *Vṛddhayavanajātaka* (3.5cd–9ab) foretells rather curiously the location, time, and character of sexual union. On the other hand, the variety of sexual behavior is generally treated in the classical India medical texts with explanations of biological nature. In the *Suśrutasaṃhitā*, four types of *klīvas* are described, and their unusual desires are the results of various forms of deficiency or the improper sexual union of the parents.[40]

The prognosis of the sex of the native is a standard topic in all *horā* texts. It is generally based on the gender of the planets and houses of the horoscopic con-

---

35 *Carakasaṃhitā* 4.3.

36 *Taishō* 397.13.169b. For the textual history of this text, see my forthcoming paper: "Silk Road Transmission of Astrological Lore to China—Indian, Chinese and Central Asian elements in the *Mahāsaṃnipātasūtra* (T397)," in *Silk Road: Interwoven History* (Cambridge, MA: Association for Central Asian Civilizations & Silk Road Studies). Note that the description is framed within a Buddhist homily on human suffering, and the embryo's recollection of past lives is also described. The relation between Indian theory of embryology and its counterpart in Chinese Buddhist texts is discussed recently in the works of Li Qinpu and Chen Ming. For an English summary, see Hsin-Yi; Lin, *Dealing with Childbirth in Medieval Chinese Buddhism: Discourses and Practices*, Ph.D. dissertation (Columbia University, 2017), 127n244, 151–152.

37 YJ *passim*. In total, there are no less than a hundred such references in the text. It is not known why YJ appears to have more sexual references than other *horā* texts. A thorough comparative study may yield some interesting results.

38 YJ 29.31.

39 YJ 7.12 13.22, 17.6. The term *klīva* (or *klība*) implies emasculation and, in some cases, refers to the neuter gender, thus not necessarily homosexuality. I follow Pingree's interpretation here.

40 *Suśrutasaṃhitā* 3.2.36–45.

TABLE 20.3 Theory of ten-month gestation in VYJ, BJ, and YJ

Month	VYJ 3.9cd–18	YJ 5.9–13	BJ 4.16
First	Venus. Union of blood and semen (*śoṇitaśukrasaṃga*)	Mars. Incipient fetus (*kalala*)	Venus. *Kalala*
Second	Mars. Thickening (*ghana*)	Venus. Muscle (*peśī*)[a]	Mars. Thickening (*ghana*)
Third	Jupiter. Hands, feet, mouth, neck	Jupiter. Shoots (*śākha*)	Jupiter. Sprout (*aṅkura*)
Fourth	Sun. Bones, marrow, fat, flesh, blood	Sun. Bones, sinew, head	Sun. Bone
Fifth	Saturn. (gender) distinction	Moon. Marrow, skin	Moon. Skin
Sixth	Moon. Body—hair, nails, tongue, anus, genitals	Saturn. Blood, hair, nails, liver	Saturn. Hair
Seventh	Mercury. Memory, five senses, self-awareness	Mercury. Mental activities	Mercury. Mental activities
Eighth	Ascendant. Hunger, thirst	Ascendant. Thirst, hunger, taste	Ascendant
Ninth	Moon. Desire, pleasure	Moon. Touch, awareness, pleasure	Moon
Tenth	Sun. Birth	Sun. Birth	Sun

a Pingree translated as "budding."

childbirth include twins and triplets (3.38–40). As in the previous cases, though similar subtopics are treated in the *Bṛhajjātaka* and the *Yavanajātaka*, no exact parallels are found, and the underlying principles behind each set of prognoses differ significantly. When compared with the medical texts, the latter's predilection for biological explanation may be observed, combined often with other factors such as the thoughts of the parents, the nature of the sexual act, and sins and virtues resulting from actions in the past lives.[48] Thus, the methods and mechanisms for explanations in the Jyotiṣa and Āyurveda texts are vastly different from each other, though one may rather superficially connect the two through a form of soft determinism.

3.5 *Interface with Mathematics and Astronomy*

As with the earlier chapters, the mathematical and astronomical concepts displayed in these two chapters of the *Vṛddhayavanajātaka* contain unique elements found in both Hellenistic and Vedic astral science. An important concept, "purity in six classes" (*ṣaḍvargaśuddha*), which is described in verse 2.13 and refers to the six divisions of a sign, that is, a half (*horā*), a third (*dreṣkāṇa*), a seventh (*saptāṃśa*), a ninth (*navāṃsa*), a thirtieth (*triṃśāṃśa*), and a sixtieth (*ṣaṣṭi-vibhāga*), is also featured in our chapter on birth.[49] The decan (*dreṣkāṇa*), or a third of a sign, an important astral concept in Hellenistic horoscopy that is ultimately of Egyptian origin, is featured once in our text (4.16). More important is the division *navāṃśa*, which is given particular importance over other divisions.[50] This unique division was introduced earlier in verse 1.21, with the understanding that one celestial revolution consists of 108 *navāṃśas*, which is also the smallest common multiple of the twelve signs and twenty-seven *nakṣatras*.[51] As the division of a ninth of a sign is not known in Greco-Babylonian horoscopy, the *navāṃśa* is likely a product of the encounter between the Hellenistic and Vedic astral sciences.

Medical Literature, 1A, 519, for an example of a description of the causes of *upaviṣṭaka* and *upaśuṣkaka* in *Aṣṭāṅgasaṃgraha* 2.4.

48 *Suśrutasaṃhitā* 3.2.46–51.

49 VYJ 4.12, 4.16. Though the six divisions are given in 1.21–23, the definition of the "six divisions" is not given anywhere. According to BJ 1.9, *varga* is a special kind of position when a planet is in an astrological subdivision presided by that planet. The subdivisions given are *rāśi, horā, dreṣkāṇa, navāṃśa, dvadaśāṃśa*, and *triṃśāṃśa*. If this is the case also for VYJ, *ṣaḍvargaśuddha* would mean "unafflicted in the six divisions." I thank Dreyer for pointing this out to me.

50 VYJ 3.6, 3.21, 3.23, 3.31, 3.34, 4.8, 4.17. The application of *navāṃśa* appears to be even more salient, or even fundamental, in the prognoses described in BJ and YJ.

51 See Mak, "The First Two Chapters," 4, fn. 17. Note the mention of the *nakṣatras* Maghā and Āśleṣā in reference to the solar and lunar divisions of the twelve signs in VYJ 1.17.

Of special interest is the prediction of the length of the wick and the amount of oil in the lamp in the maternity chamber (4.5). The descriptions are overly laconic, though the method of calculation may be deduced from similar passages in other *horā* texts. For example, the length of the wick is assumed to be in proportion to how far the Ascendant has risen, and the fullness of the oil lamp is determined by the phase of the Moon.[52]

The *Vṛddhayavanajātaka* is characterized by a paucity of actual descriptions of mathematical procedures and astronomical observations. Mīnarāja appears to be nonetheless interested in the technicalities of spatial orientation, which plays an important role in the cosmology of all *horā* texts. This includes the symbolism implicit in the geometry of the horoscope, in particular, the Ascendant, the Descendant, and the zenith as the first, seventh, and tenth house respectively, and the cardines corresponding to the four directions of any object, such as the four corners of the bed in the maternity chamber (4.6).

4 Conclusion

Similar to the astral materials described in the first two chapters of the *Vṛddhayavanajātaka*, the materials found in the text's following two chapters on conception and birth, including references to idiosyncratic divinatory techniques and specific medical theories such as the ten-month gestation period, display unique influences from both Hellenistic and indigenous Indian sources. There is no reason to doubt Mīnarāja's statement that the text was a condensed version of a larger astral compilation by Maya, a work no longer extant. While parallel materials are found in other *horā* texts, such as the *Bṛhajjātaka* and the *Yavanajātaka*, the content of the three texts differ enough to suggest that the three works are of different provenances. These texts display a varying degree of understanding and familiarity with the latest scientific knowledge of their time, be it astronomy or medicine. The authors of the Jyotiṣa texts, just like those of the Āyurveda texts, are keen observers of the variety of human conditions and behaviors, although their interests in the exact sciences may be subjugated by the astrological and Āyurvedic schematics, which sometimes operate independently from empirical observations. In all likelihood, the centuries preceding the composition of the *Vṛddhayavanajātaka* were a time when

52 YJ 6.12, BJ 5.18. More elaborate prognoses, such as the color of the flame, are given in Maṇittha and cited by Utpala on BJ 5.8. See Pingree, *The Yavanajātaka*, vol. 2, 267.

such an amalgamation of the foreign and local, the scientific and superstitious, and the new and old took place, bearing in mind that boundaries of knowledge continued to blur as authors synthesized new systems with every work they created. Despite the misleading reference to the Greek in the title of the text, works with a purported foreign origin such as the *Vṛddhayavanajātaka* may, in essence, be considered attempts to appropriate a foreign body of knowledge, first by the Greco-Indians and later by the Indians themselves. Rather than a straightforward translation of a Greek exemplar into Sanskrit, these *horā* texts display not only an adaptation of new concepts using existing vocabulary but also great creativity, resulting in the richness and variety of ideas and expressions we see in these texts.

Sanskrit Edition

Editorial note: The text here is based on Pingree's edition. Variants are indicated in a footnote only where the reading differs from the manuscripts' or Pingree's emendation (*ped*). Hypermetrical lines or hemistichs are marked with [h.m.].

Chapter 3: Ādhāna

3.1ab ādhānapṛcchodbhavasāmyayuktaṃ phalaṃ yatas tasya parīkṣaṇārtham /
3.1cd yogān vicitrān pravadāmy ato 'haṃ cihnair yathā jantuviniścayaḥ syāt //
3.2ab lagnaṃ yadā paśyati sūryasūnur nīcāśritaḥ saumyadṛśā vihīnam /
3.2cd tadāntyajātaṃ pravadanti martyaṃ sūryasya vīryeṇa divā prasaṅgāt //[a]
3.2e somo 'thavā rātriratena nityaṃ[b]
3.3ab bhaumo yadā vaiśyasamudbhavena sakṣīṇacandro nṛpasaṃbhavena /
3.3cd astaṃgato jño dvijavarṇayogāt sarvair atha mlecchasamudbhavena //
3.4ab eṣāṃ mṛtir dāsajanādinīcair dvābhyāṃ vadhūbhir vikṛtir naraiś ca /
3.4cd catuṣpade mūrtiṣu talpam etvā āpye kṣitau prāntarake nṛsaṃjñe //
3.5ab kīṭe vilomaṃ pravadanti bhāvaṃ vācyaṃ vilagnādhipateḥ svabhāvāt /

---

[a] tadāntyajātaṃ] emend., tadānyajātaṃ Σ. [b] LQW only.

English Translation

Chapter 3: Conception

3.1 In order to examine the [astrological] effects associated with the queries, occasion, and likeness of the conception [of a child], I will henceforth describe the various [astrological] configurations with the signs (*cihna*) as [described in] the Disquisition on Offspring (*jantuviniścaya*).

[Conceptions of different social classes]

3.2 If Saturn in debilitation aspects the Ascendant without the aspect of benefics, then [the sages] predict the [conception] by a person of the last caste (i.e., Śūdra). [The same results] from the union during the day when the Sun is powerful, or when the Moon [was waxing] during sex at night.

3.3 If Mars [aspects the Ascendant], then there would be a conception by Vaiśyas. If [the Ascendant is aspected] with the waning Moon, then a conception by kings (i.e., Kṣatriyas). If [aspected by] Mercury in the Descendant, then a union of Brahmins. If [aspected by] all [planets], then a conception by foreigners (*mleccha*).[53]

3.4ab For those [conceptions under such configuration] by both women and men of low [castes], such as slaves, there will be death and deformation (*vikṛti*).

[Nonhuman birth][54]

3.4cd–3.5ab Having had sexual intercourse, [beings conceive offspring] in the form of quadrupeds, aquatics, the earthly, and the silvan in the male form; in the case of insects, the opposite nature is predicted. [The nature of the conception] is predicted from the nature of the planetary lord of the Ascendant.

53 A brief discussion on this topic is given in Pingree's commentary on 4.2–3 (DEP 56).
54 Cf. BJ, ch. 3 titled "Viyonijanman."

3.5cd kalatratah saṅgavidhiḥ pradiṣṭo vikāravaikṛtyasamo grahāc ca //
3.6ab yadā ravir nekṣati janmalagnaṃ tadāndhakāre surataprayogaḥ /
3.6cd tasminn udaksthe svanavāṃśabhāge divā prasaṅgaḥ śaninā ca rātrau //
3.7ab sarvair adṛṣṭe pravaded araṇye jalasya madhye jalalagnasaṃsthe /
3.7cd kalatragaḥ śītamayūkhamālī yadā ravir mūrtigatas tadā syāt //
3.8ab sandhyā prayogo vyayage 'tha bhaume ūrdhvaṃ praśastaṃ ravinandanena /
3.8cd sītkāramiśraṃ surapūjitena śaśvat kṣatakrāntam urupracaṇḍam //
3.9ab śukreṇa līnaṃ śaśinā vidagdhaṃ syāt somajenaiva nitāntadīrgham /
3.9cd ādyasya māsasya bhṛgur vidhātā tasmin bhavec choṇitaśukrasaṅgaḥ //
3.10ab tadrūpaceṣṭābalahānidīptyā garbhasya vācyaṃ svaphalaṃ jananyāḥ /
3.10cd dvitīyamāsādhipatiḥ kujaś ca tasmin ghanaṃ tasya bhavet samantāt //
3.11ab jīvas tṛtīyasya karāṅghrivaktragrīvādikaṃ tatra bhavet samagram /
3.11cd sūryaś caturthasya patiḥ pradiṣṭaḥ asthīni tatra prabhavanti puṃsāṃ //
3.12ab majjā ca medaś ca sumāṃsaraktaṃ vyaktiṃ samāyāti vibhāgataś ca /

[Location and nature of sexual union]

3.5cd+3.6 The rules for the union with the wife are said to be in accord with the change (*vikāra-*) and deviation (*vaikṛtya-*) resulting from the planets. If the Sun does not aspect the birth Ascendant, then the sexual union [takes place] in the dark. If [the Sun] is in its own *navāṃśa* in the tenth house (lit. above, *udak*), [then] the sexual union [takes place] during the day. If it is Saturn [instead of the Sun], [the sexual union takes place] at night.

3.7–3.8a If [the Sun or the Moon] is not aspected by any [planet], then [the sexual union] is predicted [to take place] in the forest. If [the Sun or the Moon] is located in the Ascendant in an aquatic sign, [then the sexual union takes place] in the middle of water. If the Moon is in the seventh house and the Sun in the Ascendant, the sexual union takes place during twilight.

3.8 If Mars is in the twelfth house, and Saturn in the tenth house, [the sexual union] would be mixed with moaning. If Jupiter [is in the tenth house], [the sexual union] would certainly be characterized by wounds and be excessively violent.

3.9ab [If the tenth house is occupied] by Venus, [the sexual union would be] lying down. [If it is occupied] by the Moon, it would be artful. If by Mercury, it would be long lasting.

[Fetal development in ten months]

3.9cd–10ab Venus is the bestower of the first month [of fetal development]. At that time there would be the union of blood and semen. The prediction of the fetus in terms of its form, behavior, strength, damage, and beauty would follow the mother's own [horoscopic] reading.

3.10cd The lord of the second month is Mars. At that time, there may be thickening (*ghana*) on every side.

3.11ab Jupiter is [the lord] of the third month. The hands, feet, face, neck, and so on would be complete then.

3.11cd–12ab The Sun is said to be the lord of the fourth month, the bones would appear. Bone marrow and fat are thoroughly reddened with flesh. [The fetus] becomes sexually distinct for males and is proportioned.

3.12cd tasmin sa sauriḥ kila pañcamasya tvacā samantākṛtim āvṛṇoti // 12[a]
3.13ab prāpṇoti puṣṭiṃ vividhāṃ ca garbhavyaktiṃ samāgacchati kāyajātāṃ /
3.13cd ṣaṣṭhasya chandro vibhutām upaiti romāṇi tatra prabhavanti gātre //
3.14ab nakhāś ca jihvā gudarandhrabhāve guptaṃ svaraṃ brahma bhaved dhi tasya /[b]
3.14cd syāñ candrasūnuḥ kila saptamasya tasmin smṛtiḥ syāt satataṃ narāṇāṃ //
3.15ab pañcendriyatvaṃ ca vivekitā ca ko 'haṃ kuto 'trāśrayam abhyupetaḥ /
3.15cd lagnādhināthas tv athavāṣṭamasya māsasya tasmin pracurā bubhukṣā //
3.16ab bhaven manuṣyasya tataḥ sutṛptiḥ bhuṅkte janitryā rasabhāvasaṅgāt /
3.16cd nakṣatranātho navamasya nāthas tasmin viraktir vividhā narāṇāṃ //
3.17ab garbhāśrayā duḥkham anantam ekaṃ kṛtaṃ smṛtiḥ pūrvaśubhāśubhasya /
3.17cd divākaras tatparato 'dhināthas tasmin pradiṣṭaḥ prasavo narāṇām //
3.18ab –
3.18cd tasmin yadi syād vyayagaḥ śaśāṅkaḥ syād aṣṭamo bhūtanayaḥ prasūtiḥ //
3.19ab yo yasya māsasya bhaved dhi nāthaḥ saṃbhogam enaṃ tu yadā prayāti /
3.19cd arvāg vidhatte sa tadāsya janma vīryeṇa hīno yadi vātmajena //

---

[a] āvṛṇoti] Σ, ātṛṇoti AB, ākṛṇoti *ped.* [b] brahma bhaved dhi tasmin] LQW, brahmabhave hi tatra Σ.

3.12cd–13ab Saturn is [the lord] of the fifth [month]. It covers the entire form [of the fetus] with skin. It reaches maturity in its various forms and develops the appearance of a fetus which is born of the body.[55]

3.13cd–14ab The Moon is [the lord] of the sixth month. It reaches final maturity and hairs appear then on the body, besides nails and tongue. The concealed [organs] are in the form of genitals and anus. The Brahman therein verily becomes the hidden sound.

3.14cd–15ab Mercury is the lord of the seventh month. There would always be the memory [of past lives] among men. There will be the functioning of five senses and the state of discrimination. He would come to understand causal relations such as who am I and from where do I come here.

3.15cd–16ab The Ascendant is the lord of the eighth month. There would be great desire to eat then. There would be great contentment for a human being. He enjoys being with the mother due to the union of nutritive fluid (*rasa-*) and passion (*bhāva-*).

3.16cd–3.17ab The Moon is the lord of the ninth [month]. There would be different kinds of changes of disposition (*virakti*) among people. Only boundless suffering is the memory of the good and the bad [things] from the past [lives].

3.17cd After that (i.e., the tenth month), the Sun is lord, then the birth of males is indicated.

3.18 When the Moon is in the twelfth house and Mars in the eighth, the birth …

3.19 The planetary lord of each of the month proceeds throughout its allotted period during pregnancy (*saṃbhoga*). [Each of them] contributes to the birth [of the native] hitherward. When either [the planetary lord] is deprived of strength or offspring …[56]

---

55 Pingree's edition reads *tasmin samantāt kṛtim ākṛṇoti* and is summarized as "strength" (Pingree, *The Yavanajātaka*, vol. 2, 260), which makes no sense to me. When compared with YJ and BJ, the development of skin is expected, but it is missing here. Note, the reading of *āvṛṇoti* from the mss. CILOQSTW is appropriate here given the concealing nature of Saturn.

56 The reading here is corrupt, with the import of each monthly planet possibly causing affliction to the pregnancy and possibly abortion. Cf. BJ 4.9.

3.20ab yogo yadādhānagataḥ sutāridvitīyaduścimyagataiḥ samastaiḥ /
3.20cd bhavet prasūtir daśame pareṇa tadā narāṇāṃ na ca yoṣitānāṃ //
3.21ab sūryeṇa tāsāṃ pravadanti śeṣaṃ yogaṃ nṛṇāṃ vandhyam anekarūpaṃ /[a]
3.21cd klīvodaye klīvayute navāṃśe klīvasya janma prabhaven narāṇām //
3.22ab evaṃ pumāṃsaḥ prabhavanti nāryaḥ strībhāvayogād bahudhā viśeṣaḥ //
3.23ab navāṃśanāthaś ca yadā vilagnaṃ paśyet suhṛtkendragataḥ sudīptaḥ /
3.23cd tadātmavargeṇa karoti janma pradhānabhūtiṃ manasā narāṇām //
3.24ab duścimyajāmitragatāḥ pracaṇḍāḥ saumyā yadā kośatrikoṇasaṃsthāḥ /
3.24cd puṃjanmadās tatra bhavanti dṛṣṭāś candreṇa jīvena śaśāṅkajena //
3.25ab mandārayoḥ saptamarāśisaṃsthayor yadā niṣeko maraṇaṃ tadā pituḥ /
3.25cd raveḥ śaśāṅkāt tv atha tajjanatyā ekāntarodhaḥ puruṣasya vācyaḥ //
3.26ab yadā himāṃśur vyayago divākaraś chidraṃ gato bhūtanayaś caturthaḥ / [h.m.]
3.26cd mṛtyus tadā saṃbhavati hy ubhābhyām śastreṇa saureṇa tu bandhanena //

---

[a] vandhyam] emend., vādhyam Σ.

[Sex of child]

3.20 If at the time of conception, there is a conjunction [of planets] at the fifth, sixth, second, or third house, and another [planet] in the tenth house, then there would be the conception of males, not females.

3.21ab [The sages predict] the remaining variety of horoscopic configurations with the Sun in these [houses] to be unproductive (i.e., will not lead to conception).

3.21cd+3.22ab When a neuter planet (*klīva-*)[57] is in the Ascendant, [or when] the *navāṃśa* is conjoined with a neuter planet, there would be birth of an intersex among men. In such a manner males and females appear, [the latter] the result of the conjunction with a female house, in multifold variety.

3.23 When the Sun is the lord of *navāṃśa*, being in its own division,[58] in the fourth house or [one of the other] cardines, aspecting the Ascendant, it brings forth the birth of a native, who is foremost in power and in mind among his kinsmen.

3.24 When the benefics are fierce and located in the third, seventh, second or fifth house, aspected by the Moon, Jupiter, and Mercury, they then produce the birth of a male.

[Death, and unfavorable or unusual conception]

3.25 When impregnation (*niṣeka*) takes place when Saturn and Mars are in the seventh house, then there would be the death of the father. If the Sun and Moon [are in the seventh house], then [the death] of its mother. Orphanhood of the native is predicted ...

3.26 When the Moon is in the twelfth house, the Sun in the eighth, Mars in the fourth, then there would be death of both [parents] by sword. If Saturn [is conjoined with one of the planets, there would be death] by ropes.[59]

---

57 That is, Mercury and Saturn, see VYJ 2.13, Mak, "The First Two Chapters," 28.

58 On the concept of division or *varga*, which is described as sixfold in VYJ, see ibid., 28, fn. 113.

59 Here I took *riṣpha* as the twelfth house, as noted in BJ and Utpala's commentary to BJ. In the first chapter (1.38), however, *riṣpha* is assigned to the eleventh house by verse division, but may be, though not desirably, read with the following line, taking the finite verb *pravadanti* as sentence-final. Ibid., 17, 24, fn. 89.

3.27ab mṛtyuṃkaraḥ śītakaraś ca riṣphe sukhasthitaḥ sūryasutaḥ sabhaumaḥ /
3.27cd na garbhasaṃbhūtir iha pradiṣṭā yogaiḥ sasaumyaih pravadanti kṛcchrāt //
3.28ab mūrtisthitas tīkṣṇakaraḥ kujo vā sakṣīṇamūrtir vidhuḥ riṣphago vā /
3.28cd vṛthāphalaḥ syāt suratopacāro nīcāśritais tryādibhir eva puṃsām //
3.29ab klīvasya lagne budhasaurayukte sutasthite vā tv atha riṣphage vā /
3.29cd klīvasya janma pravadanti puṃsāṃ śubhekṣite tatra yathā svarūpam //
3.30ab yadā kujaḥ saptamarāśimāśritaḥ śanaiścaro vā raviṇā samāhitaḥ / [h.m.]
3.30cd na mūrtigo devaguruḥ sito vā tadā na garbhaṃ pravadanti yoṣitāṃ // [d: h.m.]
3.31ab nabhastalastho yadi vāsurārcitas trikoṇago devaguruḥ saśītaguḥ / [h.m.]
3.31cd garbhas tadā saṃbhavati prajānāṃ navāṃśako vā himaraśmijasya //
3.32ab tṛtīyajāmitragatau sitārkau śanaiścaro lābhagato yadā syāt /
3.32cd pumāṃs tadā garbhagataḥ pravācyo jīvo 'thavā svoccagatas trikoṇe //
3.33ab puṃvargage sūryasute mahīje divākare lagnam upāśrite vā /
3.33cd garbhe pumān śītakare 'thavāmbare svatuṅgage śukradṛśā samanvite // 33 [h.m.]
3.34ab kumbhe vilagne mithune 'tha kanye napuṃsake tatsutavargage vā /
3.34cd napuṃsakākhyaṃ pravadanti garbhaṃ tadā narāṇāṃ ca navāṃśake vā //
3.35ab saumyānvite tatra yathā svarupaṃ naraṃ striyaṃ vā tanayair vihīnam /
3.35cd pāpānvite cārdham uśanti nāryā ardhaṃ narasyaiva bhaved dhi garbhe //
3.36ab strīlagnage rātrikare ca śukre samasthite sūryasute sajīve /[a]
3.36cd strīgarbham uktaṃ dharaṇīsutena svavargasaṃsthe śaśinā ca dṛṣṭe //

---

[a] rātrikare] emend., ratrikare Σ.

3.27 The Moon in the twelfth house brings death when Saturn together with Mars are in the fourth house. In such case pregnancy is not expected. [Even] with the conjunction of benefics, [the sages] expect conception to take place only with difficulties.

3.28 When either Sun or Mars is in the Ascendant, or if the Moon is in the twelfth house, [or] when the three [planets] are in debilitation, the sexual conduct would be in vain for yielding males.

3.29 [The sages] predict the [the birth] of an intersex when Mercury and Saturn conjoin in the Ascendant, the fifth or twelfth house. If aspected by a benefic, then the native would be a male in [the planet's] own form.

3.30 If Mars is in the seventh house, conjoined with Saturn or the Sun, and Jupiter or Venus is not in the Ascendant, the sages then predict no conception among the women.

3.31 If Venus is in the tenth house, Jupiter and the Moon in the fifth, or Mercury has [the position of] *navāṃśa*, then there would be conception.

3.32 If Venus and Mars are in the third or the seventh house, or Saturn in the eleventh, or Jupiter is in exaltation and in the fifth or the ninth house, then the conception of a male is predicted.

3.33 If Saturn or Mars is in the male division or if the Sun is in the Ascendant, or the Moon is in the tenth house, in exaltation and aspected by Venus, then there is the conception of a male.

3.34 If a neuter [planet] is in the Ascendant, in Aquarius, Gemini, or Virgo, or in its derived divisions or [own] *navāṃśa*, [the sages] predict the birth of a native as an intersex.

3.35 If conjoined with a benefic, then the man or woman [is predicted to be] without offspring. If conjoined with a malefic, then [the sages] declare half [of the child] would be of a woman. [The fetus] in the womb would indeed be half male.

3.36 If Moon and Venus are in the female Ascendant (i.e., in a female sign), Saturn and Jupiter are in the even houses, [or] when [the Ascendant] Mars is in its own division or if it is aspected by the Moon, the conception of a female is predicted.

3.37ab yadā himāṃśuḥ surapūjito 'thavā śukreṇa dṛṣṭaḥ samarāśisaṃsthitaḥ /
3.37cd tadābalā garbhagatā nṛṇāṃ ca śanaiścare vā ravibhāgam āśrite //
3.38ab dvidehalagne himaraśmiyukte budhe svasaṃsthe ravije ca lābhe /
3.38cd yugmaṃ vadel lagnagataṃ budhena dṛṣṭaṃ striyatvaṃ pravaded ubhābhyāṃ //
3.39ab śanaiścare mūrtigate śaśāṅke ṣaṣṭhe budhe saptamage ca śukre /
3.39cd sūrye 'ṣṭame saumyavivarjite ca trayaḥ pumāṃsaḥ pravadanti garbhe //
3.40ab evaṃ kuje saṃpravadanti nāryo napuṃsakākhyāni divākare ca //

iti śrīvṛddhayavane ādhānādhyāyas tṛtīyaḥ //

Chapter 4: Janman

4.1ab lagnaṃ yadā paśyati devamantrī sito 'thavā rātrikaraḥ prapūrṇaḥ /
4.1cd svoccāśritaḥ kendragataḥ svavarge suśuddhavīryasya bhavet prasūtiḥ //
4.2ab lagnāśrite śītakare budhe vā nīcāśrite vā guruṇā tv adṛṣṭe[a] /
4.2cd anyena jātaṃ pravadanti mānavaṃ +++++++++++ //

[a] tvadṛṣṭe] emend., tu dṛṣṭe. Cf. BJ 5.6, YJ 6.3.

3.37 If the Moon or Jupiter is located in an even sign aspected by Venus, or if Saturn is in the *navāṃśa* (*bhāgam*) of the Sun, then the conception of males would be without strength.

3.38 If the Moon is in the two-bodied Ascendant (i.e., a bicorporal sign), Mercury in the second house, the Sun in the eleventh house, one predicts twins. If the Ascendant is aspected by Mercury, then one predicts female for both [twins].[60]

3.39 If Saturn and the Moon are in the Ascendant, Mercury in the sixth house, Venus in the seventh house, the Sun in the eight house without a benefic, [the sages] predict the conception of male triplets.[61]

3.40 [They predict] the conception of female [triplets] if Mars [is in the eight house without a benefic]. If the Sun, [the sages predict] intersexes.

Chapter 4: Birth

[Condition of the native at birth]

4.1 If Jupiter, Venus, or the full Moon aspects the Ascendant, and (all three planets) are in exaltation,[62] located at the cardinal directions, and in their own division, there will be the birth of a very pure and valiant man.

4.2 If either the Moon or Mercury is in the Ascendant or its debilitation,[63] but not aspected by Jupiter,[64] [the sages] predict a human birth by another [father other than the mother's husband] ...[65]

---

60 Cf. YJ 5.29–32, association with bicorporal signs and Mercury.

61 Cf. YJ 5.33.

62 As defined earlier in 1.44c–46c.

63 Or alternatively, "if the Moon is in the Ascendant and Mercury is debilitated." The first reading is preferred as it is comparable to 4.1; the inclusion of two *vā*-s also makes the latter reading somewhat unsatisfactory. For the definition of debilitation (*nīca*), see 1.46d–1.47.

64 For various configurations involving the Ascendant being aspected by a malefic or a benefic, see also the earlier 3.3. Cf. BJ 4.7 for prediction of the death of the mother when the Moon or the Ascendant is found in between malefics on both sides and not aspected by a benefic.

65 The emendation from *tu* to *tva* is needed to bring out the condition required for illegitimate birth, namely, absence of aspect by Jupiter to the Ascendant and certain planets. Cf. BJ 5.6: *na lagnam induṃ ca gurur nirīkṣate na vā śaśāṅkaṃ raviṇā samāgatam / sapāpako 'rkeṇa yuto 'thavā śaśī pareṇa jātaṃ pravadanti niścayāt //* Similarly, YJ 6.3: *ajīvabhāge 'py anavekṣite vā jīvena candre 'tha vilagnabhe vā / jātaṃ parodbhūtam iti bruvanti vācyo janenādhibalāvalokāt //* The condition of legitimate birth is positively stated in a verse by Garga (*gārgiḥ*) cited by Utpala: *gurukṣetragate candre tadyukte vānyarāśige / taddreṣkāṇe tadaṃśe vā na parair jāta iṣyate //*

4.3ab śanaiścare lagnagate ca nīcage pāpāṃśake pāpayute ca dṛṣṭe / [h.m.]
4.3cd pāpasya lagne guruṇā vimukte pāpāt prasūtiṃ pravadanti candre //
4.4ab svanīcasaṃsthe raviṇā samanvite lagne care sūryasutena vakre / [h.m.]
4.4cd vyayasthite vā dhanage ca candre pituḥ parokṣasya bhavet prasūtiḥ //
4.5ab dvāraṃ vadet kendram upāgatāc ca tatra grahābhāvata eva lagnāt /
4.5cd dīpo 'rkato lagnavaśena vartiḥ snehaṃ śaśāṅkāt pravaded yathāvat //
4.6ab vilagnabhāvodbhavamūrtikāḥ syur nāthād vaded varṇaviśeṣam eṣām /

4.3 If Saturn is in the Ascendant, in its debilitation, in a malefic *navāṃśa*, conjoined with or aspected by a malefic, and if the Moon is in the Ascendant, which is devoid of Jupiter, [the sages] predict the birth of a bad native from a bad [father].[66]

4.4 If Mars is in its debilitation, conjoined with the Sun in the Ascendant, with Saturn in a movable sign,[67] and the Moon in the twelfth or second house, then the native will have an absent father.[68]

[Environment surrounding the native at birth]

4.5 One may predict the [direction of] the door [of the maternity chamber] based on [the planet that is] in the cardines. And in the case of the absence of a planet there (i.e., in the cardines), [the prediction may be] based on the Ascendant.[69] [The position of] the lamp [of the maternity chamber] is based on [the position of] the Sun. [The length of its] wick is [determined] by the strength of the Ascendant (i.e., how far it has risen). In such a way, one predicts the [amount of] oil [in the lamp] based on [the phase of] the Moon accordingly.[70]

4.6 The bodies [of the natives] would be the results of the conditions of the Ascendant.[71] One may predict the different colors of these [bodies] according to the lord.[72] The bed [in the maternity chamber may be predicted] according to the decans.[73] The house [may be predicted] according to the exaltation and

---

66 For parallel materials on 4.2–3, see DEP 56–59.

67 As defined earlier in 1.19, that is, *meṣa, karkaṭa, tūla*, and *makara*.

68 The verse on the absence of father is given in YJ 6.1 and BJ 5.1. The formulation here is different from the other two, which both focused solely on the Moon not aspecting the Ascendant (*anīkṣyamāṇe tu vilagnam indāu / lagnam indāv apaśyati*).

69 Cf. YJ 6.2ab: *gṛhe tu digdvāravidhir grahebhyaḥ kendropagebhyo balavac ca vācyaḥ* / BJ 5.18b: *dvāraṃ ca tadvāstuni kendrasaṃsthair jñeyaṃ grahaiḥ vīryasamanvitair vā* // Once again, the description of BJ and YJ are similar. But VYJ has the extra information on the Ascendant as a secondary criterion, which is described in LJ 6.6a (*dvāraṃ vāstuni kendropagatād grahād asati vā vilagnarkṣāt*) and Maṇittha as quoted by Utpala (*lagne yo dvirasāṃśas tadabhimukhaṃ sūtikāgṛhe dvāram*).

70 The second half of this verse on the lamp, wick, and oil is identical to BJ 5.18a in content and is sketched out in an equally laconic manner: *snehaḥ śaśāṅkād udayāc ca vartir dīpo 'rkayuktarkṣavaśāc carādyaḥ*. This topic is absent in YJ.

71 Cf. BJ 5.23a: *lagnanavāṃśapatulyatanuḥ syād.*

72 Cf. BJ 5.23bc: *vīryayutagrahatulyavapur vā / candrasametanavāṃśapavarṇaḥ.*

73 The use of decans for prediction is related to the bed (cf. BJ 5.21), but it is used for the melothesia (cf. BJ 5.24).

4.6cd śayyā tribhāgād gṛham uccanīcāt teṣāṃ varāṅgāni tathā pareṣāṃ //
4.7ab yadā śaniḥ pañcamadharmago 'thavā bhaumo 'tha bandhuvyayagaḥ kathaṃcana /
4.7cd tadāṅgabhaṅgaṃ pravadet samagraṃ dṛṣṭe balena[a] pravadanti kiṃcit //
4.8ab himāṃśumālī svanavāṃśasthitaḥ svocce 'tha kendre suranāthapūjitaḥ / [h.m.]
4.8cd tulyaṃ tadā dhātuvivṛddhagātraṃ śukreṇa badhnāti samanvitaś ca //
4.9ab bhaume 'ṣṭame sūryasute vilagne nabhastalasthe himaraśmije ca /
4.9cd vane prasūtiṃ pravadanti yoṣitāṃ sūryeṇa mārge manujair vivarjite // [h.m.]
4.10ab śanaiścare mūrtigate himāṃśuje vyayasthite nīcagate prabhākare / [h.m.]
4.10cd vilomajanma pravadanti bhūmije sabhārgave nālaviveṣṭitasya // [h.m.]
4.11ab candre 'mbarasthe śaśije ca kāme bṛhaspatau mūrtigate stanāntare / [h.m.]
4.11cd garbhasya tallāñchanam uktam ādyair viparyaye pṛṣṭhavibhāgataś ca //
4.12ab jīve vilagne bhṛguje vyayasthe śaśāṅkaje lābham upāśrite ca /
4.12cd vāmāṅgasaṃsthas tilakaḥ pravācyaḥ ṣaḍvargaśuddhau tv atha dakṣiṇasthaḥ //

[a] dṛṣṭaṃ balena] emend., dṛṣṭaṃ baleśaṃ Σ, kṛṣṭaṃ balonaṃ *ped.*

debilitation.[74] Their heads (*varāṅgāni*) [correspond to the Ascendant] and for the other [parts of the body], it would be similarly so [as the zodiacal man] (?).[75]

[Deformity, unusual birth, and bodily marks]

4.7 If Saturn is in the fifth and ninth house, or if Mars is in the fourth and twelfth house, then one would predict completely broken limbs. If it is aspected by a powerful [benefic], [the sages] may predict some [broken limbs].[76]

4.8 If the Sun and the Moon are in their own *navāṃśa*, Jupiter in exaltation or the cardine, conjoined with Venus, they similarly cause obstruction to the body through the excess of elements.

4.9 If Mars is in the eighth house, Saturn in the Ascendant, and Mercury in the tenth house, they predict the birth of females in the forest. If the Sun [is in the tenth house], then [the child will be born] on a path devoid of people.

4.10 If Saturn is in the first house, the Moon in the twelfth, the Sun in its debilitation, and Mars conjoins with Venus, [the sages] predict an inverted birth [with the fetus] entwined with the umbilical cord (*nālaviveṣṭita*).[77]

4.11 When the Moon is in the tenth house, Mercury in the seventh, and Jupiter in the first house, the excellent ones predict a special mark near the breast of the fetus. In case of the opposite, [the mark] would be proportionally on the back (?).

4.12 When Jupiter is in the Ascendant, Venus in the twelfth, Mercury sitting in the eleventh house, a bodily mark located on the left limb is predicted. If the sixfold divisions [of the sign of the planet] are pure (*ṣaḍvargaśuddhi*), then [the bodily mark] is located on the right [limb].[78]

---

74 Cf. YJ 6.5: *nīcoccatoyāstamayācarasthaiḥ kāryā gṛhāṇām avikalpanā ca.*

75 Cf. BJ 5.23d: *kādivilagnavibhaktabhagātraḥ*. The scheme of melothesia is not given here but is given in BJ 5.24.

76 Neither the main variant reading *dṛṣṭaṃ baleśaṃ* nor Pingree's emendation *kṛṣṭaṃ balonaṃ* is grammatically fitting. The emendation *dṛṣṭe balena* is supplied here on the basis of the structure of a number of verses in this chapter, with *pada* d stating an additional protasis and apodasis (4.8, 11, 12, 13).

77 Cf. BJ 5.5: *chāge siṃhe vṛṣe lagne tatsthe saure 'thavā kuje / rāśyaṃśasadṛśe gātre jāyate nālaveṣṭitaḥ //*

78 The expression *ṣaḍvargaśuddhi-* first occurred in 2.13.

4.13ab vyaye śaśī lābhagataḥ kharāṃśus trikoṇagaḥ sūryasutaḥ sapāpaḥ /
4.13cd hīnāṅgatāṃ tatra vadanti bhūmije gātrādhikatvaṃ pravadanti bhāgataḥ // [h.m.]
4.14ab śanaiścare vīryayute tu taijasaṃ sūryeṇa tāmrodbhavam eva bhūṣaṇam / [h.m.]
4.14cd candreṇa māṇikyabhavaṃ hiraṇyajaṃ saumyena śukreṇa ca rītisambhavam // [h.m.]
4.15ab jīvena nānāvidham iṣṭakāmadaṃ bhaumena vindyād bahuvastrasambhavam // [h.m.]
4.16ab gṛhaṃ śaśāṅkena sudārusambhavaṃ manojarūpaṃ vividhānvitaṃ ca / [a: h.m.]
4.16cd susūtram uktaṃ maṇikuṭṭimānvitaṃ ṣaḍvargaśuddhena ca tuṅgagena // [h.m.]
4.17ab bhaume tu dagdhaṃ svanavāṃśasthe parāṃśakasthe śiśilasvabhāvam / [h.m.]
4.17cd kuvastuyuktaṃ bahupāpadṛṣṭe saumyekṣite vā dhanadhānyayuktam //
4.18ab sūryeṇa citraṃ bahugairikādyaiḥ prabhāsametaṃ bahubhūmikaṃ ca /
4.18cd ekatra ramyaṃ tv aparatra śīrṇaṃ pāpekṣite vā tapanasvabhāvam //
4.19ab budhena ramyaṃ bahuvaṃśayuktaṃ paṭāvṛtaṃ bhūrivanāḍhyam eva /
4.19cd śucipragalbhaiḥ puruṣaiḥ sametaṃ vicitrarūpaṃ prabhayā sametam //

4.13 If the Moon is in the twelfth house, the Sun in the eleventh house, Saturn in a trine house (fifth or ninth) conjoined with a malefic, [the sages] predict then the loss of limbs. If Mars is in that [house], then [the sages] predict an additional limb or a partial [limb formation].

[Characteristics of birthplace]

4.14 If Saturn is conjoined with a strong [planet], [the decoration of the house is] brilliant. If [conjoined with] the Sun, the decoration [of the house] will be made of copper. With the Moon, it will be made of ruby. With Mercury, it will be made of gold. With Venus, it will be made of brass.[79]

4.15 [If Saturn is conjoined] with Jupiter, [the decoration will be] of various kinds which fulfill one's desires. With Mars, it will be in the form of many clothes.

4.16 [If Saturn is conjoined] with the Moon in exaltation and its sixfold division is pure, the house is predicted to be made of good wood, with an appearance [as lovely as] the love god (*manoja-*), furnished with various [things], and it has a floor with inlaid gems, and is nicely designed.

4.17 But if Mars is in its own *navāṃśa* [while Saturn is conjoined with the Moon], then [the house will be] burnt. If [Mars] is in another *navāṃśa*, then [the house] is of a cool nature. If [Saturn is] aspected by many malefics, it becomes a bad house. If aspected by the benefics, [the house] is endowed with wealth and grain.

4.18 [If Saturn is conjoined] with the Sun, [the house will be] painted with a great amount of brilliant mineral and have much land. It will be pleasant in one place but faded in another. Or, if [the Sun] is aspected by malefics, [the house] will be hot.

4.19 [If Saturn is conjoined] with Mercury, [the house] will be pleasant, with many reeds. It will be covered with fine drapes, surrounded by great forest. It will be filled with virtuous and brave men, have a charming appearance, and be brilliant.

---

79 Cf. BJ 5.8–17, YJ 6.4–12.

4.20ab jīvena nānāvidharatnayuktaṃ subhūriśālaṃ bahumaṇḍalāḍhyam /
4.20cd vitānacūlādhvajasarvadikṣu pramaṇḍitaṃ dhāturasair viśālaiḥ //
4.21ab svabhāvaramyaṃ suvidagdhalokaṃ manoramaṃ sphāṭikakāñcanāḍhyam /
4.21cd tuṅgaṃ viśālaṃ suvibhaktamārgaṃ vicitrakāṣṭhaprabhavaṃ nitāntaṃ //
4.22ab śīrṇaṃ kucelaiḥ manujaiḥ sametaṃ raktānvitaṃ strīrahitaṃ sadaiva /
4.22cd bhaved gṛhaṃ kaṇṭakasaṅghayuktaṃ raudraṃ karālaṃ vikṛtapratāpam //
4.23ab śanaiścare kendragate svanīcage sutasthite bhūtanaye sabhāskare / [h.m.]
4.23cd gṛhaṃ tṛnaiḥ sarvata eva yuktaṃ jīrṇaṃ nitāntaṃ tapanasvabhāvam //
4.24ab śanaiścarāṃśe himaraśmiputre yadā tanusthaḥ svagṛhe śaśāṅkaḥ /
4.24cd tadā gṛhaṃ nūtanavīrudhāvṛtaṃ kvacit sukāṣṭhaiḥ parito vipaścitām // [h.m.]
4.25ab jīvo yadā bandhugataḥ svabhāgago nabhastalastho bhṛgunandanas tadā / [h.m.]
4.25cd candro 'stago mūrtigataḥ śanaiścaras tadāvṛtaṃ vastravaraiḥ sumandiram // [h.m.]
4.26ab śanaiścare mūrtigate kuje 'thavā pāpāṃśakasthe bahupāpavīkṣite / [h.m.]
4.26cd gṛhāśrayaṃ janma vadec chubhāṃśake kuṭārake chidrayute virūpake // [h.m.]

iti śrī vṛddhayavane janmādhyāyaś caturthaḥ //

4.20 [If Saturn is conjoined] with Jupiter, [the house] will have different kinds of gems, will have great many halls and great surrounding land. It will be furnished with all such things as awnings, turrets, flags, and wondrous mineral materials.

4.21 [The house will be] pleasant by nature, have a bright tawny appearance,[80] charming, endowed with crystal and gold. It will be tall and spacious, its paths well proportioned. It is made of various woods and is excellent.

4.22 [If Saturn is conjoined with Mars, the house will be] shattered, associated with men in bad clothes, imbued with passion, always deprived of women. The house would have a mass of thorns. It is fierce, dreadful, and has wicked power.[81]

4.23 If Saturn is in a cardine and is in its debilitation, and Mars is in the fifth house conjoined with the Sun, then the house will be surrounded by grass everywhere. It is excessively withered and scorched.

4.24 When Mercury is in the *navāṃśa* of Saturn, and if the Moon is in its own sign and in the first house, then the house will be somewhere, covered by new creepers and [made of] good wood. There will be learned people everywhere.

4.25 If Jupiter is in the fourth house and in its own division, Venus in the tenth house, the Moon in the seventh house, Saturn in the first house, then [the house] will be covered with the best drapes and have an excellent temple.

4.26 If Saturn is in the first house, and Mars in a bad *navāṃśa*, or a good *navāṃśa* but aspected by many malefics, then one predicts the birth inside a house where the roof (*kuṭārake*?) will have holes and an unseemly appearance.

80 Or associated with brilliant clever men. Reference to a specific type of man associated with a house is also found in the next verse (4.22).

81 The *protasis* is not stated at all. The *apodasis* at any rate does not connect with the previous verse and we supply here the most negative combination of malefics, namely Saturn and Mars.

Abbreviations

BJ	*Bṛhajjātaka* by Varāhamihira (Jhā, ed.)
CESS	*Census of the Exact Sciences in Sanskrit* (Pingree, ed.)
T	*Taishō Shinshū Daizokyō*大正新脩大藏經 (Takakusu and Watanabe, eds.)
VYJ	*Vṛddhayavanajātaka* by Mīnarāja (Pingree, ed.)
YJ	*Yavanajātaka* by Sphujidhvaja (Pingree, ed.)

Bibliography

Eliot, Charles. *Hinduism and Buddhism: An Historical Sketch*. London: Edward Arnold & Co., 1921.

Geller, Markham J. *Melothesia in Babylonia: Medicine, Magic, and Astrology in the Ancient Near East*. Berlin: De Gruyter, 2014.

Karashima, Seishi. "The Triṣṭubh-Jagatī Verses in the *Saddharmapuṇḍarīka*." *ARIRIAB* (*Annual Report of the International Research Institute for Advanced Buddhology at Soka University*) 19 (2016): 193–210.

Jhā, Sītārāma, ed. *Bṛhajjātakam*. Varanasi: Ṭhākuprasāda, 1973.

Karttunen, Klaus. *Yonas and Yavanas in Indian Literature*. Helsinki: Finnish Oriental Society, 2015.

Lin, Hsin-Yi. *Dealing with Childbirth in Medieval Chinese Buddhism: Discourses and Practices*. Ph.D. dissertation. Colombia University, 2017.

Mak, Bill. "The Date and Nature of Sphujidhvaja's *Yavanajātaka* Reconsidered in the Light of Some Newly Discovered Materials." *History of Science in South Asia* 1 (2013): 1–20.

Mak, Bill. "The Last Chapter of Sphujidhvaja's *Yavanajātaka* Critically Edited with Notes." *SCIAMVS* 14 (2013): 59–148.

Mak, Bill. "The 'Oldest Indo-Greek Text in Sanskrit' Revisited: Additional Readings from the Newly Discovered Manuscript of the *Yavanajātaka*." *Journal of Indian and Buddhist Studies* 印度學佛教學研究 62, no. 3 (2014): 1101–1105.

Mak, Bill. "The First Two Chapters of Mīnarāja's *Vṛddhayavanajātaka*." *Zinbun* 48 (2017): 1–31.

Meulenbeld, G. Jan. *A History of Indian Medical Literature*. 3 vols. Groningen: E. Forsten, 1999.

Needham, Joseph. *A History of Embryology*. Second edition. Cambridge: Cambridge University Press, 1959.

Pingree, David. *Census of the Exact Sciences in Sanskrit*. 5 vols. Philadelphia: American Philosophical Society, 1970–1994.

Pingree, David, ed. *Vṛddhayavanajātaka of Mīnarāja*. 2 vols. Baroda: Oriental Institute, 1976.

Pingree, David. *The Yavanajātaka of Sphujidhvaja*. 2 vols. Harvard Oriental Series 48. Cambridge: Harvard University Press, 1978.

Neugebauer, O. "Decem tulerunt fastidia menses." *The American Journal of Philology* 84 (1963): 64–65.

Sachau, Edward C. *Alberuni's India*. 2 vols. London: Trübner, 1888.

Stol, Marten. *Birth in Babylonia and the Bible: Its Mediterranean setting*. Groningen: Styx, 2000.

Swami Sada Shiva Tirtha. *The Āyurveda Encyclopedia*. Delhi: Sri Satguru, 1998.

Takakusu, Junjirō and Kaigyoku Watanabe et al, eds. *Taishō Shinshū Daizokyō* 大正新脩大藏經. 85 vols. Tokyo: Issaikyō kankōkai, 1924–1932.

Wujastyk, Dominik. "Indian Medicine." In *Companion Encyclopedia of the History of Medicine*, edited by W.F. Bynum and Roy Porter, 755–778. London: Routledge, 1993.

Yano, Michio. "Medicine and Divination in India." *East Asian Science, Technology, and Medicine* 24, no. 4 (2005): 44–61.

Yano, Michio 矢野道雄. "Indo-no senseijutsu-tenmonsho-ni mirareru girishago-karano shakuyōgo-ni tsuite" インドの占星術・天文学書に見られるギリシア語からの借用語について [Greek Words Borrowed in Sanskrit Astronomical and Astrological Texts] (in Japanese). *The Bulletin of the International Institute for Linguistic Sciences Kyoto Sangyo University* 京都産業大学国際言語科学研究所所報 8 (1987): 74–85.

Zimmermann, Francis. "Les Aspects Médicaux du *Yavanajātaka* (traité sanskrit d' astrologie)." *Sudhoffs Archiv für Geschichte der Medizin (und der Naturwissenschaften) Archiv* 65, no. 3 (1981): 299–305.

Zysk, Kenneth G. *Asceticism and Healing in Ancient India: Medicine in the Buddhist Monastery*. New York and Oxford: Oxford University Press, 1991.

Zysk, Kenneth G. *The Indian System of Human Marks*. Leiden and Boston: Brill, 2016.

CHAPTER 21

Modern Sāṃkhya: Plurality of Sāṃkhya Traditions in Modern India

Knut A. Jacobsen

Is there a modern Sāṃkhya in India, and in that case, what is it? Was there a modern revival of Sāṃkhya in India? And did the enormous interest in modern Yoga, which for long has favored the *Yogasūtra*, a text which, together with its *bhāṣya* (*Vyāsabhāṣya* or *Yogabhāṣya*) represents a second school of Sāṃkhya philosophy called Yoga, lead also to a greater interest in modern Sāṃkhya? In a recently published booklet, *Modern Sāṃkhya*, which gives an interpretation of Sāṃkhya spirituality for a Western audience, the author states that "no one has ever attempted to reconstruct classical Sāṃkhya to be utilized as a contemporary spiritual path."[1] This erroneous statement primarily shows that the modern Sāṃkhya gurus and institutions in India who have taught and followed Sāṃkhya as a spiritual path are not well-known, and hardly known at all outside of India. It illustrates that there is a need for researchers on Sāṃkhya to pay more attention to modern Sāṃkhya teachers and institutions. My recently published book, *Yoga in Modern Hinduism: Hariharānanda Āraṇya and Sāṃkhyayoga* was a contribution to this new research.[2] The book was an analysis of a late nineteenth- and early twentieth-century Indian institutionalization of Sāṃkhya as a spiritual path, the institution of Kāpil Maṭh, its teachers and disciples and their practices and writings. In this chapter I will first introduce the Kāpil Maṭh tradition as a contemporary institutionalization of Sāṃkhya and thereafter give some brief and preliminary remarks on two modern Indian spiritual teachers from other traditions who have claimed to follow and promote Sāṃkhya. Sāṃkhya in Indian history is a diverse phenomenon, and these modern teachers do manifest some of this diversity. Furthermore, adding to the diversity, influences from encounters with the West are detectable in these modern Indian Sāṃkhya teachings. Modern Sāṃkhya

1 D.E. Osto, *Modern Sāṃkhya: Ancient Spirituality for the Contemporary Atheist* (Palmerstone North, New Zealand, 2016).

2 Knut A. Jacobsen, *Yoga in Modern Hinduism: Hariharānanda Āraṇya and Sāṃkhyayoga* (London: Routledge, 2018).

 | DOI:10.1163/9789004438224_022

illustrates both the plurality of Sāṃkhya traditions in India, historical and contemporary, and that Sāṃkhya in the nineteenth and twentieth centuries was globalized.

There have been numerous teachers of Yoga in modern India, but the number of teachers of Sāṃkhya has been limited. Consequently, many have probably been under the impression that Sāṃkhya was no longer a living tradition in modern India. Vedānta has dominated modern Hinduism and many modern interpretations of the foundational text of the Yoga *darśana*, the *Yogasūtra* have also been influenced by Vedānta. Yoga practice has been dominated by physical exercises and been influenced by Haṭhayoga and other schools of Yoga, not by Sāṃkhya. Body yoga dominates modern Yoga, but body yoga plays a very minor role in Sāṃkhya and the *Yogasūtra*.

1 Three Examples of Sāṃkhya as a Contemporary Spiritual Path

The Kāpil Maṭh tradition was the most significant institutionalization of modern Sāṃkhya in India.[3] The tradition has not been well known, for several reasons, although the English translation (by P.N. Mukerji) of the Bengali text *Kāpilāśramīya Pātañjal Jogdarśan* ("The Yoga Philosophy of Patañjali of the Kāpilāśram") of Hariharānanda Āraṇya,[4] was published in an American edition by State University of New York Press in 1983,[5] which made a new audience aware of his Yoga teaching. The *Kāpilāśramīya Pātañjal Jogdarśan* is a Bengali translation of the Sanskrit text *Pātañjalayogaśāstra* (*Yogasūtra* and its autocommentary *Yogabhāṣya* or *Vyāsabhāṣya*) and a detailed Bengali commentary with comprehensive explanatory notes.[6] The center of Kāpil Maṭh is in Madhupur in Jharkhand. The tradition has remained small, its teachers and disciples have valued solitude, and the gurus one after the other have isolated themselves in Kāpil Maṭh's artificial cave. The cave building has no entry or exit door but has a window opening into a small meditation hall, and the gurus have met the Sāṃkhya devotees a few days a month through this window opening. The founder of the Kāpil Maṭh tradition Hariharānanda Āraṇya (1869–1947) and the second guru of the institution Dharmamegha Āraṇya (1892–1985) attracted

3 See Jacobsen, *Yoga in Modern Hinduism*.

4 Hariharānanda Āraṇya, *Yoga Philosophy of Patañjali*, trans. P.N. Mukerji, third edition, revised and enlarged (Calcutta: Calcutta University Press, 1981).

5 Hariharānanda Āraṇya, *Yoga Philosophy of Patañjali*, trans. P.N. Mukerji (Albany: University of New York Press, 1983).

6 Hariharānanda Āraṇya, *Kāpilāśramīya Pātañjal Jogdarśan* (Kolkata: Praścimabai Rājya Puṣtak Parṣad, 1988 [1911]).

a few disciples, some of whom became *saṃnyāsin*s, but most of their followers remained lay disciples. Several Sāṃkhya scholars have been lay disciples of Kāpil Maṭh, and these scholars represent a significant part of twentieth century Indian scholarship on Sāṃkhya and Sāṃkhyayoga.[7] The gurus, *saṃnyāsin*s and lay disciples have published a large number of books on Sāṃkhya in Bengali, Hindi, Sanskrit and English.[8] The Kāpil Maṭh tradition considers Sāṃkhya and Yoga as two parts of one philosophical system. Sāṃkhya is the theory part and Yoga provides the practice part.

Hariharānanda Āraṇya's excellent skills in Sanskrit and his views on the life style of the *saṃnyāsin* connected him to the classical traditions of Sāṃkhya and Yoga. His translation of the *Pātañjalayogaśāstra* from Sanskrit to Bengali, with explanatory notes, was his most important work, but he also wrote many texts in Sanskrit, also Sanskrit commentaries on Sāṃkhya and Yoga texts, recalling the classical Sanskrit philosophical commentarial tradition. Āraṇya represents a form of Sāṃkhyayoga orthodoxy in the modern world, but his orthodoxy was a revival influenced by the interest of Orientalists in Sāṃkhya and the ancient Sāṃkhya and Sāṃkhyayoga texts, as I have argued in *Yoga in Modern Hinduism*. Āraṇya's knowledge of Sanskrit, his interest in, and access to the ancient texts, and his fondness for solitary life, living locked up in an artificial cave the last twenty years of his life, made him seem as if belonging to a pre-modern Sāṃkhya tradition. However, he revived the tradition based on his access to ancient Sanskrit texts and choice of an ascetic solitary life style.[9] The Maṭh survived beyond the passing away of the founder because a lineage of gurus was established. The gurus have continued the solitary life style of the founder. The current guru is the third in line.

The Kāpil Maṭh tradition has been the most significant institution of Sāṃkhya as a spiritual practice in modern India, but not the only one. A small modern Sāṃkhya tradition was the one that was promoted by the Bengali saint Śrī Anirvāṇ (1896–1978, born as Narendra Chandra Dhar) and his Western disciple Lizelle Reymond (1899–1994). But Anirvāṇ and Reymond promoted a quite different tradition of Sāṃkhya than the Kāpil Maṭh tradition.

Anirvāṇ studied Sanskrit at the University of Calcutta and lived in the *āśram* of Svāmī Nigamānanda of Sārasvat Saṅgh in Kokilamukh near Jorhat in Assam. Anirvāṇ's monastic name was Nirvāṇānanda. Swami Nigamānanda had a varied spiritual background in Śāktism, Tantrism, and Vedānta, but he had stayed

7 See Jacobsen, *Yoga in Modern Hinduism*, 189–201.
8 See Jacobsen, *Yoga in Modern Hinduism*, 189–201 and 209–213.
9 Jacobsen, *Yoga in Modern Hinduism*.

in the Theosophical Society at Adyar and was also inspired by Vivekānanda.[10] After a number of years in the *āśram* of Nigamānanda, at the age of thirty, Anirvāṇ gave up his monastic life and left. He then changed his name from Nirvāṇānanda to Anirvāṇ ("undying," "nonextinguishable"). He thereafter identified himself as a Baul. Anirvāṇ wrote a number of books, all in Bengali, the three volumes *Veda-Mīmāṃsā* is considered his *magnum opus*,[11] except for one book that was written in English, *Buddhiyoga of the Gita and other Essays*.[12] Anirvāṇ translated Aurobindo's *Life Divine* from English into Bengali.[13] During the translation of *Life Divine*, he retired to Almora for four years, and it was here Lizelle Reymond met him and stayed with him. Anirvāṇ and Reymond characterized themselves as followers of "a spiritual discipline based on the Sāmkhya teaching."[14] However, this Sāṃkhya teaching is not the philosophical Sāṃkhya and Sāṃkhyayoga *darśanas* that were promoted by the *Sāṃkhyakārikā* and the *Pātañjalayogaśāstra* and their commentarial traditions in the Sanskrit language,[15] and by the Kāpil Maṭh tradition, which was a revival of these traditions. In contrast, the teachings of Anirvāṇ and Reymond do not build on the Sāṃkhya and Sāṃkhyayoga philosophical Sanskrit commentarial texts and do not seem to have been formed by them. They recognize that there is a Sāṃkhya and Sāṃkhyayoga philosophical Sanskrit tradition, but they identify their own Sāṃkhya as a different one. Anirvāṇ's Sāṃkhya is in fact not recognizable as philosophical Sāṃkhya. Sāṃkhya concepts are used, but because of the great influence of Sāṃkhya on the Hindu religious teachings, these concepts are also found in many other traditions, classical as well as modern. Maya Burger comments that the Sāṃkhya that was most important to Anirvāṇ "encompasses the entire Indian tradition from the Vedas to

---

10 Jacobsen, Knut A., "In Kapila's Cave: A Sāṃkhya-Yoga Renaissance in Bengal," in *Theory and Practice of Yoga: Essays in Honour of Gerald James Larson*, ed. Knut A. Jacobsen (Leiden: Brill, 2015), 340; Benoy Gopal Ray, *Religious Movements in Modern Bengal* (Santiniketan: Visva Bharati, 1965), 100; Ram Swarup, "Introduction," in Anirvan, *Buddhiyoga of the Gita and Other essays* (New Delhi: Biblia Impex, 1983), vii.

11 For a list of the published works of Śrī Anirvāṇ, see Shri Anirvan, *Buddhiyoga of the Gita and other Essays* (New Delhi: Biblia Impex, 1983), 180–181.

12 Ram Swarup, "Introduction," in *Inner Yoga: Selected Writings of Sri Anirvan*, ed. Margaret Case (Sandpoint, ID: Morning Light Press, 2007), v.

13 In Bengal he was apparently known mainly among the followers of Aurobindo, because of his Bengali translation of Sri Aurobindo's *Life Divine*. See http://anantaprema.blogspot.co.uk/2008/10/anirvan-on-sri-aurobindo.html, accessed October 2, 2019.

14 Lizelle Reymond, *To Live Within*, trans. Nancy Pearson and Stanley Spiegelberg (London: George Allen & Unwin, 1971), 3.

15 Summaries of the texts of these two philosophical traditions are found in Larson and Bhattacharya 1987 and 2008.

the Tantras and Purāṇas."[16] Anirvāṇ's Sāṃkhya, writes Burger, "reveals many combined perspectives, tantric and psychological."[17] Burger considers Anirvāṇ's Sāṃkhya to be a "mystical path enshrined in the Vedas and the Upaniṣads, which in the course of centuries has found clear expressions in the Purāṇas and the Tantras, especially in the later scriptures."[18] There have also historically been many traditions of Sāmkhya, one of which has been identified as nondifferentiated Sāṃkhya and is different from the *darśana* traditions associated with the *Sāṃkhyakārikā* and the *Pātañjalayogaśāstra*, but encompasses larger part of the Indian tradition, Vedas, epics, Purāṇas, and Tantras. Nondifferentiated Sāṃkhya was found in the Upaniṣads as well as in the *Mahābhārata*, and flourished unrestrained by the development of the technical Sāṃkhya of the *Sāṃkhyakārikā* and the *Pātañjalayogaśāstra* traditions.[19] This nondifferentiated Sāṃkhya has blended with many systems of knowledge and traditions of religion throughout history. Nondifferentiated Sāṃkhya can be defined as a way of thinking about the world by means of certain Sāṃkhya concepts and seems not to be a system of philosophy, but part of Hindu religious life. In nondifferentiated Sāṃkhya the highest principles are sometimes considered to be *brahman*, *ātman*, *kṣetrajña*, *puruṣa*, Īśvara, Nārāyaṇa, Kṛṣṇa, etc. Nondifferentiated Sāṃkhya continued to flourish from the period of the theistic Upaniṣads, the *Kaṭhopaniṣad* and *Śvetāśvataropaniṣad*, and the *Mahābhārata* up to the present. One important textual tradition of Sāṃkhya different from the Sāṃkhya systems of the *Sāṃkhyakārikā* and the *Pātañjalayogaśāstra*, emerging perhaps from this undifferentiated Sāṃkhya, is found in the Sāṃkhya chapters of the *Bhāgavatapurāṇa* called the *Kapilagītā* or *Kapilopadeśa* (*Bhāgavatapurāṇa* 3.25–3.33). This Sāṃkhya is also identified with a sage by the name of Kapila.[20] Anirvāṇ's Sāṃkhya shows some kinship to the

16 Maya Burger, "Sāṃkhya in Transcultural Interpretation: Shri Anirvan (Śrī Anirvāṇa) and Lizelle Reymond," in *Yoga in Transformation: Historical and Contemporary Perspectives*, eds. Karl Baier, Philipp André Maas, and Karin Preisendanz (Göttingen: V&R Unipress GmbH, 2018), 475.

17 Burger, "Sāṃkhya in Transcultural Interpretation," 478.

18 Burger, "Sāṃkhya in Transcultural Interpretation," 474.

19 Johannes Bronkhorst, "Systematic Philosophy Between the Empires: Some Determining Features," in *Between the Empires: Society in India 300 BCE to 400 CE*, ed. Patrick Olivelle (Oxford: Oxford University Press), 287–313.

20 There are several Kapilas in the history of Sāṃkhya (see Jacobsen 2008). In addition to the *Sāṃkhyakārikā* tradition, a child Kapila was promoted in the *Bhāgavatapurāṇa*, and the *Sāṃkhyasūtra* (fifteenth century CE) was also supposedly written by a person called Kapila. Kapila seems to have been a common name. In the Pāli texts alone, twelve different Kapilas have been identified. See http://www.vipassana.info/ka/kapila.htm, accessed

nondifferentiated Sāṃkhya tradition. Ram Swarup, who has written the "Introduction" in several of Anirvāṇ's books,[21] writes about Anirvāṇ:

> He started as a *Sāṅkhya*[22] teacher and *saṅkhya* (sic) *Sādhanā*, according to him, is based on self-observation, on equal-mindedness, on opening out to new influences and experiences, on recreating a new "I," on becoming a Puruṣa, which is the master of Prakṛti. This is the *sādhanā* he taught to many pupils including Lizelle Reymond.[23]

But Anirvāṇ also used many non-Sāṃkhya concepts and ideas. Ram Swarup continues:

> Shri Anirvan also talked of the Void, of death in life, of becoming sky-like, *ākāśavat*. Void to him was the prototype of the life of a liberated man. ... Shri Anirvan often talked of "God" ... His God was derived from Hindu scriptures which has *advaita* as its background, a Soul-God.[24]

In *Letters from a Baul: Life Within Life*, Lizelle Reymond commented "He always spoke with reference to the Science of Samkhya, not the philosophical Samkhya taught by Vijnana Bhikshu in the XVe century, but that Samkhya which is the essence of life."[25] For Reymond, the Sāṃkhya of Vijñānabhikṣu here perhaps denotes the whole Sāṃkhya *darśana* tradition. Anirvāṇ's teaching seems to have been a mixture of nondifferentiated Sāṃkhya, Tantra, Vedānta, Sahajiyā, Bauls and New Age thought. An essay called "Sāṃkhya" is found in Anirvāṇ's *Letters from a Baul*.[26] Here he refers to *Bhāgavatapurāṇa* and not the Sāṃkhya *darśana* texts. He writes "There are two Samkhyas: the first is philosophical, the second, mystical."[27] The first, philosophical Sāṃkhya is "all that is known

October 2, 2019. In addition, there are many different Kapilas mentioned in Hindu texts, as well as some in the Jain texts.

21 Ram Swarup, "Introduction," in Anirvan, *Buddhiyoga of the Gita and Other essays* (New Delhi: Biblia Impex, 1983), vi–xvi; and Ram Swarup, "Introduction," in *Inner Yoga: Selected Writings of Sri Anirvan*, ed. Margaret Case (Sandpoint, ID: Morning Light Press, 2007), ii–xii.

22 In quotations the transliterations of terms from Indian languages follow those used by the authors.

23 Ram Swarup "Introduction," in Anirvan, *Buddhiyoga of the Gita and Other essays*, xii.

24 Ram Swarup "Introduction," in Anirvan, *Buddhiyoga of the Gita and Other essays*, xii.

25 Reymond in Anirvan, *Letters from a Baul*, i.

26 Shri Anirvan *Letters from a Baul: Life Within Life*, compiled by Lizelle Reymond (Calcutta: Sri Aurobindo Pathamandir, 1983), 11–28.

27 Anirvan, *Letters from a Baul*, 17.

of Samkhya in the West," he thinks. The mystical Sāṃkhya, he claims, is the Sāṃkhya known in the Vedas, Upaniṣads and Purāṇas and especially in the Tantras.[28] He argues that "the whole Tantric way of life is none other than a Saṃkhya whose sources can be discovered in the most ancient Vedic verses."[29] That there are two Sāṃkhyas, relate to the tradition that there were two Kapilas, one the promoter of Sāṃkhya found in the *Sāṃkhyakārikā* tradition and another in the *Bhāgavatapurāṇa*. In the *Bhāgavatapurāṇa*, Sāṃkhya and Yoga is taught by a child by the name of Kapila to his mother Devahūti, and the purpose is to prepare for meditation on Kṛṣṇa. The founder of ISKCON, Bhaktivedānta Prabhupāda, also from Bengal, in a book on Kapila in the *Bhāgavatapurāṇa*, also promoted the idea that there were two Kapilas, and therefore two Sāṃkhyas, one known in the West, and the other one the Sāṃkhya of *Bhāgavatapurāṇa*,[30] using some of the same arguments as Anirvāṇ. For Anirvāṇ though, the real Sāṃkhya was Tantra.[31]

The nondifferentiated Sāṃkhya has throughout history blended with many systems of knowledge and religion, and Anirvāṇ is an example of this. Anirvāṇ argued against philosophical Sāṃkhya. He argued that, contrary to philosophical Sāṃkhya, there is a higher *prakṛti* beyond the complex of the three *guṇas*, which is *śuddhisattva* totally untouched by *rajas* and *tamas*.[32] This, he wrote, is the "concept of *sat-cit-ānanda* common to the mystical philosophies of Samkhya and Vedanta."[33] Anirvāṇ also believed that divine energy (*śakti*) is "the most subtle element" of *prakṛti*. On suffering, Anirvāṇ writes:

> There is no question of fleeing from pain, but rather of experiencing it as the embrace in which Kali enfolds Shiva prostrate beneath her. This is technically described in the Tantras by the word *viparitarati* (or the inverted coitus), in which the Puruṣa is passively accepting even death and destruction from then active Prakriti. Evil and pain, from which the world-negating Vedanta assiduously turns away, are transformed here into bliss (Ananda).[34]

---

28 Anirvan, *Letters from a Baul*, 18.

29 Anirvan, *Letters from a Baul*, 18.

30 A.C. Bhaktivedanta Swami Prabhupada, *Teachings of Lord Kapila: The Sone of Devahūti* (New York: Bhaktivedanta Book Trust, 1979).

31 Anirvan, *Letters from a Baul*, 18.

32 This view agrees with the idea of Īśvara in the *Pātañjalayogaśāstra*. In the *Pātañjalayogaśāstra*, Īśvara is a particular *puruṣa* who has never been bound to *saṃsāra* and is untouched by *rajas* and *tamas* but is associated with a particular pure dimension of *sattva*.

33 Anirvan, *Letters from a Baul*, 19.

34 Anirvan, *Letters from a Baul*, 18.

In Anirvāṇ's texts there are detectable influences from the Bengali Tantric traditions especially of the Sahajiyās and Bauls. Anirvāṇ uses the term *sahaja* as the ultimate goal, a term used by the Sahajiyās, to connote the ultimate state of salvific liberation. Anirvāṇ called himself a Baul, and the Bauls are considered a branch of modern Sahajiyās.[35] "Sāṃkhya is a technique to realize the expansion of *sahaja*," wrote Anirvāṇ,[36] and *sahaja* is "a state in which the mind is freed from all duality."[37]

We see in Anirvāṇ different traditions blended with nondifferentiated Sāṃkhya, and the integration of Sāṃkhya, Vedānta, Tantra, Sahajiyā, etc. and possibly some similarities to Aurobindo, but his Sāṃkhya does not connect to the Sāṃkhya philosophical tradition of the *Sāṃkhyakārikā* and its classical commentaries. Sāṃkhya is interpreted in a general sense as an attitude of dispassion and cultivation of witness consciousness:

> the ideal of Samkhya is to learn how to stand back, and the ideal of *yoga* is *vairāgya*, which means to learn how to observe oneself dispassionately and without judgment.[38]

Anirvāṇ's Sāṃkhya seems also to have contacts with Western Esotericism. Reymond states that "*Karma* is the occult force of *prakriti* (Great nature) which brings people together for some secret purpose."[39] The terms "occult force" and "secret purpose" seem quite removed from philosophical Sāṃkhya. The influence on Anirvāṇ and Reymond from Gurdjieff, a Russian who had a significant influence on Western Esotericism and New Age traditions, and from Ouspensky, one of his students, has recently been discussed by Maya Burger.[40] Burger suggests that "Sāṃkhya is Gurdjieff in the spiritual path of Lizelle Reymond."[41] Burger has shown that Reymond also inserted references to Sāṃkhya in her editing and translations of Anirvāṇ's texts in order to Indianize the philosophy, i.e. make her rendering of Anirvāṇ appear more traditional.[42] Anirvāṇ in

35 Glen A. Hayes "Sahajiyās," in *Brill's Encyclopedia of Hinduism*, eds. Knut A. Jacobsen, Helene Basu, Angelika Malinar, and Vasudha Narayan, vol. 3 (Leiden: Brill, 2011), 507–513.
36 Anirvan, *Letters from a Baul*, 4.
37 Anirvan, *Letters from a Baul*, 3.
38 Anirvan, *Letters from a Baul*, 6.
39 Reymond, *To Live Within*, 40.
40 Burger, "Sāṃkhya in Transcultural Interpretation."
41 Burger, "Sāṃkhya in Transcultural Interpretation," 478.
42 Burger, "Sāṃkhya in Transcultural Interpretation."

the texts edited by Reymond quoted approvingly Gurdjieff and Ouspensky.[43] A detailed study of Anirvāṇ is outside of the scope of this article.

A third modern Indian tradition of Sāṃkhya as a spiritual practice that will be briefly mentioned here is the teaching of Brahmarṣi Viśvātmā Bāvrā, which differs both from the Kāpil Maṭh tradition as well as from Anirvāṇ's tradition. Western students of Brahmarṣi Viśvātmā Bāvrā have been active in publishing English translations of his texts. Bāvrā was a life long *saṃnyāsin* with followers especially in Haryana, but he also visited Canada and many countries in Europe. He founded and led the International Brahmrishi Mission, India. He taught Sāṃkhya and, in contrast to Anirvāṇ, he taught the Sāmkhya and Sāṃkhyayoga *darśana* texts. Some of his disciples seem to consider themselves Sāṃkhyans and they have published English translations of Sāṃkhya texts and commentaries such as *Tattvasamāsa*, *Pañcaśikhasūtra* and *Sāṃkhyakārikā* with the *Gauḍapādabhāṣya* based on tapes and notes of Bāvrā's lectures and have included their own interpretations.[44] As in the case with Anirvāṇ, Western disciples have contributed to promoting his teaching internationally. William and Margot Milcetich have compiled and edited his publications of *Tattvasamāsa*, *Pañcaśikhasūtra* and *Sāṃkhyakārikā* with the *Gauḍapādabhāṣya* and the *Yogasūtra*. Bāvrā understood that Sāṃkhya was at the foundation of Yoga, and his disciples refer to "Swami Bawra's revival of Samkhya as the basis of Yoga."[45]

Bāvrā's book *Kapil's Samkhya and Patanjali's Yoga* contains lecture notes on the *Tattvasamāsa* and the *Yogasūtra*,[46] while the book *Samkhyakarika with Gaudapadacarya Bhasya* contains lecture notes on *Pañcaśikhasūtra* and the *Sāṃkhyakārikā* with the *Gauḍapādabhāṣya*.[47] These are texts of the Sāṃkhya

43 Reymond, *To Live Within*, 162–163, 180, 188, 230, 231.

44 Brahmrishi Vishvatma Bawra, *Samkhya Karika with Gaudapadacarya Bhasya with Commentary by Brahmarishi Vishvatma Bawra*, compiled and edited by William F. Milcetich, rev. ed. (United States of America: Brahmarishi Yoga Publications, 2018); Brahmrishi Vishvatma Bawra, *Kapil's Samkhya Patanjali's Yoga: Commentary Inspired from lectures of Brahmarishi Vishvatma Bawra*, compiled and edited by William and Margot Milcetich (United States of America: Brahmarishi Yoga Publications, 2007).

45 William Milcetich and Margot Milcetich, "Preface," in Brahmrishi Vishvatma Bawra, *Kapil's Samkhya Patanjali's Yoga: Commentary Inspired from lectures of Brahmarishi Vishvatma Bawra*, compiled and edited by William and Margot Milcetich (United States of America: Brahmarishi Yoga Publications, 2007), viii.

46 William Milcetich and Margot Milcetich, "Preface."

47 William Milcetich, "Editor's Revision," in Brahmrishi Vishvatma Bawra, *Samkhya Karika with Gaudapadacarya Bhasya with Commentary by Brahmarishi Vishvatma Bawra*, compiled and edited by William F. Milcetich, rev. ed. (United States of America: Brahmarishi Yoga Publications, 2018).

darśana, but the interpretations offered by Bāvrā differ from the classical Sāṃkhya and has kinship to Vedānta. He argued that in its unmanifest state there is only one *puruṣa*, and he denied the fundamental Sāṃkhya doctrine of *puruṣabahutva*, the plurality of *puruṣas*. Bāvrā thought that as there is one *prakṛti* in its unmanifest state there is also only one *puruṣa* in its unmanifest state: "In the formless state, there is one Nature and one Spirit, one *purusha*. But when nature evolves, projecting the universe, then purusha manifests in numerous forms."[48] Here there are two stages of *puruṣa*, an unmanifest stage and a manifest stage. To solve the issue of one and many *puruṣas* he argued that the knower is one and the indwellers are many. He wrote:

> Is *purusha* one or many? In Sāmkhya there are two separate stages of the Knower. The first is *jñā*, the Knower, and the second is *purusha*, the indweller and enjoyer. In Īsvarakrsna's Kārikā we see two distinct aspects. First, the Knower was introduced as *jñā*, pure Knower. This Knower is alone, only a seer. It is pure consciousness, one without a second, all-pervading and beyond the three attributes. But when this same *jña* manifests and, due to its all-pervading quality, indwells numerous, countless mediums or *chittas*, it appears as many enjoyers, rather than one. … The Knower is one, but the enjoyer is many. *Jñā* is one, but *purushas* are many. … When purusha achieves emancipation or liberation, realization is of one Spirit, one seer that is the self of all.[49]

His denial of ultimate plurality goes against a fundamental tenet of Sāṃkhya. Bāvrā claims that Kapila believed in only one *puruṣa*, and that Pañchaśikha changed the teaching into many *puruṣas*.[50]

Bāvrā also denies the ultimate dualism of Sāṃkhya. In his view, "When the original oneness of Brahman manifests, both nature and spirit appear in causal form in a stage of manifestation called cosmic intellect or cosmic *chitta*."[51] In the Sāṃkhya *darśana* there is no original oneness from which *prakṛti* and *puruṣa* emerge, but the dualism of two independent principles is ultimate. William Milcetich, the disciple of Bāvrā who has compiled and published notes from Bāvrā's lectures, writes that "academics began to classify Sāṃkhya as atheistic" because there is no indication in Sāṃkhya of *prakṛti* and *puruṣa* "being derived from a higher unified principal" or of "any God as a creator of the uni-

48 Bawra, *Samkhya Karika with Gaudapadacarya Bhasya*, 99.

49 Bawra, *Samkhya Karika with Gaudapadacarya Bhasya*, 39–40.

50 Bawra, *Kapil's Samkhya Patanjaili's Yoga*, 20.

51 Bawra, *Kapil's Samkhya Patanjaili's Yoga*, 131.

verse."[52] But, thinks Milcetich, "pre-classical texts highlight the existence of a *Purushottama* or Supreme Being as the ultimate reality, and *purusha* and *prakriti* are phases of this highest Lord."[53] He claims "Prakriti is the dynamic power evolving this universe, performing under the influence of the Supreme Lord."[54] The author refers the reader to the *Bhāgavatapurāṇa*, and claims that "Early and valuable ideas not included in the Sāṃkhya Kārikā are presented there."[55] Milcetich thinks that also the *Pañcaśikhasūtra* is a documentation of this early Sāṃkhya.

Bāvrā goes against Vedānta on the view of the reality of the world: "The universe is not a futile or fanciful illusion, as some have taught. It is the projection of divine nature in light of the supreme Spirit. We live in a real universe. Lord Kapila teaches there is no illusion."[56] According to Bāvrā, Kapila in the Sāṃkhya philosophy "articulated the essence of the eternal Veda."[57] Bāvrā's disciple Milcetich identifies Sāṃkhya with the teaching of the child Kapila in the *Bhāgavatapurāṇa* who taught Sāṃkhya, Yoga and *bhakti* to his mother Devahūti. Bāvrā promoted and commented in detail on the Sāmkhya *darśana* in lectures and read with disciples Sāmkhya *darśana* texts such as the *Sāṃkhyakārikā*, *Tattvasamāsa*, *Pañcaśikhasūtra* and even the *Gauḍapādabhāṣya* on the *Sāṃkhyakārikā*, but his views on Sāṃkhya differ from those of the *Sāṃkhyakārikā* tradition. However, he nevertheless exemplifies the attempt to reconstruct Sāṃkhya as a contemporary spiritual path.

2 Conclusion

The Kāpil Maṭh institution builds on detailed studies of the classical Sāṃkhya and Sāṃkhyayoga texts and can be considered a revival of a perceived Sāṃkhya orthodoxy.[58] Anirvāṇ's teaching is also called Sāṃkhya but represents a blending of multiple traditions, in particular nondifferentiated Sāṃkhya, Tantra, Gurdjieff, and Vedānta. The Sāṃkhya teaching of Brahmarṣi Viśvātmā Bāvrā uses the Sāṃkhya philosophical texts, but his interpretations seem influenced by the *Bhāgavatapurāṇa*, Vedānta and nondifferentiated Sāṃkhya. The Sāṃ-

52 Milcetich, "Editor's Revision," iii.
53 Milcetich, "Editor's Revision," iii.
54 Milcetich, "Editor's Revision," iii–iv.
55 Milcetich, "Editor's Revision," iv.
56 Bawra, *Samkhya Karika with Gaudapadacarya Bhasya*, 99.
57 Bawra, *Samkhya Karika with Gaudapadacarya Bhasya*, 1.
58 Jacobsen, *Yoga in Modern Hinduism*.

khya teachings of the Kāpil Maṭh institution, Śrī Anirvāṇ, and Brahmarṣi Viśvātmā Bāvrā are strikingly different, and illustrate different forms of Sāṃkhya as a spiritual practice in the modern world. The institutions and gurus of modern Sāṃkhya in India discussed in this article saw Sāṃkhya as an inherent part of Yoga. They taught Sāṃkhya as a spiritual practice and created modern spiritual traditions. Their examples demonstrate that Sāṃkhya in different forms has continued to give rise to and influenced traditions of religion and spirituality in modern India. In modern India, Sāṃkhya has most certainly been reconstructed to give rise to contemporary spiritual paths.

References

Anirvan, Shri. *Buddhiyoga of the Gita and other Essays*. New Delhi: Biblia Impex, 1983.

Anirvan, Shri. *Letters from a Baul: Life Within Life*, compiled by Lizelle Reymond. Calcutta: Sri Aurobindo Pathamandir, 1983.

Āraṇya, Hariharānanda. *Yoga Philosophy of Patañjali*, translated by P.N. Mukerji. Third edition, revised and enlarged. Calcutta: Calcutta University Press, 1981.

Āraṇya, Hariharānanda. *Yoga Philosophy of Patañjali*, translated by P.N. Mukerji. Albany: University of New York Press, 1983.

Āraṇya, Hariharānanda. *Kāpilāśramīya Pātañjal Jogdarśan*. Kolkata: Praścimabai Rājya Puṣtak Parṣad, 1988 [1911].

Bawra, Brahmrishi Vishvatma. *Kapil's Samkhya Patanjali's Yoga: Commentary Inspired from Lectures of Brahmarishi Vishvatma Bawra*, compiled and edited by William and Margot Milcetich. United States of America: Brahmarishi Yoga Publications, 2007.

Bawra, Brahmrishi Vishvatma. *Samkhya Karika with Gaudapadacarya Bhasya with Commentary by Brahmarishi Vishvatma Bawra*, compiled and edited by William F. Milcetich. Revised edition. United States of America: Brahmarishi Yoga Publications, 2018.

Bronkhorst, Johannes. "Systematic Philosophy Between the Empires: Some Determining Features." In *Between the Empires: Society in India 300 BCE to 400 CE*, edited by Patrick Olivelle, 287–313. Oxford: Oxford University Press, 2006.

Burger, Maya. "Sāṃkhya in Transcultural Interpretation: Shri Anirvan (Śrī Anirvāṇa) and Lizelle Reymond." In *Yoga in Transformation: Historical and Contemporary Perspectives*, edited by Karl Baier, Philipp André Maas, and Karin Preisendanz, 465–483. Göttingen: V&R Unipress GmbH, 2018.

Hayes, Glen A. "Sahajiyās." In *Brill's Encyclopedia of Hinduism*, edited by Knut A. Jacobsen, Helene Basu, Angelika Malinar, and Vasudha Narayanan, vol. 3, 507–513. Leiden: Brill, 2011.

Jacobsen, Knut A. "In Kapila's Cave: A Sāṃkhya-Yoga Renaissance in Bengal." In *Theory and Practice of Yoga: Essays in Honour of Gerald James Larson*, edited by Knut A. Jacobsen, 333–350. Leiden: Brill, 2005.

Jacobsen, Knut A. *Kapila: Founder of Sāṃkhya and Avatara of Viṣṇu*. New Delhi: Munshiram Manoharlal, 2008.

Jacobsen, Knut A. *Yoga in Modern Hinduism: Hariharānanda Āraṇya and Sāṃkhyayoga*. London: Routledge, 2018.

Larson, Gerald James and Ram Shankar Bhattacharya, eds. *Sāṃkhya: A Dualist Tradition in Indian Philosophy. Encyclopedia of Indian Philosophies*. Vol. 4. Delhi: Motilal Banarsidass, 1987.

Larson, Gerald James and Ram Shankar Bhattacharya, eds. *Yoga: India's Philosophy of Meditation. Encyclopedia of Indian Philosophies*. Vol. 12. Delhi: Motilal Banarsidass, 2008.

Milcetich, William. "Editor's Revision." In Brahmrishi Vishvatma Bawra, *Samkhya Karika with Gaudapadacarya Bhasya with Commentary by Brahmarishi Vishvatma Bawra*, compiled and edited by William F. Milcetich, revised edition, i–iv. United States of America: Brahmarishi Yoga Publications, 2018.

Milcetich, William and Margot Milcetich. "Preface." In Brahmrishi Vishvatma Bawra, *Kapil's Samkhya Patanjali's Yoga: Commentary Inspired from lectures of Brahmarishi Vishvatma Bawra*, compiled and edited by William and Margot Milcetich, vi–viii. United States of America: Brahmarishi Yoga Publications, 2007,

Osto, D.E. *Modern Sāṃkhya: Ancient Spirituality for the Contemporary Atheist*. Palmerstone North, New Zealand, 2016.

Prabhupada, A.C. Bhaktivedanta Swami. *Teachings of Lord Kapila: The Sone of Devahūti*. New York: Bhaktivedanta Book Trust, 1979.

Ray, Benoy Gopal. *Religious Movements in Modern Bengal*. Santiniketan: Visva Bharati, 1965.

Reymond, Lizelle. *My Life with a Brahmin Family*. London: Rider, 1958.

Reymond, Lizelle. *To Live Within*, translated from French by Nancy Pearson and Stanley Spiegelberg. London: George Allen & Unwin, 1971.

Swarup, Ram. "Introduction." In Anirvan, *Buddhiyoga of the Gita and Other essays*, vi–xvi. New Delhi: Biblia Impex, 1983.

Swarup, Ram. "Introduction." In *Inner Yoga: Selected Writings of Sri Anirvan*, edited by Margaret Case, ii–xii. Sandpoint, ID: Morning Light Press, 2007.

The Works of Kenneth G. Zysk

Books and Monographs

w/ Kamal Sheel and Charles Willemen, eds. *From Local to Global: Papers in Asian History and Culture. Prof. A.K. Narain Commemoration Volume*. 3 vols. New Delhi: Buddhist World Press, 2017.

The Indian System of Human Marks. 2 vols. Sir Henry Wellcome Asian Series 15. Leiden: Brill, 2015.

Siddha Medicine in Tamil Nadu. Tranquebar Initiativets Skriftserie 4. København: Nationalmuseet, 2008.

Conjugal Love in India: Ratiśāstra and Ratiramaṇa. Text, Translation, and Notes. Sir Henry Wellcome Asian Series 1. Leiden: Brill, 2002.

Religious Medicine: The History and Evolution of Indian Medicine. New Brunswick: Transaction Publishers, 1992.

Asceticism and Healing in Ancient India: Medicine in the Buddhist Monastery. New York: Oxford University Press, 1991.

Basham, Arthur L. *The Origin and Development of Classical Hinduism*, edited, revised, and completed by Kenneth G. Zysk. Boston: Beacon Press, 1989.

Basham, Arthur L. *The Sacred Cow: The Evolution of Classical Hinduism*, edited and annotated by Kenneth G. Zysk. London: Rider, 1989.

Religious Healing in the Veda, with Translations and Annotations of Medical Hymns from the Ṛgveda and the Atharvaveda and Renderings from the Corresponding Ritual Texts. Transactions of the American Philosophical Society 75, part 7. Philadelphia: The American Philosophical Society, 1985.

Articles and Chapters

"Mesopotamian and Indian Physiognomy." In *Visualizing the Invisible with the Human Body: Physiognomy and Ekphrasis in the Ancient World*, edited by J. Cale Johnson and Alessandro Stavu, 41–60. Berlin and Boston: De Gruyter, 2019.

"Mesopotamian Physiognomic Omens in India." *Zeitschrift der Deutsche Morgenländischen Gesellschaft* 169, no. 2 (2019): 379–394.

"Varāhamihira's Physiognomic Omens in the *Garuḍapurāṇa*." *History of Science in South Asia* 7 (2019): 72–81.

"Greek and Indian Physiognomics." *Journal of the American Oriental Society* 138, no. 2 (2018): 313–325.

w/ Tsumoto Yamashita, "Sanskrit Medical Scholasticism: Jajjaṭa's *Nirantarapadavyā-*

khyā and Other Commentaries on the *Carakasaṃhitā*, Cikitsāsthāna 2.1." *eJournal of Indian Medicine* 10, no. 1 (2018): 1–113.

"The Human Character Types in Ancient India: A Study in the Transmission of Knowledge between Genres in Early Sanskrit Literature." *Indo-Iranian Journal* 61, no. 3 (2018): 218–261.

w/ Marko Geslani, Bill Mak, and Michio Yano, "Garga and Early Astral Science." *History of Science in South Asia* 5, no. 1 (2017): 151–191.

"The History of Indology in Denmark." In *History of Indological Studies: Papers of the 12th World Sanskrit Conference*. Vol. 11, no. 2, edited by Klaus Karttunen, 175–186. Delhi: Motilal Banarsidass, 2015.

"An Indologist Looks at Siddha Medicine in Tamilnadu." In *Medical Texts and Manuscripts in Indian Cultural History*, edited by Dominik Wujastyk, Anthony Cerulli, and Karin Preisendanz, 181–194. New Delhi: Manohar, 2013.

w/ Tsumoto Yamashita, "Jajjaṭa's *Nirantarapadavyākhyā* on the *Carakasaṃhitā*, Cikitsāsthāna 2.1.1–4ab." In *Medical Texts and Manuscripts in Indian Cultural History*, edited by Dominik Wujastyk, Anthony Cerulli, and Karin Preisendanz, 221–242. New Delhi: Manohar, 2013.

"The Use of Manuscript Catalogues as Sources of Regional Intellectual History in India's Early Modern Period." In *Aspects of Manuscript Culture in South India*, edited by Saraju Rath, 253–287. Brill's Indological Library 40. Leiden and Boston: Brill, 2012.

"Carsten Niebuhr og Rasmus Rask i Indien: Sådan kom indologien til Danmark." In *Tværkultur. Om forskere på rejse: 250-året for Carsten Niebuhrs rejse til Det Lykkelige Arabien*, edited by Annemarie Hede-Andersen, Claus Valling Pedersen, and Ditte Marie Søgaard, 19–26. Årbog for ToRS (University of Copenhagen) 2010. København: KUA Grafisk, 2011.

"Energy-Based Bodywork Therapy: Shiatsu and Ayurveda." In *Massage Therapy: Principles and Practice*, edited by Susan G. Salvo, fourth edition, 314–345. St. Louis, Missouri: Saunders Elsevier, 2011.

"Traditional Medicine in India: Ayurveda and Siddha." In *Fundamentals of Complementary and Alternative Medicine*, edited by Marc S. Micozzi, fourth edition, 455–467. St. Louis, Missouri: Saunders Elsevier, 2011.

"Citations in Jajjaṭa's *Nirantarapadavyākhyā*." *eJournal of Indian Medicine* 3, no. 2 (2010): 69–99.

"Kāma." In *Brill's Encyclopedia of Hinduism. Volume II: Sacred Texts, Ritual Traditions, Arts, Concepts*, edited by Knut A. Jacobsen, Helene Basu, Angelika Malimar, and Vasudha Narayanan, 772–777. Leiden: Brill, 2010.

"Animal Usage in the Sanskrit Traditions of Lovemaking, Lawful Conjugal Love, and Medicine." In *Penser, dire et représenter l'animal dans le monde indien*, edited by Nalini Balbir and Georges-Jean Pinault, 287–304. Bibliothèque de l'École des Hautes Études, Sciences historiques et philologiques 345. Paris: Honoré Champion, 2009.

"Purāṇ: upakathā evaṃ bharatīya cikitsāvidyār vrāhmaṇāyan: agoṃḍāder śāstramate ātmasthakaraṇ." (Translated into Bengali by Jayanta Bhattacharya.) In *Bhārater patabhumite cikitsāvijñāner itihās: saṃkṣipta paryalocanā*, edited by Jayanta Bhattacharya, 36–52. Kolkata: Avabhās, 2009.

"Sanskrit Commentaries on the *Carakasaṃhitā*, with Special Reference to Jajjaṭa's *Nirantarapadavyākhyā*." *eJournal of Indian Medicine* 2, no. 3 (2009): 83–99.

"Some Reflections on Siddha Medicine in Tamilnadu." *Indian Journal of History of Science* 44, no. 2 (2009): 199–208.

"Akattiya and the origins of Siddha Medicine in Tamilnadu." Galathea 3, Nationalmuseets Tranquebar Initiativ, 2010. www.galathea3.dk.

"Alchemy in Citta Medicine." Galathea 3, Nationalmuseets Tranquebar Initiativ, 2010. www.galathea3.dk.

"Citta Surgery: Varman-Therapy." Galathea 3, Nationalmuseets Tranquebar Initiativ, 2010. www.galathea3.dk.

"First Report from India." Galathea 3, Nationalmuseets Tranquebar Initiativ, 2010. www.galathea3.dk.

"Fortune-telling in Pondicherry." Galathea 3, Nationalmuseets Tranquebar Initiativ, 2010. www.galathea3.dk.

"'Health Tourism' in Kerala." Galathea 3, Nationalmuseets Tranquebar Initiativ, 2010. www.galathea3.dk.

"The Principles of Disease and Diagnosis in Citta Medicine." Galathea 3, Nationalmuseets Tranquebar Initiativ, 2010. www.galathea3.dk.

"The Principles of Treatment in Citta Medicine." Galathea 3, Nationalmuseets Tranquebar Initiativ, 2010. www.galathea3.dk.

"Traditional South Indian Medicine Surrounding Tranquebar." Galathea 3, Nationalmuseets Tranquebar Initiativ, 2010. www.galathea3.dk. (Translated as "Traditionel sydindisk lægekunst i området omkring Tranquebar" at www.galathea3.dk.)

"The Bodily Winds in Ancient India Revisited." *Journal of the Royal Anthropological Institute* 13 (2007): 105–115.

"Indian Traditions of Physiognomy: Preliminary Remarks." In *Theory and Practice of Yoga: Essays in Honour of Gerald James Larson*, edited by Knut A. Jacobsen, 425–443. Studies in the History of Religions 110. Leiden and Boston: Brill, 2005.

"Potency Therapy in Classical Indian Medicine." *Asian Medicine: Tradition and Modernity* 1, no. 1 (2005): 101–118.

"Medicine." In *Encyclopedia of Buddhism*. Vol. 2, edited by Robert E. Buswell, 518–520. New York: Macmillan Reference USA, 2004.

"Medical Sanskrit: An Exercise in Translating and Interpreting Sanskrit Medical Literature." *Orientalia Suecana* 51–52, no. 1 (2002–2003): 491–502.

"Classical Ayurveda and Its Modern Practice: Study and Fieldwork of Ayurveda in India and in America." In *Ayurveda at the Crossroads of Care and Cure: Proceedings of the*

Indo-European Seminar on Ayurveda Held at Arrábida, Portugal, in November 2001, edited by Ana Salema, 218–222. Lisbon and Pune: Centro de História de Além-Mar, Universidade Nova de Lisboa, 2002.

"New Age Āyurveda or What Happens to Indian Medicine When It Comes to America." *Traditional South Asian Medicine* 6 (2001): 10–26.

"Did Ancient Indian Medicine Have a Theory of Contagion?" In *Contagion: Perspectives from Pre-Modern Societies*, edited by Lawrence I. Conrad and Dominik Wujastyk, 79–95. Aldershot: Ashgate, 2000.

"Mythology and the Brahmanization of Indian Medicine: Transforming Heterodoxy into Orthodoxy." In *Categorisation and Interpretation: Indological and Comparative Studies from an International Indological Meeting at the Department of Comparative Philology, Göteborg University*, edited by Folke Josephson, 125–145. Meijerbergs Arkiv för Svensk Ordforskning 24. Göteborg: Adoptionscentrum, 1999.

"*Aṣṭāṅgasaṅgraha*, Kalpasthāna IV: Translation and Notes." *Journal of the European Āyurvedic Society* 5 (1997): 25–45.

"A Working Glossary of Pāli Plant Names." In *Researches in Ayurveda, Past and Present: Felicitation Volume in Honour of Prof. Vaidya Priyavrata Sharma*, edited by Banwari L. Gaur and Satyanarayan Sharma, 35–38. Jaipur: Publication Scheme, 1997.

"Traditional Āyurveda." In *Fundamentals of Complementary and Alternative Medicine*, edited by Marc S. Micozzi, 233–242. New York: Churchill Livingston, 1996.

"*Aṣṭāṅgasaṅgraha*, Kalpasthāna III: Translation and Notes." *Journal of the European Āyurvedic Society* 4 (1995): 26–54.

w/ James F. Hartzell, "Columbia University Dharam Hinduja Indic Research Center Conference: Health, Science, and the Spirit: Veda and Ayurveda in the Western World." In *The Journal of Alternative and Complementary Medicine* 1, no. 3 (1995): 297–301.

"New Approaches to the Study of Early Buddhist Medicine: Use of Technical Brāhmaṇic Sources in Sanskrit for the Interpretation of Pāli Medicial Texts." *Pacific World*, New Series, no. 11 (1995): 143–154.

"Vital Breath (*prāṇa*) in Ancient Indian Medicine and Religion." In *The Comparison between Concepts of Life-Breath in East and West*, edited by Yosio Kawakita, Shizu Sakai, and Yasuo Otsuka, 33–65. Tokyo: Ishiyaku EuroAmerica, 1995.

"Magic, Myth, Mysticism, and Medicine." In *Heinrich Zimmer: Coming into His Own*, edited by Margaret Case, 87–103. Princeton: Princeton University Press, 1994.

"*Aṣṭāṅgasaṅgraha*, Kalpasthāna II: Translation and Notes." *Journal of the European Āyurvedic Society* 3 (1993): 319–351.

"The Science of Respiration and the Doctrine of the Vital Breaths in Ancient India." *Journal of the American Oriental Society* 113, no. 2 (1993): 198–213.

"Reflections on an Indo-European System of Medicine." In *Perspectives on Indo-European Language, Culture and Religion: Studies in Honor of Edgar C. Polomé*. Vol. 2, edited

by Roger Pearson, 321–336. Journal of Indo-European Studies Monograph Series 9. McLean, Virginia: Institute for the Study of Man, 1992.
"*Aṣṭāṅgasaṅgraha*, Kalpasthāna 1: Translation and Notes." In *Medical Literature from India, Sri Lanka, and Tibet: Panels of the VIIth World Sanskrit Conference*. Vol. 7, edited by Gerrit J. Meulenbeld, 113–136. Leiden: Brill, 1991.
"Indian Ascetic Traditions and the Origins of Āyurvedic Medicine." *Journal of the European Āyurvedic Society* 1 (1990): 119–124.
"Mantra in Āyurveda: A Study of the Use of Magico-Religious Speech in Ancient Indian Medicine." In *Mantra*, edited by Harvey Alper, 123–143. Albany, New York: SUNY Press, 1989.
"Buddhist Healing and Āyurveda: Some General Observations." In *Studies in Orientology: Essays in memory of Prof. A.L. Basham*, edited by Sachindra K. Maity, Upendra Thakur, and Awadh K. Narain, 124–135. Agra: Y.K. Publishers, 1988.
"Arthur Llewellyn Basham: 24 May 1914–27 January 1986." *Journal of the American Oriental Society* 106, no. 2 (1986): 411–412.
"The Evolution of Anatomical Knowledge in Ancient India, with Special Reference to Cross-Cultural Influences." *Journal of the American Oriental Society* 106, no. 4 (1986): 687–705.
"Towards the Notion of Health in the Vedic Phase of Indian Medicine." *Zeitschrift der Deutschen Morgenländischen Gesellschaft* 135, no. 2 (1985): 312–318.
"An Annotated Bibliography of Translations into Western Languages of Principal Sanskrit Medical Treatises." *Clio Medica* 19, nos. 3–4 (1984): 281–291.
"Fever in Vedic India." *Journal of the American Oriental Society* 103, no. 3 (1983): 617–621.
"Some Observations on the Dissection of Cadavers in Ancient India." *Ancient Science of Life* 2, no. 3 (1983): 187–189.
"Studies in Traditional Indian Medicine in the Pāli Canon: Jīvaka and Āyurveda." *Journal of the International Association of Buddhist Studies* 5 (1982): 70–86.
"A Note on Ancient Indian Medicine." *Bulletin of the Indian Institute of History of Medicine* 8 (1977): 14–23.

Book Reviews

Mathematics and Medicine in Sanskrit: Papers of the 12th World Sanskrit Conference. Vol. 7, edited by Dominik Wujastyk. *Acta Orientalia* 71 (2010): 342–346.
Kleine Schriften zur traditionellen Medizin Südasiens, by Reinhold F.G. Müller. *Indo-Iranian Journal* 52, no. 1 (2009): 75–77.
Traditional South Asian Medicine. Vol. 8, edited by Rahul P. Das. *Orientalistische Literaturzeitung* 105, no. 1 (2010): 104–105.
The Self-Possessed: Deity and Spirit Possession in South Asian Literature and Civilization,

by Frederick M. Smith. *Journal of the American Academy of Religion* 76 (2008): 246–249.

A History of Indian Medical Literature, edited by Gerrit J. Meulenbeld. 3 vols. *Indo-Iranian Journal* 45, no. 4 (2002): 358–361.

Buddhism and Traditional Medicine in Sri Lanka, by Jinadasa Liyanaratne. *Bulletin of the School of Oriental and African Studies* 64, no. 1 (2001): 133–134.

Vāgbhaṭa's Aṣṭāṅgahṛdayasaṃhitā: The Romanised Text Accompanied by Line and Word Indexes, compiled and edited by Rahul P. Das and Ronald E. Emmerick. *Journal of the Royal Asiatic Society* 11, no. 1 (2001): 90–91.

Census of the Exact Sciences in Sanskrit. Series A, vol. 5, by David Pingree. *Journal of the American Oriental Society* 116, no. 3 (1996): 607–608.

Classifying the Universe: The Ancient Indian Varṇa System and the Origin of Caste, edited by Brian K. Smith. *Journal of Asian Studies* 55, no. 3 (1996): 770–771.

Health and Medicine in the Hindu Tradition, by Prakash N. Dasai. *Journal of the American Oriental Society* 111, no. 3 (1991): 597–598.

Galle: Geomedizinische Analyse einer Küstenstadt auf Sri Lanka (Ceylon), by Hella Wellmer. *Journal of the American Oriental Society* 112, no. 1 (1992): 175–176.

Jewels of the Doctrine: Stories of the Saddharma Ratnāvaliya, translated by Ranjini Obeyesekere. *Tricycle: The Buddhist Review* 1 (1991): 93–94.

Das Wissen von der Lebensspanne der Bäume: Surapālas Vṛkṣāyurveda, by Rahul Peter Das. *Indo-Iranian Journal* 34, no. 4 (1991): 261–220 and *Journal of the American Oriental Society* 110, no. 4 (1990): 743–744.

Viśvarūpa, by Thomas S. Maxwell. *Religious Studies Review* 15, no. 2 (1989): 184.

The Jungle and the Aroma of Meats: An Ecological Theme in Hindu Medicine, by Francis Zimmermann. *Indo-Iranian Journal* 33, no. 3 (1990): 205–209.

Studies on Indian Medical History: Papers Presented at the International Workshop on the Study of Indian Medicine Held at the Wellcome Institute for the History of Medicine, 2–4 September 1985, edited by Gerrit J. Meulenbeld and Dominik Wujastyk. *Indo-Iranian Journal* 32, no. 4 (1989): 322–327.

Philosophy in India: Traditions, Teaching, and Research, by Kotta S. Murty. *Journal of the American Oriental Society* 109, no. 1 (1989): 172–173.

Guide to Hindu Religion, by David J. Dell et al. *Journal of the American Oriental Society* 109, no. 1 (1989): 172–173.

Das Kāṭhaka-Gṛhya-Sūtra: mit Vivaraṇa des Ādityadarśana, Bhāṣya des Devapāla, Gṛhyapañcikā des Brāhmaṇabala; kritische Edition mit Anmerkungen. Teil 1: Kaṇḍikā und Sandyopāsanamantrabhāṣya des Devapāla, by Caren Dreyer. *Journal of the American Oriental Society* 108, no. 4 (1988): 639–640.

Science and Technology in South Asia: Proceedings of the South Asia Seminar. Vol. 2, edited by Peter Gaefke and David Utz. *Journal of the American Oriental Society* 107, no. 4 (1987): 838.

Kauṭilya's Arthaśāstra: A Comparative Study, by Narasingha Prosad Sil. *Journal of the American Oriental Society* 107, no. 4 (1987): 838–839.

Structural Depths of Indian Thought, by Poola T. Raju. *Journal of the American Oriental Society* 107, no. 3 (1987): 521–522.

Buddhism and Life: The Anthropological Study of Religion and the Sinhalese Practice of Buddhism, by Martin Southwold. *Journal of the American Oriental Society* 107, no. 1 (1987): 206–207.

Ḍalhaṇa and His Comments on Drugs, by Priya V. Sharma. *Journal of the American Oriental Society* 106, no. 4 (1986): 864–865.

Proceedings of the International Workshop on Priorities in the Study of Indian Medicine, edited by Gerrit J. Meulenbeld. *Journal of the American Oriental Society* 106, no. 4 (1986): 865.

An Untouchable Community in South India: Structure and Consensus, by Michael Moffatt. *Journal of the American Oriental Society* 106, no. 4 (1986): 887–888.

Buddhist Cosmology (From Single World System to Pure Land: Science and Theology in the Images of Motion and Light), by Randy Kloetzli. *Journal of the American Oriental Society* 106, no. 4 (1986): 888.

Die Kulturen Kontinental-Südostasiens, by Manuel Sarkisyanz. *Journal of the American Oriental Society* 106, no. 4 (1986): 888–889.

Surabhi: Sreekrishna Sarma Felicitation Volume, edited by Kopalle S. Ramamurthi et al. *Journal of the American Orientla Society* 106, no. 4 (1986): 889.

Medicine: Its Magico-Religious Aspects According to the Veda and Later Literature, by Ganesh U. Thite. *Journal of the American Oriental Society* 105, no. 4 (1985): 808.

Varāhamihira's Bṛhat Saṃhitā, translated by M. Ramakrishna Bhatt. *Journal of the American Oriental Society* 105, no. 4 (1985): 790–791.

Jyotiṣa: Das System der indischen Astrologie, by Hans-Georg Tüstig. *Journal of the American Oriental Society* 105, no. 4 (1985): 790–791.

A Comprehensive Śulbasūtra Word Index, by Axel Michaels. *Journal of the American Oriental Society* 105, no. 4 (1985): 807.

Modern Hindu Mysticism, by Kamakhya P.S. Choudhary. *Journal of the American Oriental Society* 105, no. 4 (1985): 807–808.

Contemporary Buddhism in Bangladesh, by Sukomal Chaudhuri. *Journal of the American Oriental Society* 105, no. 1 (1985): 180–181.

Aryan Medicial Science: A Short History, by H.H. Bhagvat Singhjee. *Journal of the History of Medicine and Allied Sciences* 40, no. 1 (1985): 111–113.

Haribhadra's Yoga Works and Psychosynthesis, by Shantilal M. Desai. *Journal of the American Oriental Society* 104, no. 4 (1984): 788.

Atīśa and Tibet, by Alaka Chattopadhyaya. *Journal of the American Oriental Society* 104, no. 4 (1984): 783.

Poetry and Speculation in the Ṛgveda, by Willard Johnson. *Journal of the American Oriental Society* 104, no. 4 (1984): 783–784.

Ādiśeṣa: The Essence of Supreme Truth (Paramārthasāra), edited and translated by Henry Danielson. *Journal of the American Oriental Society* 104, no. 4 (1984): 784.

Yogaśataka: texte médical attribué à Nāgārjuna; textes sanskrit et tibétain, edited and translated by Jean Filliozat. *Indo-Iranian Journal* 23, no. 4 (1981): 309–313 and *Journal of the American Oriental Society* 104, no. 4 (1984): 782.

The Ṣaṭsāhasra Saṃhitā, edited and translated by Jan A. Schoterman. *Journal of the American Oriental Society* 104, no. 3 (1984): 606.

The Cult of the Serpent: An Interdisciplinary Survey of Its Manifestations and Origins, by Balaji Mundkar. *Journal of the American Oriental Society* 104, no. 3 (1984): 605.

Shamans, Mystics, and Doctors: A Psychological Inquiry into India and Its Healing Traditions, by Sudhir Kakar. *Journal of the American Oriental Society* 103, no. 4 (1983): 736.

The Cult of Tārā: Magic and Ritual in Tibet, by Stephen Beyer. *Religious Traditions* (*Australia*) 3 (1980): 59–60.

Classical Sāṃkhya. Second edition, by Gerald J. Larson. *Asian Studies Association of Australia. Review* 4, no. 2 (1980): 88–90.

Agniveśa's Charaka Saṃhitā. Vol. 1, translated by Ram K. Sharma and Vaidya B. Das. *Asian Studies Association of Australia. Review* 3, no. 1 (1979): 71–73.

Index

www.ingramcontent.com/pod-product-compliance
Lightning Source LLC
Chambersburg PA
CBHW071328230926
51849CB00002B/337
* 9 7 8 9 0 0 4 4 3 6 6 9 5 *